Creo Parametric 12.0 Tutorial

Roger Toogood, Ph.D.
Edited by David Rocheleau, Ph.D., P.E.

SDC
PUBLICATIONS

SDC Publications
P.O. Box 1334
Mission, KS 66222
913-262-2664
www.SDCpublications.com
Publisher: Stephen Schroff

ISBN-13: 978-1-63057-837-4
ISBN-10: 1-63057-837-1

Printed and bound in the United States of America.

Preface

This tutorial was created to introduce new users to Creo Parametric 12.0 and covers the major concepts and frequently used commands required to advance from a novice to an intermediate user level. Major topics include part and assembly creation, and creation of engineering drawings. The functions that make Creo a parametric solid modeler are illustrated with numerous hands-on examples and exercises.

The commands are presented in a click-by-click manner. In addition to showing the command usage, an effort has been made to explain *why* certain commands are being used. This includes the relationship of feature selection and construction to the overall part design philosophy, that is the consideration of *design intent*. Simply knowing where commands can be found is only half the battle. As is pointed out numerous times in the text, creating useful and effective models requires advance planning and forethought. The analogy made frequently is to playing a strategy game like chess, where skillful players are thinking many moves ahead in the game. Moreover, since error recovery is an important skill, some time is spent exploring the created models (in fact, sometimes intentionally inducing errors), so that users will become comfortable with the "debugging" phase of model creation.

Users of this book come with a broad range of backgrounds - some have previous CAD experience, others may have only an introductory programming course, and others may have neither. Students also have a wide range of abilities both in spatial visualization and computer skills. The approach taken here is meant to allow accessibility to learners of all levels. These lessons, therefore, were written for new users with no previous experience with CAD, although some familiarity with computers is assumed.

This book is ***NOT*** a complete reference for Creo Parametric. Coverage of all that Creo has to offer within a single (even not-so-thin) volume is quite impossible. All the several thousand pages of reference manuals and official documentation are available on-line. The Help Center offers good search tools and cross-referencing to allow users to find relevant material quickly and in much more detail than can be presented here. The on-line help also contains a number of additional "how to" documents and examples; however, these typically assume some knowledge of the program.

The lessons in this tutorial are meant to be covered sequentially. Discussion of commands is, for the most part, restricted to their use within the context of the lesson (a Just-in-Time delivery!). For this reason, although they may be mentioned, many options to commands are not dealt with in detail all in the same place in the text, as is done in the on-line reference material. Such a discussion would interrupt the flow of the work. Although the index provides locations within the text where a command is used, this is not exhaustive but rather meant to act as a quick reference or reminder.

Notes for the *Creo Parametric 12.0* Edition

The organization of the lessons remains the same as the previous editions. Most new functions introduced with Creo 12.0 are outside the scope of this book, which continues to focus on the core concepts and tools.

Additions to previous releases of Creo continue to be the refinement of multibody modeling functions. Multibody functions are mentioned briefly here and a short section introducing how they are created is included. However, to make intelligent use of multiple bodies, the basic material covered here is necessary first. Therefore, to keep the size of this Tutorial manageable, a broader discussion of how and when to use multiple bodies is postponed to the *Advanced Tutorial*. For users who want to explore these functions immediately, there are some lessons included in the on-line Help pages.

Earlier releases of Creo introduced detachable Design and Layer Trees, detachable dashboard panels, and transparency settings for separate bodies (independent of appearance settings). A new tool for examining model structure (snapshots) was added. Some other visual changes in the user interface were introduced (e.g. shaded datum planes). A new sketch-based hole pattern was introduced.

Some interesting and useful additions include better model tree interaction and organization for understanding design intent and managing complex assemblies. Usability upgrades include streamlining workflows that previously required more steps in earlier versions.

Note to Instructors

The tutorials consist of the following:

- 1 lesson introducing the program and its operation
- 6 lessons on features used in part creation
- 1 lesson on modeling utilities
- 1 lesson on creating engineering drawings
- 2 lessons on creating assemblies and assembly drawings

Each of these will take between 2 to 4 hours to complete (thus requiring some time spent outside of the regularly scheduled computer lab time). The time required will vary depending on the ability and background of the student. Moreover, additional time would be beneficial for experimentation and exploration of the program. Most of the material can be done by the student on their own; however, there are a few "tricky" bits in some of the lessons. Therefore, it is important to have teaching assistants available (preferably right in the computer lab) who can answer special questions and especially bail out students who get into trouble. Most common causes of confusion are due to not completing (or even doing!) the lessons or digesting the material. This is not surprising given the volume of new information to absorb, the lack of time in students' schedules, and maybe the propensity of some to want to learn by trial-and-error! However, I have found that most student questions are answered within the lessons.

In addition to the tutorials, some class time (two to three hours) over the duration of the course will be invaluable in demonstration and discussion of some of the broader issues of feature-based modeling. It takes a while to accept that just creating the geometry is not sufficient for a design model that will inevitably be involved in design revisions, usually by other people.

It is important for students to keep up the pace with the lessons through the course. To that end, laboratory exercises have involved short quizzes (students produce written answers to questions chosen at random from the end of each lesson), creating models of parts sketched on the whiteboard in isometric or multiview, or brought into the lab (usually large models made of styrofoam or smaller models made on a rapid prototype machine that they can take to their desk). Of these, the latter two activities seemed to have been the most successful. It appears that many students, after having gone through the week's lesson (usually only once, and very quickly) do not absorb very much. The second pass through the lesson usually results in considerably more retention. Students really don't feel comfortable or confident until they can make parts from scratch on their own. Each lesson concludes with a number of simple exercise parts that can be created using new commands taught in that lesson. In addition to these, a project is also included that consists of a number of parts that are introduced with the early lessons and finally assembled at the end. The Panavise remains my favorite choice for this project, since its parts have a wide variety of feature types and degree of difficulty. It would be most beneficial if students could have at their disposal a physical model (of the Panavise or some other object) which they can reverse engineer.

Acknowledgments

The inspiration for these lessons was based on the web pages produced in 1996 for Pro/ENGINEER Release 16 by Jessica LoPresti, Cliff Phipps, and Eric Wiebe of the Graphic Communications Program, Department of Mathematics, Science and Technology Education at North Carolina State University. Permission to download and modify their web pages is gratefully acknowledged. Since that time (July, 1996) the tutorials have been rewritten/updated over twenty times: initially to accommodate our local conditions and then for Releases 18, 20, 2000i (with another foray into PT/Modeler in between), 2000i^2, 2001, Wildfire 1.0 - 5.0, Creo Parametric 1.0 - 9.0, and now Creo Parametric 12.0.

Some of the objects and parts used in these tutorials are based on illustrations and problem exercises in **Technical Graphics Communication** (Irwin, 1995) by Bertoline, Wiebe, *et al.* This book and its newer editions are excellent sources for examples and additional exercises in part and assembly modeling, and drawing creation.

The Panavise project in this tutorial is based on a product patented by Panavise Products, Inc. and is used with the written permission of Panavise Products, Inc., Reno, Nevada. The name Panavise is a registered trademark of Panavise Products, Inc., Reno, Nevada, and is used with the written permission of Panavise Products, Inc. Such permission is gratefully acknowledged.

Any other similarity of objects, parts, and/or drawings in this tutorial is purely coincidental and unintentional.

These tutorials (for Release 16) were first written as web pages and released in September, 1996. Since then, a number of users (students, instructors, industrial users, even a patent lawyer!) and reviewers have returned comments on the lessons, which are gratefully acknowledged. Constructive comments and suggestions continue to arrive regularly from readers all around the world. Many thanks to all these people, who are too numerous to list here. Notwithstanding their assistance, any errors in the text or command sequences are those of the author!

Acknowledgment is due to Stephen Schroff and the staff at SDC (especially Zach, Karla, Tyler, and Megan) for all their efforts.

Thank you to Trent Miller, University of South Carolina, for creating the model on the front cover.

To users of this material, I hope you enjoy the lessons.

RWT
Edmonton, Alberta
24 May 2022

David Rocheleau
Columbia, SC
10 January 2026

TABLE OF CONTENTS

Introduction to Creo Parametric

Lesson 1: User Interface, View Controls and Model Structure

Lesson 2: Creating a Simple Object (Part I)

Lesson 3: Creating a Simple Object (Part II)

Lesson 4: Revolved Protrusions, Mirror Copies, Model Analysis

Lesson 5: Modeling Utilities

Lesson 6: Datum Planes and Sketcher Tools

Lesson 7: Patterns and Copies

Lesson 8: Engineering Drawings

Lesson 9: Assembly Fundamentals

Lesson 10: Assembly Operations

Lesson 11: Sweeps and Blends

Appendix: Creo Parametric Customization

INTRODUCTION to

Creo Parametric 12.0

A Few Words Before You Dive In...

This tutorial was written for new users who are getting started with Creo® Parametric 12.0 (<**www.ptc.com/en/products/creo**>). The lessons in this book will introduce you to the basic functionality of the program and are meant to be used alongside the running Creo software. Description of command sequences is accompanied by a discussion of where the commands fit into an overall modeling strategy. In addition to learning *what* each command or function does, it is important to understand *why* it is used. We will sometimes make intentional errors so that we can discover how Creo responds. **Therefore, just clicking through the command sequences given here is not enough; you will learn the material best if you take time along the way to read the text carefully and think about what you are doing and observing what happens**. You will also learn considerably more by exploring the program on your own and experimenting with the commands and options.

You are about to learn how to use the most sophisticated and powerful solid modeling program available. It may be the most complex software you will ever use. Its power derives from its extremely rich command set, that understandably requires quite a long time to master. Creo's learning curve has the reputation of being very steep although this is diminishing and is now quite undeserved. The goal of this tutorial is to help you with this learning as effectively and efficiently as possible. Learning Creo is a challenging task, but not impossible. Do not be discouraged, as you will find it well worth the effort.

Please note that this book is not a reference manual. Not all the available commands are covered, nor will a comprehensive discussion of the myriad available options be attempted. The tutorial is meant to give you a good start and to give you a solid (no pun intended!) foundation on which to build further knowledge. On completion of these lessons (in about 30 or 40 hours from now), you should:

1. be able to create models of relatively complex parts and assemblies.
2. know how to produce the related detailed engineering drawings.
3. be prepared for further exploration of Creo and its related modules.
4. understand the terminology used in the program, and solid modeling in general.
5. understand the design philosophy and methods embedded in Creo.

The last two are important so that you can understand the on-line reference documentation and explore other commands and functions in the software.

In the early lessons and as each new function is introduced, commands are presented in considerable detail to explain what is going on and why. As you progress through the lessons, you will be given fewer details about commands that have been covered previously. For example, in Lesson #2 we find out how to create a two-dimensional sketch, mouse click by mouse click. Later on, you will be asked to "Create the sketch shown in the figure" assuming that you know how to do that. Thus, the lessons build on each other and are meant to be done in the order presented. It is important for you to go through the lessons in sequence and to have a good understanding of the material before you go on to the next lesson.

You may have to go through each lesson (or some portions) more than once to gain an adequate level of understanding. Each lesson has some questions and exercises at the end to allow you to check your knowledge of the concepts and commands. No answers are given here for these questions - you will learn the material best if you have to dig them out for yourself, not to mention what you will discover accidentally along the way! And don't worry if you don't get all the answers right away. Some questions are posed intentionally to encourage your continued and deeper exploration of the program. This sort of "on-the-fly" discovery is a never-ending activity even with experienced users because Creo is such a big program. Finally, each lesson concludes with a project activity that will result in the creation of a simple assembly.

The images presented here should correspond with those obtained in the Creo windows, and can be used to check your work as you proceed through the lessons. Figures in the printed hard copy document, however, are only available in black-and-white, whereas color plays an important role on the Creo screen. In addition to distinguishing between different parts, color is used to indicate the meaning of lines (edge, axis, datum curve, hidden line, and so on). Where a line interpretation may be ambiguous in the black and white version here, the color is indicated in the text. Also, some modifications have been made to the default system font in order to make the figures clearer.

These lessons were developed using the Windows-64 version of the software. The version used was 12.4.0.0. Minor variations might be expected with later builds, although these will likely not affect the material presented here.

What *IS* Creo Parametric?

Actually, Creo is a suite of programs that are used in the design, analysis, and manufacturing of a virtually unlimited range of products. Its field of application is generally mechanical design, although recent additions to the program are targeted at ship building and structural steel framework as well[1]. Creo Parametric is one of a series of related software packages (including Creo Direct, Creo Simulate, Creo Schematics, Creo Layout, and others). In these lessons, we will be dealing only with the major

---

[1] People who work in general architectural or civil engineering design (like highway design) would most likely not use Creo, as its design and functions do not lend themselves directly to those activities.

front-end module, known as Creo Parametric. This program (previously known as Pro/ENGINEER and Wildfire) is used for part and assembly design, and production of engineering drawings. There are a wide range of additional/optional modules available to handle tasks ranging from sheet metal operations, piping layout, mold design, wiring harness design, NC machining, and other functions. Sensitivity studies and design optimization based purely on geometry are handled by a module called Behavioral Modeling Extension (BMX). Mechanism design, kinematics, dynamics, and animation is accomplished using the Mechanism Dynamics Extension (MDX). Creo Simulate (previously known as Pro/MECHANICA, also from PTC)[2], integrates with Creo to perform structural analysis (static stress and deformation, buckling and fatigue analysis, vibration), thermal analysis, and dynamic motion analysis of mechanisms. Creo Simulate can also do sensitivity studies and design optimization, based on the model created in Creo.

In a nutshell, Creo Parametric is a *parametric, feature-based solid modeling* system.

"Feature-based" means that you create your parts and assemblies by defining high level and physically meaningful features like extrusions, sweeps, cuts, holes, slots, rounds, and so on, instead of specifying low-level geometry like lines, arcs, and circles. This means that you, the designer, can think of your computer model at a very high level, and leave all the low-level geometric detail for Creo to figure out. Features are specified by setting values and attributes of elements such as reference planes or surfaces, direction of creation, pattern parameters, shape, dimensions, and others. Features can either add or subtract material from the model, or be simple non-solid geometric entities like references axes and planes. Creo Parametric is a "history based" modeler, which means that the order of feature creation is important. Its cousin, Creo Direct, is a "direct" modeler where the order of feature creation is irrelevant.

"Parametric" means that the physical shape of the part or assembly is driven by the values assigned to the attributes (primarily dimensions) of its features. You may define or modify a feature's dimensions or other attributes at any time (within limits!). Any changes will automatically propagate through your model. You can also relate the attributes of one feature to another. For example, if your design intent is such that a hole be centered on a block, you can relate the dimensional location of the hole to the block dimensions using a numeric formula, appropriately called a *relation*; if the block dimensions change, the centered hole position will be re-computed automatically.

"Solid Modeling" means that the computer model you create contains all the "information" that a real solid object would have. It has volume and therefore, if you provide a value for the density of the material, it has mass and inertia. Unlike a surface model, if you make a hole or cut in a solid model, a new surface is automatically created and the model "knows" which side of this surface is solid material. The most useful thing about solid modeling is that it is impossible to create a computer model that is

[2] A companion book, *Creo Simulate Tutorial*, is also available from the publisher, SDC Publications.

ambiguous or physically non-realizable, such as the "object" shown in the figure at the right. This figure shows what appears to be a three-pronged tuning fork at the left end, but only has two square prongs coming off the handle at the right end. With solid modeling, you cannot create a "model" such as this that could not physically exist. This type of ambiguity is quite easy to do with just 2D, wireframe, or sometimes even surface modeling.

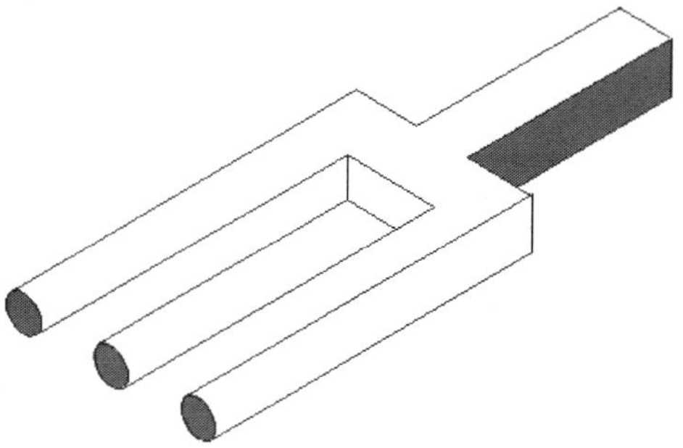

The 3-Pronged Blivot -
A Non-realizable Object
(*NOT* created with Creo!)

Although the emphasis in this book is on creation of solid features, Creo has a very extensive command set for creating and manipulating surfaces (called "quilts"). This advanced surface modeling is important in, for example, the design of consumer products. Once the surfaces are created, the part model can be "solidified" so that the normal solid features (like standard holes) can be applied.

Whether or not the part could actually be manufactured is another story. Here is a cut-away view of a physically possible part, but don't take this to the machine shop and ask them to machine the cavity inside the part! Creo will let you make this model, but concerns of manufacturability are up to you.

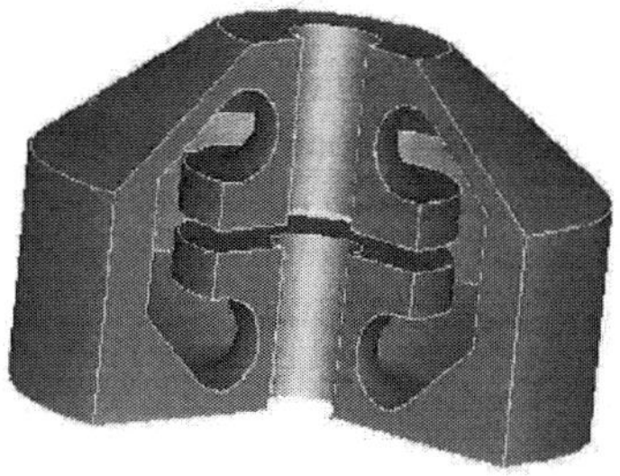

Could your machine shop make this?

An important aspect of feature-based modeling in Creo, due to its history-based modeling paradigm, is the concept of **parent/child relationships**.

Without going into a lot of detail at this time, a child feature is one that references a previously created (parent) feature. For example, the surface of a block might be used as a reference plane to create a slot. A change to the parent feature (the block) will potentially affect the child (the slot). For example, deleting a parent feature will delete all its children since one or more references required to create the children would no longer exist. Creo has special functions available to manage parent/child relationships. This can get pretty complicated with a complex model (a good reason to try to keep your models simple!), so we will leave the details for later. However, you should keep parent/child relations in mind when you are specifying feature references for a new feature you are creating: If the parent feature is temporary or is likely to change, what effect will this have on the children? Will the references still correctly capture your design intent?

Once your model is created, it is very easy to get Creo to produce fully detailed standard format **engineering drawings** almost completely automatically. In this regard, Creo also has **bidirectional associativity** - this means you can change a dimension value on the drawing and the shape of the model will automatically change, and vice versa.

Of course, few parts live out their existence in isolation. Thus, a major design function accomplished with Creo is the construction of assemblies of parts. Assembly is accomplished by specifying physically-based geometric constraints (insert, mate, align, and so on) between part features. With assemblies you can see how the different parts will fit together or interfere with each other, or see how they move with respect to each other, for example, in a linkage assembly. Assembly models are also associative: in a properly made assembly model, changes can propagate through to other parts in the assembly automatically. To a new user of the program, this is almost magic! And, of course, drawings of assemblies can also be created.

Creo uses standard Windows interface methods and contains a mode of operation called "Direct Modeling" that makes feature creation very simple[3]. At the same time, for power users, there are a large number of shortcuts (think "right mouse button") which can speed up your work quite a lot. These make the program easy to use and add a lot of visual excitement to working with the program. Another key aspect of the program is its readiness for collaboration of users over the internet. We will not be delving into these tools here but instead concentrate on stand-alone usage to create models and assemblies.

Many capabilities introduced into Creo in recent years are beyond the scope of this tutorial. These include Model-Based Definition (MBD) in which all the information required for part manufacturing is directly included in the 3D model. More recently there have been enhancements and added capabilities in Augmented Reality (AR) and Additive Manufacturing. A major enhancement is the use of multiple bodies within the same part model and tools for incorporating simulation in early design models.

This sounds like it's pretty complicated!...

It is important to realize that you won't be able to master Creo overnight. Its power derives from its flexibility and rich set of commands. It is natural to feel overwhelmed at first! With not too much practice, you will soon become comfortable with the basic operation of the program. As you proceed through the lessons, you will begin to get a feel for the operation of the program, and the philosophy behind feature-based design. Before you know it, you'll feel like a veteran and will gain a tremendous amount of personal satisfaction from being able to competently use it to assist you in your design tasks.

Using Creo is quite different from previous generation CAD programs. This is a case where not having previous CAD experience might even be an asset since you won't have to unlearn anything! For example, because it is a solid modeling program, all your work is done directly on a 3D model rather than on 2D views of the model. Spatial visualization is very important and, fortunately, the display is very easy to manipulate. Secondly, as with computer programming, with Creo you must do a considerable amount of thinking and planning ahead (some fast free-hand sketching ability will come in handy

[3] Although it has minor similarities, the direct modeling operations in Creo Parametric should not be confused with its cousin Creo Direct.

here!) in order to create a clean model of a part or assembly. Don't worry about these issues yet - they will not interfere with your learning the basic operation of the program. As you become more adept with Creo, you will naturally want to create more complex models. It is at that time that these high-level issues will assert themselves. In the meantime, have fun and practice, practice, practice.

Overview of the Lessons

A brief synopsis of the lessons in this tutorial is given below. Each lesson should take at least 2 to 3 hours to complete. If you go through the lessons too quickly or thoughtlessly, you may not understand or remember the material. ***Do not confuse recognition with comprehension***. Like sports, watching someone else do something doesn't mean you can do it (yet!). For best results, it is suggested that you scan/browse ahead through each lesson completely before going through it in detail. You will then have a sense of where the lesson is going, and not be tempted to just follow the commands blindly. You need to have a sense of the whole forest when examining each individual tree!

In order to complete some of the lessons, you will need to install some tutorial files on your hard disk. These are available for download from the following site:

http://www.sdcpublications.com/downloads/978-1-63057-837-4

Look for Commercial Edition Exercise Files. Further instructions for this are in Lesson #1.

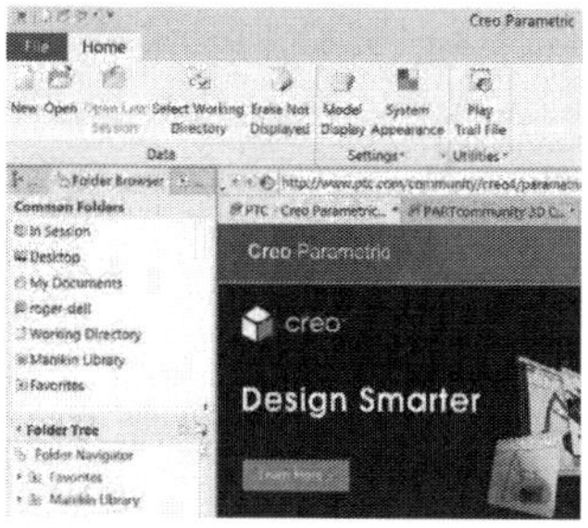

Lesson 1 - User Interface, View Controls and Model Structure

How to start Creo; representation of Creo command syntax; command flow in Creo; special mouse functions; Creo windows; controls for managing the view and display of objects; the model tree; how parts and assemblies are structured.

Lesson 2 - Creating a Simple Object (Part I)

Creating a simple part using sketched features; datum curves; Sketcher and Intent Manager are introduced; sketching constraints, alignments, and procedures; feature database functions are introduced; part templates.

Lesson 3 - Creating a Simple Object (Part II)

Placed features (hole, chamfer, round) are added to the block created in Lesson #2; listing and naming features; modifying dimensions; adding relations to control part geometry; more Sketcher tools; implementing *design intent*.

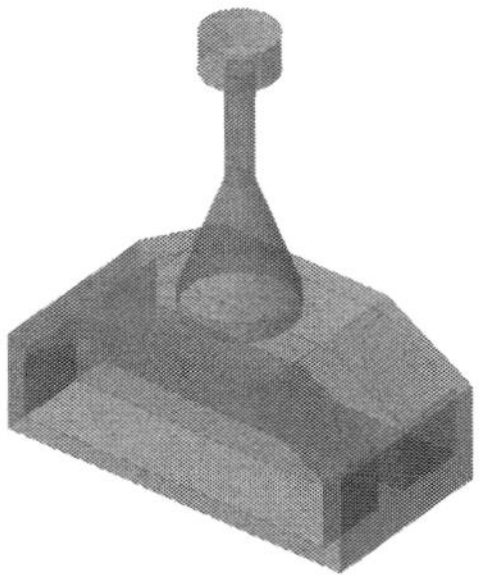

Lesson 4 - More Features for Creating Parts

A new part is modeled using a number of different feature creation commands and options: both sides protrusions, an axisymmetric (revolved) protrusion, a cut, rounds, and chamfer. More Sketcher tools. Edge sets. Mirrored features. Model analysis tools. We will intentionally make some modeling errors to see how Creo responds. A short introduction to multibody modeling.

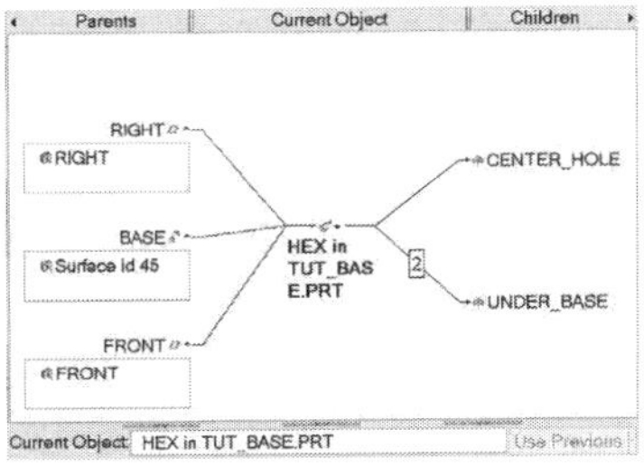

Lesson 5 - Modeling Utilities, Parent/Child Relations, and the 3 R's

These utilities are used to investigate and edit your model: changing references, changing feature shapes, changing the order of feature regeneration, changing feature attributes, and so on. Suppressing and resuming features. If your model becomes even moderately complex, you will need to know how to do this!

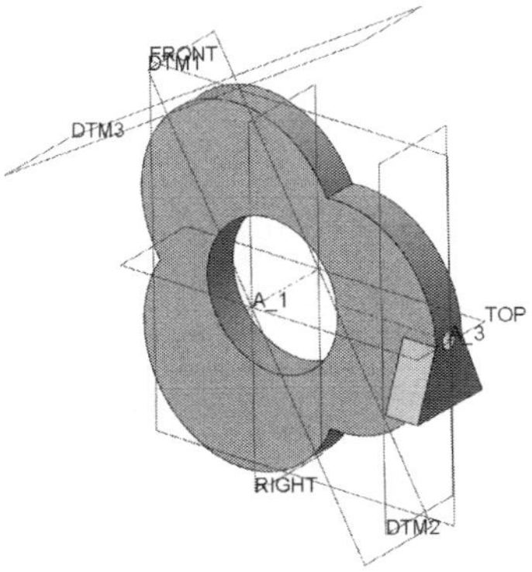

Lesson 6 - Sketcher Tools and Datum Planes

More tools in Sketcher are introduced, including sketching relations. The mysteries of datum planes and make datums are revealed! What are they? How are they created? How are they used to implement design intent? Sketch regions.

Lesson 7 - Patterns and Copies

Creating a counterbored hole and hole notes. Patterns (one-dimensional or two-dimensional); radial patterns of placed and sketched features. Pattern groups. Copies using translation, rotation, or mirroring.

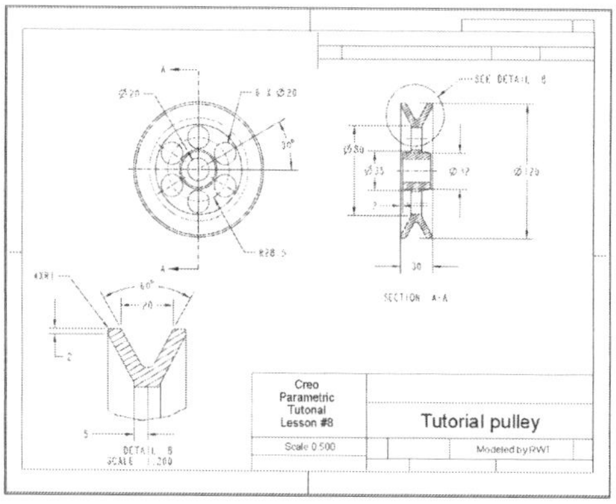

Lesson 8 - Creating an Engineering Drawing

Procedures for making dimensioned engineering drawings. Two new parts are created (both parts will also be used in Lesson #9 on assemblies). Much of the work in creating the drawing is done by Creo, although a fair amount of manual labor must go into improving the cosmetics of the drawings.

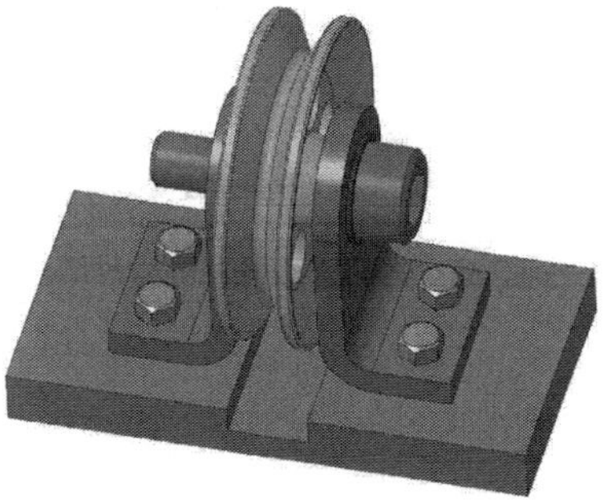

Lesson 9 - Assembly Fundamentals

How to create an assembly from previously created parts. This involves creating placement constraints that specify how the parts are to fit together. Assigning appearances (colors).

Lesson 10 - Assembly Operations

Modifying the assembly created in Lesson #9. This includes changing part dimensions, adding assembly features, suppressing and resuming components, creating exploded views, and creating an assembly drawing. Display styles.

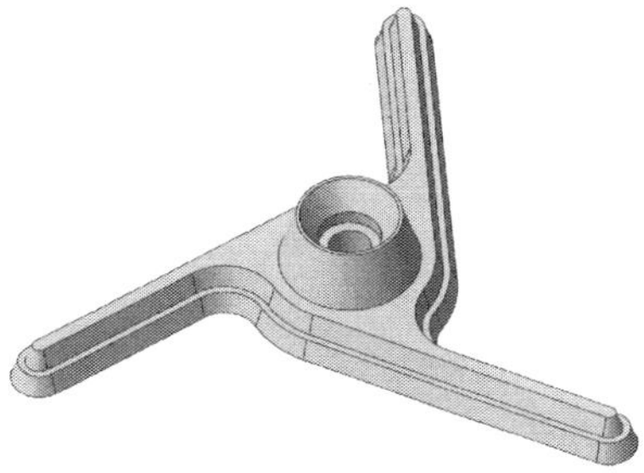

Lesson 11 - Sweeps and Blends

These are the most complicated (that is, flexible and powerful) features covered in these lessons. They are both types of solid protrusions but can also be used to create cuts and slots.

Once again, as you go through these lessons, take the time to explore the options available and experiment with the commands. You will learn the material best when you try to apply it on your own ("flying solo"), perhaps trying to create some of the parts shown in the exercises at the end of each lesson.

On-Line Help

Should you require additional information on any command or function, Creo comes with extensive on-line help. This contains the complete text of *all* reference manuals for the software. There are several ways you can access the on-line help. These are presented in Lesson #1.

To those of you who have read this far: Congratulations! You are probably anxious to get going with Creo. Let's get started...

This page left blank.

Lesson 1

User Interface, View Controls and Model Structure

Synopsis

Starting Creo Parametric; command syntax; mouse functions; view and display controls; structure of parts and assemblies; the model tree; obtaining hard copy; on-line help.

Overview of this Lesson

We are going to cover a lot of introductory ground in this lesson with two main objectives. The first is to introduce you to the Creo Parametric user interface, some of the terminology used, and commands for controlling the display of parts and assemblies. You need to be comfortable with these controls so that later on we can concentrate on commands for actually creating objects or dealing with other issues of running the program. Our second objective is to explore how models of parts and assemblies are structured: what features are used, how they are ordered, and so on[1].

Keep in mind that creating a Creo Parametric model involves much more than simply producing geometry or creating pretty pictures. A model can have a number of purposes: engineering simulation and analysis, visualization, production of drawings, manufacturing and production planning, marketing, and so on. More importantly, the model must be easily understood by the people who will use it. If the model is used in design of a new product (which is typically a very iterative process), then we must ensure that it is simple, flexible, and robust to the inevitable modifications that will occur as the design evolves. Plus, there are often several different ways to create the desired geometry. Obviously, if you only know one way of doing something, your options will be limited! If you know several, which one should you use, and why? Furthermore, since a great deal of design these days is done in groups and teams, it is inevitable that you will

---

[1] An old saying, attributed to Lao Tzu (among others!), goes something like this: "Give a man a fish and you feed him for a day. Teach him how to fish and you feed him for a lifetime." Notwithstanding the gender discrimination here, perhaps the best way to get started is just to have a look at the fish!

be passing your models on to someone else. It must be easy for them to figure out how your model was made. *Simple .. flexible .. robust.* These goals are not that much different from writing a computer program. As with computer programming, your job will be easier if you take some time to plan your model before you sit down at the computer. In other words, do not use "Ready!...Fire!...Aim!" as your modeling strategy. This will inevitably lead to frustration with a poorly designed model.

The point of these comments is that modeling is not a trivial task. In this tutorial you must try to connect the "what, how, where" of issuing commands with the "when" and "why" in order to develop good modeling skills.

We will go at quite a slow pace at the start[2]. This should leave you sufficient time for experimenting on your own, which you are strongly encouraged to do. Here's what we will cover in this lesson:

1. Layout of the screen and user interface
    - How commands are entered
    - How this tutorial will represent the command sequence
    - Files and directories
2. View and display controls
3. Exploring the data structure for a part, including multibody modeling
4. Exploring the data structure for an assembly
5. How to get printed hard copy
6. How to get on-line Help

We will spend most of our time on sections 2 through 4. It will be a good idea to browse ahead through each section to get a feel for the direction we are going, before you do the lesson in detail. A few words of caution: Take your time through the lessons. Resist the temptation to skip over the discussion and just execute the commands. There is a lot of material here which will be useful later, and not much that you can ignore without eventually paying for it. It is likely that you won't be able to absorb everything with a single quick pass-through.

Helpful Hint

You may find it helpful to work with a partner on some of these lessons because you can help each other with the "tricky bits." Split the duties so that one person is reading the tutorial out loud while the other is doing the keyboard and mouse stuff, and then switching duties periodically. It will be handy to have two people scanning the menus for the desired commands and watching the screen. Creo Parametric uses a lot of *VERY* subtle visual cues to alert you to what the program is doing or requires next. Having two people watching out for this may be helpful.

---

[2] Lao Tzu also apparently said "A journey of a thousand miles begins with a single step" although given the time and place it is doubtful that he actually used "miles"!

Starting Creo Parametric

To start Creo Parametric there may be an icon right on the desktop or you may have to look in the **Start** menu at the bottom left of the screen on the Windows taskbar. The startup is complete when your screen looks like Figure 1. The screen shown in the figure is the bare-bones, default Creo Parametric screen. If your system has been customized[3], the menu bars and window contents of your interface may look slightly different.

Figure 1 The Creo Parametric start-up screen

On the left is a multi-purpose area called the **Navigator**, with some associated control tabs at the top. In the default start-up, the Navigator shows you the *Common Folders* structure on your machine, among which is your current *Working Directory*. The working directory is the default location where Creo Parametric will look for and save your files. Other tools in the Navigator allow you to organize folders and web sites into groups of favorites. These are accessed using the tabs at the top. At the bottom of the Navigator pane is the ***Folder Tree*** button which if selected will expand to show the directory structure of your computer. To the right of the Navigator is the **Browser**, which is an integrated web browser[4]. A major focus of Creo Parametric is connectivity between/among users. To that end, many Internet communication tools are embedded

---

[3] Customizing the operation and interface of Creo Parametric is discussed briefly in the Appendix. These topics are discussed at length in the *Creo Parametric Advanced Tutorial* available from SDC.

[4] Depending on the customization of your installation, the Browser may not be displayed (see button at lower left), or might be showing different content. By default, Creo uses an embedded Chromium browser.

within the program. In addition to communicating with other users, the Browser allows you to launch regular web pages and, for example, download part files from the internet directly into your session (see the *3D Modelspace* tab). The browser functions in the same way as normal web browsers, including history and printing functions.

The Navigator and Browser areas can be resized by dragging left/right on the vertical right borders. Each area can be minimized by clicking on the buttons at the lower left corner of the screen - do that now. You are left with the screen shown in Figure 2.

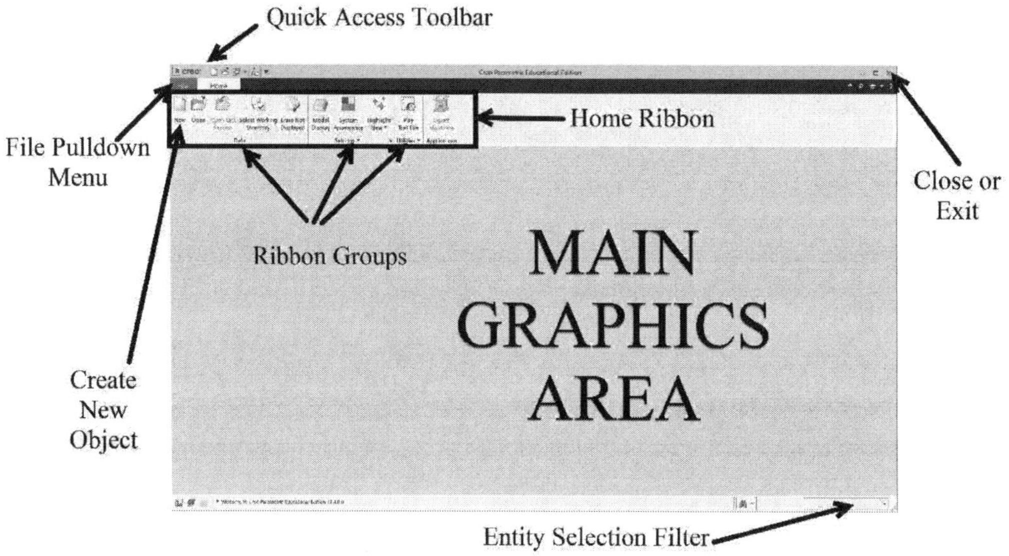

Figure 2 The Creo Parametric screen

The main user interface element in Creo Parametric is the Command Ribbon. There are many different possible ribbons and these are context sensitive - that is, the ribbon will change depending on what you are doing. Figure 2 shows the default **Home** ribbon that appears on startup, or when no objects are loaded into the current session. In Figure 2, the main graphics area is, of course, where most of the action will take place. As you move the mouse (slowly) across the buttons and interface items, a tool tip box will pop up. Occasionally a button or option will be grayed out, meaning they are inactive at this time. The prompt/message area below the graphics area shows brief system messages (including errors and warnings) during command execution. Creo Parametric is usually set up to show only the last line of text in this message area, but you can resize the area by dragging on its upper horizontal border. When the mouse cursor is in the message area, you can use the scroll wheel to review the message history.

Before we load an object into the program, let's explore the interface a bit more. Along the way we'll discuss how this tutorial will deal with command entry.

How commands are entered into Creo Parametric

There are a number of ways that you will be interacting with the program: menu picks, shortcut buttons, keyboard entry, and several special mouse functions.

The Quick Access and Graphics Toolbars

These are sets of command buttons that are most frequently used, and therefore are on the screen more-or-less full time. We will see the graphics toolbar when we load a part or assembly. You can add your own buttons and toolbars to customize either of these areas[5].

The Ribbon User Interface

The main user interface element is the context sensitive command ribbon. Different ribbons will appear depending on your current activity (called the *mode,* as in part mode or assembly mode) and your current task. For example, in part mode there are ribbons for modeling, analysis, rendering, viewing, and so on. Within each ribbon, commands are arranged in groups (see Figure 2) and represented by an icon, text, or both. Commonly used commands are usually represented by larger icons. If there are additional group commands, the group may have an overflow area that can be displayed by clicking on the group name. We are going to spend a lot more time exploring the various ribbons throughout the remainder of this Tutorial.

Automatic Pop-up Menus

Probably the easiest way to give commands is by selecting from a pop-up menu that may appear at the cursor location or beside a selected entity. These menus are context sensitive - that is, the commands in the pop-up depend on what has been selected. The commands are also available in the ribbons, but by showing them at the current point of interest (i.e. the cursor location), it means you can stay focused on what you are doing instead of searching through the ribbons. The pop-up menu goes away if you click on the background of the graphics window. When a pop-up menu is open, you can access additional commands (like customizing the pop-up menu) by clicking the right mouse button.

Right Mouse Button (RMB) Pop-up Menus

These menus are available in a number of operating modes by clicking and sometimes holding down the *right* mouse button. This menu is also context sensitive. Often the alternate location for these pop-up commands is several levels deep in the menus so it is much quicker to get to them using the pop-up. These menus also go away if you click on the graphics background.

---

[5] Some simple customization methods are shown in the Appendix. HINT: *Right click* the mouse on a toolbar.

Helpful Hint

While you are learning the interface, it won't hurt to periodically execute a right-click. This will usually do no harm, and will let you get familiar with the commands available in this way. Remember that the automatic and RMB pop-up menus are context sensitive, so they will change depending on what you are doing!

Pull-Down Menus

Although the ribbon dominates the command layout, occasionally some pull-down menus are presented across the top of the Creo Parametric window. For example, click on the ***File*** menu to open it and scan down the list of available commands. Many of these have direct analogs and similar functions to familiar Windows commands. Commands unavailable in the current context are always grayed out. The available menu choices will also change depending on the current operating mode. Many are grayed out at this time since we have no object loaded to work on. We will introduce these as they become available and on an "as-needed" basis as we go through the lessons. Some menu commands will open up a second level menu (these have a '›' symbol to their right).

Helpful Hint

Each time you come to a new dialog window or menu get in the habit of quickly scanning the listed commands and noting the pop-up messages. Such exploration builds a familiarity with the location of the commands.

Dialog Windows

Many versions ago, Creo Parametric moved away from a cascading menu style (a legacy of its Unix heritage), and adopted a conventional Windows-style interface almost everywhere. Dialog windows play a major role. In these windows you must enter text or numeric data, select options from pull-down lists, set toggles (radio buttons), check options, and so on - pretty familiar stuff. The idea is to set/select data in the dialog window and then select ***OK.*** Occasionally you will have to select ***Apply*** in order for the settings to "stick" - if you select ***Cancel***, all changed dialog settings are ignored. You can usually set the options in any order although sometimes the layout and data entry fields of the dialog window will change when particular options are selected (in which case, obviously, order matters).

Menu Picks

The conversion of the Creo Parametric interface to eliminate cascading menus is not quite complete. There are a very few places in the program where remnants of this old style still exist[6]. In these cases, the text commands (and command options) are initiated using picks on menus that will pop up at the time they are needed, with commands and options arranged in a vertical list. As you move the mouse pointer up and down within the list, each command will highlight and a one-line message describing the command under the pointer will appear in the message area. Since these occurrences are rare, an example is shown in Figure 3.

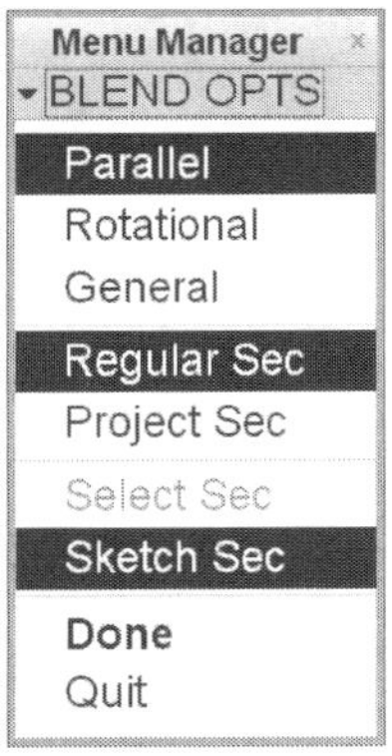

Figure 3 Example of old style cascading menu

In the cascading menus, the pre-selected default command in a menu is highlighted (white on black). This default can often be accepted by clicking the middle mouse button (a *middle-click*) with the mouse positioned anywhere in the graphics window. With the exception of this default command, you execute a menu command by picking it using the *left* mouse button. Menu choices that are "grayed-out" are either not available on your system or are not valid commands at that particular time. Generally, you go through these menus top to bottom and when you pick a command or option, other menus may pop open below the current one. When these represent options for the current command, the default option will again be highlighted. You can select another option by left-clicking on it. There may be several groups of options on a single menu separated by horizontal lines. When all the options in a menu are set the way you want (by *left-clicking*), click on ***Done*** at the bottom of the option menu window (or *middle-click*).

Helpful Hint

Clicking the *middle* mouse button is often synonymous with selecting ***Done*** or pressing the ***Enter*** key on the keyboard.

You can often back out of a command menu by pressing an available ***Done-return*** or ***Quit*** command, or by pressing a command on a higher menu. At times, you will be given a chance to ***Cancel*** a command. This usually requires an explicit confirmation, so you don't have to worry about an accidental mouse click canceling some of your work.

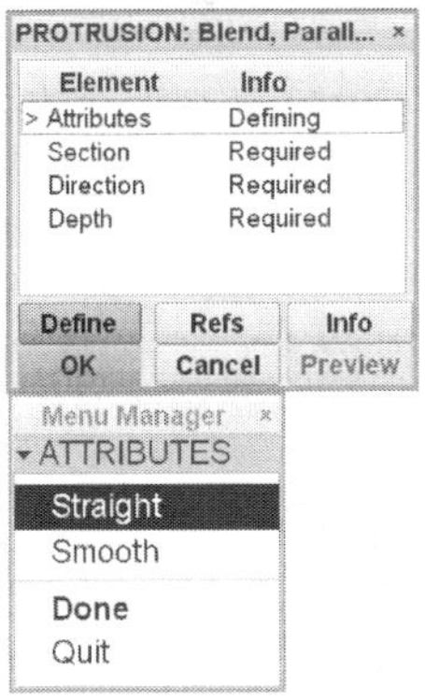

Figure 4 Example of the **Feature Elements** window

Frequently these cascading menus are associated with another holdover from the past - the **Feature Elements** window. Although we will not likely encounter any of these in this tutorial, they are mentioned here for completeness. Figure 4 shows an example that used to occur when creating a blend. In these windows, various feature elements can be selected in the first column and then specified using the ***Define*** button. When first entering this window,

[6] These sometimes also occur when working with legacy parts, that is, parts created in a previous version of Creo Parametric. See, for example, the variable section sweep feature (#5) in the part *creo_handle* that we will load a bit later this lesson.

the feature creation process proceeds automatically to specify the required elements. Prior to acceptance (with ***OK***), options can be set and/or the feature can be ***Preview***ed.

Keyboard Entry

Generally, you will only use the keyboard to enter alphanumeric data when requested, such as object or file names, numerical values, and so on, into dialog windows or pop-up prompts that open as required. The keyboard is also used to launch *mapkeys*, which are special keyboard inputs that will launch a pre-programmed sequence of commands (commonly called macros). See the *Advanced Tutorial* for further information.

You will have to get used to being aware of several areas on the screen: the ribbon at the top, the Navigator window, the graphics window, and the command/message area at the bottom. At the start, this will get a little hectic at times. Until you become very familiar with the menu picks and command sequence, keep an eye on the one-line message description in the message window. There is often enough information there to help you complete a command sequence. (Also, read the Hint on page 2 about working with another person to start with.)

Helpful Hint

Although it is a rare occurrence, if your mouse ever seems "dead" (menus, toolbars, and so on won't respond to mouse clicks) then check the message window; Creo Parametric is probably waiting for you to type in a response somewhere, perhaps in a dialog window that has moved behind the visible one.

Mouse Functions

The mouse is by far the most important input device. Creo Parametric is designed to be used with a 3-button mouse. If it has a middle scroll wheel, all the better. The mouse buttons are sometimes used in combination with keyboard keys (Shift, Alt, and Ctrl). We will assume here that you have the default mouse set-up (with apologies to left handed users who may have re-mapped the mouse buttons). As you will have anticipated, most selections of menu commands, shortcut buttons, and so on, are performed by clicking with the left mouse button (LMB). In this book, whenever you select, click, or pick a command or entity, this is done with the LMB unless otherwise directed. A middle- or right-click will always be explicitly stated as MMB or RMB, respectively. Entity selection with the LMB also involves the Entity Selection Filter at the bottom right of the graphics window, as shown in Figure 2.

The functions controlling the view of the object in the graphics window are all associated with the middle mouse button (MMB) and scroll wheel (if the mouse has one). These are the important Spin, Pan, Zoom functions as shown in Table 1.1. Some of these are used in combination with the Shift and Control keys on the keyboard. The action of the mouse is also affected by the selection of a View Mode. All dynamic view operations involve dragging the mouse. The more comfortable you get with these mouse functions, the quicker you will be able to work. They will become second nature after a while. We will open a part in a few minutes so that you can investigate view control functions and there is a special part for you to play with at the end of the lesson to practice dynamic rotations.

The dynamic view controls in Table 1-1 refer to display of 3D objects. When viewing 2D objects such as drawings, some of the mouse functions change slightly (primarily the spin command). Mouse functions associated with the right mouse button (RMB) will also be introduced a bit later in the lessons. The main function associated with the RMB is to launch context sensitive pop-up menus as described above.

How this tutorial will represent the command sequence

Creo Parametric generally operates on the assumption that you are an experienced user, so does not blatantly display a lot of prompting and/or unnecessary information that would slow it (or you) down. Prompts and cues that it gives are short, crisp, and sometimes quite subtle (like the color of a line or shape of a small icon on the screen). Not much hand-holding here. And certainly no cute but annoying wizards! This is great for power users, but may be intimidating for new users. Just remember that even power users started out from scratch at some point.

We will try to discuss each new command as it is entered. Eventually, you may be told to enter a long sequence of commands that may span several menus and/or require keyboard input. Fortunately, since Creo generally uses standard Windows interaction methods, it is pretty easy to figure out how to tell Creo Parametric what you want it to do. You will know that you are beginning to understand the interface when you can enter an area of the program you have not seen before and correctly anticipate the required input.

/* /* /* **Caution** *\ *\ *\

This tutorial tries to present the command sequence as accurately and concisely as possible. This is a difficult task due to the different ways you will be interacting with the program, the many available shortcuts, and the diverse nature of the presented information (text, graphics, tool icons, line colors, menus, dialogs, etc.). Also, many functions are highly automated with the use of numerous defaults. In these cases, an inadvertent mouse click on the wrong thing or at the wrong time can sometimes lead you far off the tutorial path. So, in the early lessons, pay close attention and try not to jump ahead. You can expect a few cases where you may have to backtrack to get back on the right path.

Table 1-1 Common Creo Parametric Mouse Functions (3D)

Function		Operation	Action
Selection (click left button)		LMB	entity or command under cursor selected
Direct View Control (drag holding middle button down)		MMB	Spin
		Shift + MMB	Pan
		Ctrl + MMB (drag vertical)	Zoom
		Ctrl + MMB (drag horizontal)	Rotate around axis perpendicular to screen
		Roll MMB scroll wheel (if available)	Zoom
Pop-up Menus (click right button)		RMB	launch context-sensitive pop-up menus

The following notation will be used to represent command input to the program. Commands are always shown in ***bold italics***. Dialog window titles are shown in **UPPERCASE BOLD**. Otherwise, the first couple of times a command is used:

- If a command is launched using a ribbon button, that will be stated in the text. The ribbon tab will be indicated in braces, as in ***{ Model }***, with the command text given, often with the button icon shown to help you identify it. For example:

  { Model }:Regenerate 

  If the command is in a ribbon group overflow area, that will be indicated in the text.

- If you are to select an option from a pull-down list, you will see the name of the option with the desired list member in parentheses as follows:

  { View }:Saved Orientation  ***(Front)*** 

- For commands picked from the Quick Access Toolbar or Graphics Toolbar, you will see the name and button of the command (with options in parentheses) as follows:

  Repaint 

- If a setting is a simple toggle, you will see the name of the option with the toggle setting (On or Off, Yes or No) in parentheses as follows:

  Bell (Off)

- If you are to enter data through the keyboard, you will see the notation using square brackets "*[...]*" as follows:

  [block]

  In this case, just enter the characters inside the square brackets.

- If you select a command (usually from the pull-down menus) that starts up another menu or window, followed by a selection from the new menu, you will see the notation using the "➤" sign as follows:

  menu1_command ➤ menu2_command

- If a number of picks are to be made from the same menu or window you will see the notation using the "|" sign as follows (these are generally listed in a top-to-bottom order in the menu but can be chosen in any order in dialog windows):

  option1 | option2 | option3

  Be aware that in some dialog windows, the contents, layout, and data entry fields may change substantially if some options are chosen.

Thus you might see a command sequence in a lesson that looks like this:

{ File }:Options ' Entity Display ' Default Geometry Display (Shading) | Show Colors (On) | Show Datum Planes (Off) | OK

Fortunately, such complicated command instructions are rare!

How to get Online Help

As you go through these lessons, you might want to consult additional reference material. Extensive online help is available. The help documentation, consisting of the entire Creo Parametric user manual set (many thousands of web pages), is viewed using your default browser. There are several ways to access the help files:

1. Select (Do this now!)

   {File}:Help ➤ Creo Parametric Help

   You will see a new browser window open as in Figure 5. Close this window for now, but come back later for more exploration.

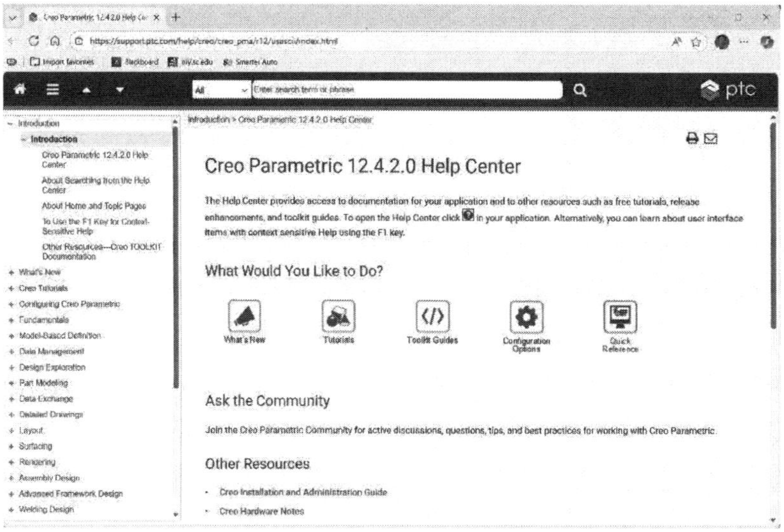

Figure 5 Launching the Creo Parametric Help Center

2. Click the ***Help*** button towards the right end of the top toolbar. This opens the same window as Figure 5. Close it again.

3. For context sensitive help, hold the mouse cursor over a button or command and press the ***F1*** key. For example, hover the cursor over the command ***{Home}:Select Working Directory*** and press F1. You should see Figure 6. Close this browser window.

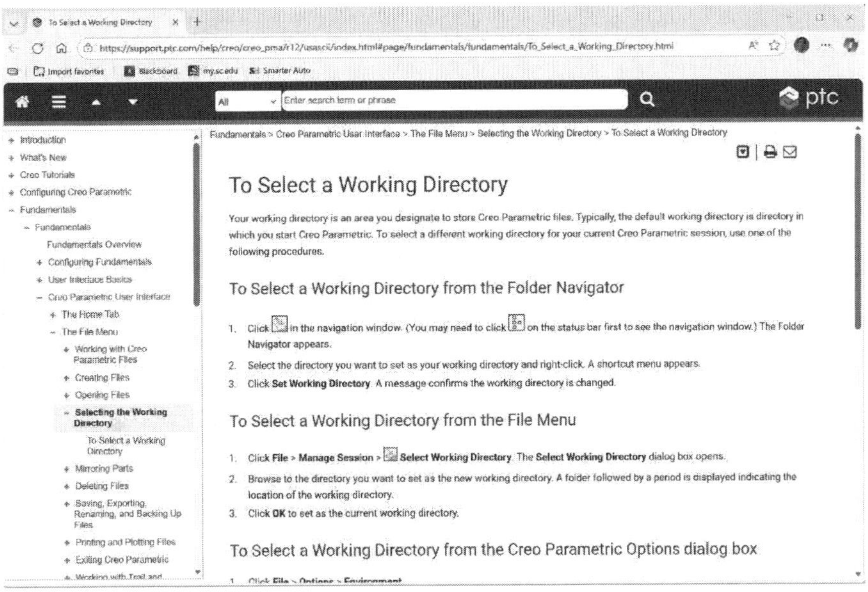

Figure 6 Getting context sensitive help with F1

As a sampler of the extensive on-line help available, click the ***Help*** button  towards the right end of the top toolbar. In the Navigation panel in the **Help** window on the left pick ***Fundamentals***.

You should see the window shown below in Figure 7. In the panel on the left side select the link ***Fundamentals Overview***. See Figure 8.

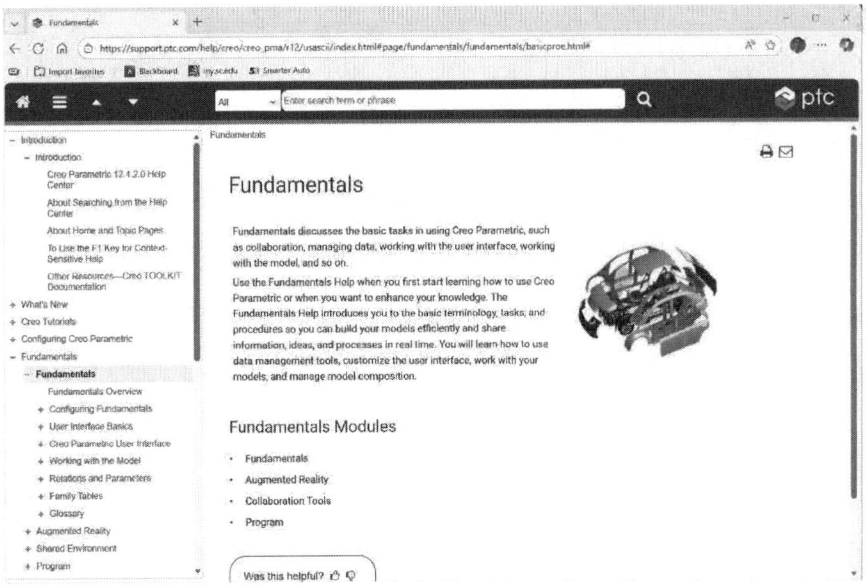

Figure 7 Opening the on-line Help Center

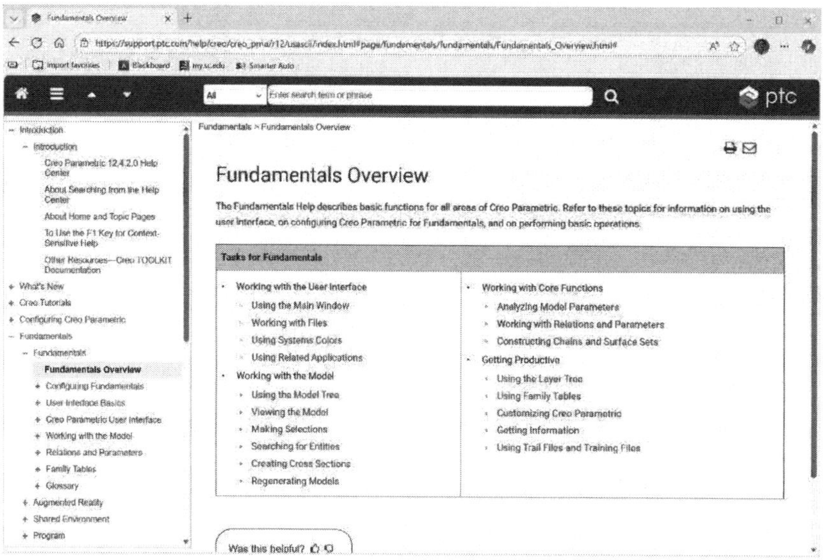

Figure 8 Creo Parametric Fundamentals help area

In the table of contents displayed in the left panel, entries will expand to show subheadings and topics. See if you can locate the topic shown in Figure 9.

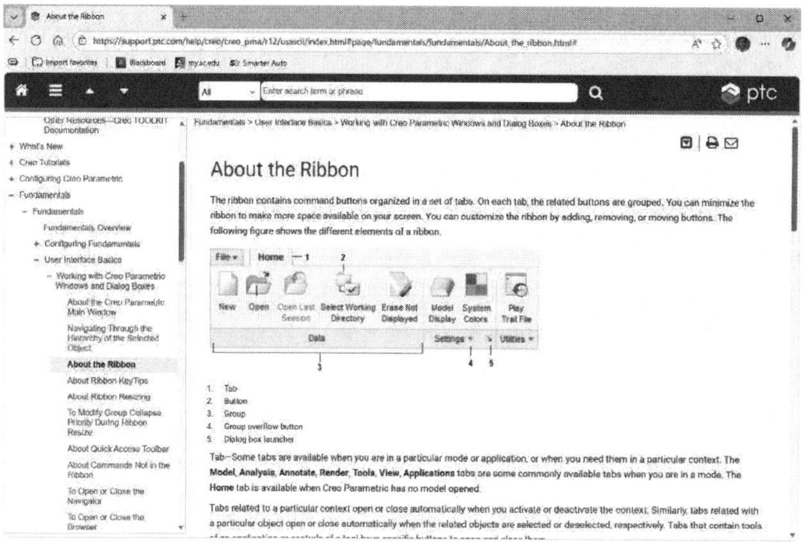

Figure 9 Exploring the Help pages

If you have a few minutes to spare now and then, browse through the manuals (especially the *Fundamentals* section).

Enter the term ***[helical sweeps]*** (remember this is without the square brackets) in the text box beside the ***Search*** button under the top banner. A number of topics will be listed - pick ***About Helical Sweeps***. You should see the contents shown in Figure 10. Notice that the Help pages include numerous examples and step-by-step instructions.

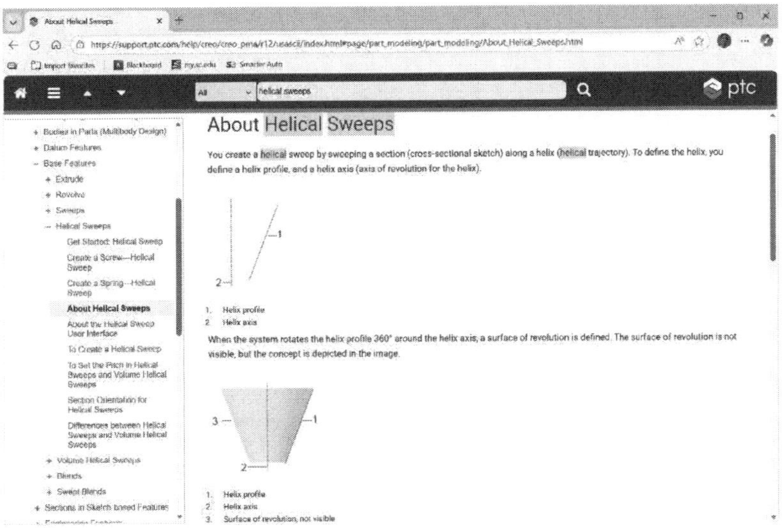

Figure 10 Using the Search function in the Help Center

In the beginning, it will be a rare event when you explore the Help pages and don't pick up something useful. If you desire and have the local facilities, you can obtain hard copy of these pages using the print button at the top of the Help window. Be aware of the cost and time involved in printing off large quantities of documentation that you may only need to read once (and think about the trees, too!). Close the **Help** window.

Finally, at the top right of the window is the set of buttons shown in Figure 11. The one at the far right will launch the Help system. The **Command Search** button in the middle will allow you to enter a command and show you where it appears in the interface. This is obviously handy if you know the command exists but have forgotten where to find it[7]. Finally, the **Minimize Ribbon** button will let you toggle the display of the ribbon groups on and off just by selecting the group tab. You can experiment with these buttons as we proceed through the rest of the lesson.

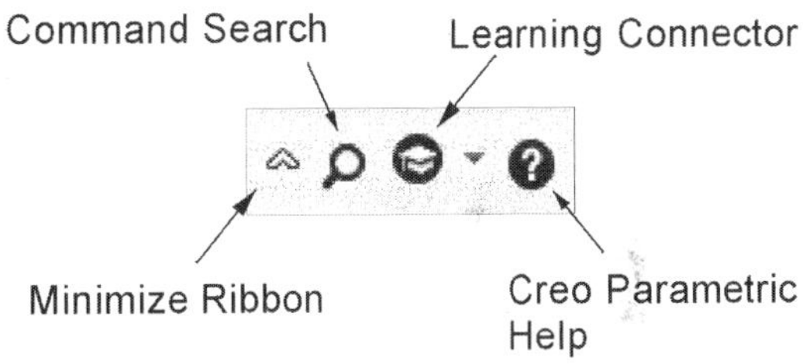

Figure 11 Utility menu

Close all the Help windows and proceed on to the next section.

Tutorial Files and Working Directory

We're finally getting close to loading a model! A number of files used in this Tutorial (parts, assemblies, drawings, and support files) are available for download from the

[7] It can also find and launch legacy commands that are not included in the ribbon interface.

publisher (see the Introduction). Your instructor may have made local copies of the files available.

These files will have to be stored somewhere on your local hard disk. Note that they must all be in the same directory. This should be your default working directory for Creo Parametric. To find out where this working directory is, since Creo Parametric is up and running at this point, all we have to do is try to open a file. Creo Parametric will automatically look first in the current working directory. Give the following commands:

File ➤ Open or ***{Home}: Open***

or use the Quick Access Toolbar button, or shortcut (***Ctrl+O***). In the left pane of the **File Open** dialog window, select the ***Working Directory*** folder (symbol). The full path to the current working directory is shown near the top of the dialog window (see Figure 12)[8]. This uses the Windows Vista style for displaying the path.

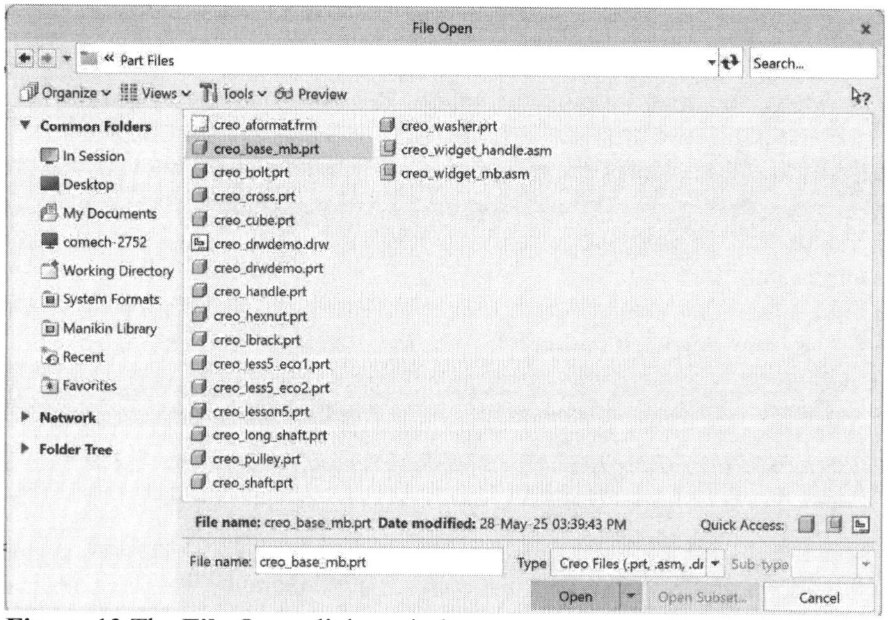

Figure 12 The **File Open** dialog window

An alternative to the ***File Open*** command we used above is to directly select the ***Working Directory*** in the Navigator (under the **Folder Browser** tab). This will open the directory in the Browser. We will experiment with this alternative a bit later.

If your *File Open* dialog window contains the files listed in Figure 12 (you may have to scroll down the list), you can proceed on to the next section on view and display controls,

---

[8] On Windows systems, the start-up working directory can be set in the shortcut that launches the program. For a desktop shortcut, right click on the icon and select ***Properties ➤ Shortcut*** and enter the path to the desired working directory in the *Start In* field.

although the following discussion of options for file management is important.

If the window does not contain the files listed in Figure 12, then either they have not been installed or they are in another directory. Select ***Cancel***, then depending on your circumstance, pick from either option A or B below.

OPTION A - Tutorial files not installed

Minimize Creo Parametric. The tutorial files are available for download from the following site (Look for Commercial Edition Exercise Files):

http://www.sdcpublications.com/downloads/978-1-63057-837-4

Download the zip file to a convenient directory. Unzip and make sure the files are being put into your working directory. You can then restore Creo Parametric and proceed to the next section on view controls.

OPTION B - Tutorial files are in another directory (changing the working directory)

One solution here is to copy the files from wherever they are to the current working directory. Presumably, however, they are in the other directory for a reason. So, you have to tell Creo Parametric to change its working directory by using the commands

{Home}: Select Working Directory

then navigate using the standard Windows operations until the desired directory is shown in the path near the top of the window. Accept the dialog with ***OK*** (or remember that a middle click is a shortcut for accepting a dialog window) and your working directory is now changed (see the message area). You can also change the working directory by opening the Folder Tree in the Navigator pane, selecting the desired directory, and using the RMB pop-up menu to select ***Set Working Directory***. The ***{Home}: Open*** command should now bring up a list of the proper files.

Controlling the Screen: View and Display Commands

Opening a Part File

If they aren't already, close the Browser and Navigator windows by clicking on the buttons at the lower left. Select

{Home}: Open

The dialog box shown in Figure 12 will appear, showing a list of part, drawing, and assembly files in the current working directory. These can be identified by the filename extensions: **prt**, **drw**, and **asm**, respectively. Click on the file **creo_base_mb.prt** and select ***Open*** or just middle click. The object shown in Figure 13 should appear, possibly at a slightly different orientation.

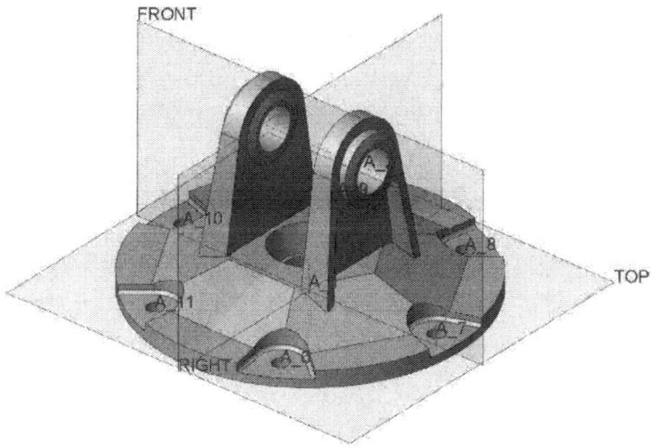

Figure 13 The part *creo_base_mb.prt*

Now that there is a part loaded, the main part mode ribbon and another toolbar have appeared. Check out the various ribbon tabs (**Model**, **Analysis**, **Annotate**, etc.) and groups paying attention to the pop-up tips. Note that many of the ribbon groups have overflow areas indicated by the ▼ symbol. These contain commands that are used infrequently. The command button icons can be different sizes, and can be displayed with or without accompanying text - users can customize these settings.

The new group of buttons in the Graphics Window is the **Graphics Toolbar**. See Figure 14. Again, use the pop-up tool tip to identify the function of these buttons and drop-down lists. This toolbar can also be easily customized[9] - hold down the RMB with the cursor on the toolbar. Note that you can position the Graphics Toolbar elsewhere on the screen and/or change its size.

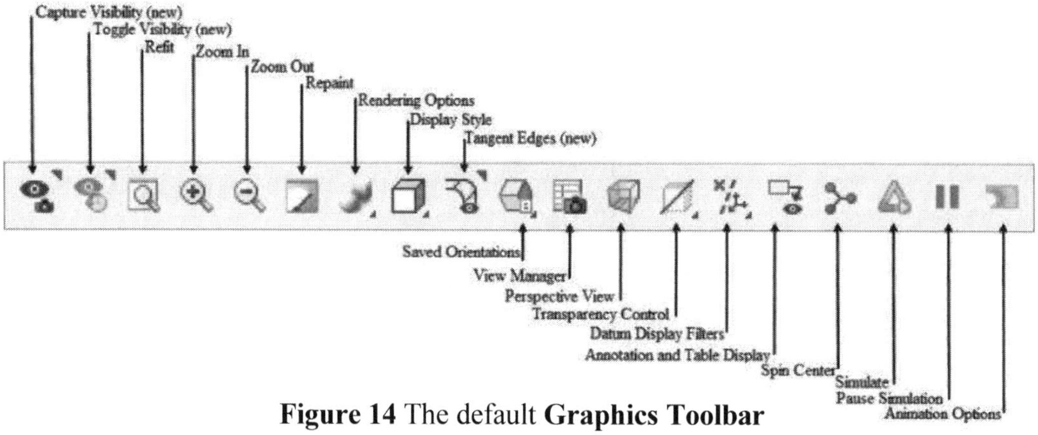

Figure 14 The default **Graphics Toolbar**

Scan these ribbon commands and buttons but resist the temptation to select any of them just yet! If you do, you can usually back out of the command with ***Cancel*** or ***Quit*** or selecting the red ✕ button that might appear at the top of a dialog window.

---

[9] Your **Graphics Toolbar** may look different from what is shown in Figure 14. Again, you can customize it any way that suits you.

Figure 15 Datum display

Also visible on the part are a number of brown (or black) lines that represent non-solid geometric construction entities called *datums* (planes and axes). We will be learning lots more about datums throughout these lessons. For now, note that all datums have labels, or *tags*, like TOP, FRONT, A_1, A_6, and so on but their display may be turned on or off. To do that go to the ***{ View }: Show*** group and find the buttons that toggle the tag display (see Figure 15). Note that there are similar looking buttons that turn the display of the datums on and off (these are also in the **Graphics Toolbar**). You will also discover that datum planes have a positive side and a negative side.

Now that we have a part loaded, we will examine a number of ways to control how Creo displays objects on the screen. The **{View}** tab and buttons on the **Graphics Toolbar** contain commands for controlling the visual appearance of the model - its orientation, size, and placement on the screen. This includes commands for zooming in and out, rotating the object, panning across an object, or selecting a pre-defined viewing direction (like TOP, FRONT, etc.).

Display controls determine how the object is represented on the screen. Options here include wireframe, hidden line, or shaded displays. All displays can be with or without color turned on. You can also control the display style of some special line types (for example the common edge of two surfaces that are tangent to each other - called tangent edges). Let's see how these controls work.

View Controls using the Mouse

The operation of the mouse for controlling the view depends on two related view controls: the spin center and the orient mode. We'll deal with the spin center first.

Spin Center On (Default)

The display of the spin center is controlled by a button  on the Graphics Toolbar. Make sure the spin center button is on (pressed in). In the approximate center of the part you should see a small red-green-blue triad. This is the spin center.

The main mouse functions for view control of 3D objects are shown back in Table 1-1 on page 10. The dynamic view controls for spin, pan, and zoom are all performed by dragging the mouse while holding down the middle mouse button (MMB). Try the following:

SPIN Hold down the MMB and drag the mouse. The object will spin, with the spin center staying fixed on the screen.

ZOOM Hold down the Control key (Ctrl) and MMB. Drag the mouse towards and away from you. This will cause your view to zoom in and out on the object. Try this with the cursor at different points on the object or screen. The center of the zoom is at the initial location of the mouse cursor.

PAN Hold down the Shift key and MMB. Dragging the mouse now translates (pans) your view across the object.

ZOOM If you have a mouse with a scroll wheel, try turning that to zoom in and out on the object. Once again, the zoom is centered on the initial location of the mouse when you start scrolling[10].

ROTATE Hold down the Ctrl Key and MMB. Drag the mouse left to right. The object will rotate around an axis through the initial location of the mouse cursor and perpendicular to the screen.

Spend some time experimenting with these, especially the spin command which may take some getting used to. Since you will be working in a complex 3D environment, you will be using these more than any other commands in Creo Parametric, so you should be very comfortable with them.

For added practice with the dynamic view controls, there is a special exercise (#3) at the end of this lesson.

See if you can obtain (approximately!) the three part orientations shown in Figure 16 at the right. These are the standard engineering views (top, front, right). These views are so commonly used that there is a shortcut to obtain them that we will see in a minute or two.

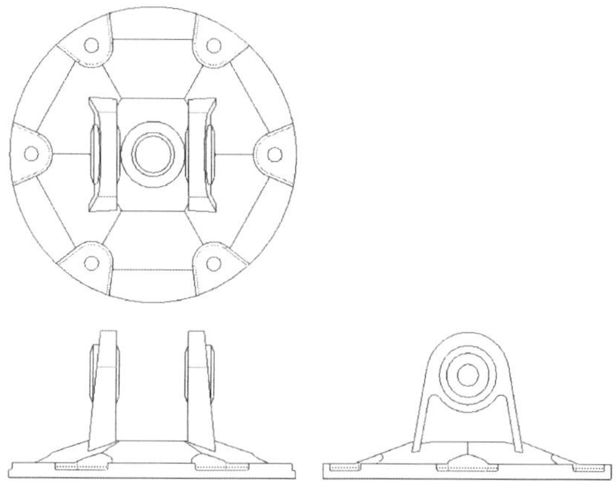

Figure 16 Standard engineering view orientations (top, front, right)

Spin Center Off

Now turn the spin center off using the button in the Graphics Toolbar and see what happens with the mouse controls. The pan and zoom controls work as before. However, using the MMB, the spin occurs around the point on the screen where the cursor is located rather than the spin center. You will find this a useful function when you are zoomed way in on the model and want to spin the part around a specific point. If you try this with the spin center turned on, you are liable to spin the object right out of view. So, remember that spin center OFF means that spin occurs around the cursor; spin center ON means spin occurs around the spin center. Leave it off for now as we explore the other view control button.

[10] Yet another option is ***Zoom To Selected*** on the RMB popup menu.

Orientation Modes

To see the second control that works with the mouse (with Spin Center off) select the following (in the ***Orientation*** group overflow)

{View}:Orientation* ➤ *Orient Mode

Several changes occur on the screen. First, at the location of the spin center, a small black circle appears. Second, there is a black diamond with a red center. This is the current center of the dynamic view and can be reset by clicking the MMB. Third, depending on the settings for your installation[11], the datum planes may disappear. Finally, the mouse cursor changes shape from the arrow pointer to a sort of spinning lollipop. If you drag the MMB, the model will spin around the red diamond on the screen. Experiment with the operation of the mouse (for example, holding down the Ctrl key while spinning with the MMB will cause the model to rotate around an axis normal to the screen). Check the pan and zoom functions - they work the same as before. So far, not much difference.

Now, hold down the right mouse button. You will get the pop-up window shown in Figure 17. Try selecting the ***Velocity*** option in this menu. The diamond changes to a circle. Now, drag a short distance with the MMB, and keep it pressed. You will see the object spinning at a constant speed. The speed and direction of the spin is controlled by the mouse position. Drag the mouse and the spin direction and speed will change. Spinning stops when you release the mouse button. When you are in velocity mode, both pan and zoom will occur at a constant velocity until you release the mouse button.

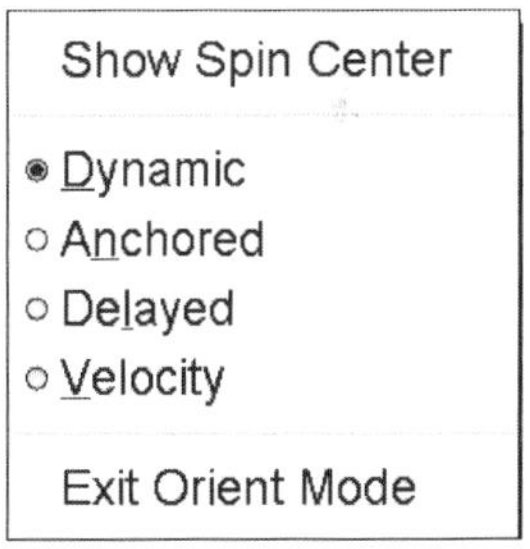

Figure 17 The ***Orient Mode*** pop-up menu

Hold down the RMB and select the ***Delayed*** option. The circle changes to a square. Any view changes are delayed until the mouse button is released. One problem with delayed mode is it is very easy to make changes that are too large (especially pan) and when the view is updated the object is completely off the screen. If that happens, use the ***Refit*** button in the Graphics Toolbar. Note that this cancels ***Orient Mode***, as does pressing any of the datum display buttons.

Finally, in the RMB pop-up, select ***Anchored***. The square changes to a red triangle. Place the cursor on any straight edge. Middle click to "grab" the edge (a very rare time that you use the MMB to select anything). Click and hold the MMB again without moving the mouse pointer, then you can drag the mouse. A portion of the chosen edge will highlight in red, and the model will spin around the highlighted line - it is anchored to it.

---

[11] There is a configuration setting (*spin_with_3D_datum_planes*, default YES) that determines whether the datum planes are displayed during some operations. When this is set to NO the datums will disappear whenever you spin the model. Another option *spin_with_part_entities* (default NO) will affect datum axes.

As you can see, there are a lot of visual cues to alert you to the current view control state, especially the spin options. If you can't remember them all now, no worries since a quick flick of the mouse with the MMB down will tell you the current mode.

The black circle means that ***Orient Mode*** is on. You can turn it off either using ***Refit***, the same ribbon toolbar icon, or select ***Exit Orient Mode*** in the pop-up menu (Figure 17).

If the spin center is turned on when you enter Orient Mode, the RMB view mode operations still work as we have just seen. In this case, though, spinning occurs relative to the spin center, as you might expect! For ***Anchored***, the model spins around an axis through the spin center parallel to the chosen edge.

Graphics Toolbar View Commands

Let's explore the buttons on the Graphics Toolbar. The ***Repaint*** button (sometimes called ***Redraw***) looks like . You can also simply press ***Ctrl+R*** (hold the Control key while you press R). This command causes a complete refresh of the graphics window, which is sometimes helpful to remove entities no longer required for display (like feature dimensions). Try out the other view buttons in this group (***Refit, Zoom In, Zoom Out***), watching the message area for prompts. The handiest of these is probably ***Refit*** since it is easier to zoom with the scroll wheel on the mouse.

Using Pre-defined Orientations

In addition to the dynamic viewing capabilities available with the mouse, you can go to predefined orientations. To view the object in the default orientation, select the ***Saved Orientations*** shortcut button and click on ***Default Orientation***. Alternatively, you can select

{View}: Standard Orientation

or press ***Ctrl+D*** (hold the Control key while you press D). Try the other named orientations (TOP, FRONT, and so on) in the saved view list - these should take you immediately to the view orientations shown in Figure 16. When a drawing is made of a part or assembly, these are exactly the views that most commonly appear. You should bear this in mind when you are creating parts - orientation matters! In the named views (TOP, FRONT, RIGHT) you are looking directly at the positive side of the datum plane of the same name.

Most parts you create will have these standard engineering views already defined, using standard part templates which we will discuss later. If you want to experiment with creating your own named views (handy for documenting the model), select the command (starting in the Graphics Toolbar)

Saved Orientations ➤ Reorient

which brings up the dialog shown in Figure 18. To expand the window, click on ***Saved Orientations***. If you want to set up your own named views, the general procedure for the ***Orient by Reference*** type (selected in the pull-down list in the middle of the window) is to select two orthogonal surfaces or datum planes and tell Creo Parametric which way they should face in the named view. These are called the view references. References can be chosen to face the top, front, left, right, bottom, or back of the screen. In Figure 18, Reference One is the RIGHT datum plane and faces front (towards you); Reference Two is the TOP datum plane and faces the top of the window. This combination of planes and directions produces the standard RIGHT engineering view (the saved name in the pane at the top of the menu). Note that you could obtain the same orientation by picking Reference Two as the FRONT datum plane facing the left edge of the screen.

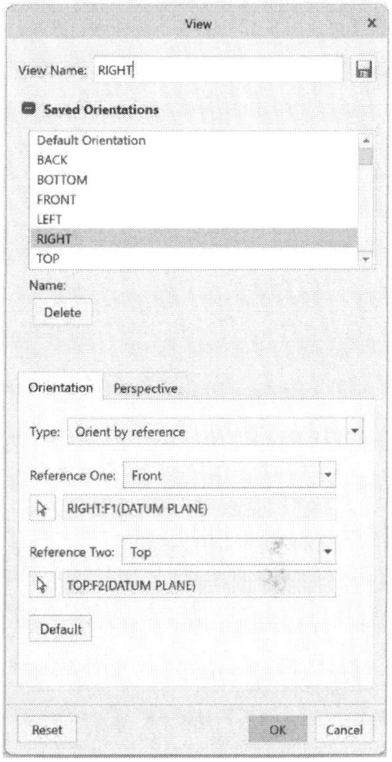

Figure 18 Dialog for creating saved/named views

Once the view orientation options have been selected, you can enter a new view name, then select the ***Save*** button to the right. If you double click a view in the **Saved Orientations** list, the model will spin to that orientation.

As a final note, you might remember that orientation references are not restricted to datum planes, as shown in Figure 18. You can use any planar surface as a reference. It is customary to use datum planes (especially the default planes) since these are less likely to be moved or otherwise changed as the model is created and modified later. Of course, if a view reference moves then the view direction will adjust automatically (assuming both references are still legal).

It is not strictly necessary to have datums situated to define a named view. Check out the **Orientation Type** drop-down list. See if you can figure out how the type ***Dynamic orient*** was used to create the named view ISO_RIGHT. It requires two specific rotations from either the existing FRONT or RIGHT views.

No doubt this all seems terribly confusing since we are using the same words ("top", "front", "right") for three purposes: naming the datum planes, specifying view directions relative to the screen, and giving names to the saved views themselves. Rest assured that this will eventually start to make sense and you will understand the statement "TOP is the FRONT reference with RIGHT facing BOTTOM"! Close the **View** window with ***OK*** (if you have made changes you want to keep) or ***Cancel***.

Object Display Commands

These buttons on the Graphics Toolbar are fairly self-explanatory. They are in a drop-down list with the default ***Display Style*** button:

Shading with reflection - shaded solid model with mirror surfaces
Shading with edges - shaded solid model with highlighted edges
Shading - shaded solid model with no highlighted edges
No Hidden - hidden edges (or portions) are not shown
Hidden line - hidden edges (or portions) are displayed in gray (dark but visible)
Wireframe - all object edges are shown as visible, even those at the back

These display modes persist until changed by selecting a different mode. Try it! These buttons do not affect the display of datums. The datums may disappear momentarily when you are changing the view, for example during spinning. The display mode is very much a matter of personal preference. Keep trying different modes in different situations to see what works best for you. Hidden line mode allows you to see hidden features (although they are dimmed). Shaded mode is useful to reinforce your mental image of the three dimensional shape, when that is sometimes unclear by looking at only the edges.

Making Bodies Transparent

A useful display is to show solids as partially transparent - sort of a compromise between wireframe and solid. Transparency can be set in an *Appearance* (discussed in Lesson 9) but there is a good shortcut for setting the transparency of individual bodies.

At the bottom right of the screen, open the **Selection Filter** (see Figure 2) and set it to ***Body***. Now click on the model. A pop-up toolbar will open beside the cursor. On the bottom row at the far right is the ***Make Transparent*** command. This works for both bodies and quilts. Click that now - the body becomes partially transparent. The default is 50% transparent which can be changed using the ***Transparency Control*** in the graphics toolbar (see Figure 14).

Datum Display Commands

The buttons in the datums group of the Graphics Toolbar (Figure 19) control the visibility (on or off) of the datums (planes, axes, points, coordinate systems). A button also controls the shading of the datum planes. Try these out now. This model does not contain any datum points or coordinate systems, so you won't notice any effect from those two buttons. Remember that these buttons do not delete these entities from the model, just turn off their display.

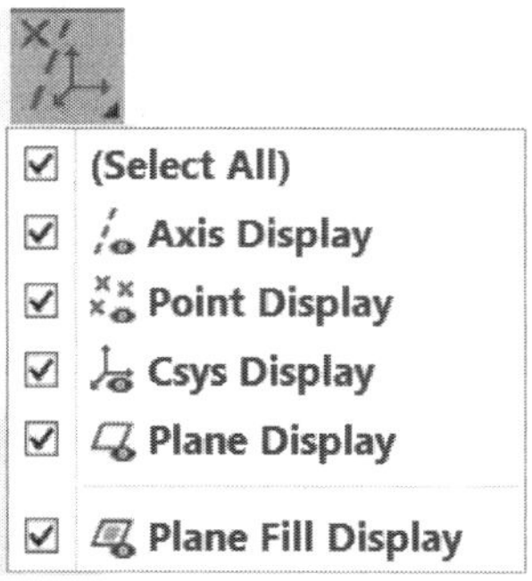

Figure 19 Datum display settings

We have seen previously how to control the display of the datum labels or *tags* using the commands in the ***{View}: Show*** group. See Figure 15. Turn off all the tag display settings.

Some Comments on Display Style

Your choice of display style is strictly personal preference. However, keep in mind the following:

- It is useful to be able to see the entire object at once. ***Hidden Line*** gives you "x-ray" vision to do just that. ***No Hidden*** hides too much, and ***Wireframe*** is sometimes ambiguous.
- On the other hand, the 3D shape of the object is most clearly shown in a shaded image. If you have trouble interpreting the line drawing image, go ahead and shade the part.
- A lot of information in Creo is portrayed using color cues. It is useful, then, to keep your model color neutral so that you won't miss anything. The default color is a neutral gray. Color is applied using *Appearances*, discussed in Lesson 9.

Let's turn the model colors off. To quickly find a function to toggle the colors, select the ***Command Search*** button (see Figure 11). In the text box that appears type ***[color]***. A drop-down list will appear (actually as you are typing in the letters) that contains commands with this keyword. This includes commands that are not in the ribbon. Select ***Model Colors*** to toggle the color display[12]. With colors turned off, explore the various display styles again. Note that transparency is not available for the default appearance if colors are turned off. Before you leave this section, turn colors back on and set the display to ***Shading*** or ***Shading With Edges*** and set transparency to 50%. Also, set the selection filter back to ***Geometry*** and turn the display of datum tags on.

We are about halfway through this lesson, so now is a good time to take a break!

Anatomy of a Part - The Model Structure

Now that we have explored most of the viewing and display commands, let's explore the model itself[13]. If the part **creo_base_mb.prt** is not loaded, do that now with ***File ➤ Open***. Open the Navigator pane and select the left-most control tab (***Model Tree***). Go to the **View** tab and turn on the display of all datum plane and axis tags. The screen should look like Figure 20 below. Let's examine the information in the model tree in the left pane. This contains a list of the features in the part database. The features were created in a top-down order in this list, called the *regeneration sequence*. Each of these features has a name - some are the default names and others have been specified. Beside each

---

[12] This is actually a very handy command to have in the Quick Access toolbar.

[13] It has taken a while, but this is where you finally get to see the fish!

feature is a small icon to help you identify the type of feature. We will discover more about these icons in a couple of minutes. As you click on any of the features listed in the model tree, edges of the selected feature are highlighted on the part in the graphics window using bright green. Try that now.

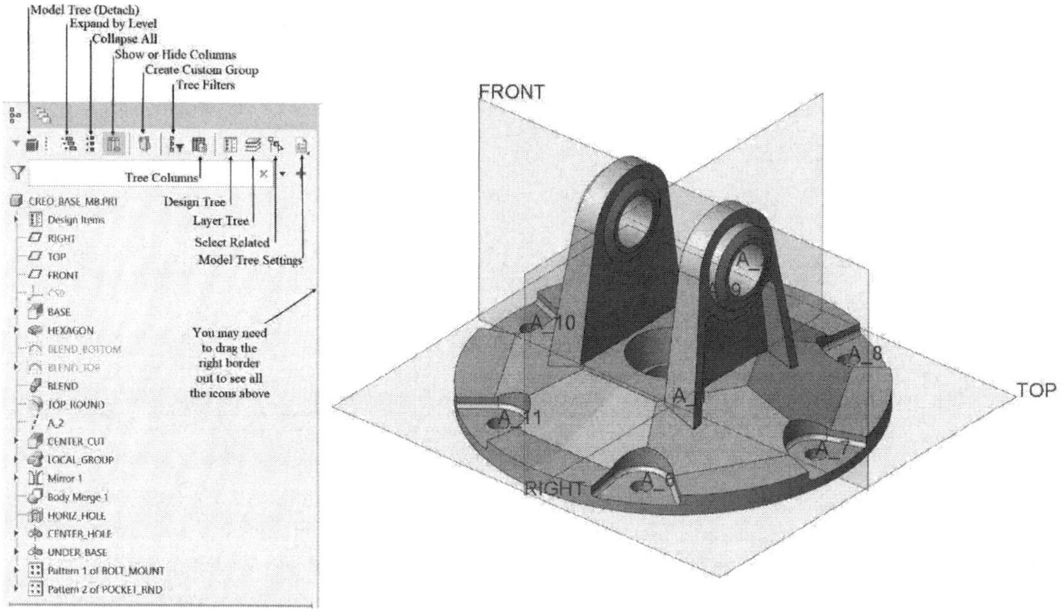

Figure 20 Part ***creo_base_mp*** with model tree (top level)

Across the top of the model tree are some buttons to control what and how model information is included in the tree display. The first button can be used to detach the tree window, that is set it up to float anywhere in the graphics window. We'll discuss some of the other buttons below.

Preselection Highlighting

Preselection highlighting is how Creo Parametric tells you what will be selected on the model if you pick with the left mouse button. This tool is part of what is called the *object/action* command structure. Creo also allows the reverse (*action/object*) style of command. In the object/action structure, you select an object first and then specify the action to be performed on it. For many people, this is a natural way of doing things. To accommodate this mode, Creo allows you to pick most objects directly in the graphics window by clicking on them[14]. When the display gets very cluttered (or the object you want is behind something else), it may be hard to pick on exactly what you want with the first click of the mouse. Thus, preselection highlighting is a method to make sure you are picking what you want in a complicated model.

---

[14] A function called ***Query Select*** is available to do this as part of the action/object selection process. The preselection method is much simpler.

To see how preselection highlighting works[15], first have a look at the ***Selection*** filter list at the bottom right of the screen, as in Figure 21. This filter lets you select what the mouse will detect. Select ***Feature***.

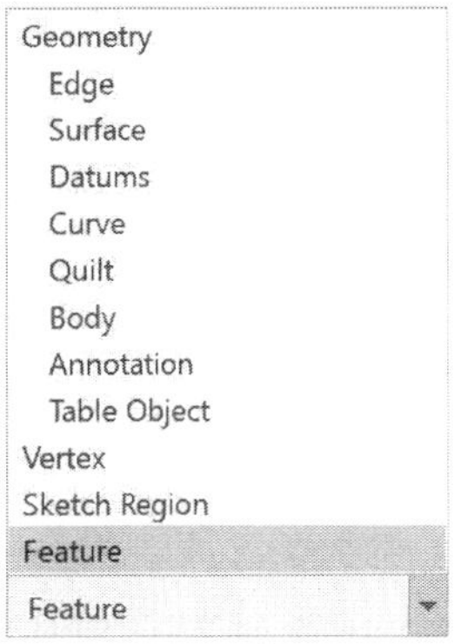

Figure 21 The Selection filter list

Move your cursor over the surfaces of the model, but do not click just yet. Some surfaces of the model will highlight in orange, and a small box will pop open beside the cursor. The box and highlighted edges show you which feature creates the highlighted surface. As you drag the cursor (slowly!) around on the model, the different features that have created different surfaces are identified by color and a pop-up box.

If a feature is highlighted and you pick on it with the left button, the highlighted edges will turn bright green and the feature name will be highlighted in the model tree. If the selected feature is inside a group in the model tree, the group will be expanded in the model tree. The highlighted feature is now selected as the object to be acted upon by the next command. To go deeper into the geometry of that feature, set the selection filter to ***Geometry*** (the default) and slowly move the cursor over the highlighted surfaces. Now the filter will identify individual surfaces and edges formed by the feature. Left click to select a highlighted surface. Now, you can left click again to select an edge (orange to green). The entity currently highlighted in green during this selection process is the active entity. As soon as you have the one you want, you can leave the selection and proceed on to perform some action on this entity (for example by using a pop-up menu).

If you set the selection filter to (***Datums***), only datum planes and axes are trapped by the filter. The selection filter (***Geometry***) is the most detailed - it traps surfaces, edges and vertices and shows who they belong to. Before we proceed, if the **Design Items** list near the top of the tree has expanded, shrink it back by selecting the ▼ in front of the icon. We'll have a look at this a bit later. The tree should look like Figure 20.

Expanding the Model Tree

We would like to set up the model tree to show a bit more information. We will find out how to do this from scratch in a later lesson. For now, we are going to read in a model tree configuration file that has been created for you. This should be in your working directory, along with all the part files downloaded from the web. To load it, select the **Model Tree Settings** button  at the top-right corner of the model tree. Then pick

Settings ➤ ***Import Tree Settings***

---

[15] Make sure this is turned on using ***File*** ➤ ***Options*** ➤ ***{Global}: Selection*** and option ***Enable preselection highlighting*** is turned on (checked). This is the default.

In the **Open File** dialog, make sure you are in the current working directory and select the file **Tut_ModTree.ui** and click ***Open* ➤ *Import* (Import for active mode only)**. This will add some columns on the right of the model tree (see Figure 22). You may have to drag the right border out a bit to see all the columns. The columns can all be resized by dragging on the vertical separators at the top. The feature number indicates the order of feature creation (*regeneration sequence*). The feature types are datum planes, protrusions, cuts, holes, and so on.

	Feat #	Feat Type	Feat Subtype	Contributes To
CREO_BASE_MB.PRT				
Design Items				
RIGHT	1	Datum Plane		
TOP	2	Datum Plane		
FRONT	3	Datum Plane		
CS0	4	Coordinate System		
BASE	5	Protrusion	Extrude	BASE_PLATE
HEXAGON	7	Protrusion	Sweep	BASE_PLATE
BLEND_BOTTOM	8	Curve	Sketched Curve	
BLEND_TOP	10	Curve	Sketched Curve	
BLEND	11	Protrusion	Blend	UPRIGHTS
TOP_ROUND	12	Round	General	UPRIGHTS
A_2	13	Datum Axis		
CENTER_CUT	14	Cut	Extrude	UPRIGHTS
LOCAL_GROUP	15	Group Head		
Mirror 1	19	Mirror		
Body Merge 1	24	Boolean Operations		BASE_PLATE, UPRIGHTS
HORIZ_HOLE	25	Hole		BASE_PLATE
CENTER_HOLE	26	Cut	Revolve	BASE_PLATE
UNDER_BASE	27	Cut	Revolve	BASE_PLATE
Pattern 1 of BOLT_MOUNT	28	Pattern		
Pattern 2 of POCKET_RND	47	Pattern	PATTERN	

Figure 22 Model tree with added columns

The subtypes give further classification information about each feature. For example, feature #5 is an extruded protrusion, feature #7 is a swept protrusion, #14 is an extruded cut, #26 is a revolved cut, and so on. We will be spending a lot of time in these tutorials finding out what these features are and how they are created. We will see in a short while what the names in the last column are all about.

Most of the features in this part have been named. This is optional, but a very good idea. Features that have not been named will appear with only the feature type (protrusion, cut, hole, etc.) in the first column of the model tree. In a large model, with dozens or even hundreds of features, this becomes very confusing and not very helpful. You should get in the habit of naming at least the key features in your models.

Notice that there seems to be a few features missing (notice the gaps at 15 - 19, 19 - 24, and 28 - 47). Click on the small triangle beside feature #15 (on the left). This *Group* contains two features: the BOSS and SIDE_POCKET on the right vertical upright. The group is then mirrored (feature 19) through the RIGHT datum plane to the other side.

In this group, expand feature #18. A datum plane (feature #17, DTM2) is grayed out, meaning it is hidden – it is not visible in the graphics window. Select DTM2 in the model tree and in the pop-up menu, select ***Show*** . It now appears in the graphics window. This datum was used to create the 2D sketch for the extruded cut SIDE_POCKET. Select the datum DTM2 in the graphics window and in the pop-up menu, pick ***Hide*** .

Close the groups at features #15 and #19. If there is a bunch of extra stuff on the screen, use ***Repaint***. Find out what is in the Pattern starting at feature #28 by expanding the model tree and use preselection to highlight the groups. Technically, this is called a *radial or axis pattern* since the pattern (called *instances*) was created by incrementing an angle around a central axis. The pattern instances are groups of features.

The last pattern (starting at feature #47) consists entirely of rounds. These rounded edges were kept separate from the previous group pattern because rounds are normally considered cosmetic features. Rounds often cause problems with downstream modeling operations such as creating drawings and performing finite element analysis. Adding them last and keeping the rounds separate means they can be easily selected and temporarily removed from the model (or *suppressed*, see Lesson #5) without disrupting other model geometry.

Compare the feature types and subtypes listed in the model tree with the feature icons on the left. You should be able to easily identify datum planes and axes, and extruded, revolved, and swept features, patterns and groups, and hidden features from their icons.

Note that, curiously, features 5, 7 and 11 are all protrusions but have different icons. The icons depend on the subtype. Compare the icons for features 5 (a protrusion) and 14 (a cut) – both are extrusions. When creating features, you actually select the subtype first (extrusion, revolve, sweep, ...) and then choose whether to add material (protrusion) or remove material (cut). Once you can recognize all the icons, you could remove the subtype column from the model tree and gain back some graphics window real estate.

The Design Tree

On top of the Model Tree, on the right, you should see the ***Design Tree*** icon. If you do not see it, drag the border out to see all the icons. Alternatively, at the top of the Model Tree you can expand it using the double right arrows » to see it.

Select the ***Design Tree*** icon. This moves the design tree into its own window. Even better, grab the Model Tree Detach icon at the top-left and you can detach it, too, to place it anywhere on the screen. See if you can make the windows look like Figure 23. If you drag the design tree window around, you will find it will automatically re-attach itself either to the side or above the model tree.

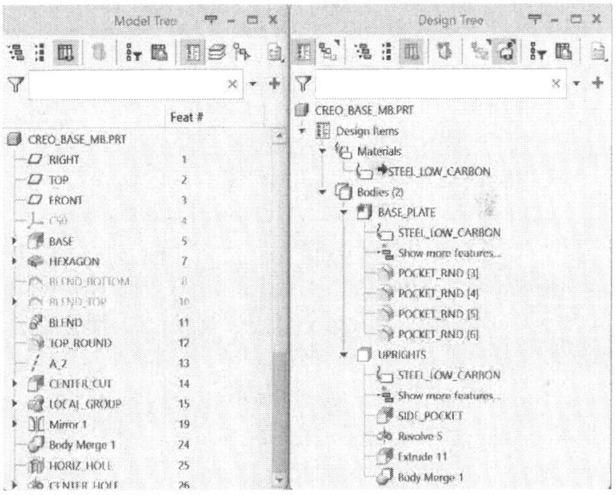

Figure 23 Detached Model and Design Trees

Detaching the Model and/or Design Trees from the Navigator is really handy if you are lucky enough to have a second monitor since you can drag either the model or design tree onto the second monitor so that they do not take up any space in the graphics window.

In the Design Tree window, pick ***Tree Filters***. This lets you choose which items will be visible in the tree window. Select the **Body/Quilt** tab. Observe the icons used to indicate various body states. Make sure the boxes beside ***Consumed Bodies*** and

Consumed Quilts are checked. We'll find out what these are about in the next section. Select ***Apply* ➤ *OK***.

The design tree for this model shows an assigned material and two bodies. It is also possible to add columns to the design tree just as we did for the model tree. Back in the model tree window, turn off the **Design Tree** using the button at the top. It now returns to its previous position at the top of the model tree (with the columns already defined there).

Note that the design tree is available only in part mode.

Introduction to Multibody Modeling

When creating solid geometry, Creo allows new features to be assigned to different bodies within the part. The bodies can then be used in Boolean operations (***Merge***, ***Intersect***, ***Subtract***) that come from a branch of modeling called Constructive Solid Geometry. It is basically a way of organizing features in closely related groups. Note that only solid geometry is involved - no datum features. As features are created, they can use previously created features in any body in the part for references. Multibody modeling is similar to creating new parts within an assembly but without having to worry about complicated assembly constraints or references. Using multibody modeling is a way to organize feature creation that offers additional flexibility in model creation. There are a number of other functions available involving bodies and situations where using multibodies makes modeling much easier.

To see how multiple bodies are used in the current part, expand the **Design Tree** listing for **Bodies (2)** near the top of the model tree. The currently active body (BASE_PLATE) has a blue star on its icon. Expand the listing for each body (BASE_PLATE and UPRIGHTS). Both have materials assigned; in this case, they are the same although that is not mandatory. The solid features contributing to each body are listed. They are also listed on the regular tree. Using **Tree Columns**, display *Feat#* and *Feature Status*. The feature status is shown in the last column. In this case, BASE_PLATE is shown as containing geometry while UPRIGHTS is shown as "Consumed". This status is also indicated in the icon for the body, which is different from the other body. Note the final feature listed for UPRIGHTS is a *Body Merge* feature (#24) involving both bodies. This is the Boolean operation that is used to join two bodies into a single solid.

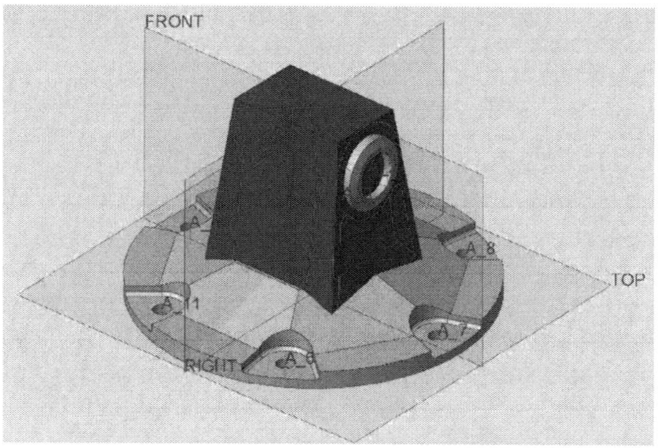

Figure 24 Snapshot showing blend feature (#11)

Let's see how you can visualize the features that were used to create the consumed body UPRIGHTS. In the Design Items list, expand the listing for the body. Click on ***Show More Features*** to get the full list.

Click on the BLEND feature (#11). Some edges on the model highlight in green, but these don't really help visualize the geometry of the blend feature. Now select the same feature, hold down the RMB and select ***Show Snapshot***. This now shows the full shape of the blend. See Figure 24. A snapshot, in the context of multibody modeling, shows you the status of a body part way through the regeneration sequence. The body will highlight in purple with the selected feature highlighted in green.

We will be discussing multibody modeling and how it can be used a bit later. Since it is a major topic but not essential for basic modeling and due to space constraints, extensive treatment is postponed to the *Advanced Tutorial*[16].

The Model Player

Just knowing the feature types and creation order may not give you a really clear idea of how this model was created. The ***Model Player*** offers a useful tool if (like now!) you have obtained a model from someone else and want to explore its structure. This is particularly useful with multibody modeling since, as you can see, there is no hint from the final geometry how the features have been organized into bodies.

Make the Navigator pane smaller so that you can see the feature names, numbers, and other columns but have a good view of the model in the graphics window. Now select

{Tools}: Model Player 

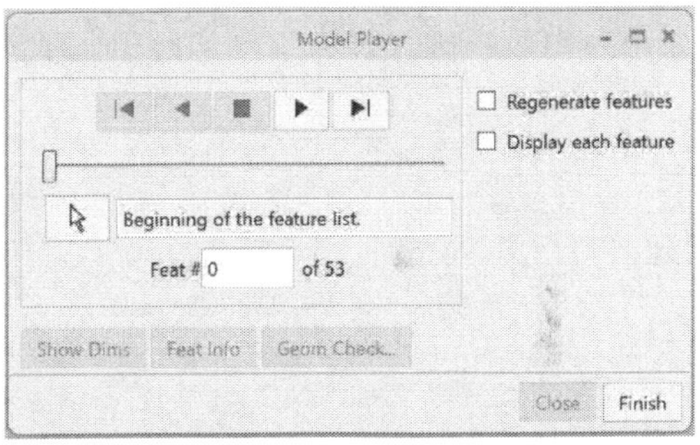

Figure 25 The Model Player window

A new dialog window opens (Figure 25). This operates much like a tape player, with buttons at the top to let you move forward and backward through the regeneration sequence. Let's start by picking on the button on the far left to get to the beginning of the sequence. The display shows we are feature 0, with nothing shown on the screen.

Exploring the Structure of a Part

We're going to step through the creation of this part to give you a feel for how models are created and how the features work together to create the geometry. This very simple part illustrates quite a number of interesting and important ideas about modeling with Creo Parametric. In the following, feel free to change model orientation and display style whenever you want.

In the Model Player, click on the button (second from right) to step forward a single feature. This brings feature #1 into the model (the RIGHT datum plane). Continue

---

[16]There are some tutorials available in the online help that illustrate operations with multibodies. Search for "Multibody Tutorials".

clicking this button until you are at feature #5. While you do this, observe the changes in the model tree as each feature is added. This is the first solid feature of the part. It is an *extruded protrusion*, created by sketching a circle on the TOP datum plane and extruding it perpendicular to the sketch. A *protrusion* adds solid material to a model (a *cut* takes it away). It has been assigned to the body BASE_PLATE (see the model tree).

Click on the ***Show Dims*** button in the Model Player window. An orange circle and a couple of dimensions appear on the model (Figure 26). This was the sketch used to define the shape. The circle has a diameter (note the symbol Ø) of 200. The sketch was then extruded a distance 10 to form the solid feature. These are the *parametric dimensions* of the feature.

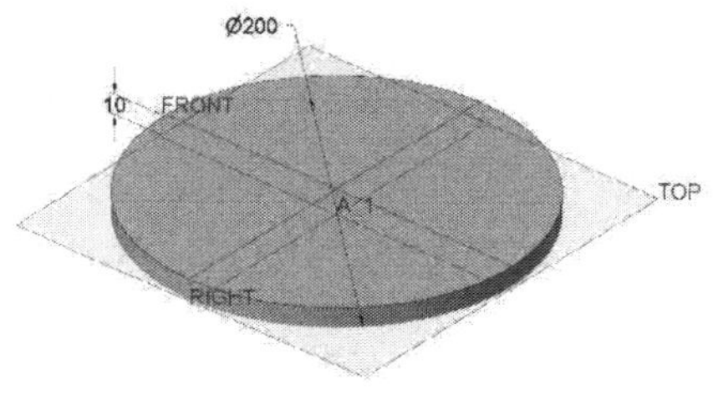

Figure 26 Feature #5 with ***Show Dims***

Move on to feature #7 HEXAGON, a swept protrusion, and expand this on the model tree. Feature #6 Sketch 1 was used to define the trajectory of the sweep - a regular hexagon. After the feature was created, the sketch was hidden in the graphics window - note it is grayed out in the model tree[17]. You can see the section used to create the sweep by selecting ***Show Dims***. Sweeps can be used to create cuts or protrusions. They come in many varieties and are among the most complex features to create. We will look at the simpler versions later in this tutorial[18].

The next features (#8, #9, #10) are all hidden. Selecting them in the model tree will show them in the graphics window in green. These rectangular sketches will be used to create feature #11.

Advance the Model Player on to feature #11 (see Figure 27). It appears in a different color which indicates that it is assigned to a different body - this is similar to the view we saw previously with the snapshot. The blend feature allows you to specify different cross sections along the length of a protrusion. This lets the upright taper from the large rectangle at the bottom to the small one at the top. Click on those in the model tree. We will deal with blends later - they are quite an advanced feature and can do much more!

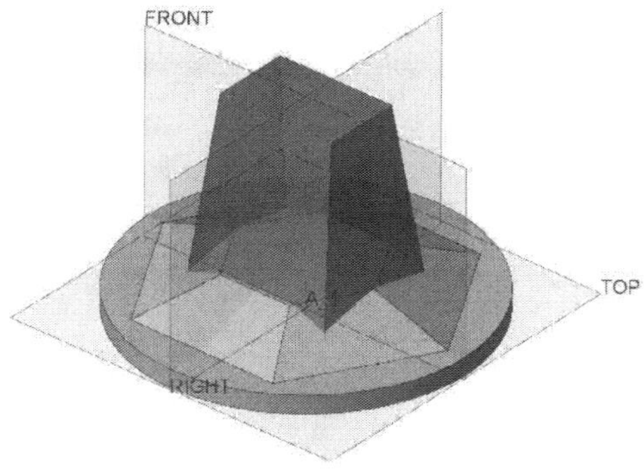

Figure 27 *Blend* feature added

---

[17] This is one of the very few times that you will see the feature numbers out of order in the model tree - note that feature #6 is listed after feature #7. This is actually because the sketch for the trajectory was created first.

[18] An entire lesson in the *Advanced Tutorial* is devoted to more advanced versions: helical sweeps and several kinds of variable section sweeps.

Click the ***Feat Info*** button in the Model Player. This opens the Browser pane and presents a great deal of information about the feature: its feature number, an internal ID number, the number and ID of parent and child features, all its attributes and dimensions, and so on. This information will make more sense to you as we proceed through the lessons. The concept of parents and children is quite important. A *parent* feature is one that supplies references necessary for the creation of a subsequent feature, which then becomes a *child* feature of the parent. The two sketches (features #8 and #10) are parents since they provided the section shapes for the blend. The children of the BLEND use one or more of its surfaces as a reference. At the bottom of the browser is a list of all the dimensions used to specify a feature (not including dimensions of its references). ***Close*** the Browser pane by picking the X button at the upper right.

Continuing on.....Feature #12 is a round that removes the top surface of the blend. Technically, this is called a **Full Round** (use ***Show Dims*** to see that it has no dimensions). Feature #13 is a Datum Axis which is defined down the center of the round. Feature #14 is an extruded cut that removes the central portion of the body.

Continue on to feature #16, the BOSS. This is a revolved protrusion, created by sketching a 2D shape on the FRONT datum plane and revolving it around axis A_2. This was the reason for creating this datum axis - an example of planning ahead. Go to the default view, turn on the datum planes, click on ***Show Dims*** and zoom in on the model to see the sketch.

Proceed to feature #18, which is another extruded cut used to hollow out the upright. Click on ***Show Dims*** and zoom in to see the sketch for this feature. What plane is it sketched on? This is tricky - there was no surface available to define the sketch at the right distance (5 units) from the left vertical surface. So, a special datum plane was used. In previous releases, this would have been called a *make datum* (or a *datum-on-the-fly*)[19]. This is DTM2 (#17) which you can find by opening the group in the model tree. Once the sketch was defined, the make datum disappears from the display (but is listed in the model tree as a hidden feature). Make datums are very powerful tools and we'll find out more about them in lesson 6.

Features #15 through #18 are now mirrored through the RIGHT datum plane to the other upright. The mirrored features form a group (feature #19). Note the group name in the model tree. Continue through to Feature #23. See Figure 28.

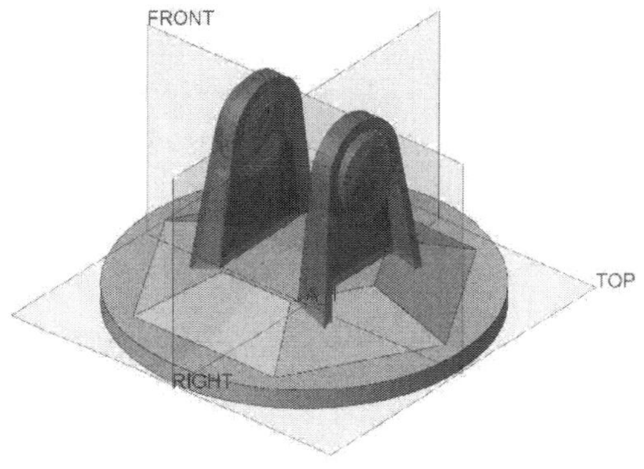

Figure 28 UPRIGHTS body complete and ready for ***Merge*** operation

---

[19] These are also called *asynchronous datums*. Kind of a mouthful!

Note that all solid features starting at #11 have been added to the UPRIGHTS body. Up in the design tree, note the star on its icon that indicates it is the active body.

Notice that it is quite legal and common to have two or more bodies overlap each other. In fact, that is one of their purposes. When solid features are added to the same body, they automatically merge together. In fact, a warning message is created if they don't overlap (something like "protrusion is entirely outside the model"). A solid feature that completely encloses another solid feature in the same body is said to "bury" it. Buried features are often a sign of a poorly thought-out model.

We now move on to a feature (#24 Body Merge) that joins the two bodies together into a single body. This is the Boolean operation ***Merge***. In this case, the UPRIGHTS body has been merged to the BASE_PLATE body and, in the process, has changed its status to Consumed. The model tree shows that it contains no geometry after the merge. This is a default action of the Merge operation.

The next feature (#25) is the horizontal hole going through the two bosses. This hole is defined on the RIGHT datum plane, is coaxial with A_2, and extends through everything in both directions (technically called a *Thru All, Both Sides* hole). Examine the Feature Element Data list in the browser panel using the ***Feat Info*** button in the model player. The only dimension required for this hole is its diameter.

Features #26 and #27 are both revolved cuts in the base of the part. Notice they occur after the Merge (why?). It would be technically possible to combine both these features into a single revolved cut. This would make the sketched shape considerably more complicated (and hence more difficult to create) and would also give less flexibility to the model. For example, feature #27 UNDER_BASE can be easily deleted from the model without affecting feature #26 CENTER_HOLE. If these were both parts of the same revolved feature, this would be more difficult.

Helpful Hint

When choosing the features to make up a model, there is always the question of maintaining a balance between a large number of simple features and a small number of complicated ones. When first starting out in Creo Parametric it is probably better to lean toward the former approach. This is what we'll do in this tutorial, following the "Keep It Simple" principle.

Looking in the model tree, observe that immediately below feature #27 there is an entry for a *pattern*. Expanding this to the next level shows a number of groups. Expanding the first group (#29) shows two features: a hole and a cut. Advance the model player to feature #31. Left click on each of the last few features to see them highlighted in the model. The group is then duplicated around the circumference of the base. This is called a *radial pattern* which is produced by incrementing an angular dimension parameter associated with the first group (the pattern leader). There are many other different ways of forming patterns - we will see several of these later in lesson 7.

Advance the model player all the way to the end of this pattern by entering **46** in the **Feat** # text box in the model player. This is the last feature in the last group of the pattern.

This almost concludes the regeneration sequence of the part using the model player. Note that you can step backwards through the sequence or enter the desired feature number directly in the window. Some other optional functions can be explored on your own.

The last six features in the part are rounds. Rounds are considered to be cosmetic features and are usually added to the model last. Bring them into the model by pressing the ***Finish*** button in the Model Player.

Close up all the levels in the model tree so that only the first level is showing as in Figure 20.

Modifying Dimensions

There is lots more we can do with the information in the model tree. For example, here are three ways you can modify the dimensions of a feature. So that we can watch what is happening, in the ***{Model}:Operations*** group select ***Regenerate*** and in the drop down menu make sure ***Auto Regenerate*** is ***Off*** (not selected).

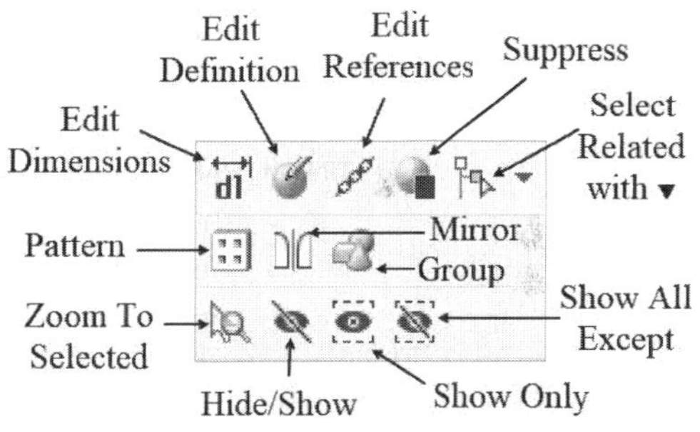

Figure 29 Typical pop-up menu

First, select feature #5 (BASE) in the model tree. Right beside the cursor, a pop-up menu will appear as in Figure 29. This is typical of the types of menus that will appear from time to time. As described previously, the contents are context sensitive. The buttons are arranged in rows in related groups. For this feature, the top row is for manipulating the feature; the second row is for creating features based on this one (by making a pattern or mirror copy in this case); the third row is for controlling the display. To demonstrate context sensitivity, compare the menus for several features. These are all different because with each feature selected (the "object"), a different set of operations is possible (the "action"). When pop-up menus like this are open, a RMB click will open another menu of related but perhaps less common commands. This includes the ability to customize the pop-up menu.

For feature #5 BASE, select ***Edit Dimensions***. The feature dimensions appear in dark blue on the screen. Double-click directly on the thickness dimension (currently 10). A small data entry box will appear at the dimension location. Type in ***25***, press Enter, and click on the background. Nothing happens on the screen except that the dimension now shows in light blue. This indicates that although the value of the parameter is changed, the geometry has not been updated to reflect the new value. To do this, we must *regenerate* the model. The command to do this is either (starting in the pull-down menu):

{Model}: Regenerate

or (this is quicker) use the shortcut button in the Quick Access toolbar at the top or (even quicker) press ***Ctrl+G***. Observe the message window. You should be informed that the regeneration was successful. Look at the thickness of the model. See Figure 30.

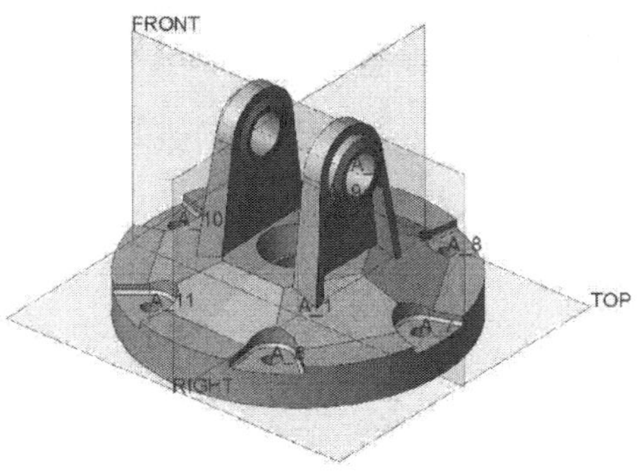

Figure 30 Part with modified base thickness

Let's change the thickness again but using a different way to launch the command. Make sure the selection filter is set to **Feature**. Move the cursor across the model until the base feature is pre-selected, then left click to select it. The same pop-up menu appears in the graphics window. In this menu, select ***Edit Dimensions***.

What happens if we make a mistake here? Change the thickness of the BASE from 25 to *75*. Can you predict what will happen? Once again, you will have to select the ***Regenerate*** command. A **Notification** window appears that indicates there is a problem. Expand the model tree to see that the regeneration has failed on the cut feature #27 (UNDER_BASE) shown in red in the model tree. All the features that depend on this one (its children) are also shown in red. There are some tools available to discover exactly why this feature failed[20]. If the model tree was closed, you would still know something was wrong (in addition to the weird model appearance) because of the small red flag that has appeared in the message line. We will be looking at those tools in lesson 5. For now, select the ***Undo*** button in the Quick Access toolbar (note the shortcut ***Ctrl+Z***). This returns the thickness back to 25.

A final way to change the value of a dimension is very easy. Just double-click on any surface of the base feature. Make sure you don't accidentally pick a surface of HEXAGON or one of the cuts. If you get the disk, the sketch will appear with dimensions. Change the thickness dimension back to ***10*** and regenerate the model. Double-clicking on any sketched feature will bring up its sketch and dimension values - no need to go looking for a menu command.

Go to ***{Model}: Operations*** and turn ***Auto Regenerate*** back on. Repeat the previous three ways of editing a dimension value. Observe the difference in the color of the edited dimension value and the shape of the mouse cursor. You may have to invoke the RMB pop-up menu to select ***Exit Edit Dimension***, and/or double click on the background.

---

[20] The change in base thickness has caused the sketch for the UNDER_BASE feature to cross through the axis of the revolve - this is not allowed!

Parent/Child Relations

As mentioned, a parent provides a reference for the creation of a child. To illustrate this, select HEXAGON (either in the model tree or using preselection in the graphics window), right click and select ***Information➤ Reference Viewer*** (see Figure 31). This brings up the reference information window shown in Figure 32. The pane on the left controls (using Filters) which information to display in the pane on the right. Turn off (uncheck) the ***System*** reference type so that you will see only direct feature reference types as shown in the Figure. Otherwise, we will accept all the defaults, so you can close the filters pane using the textured button on the vertical sash.

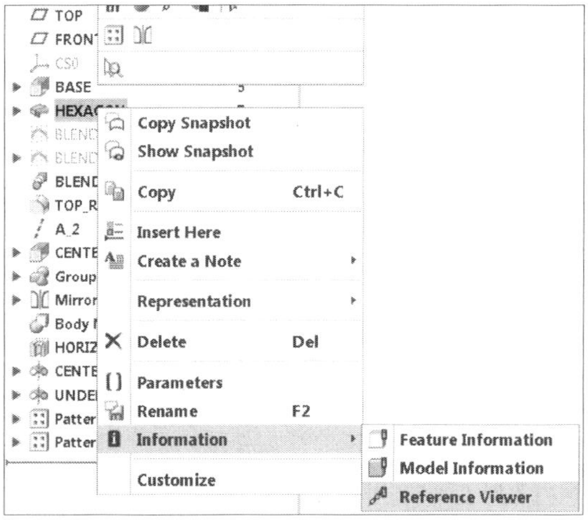

Figure 31 Launching the Reference Viewer

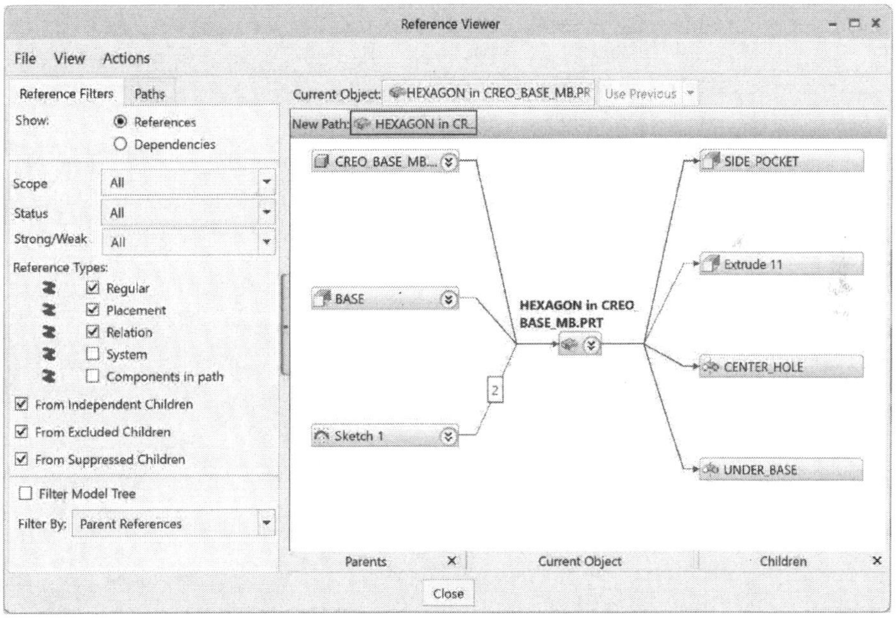

Figure 32 The **Reference Viewer** for exploring parent/child relations

The pane shows a graphical display of the relationships between features, with parent features on the left and child features on the right. In an assembly, this display also shows relations between components. The HEXAGON feature is indicated as the current object. Note the icons for each feature that indicate their type (datum plane, extrude, sweep). If necessary, expand this window so that the other feature names are listed, as in Figure 32, and move it so that the model is visible in the graphics area. As you pass the mouse over these features, they will be highlighted on the model.

The feature BASE is a parent of HEXAGON. Clicking the small down arrowhead beside BASE shows that an axis of BASE is involved. How? If you hold your mouse on the line in the Reference Viewer connecting BASE to HEXAGON, a pop-up message will show that the axis of BASE was a dimensioning reference for HEXAGON (see *Used As*).

What other children does BASE have? If you select BASE in the **Reference Viewer** and click the RMB, another pop-up menu appears. In this menu, select ***Set as Current***. This changes the current object to BASE. As expected, the child list contains HEXAGON but also numerous other features. Since BASE was the first solid feature in the part, virtually everything else is a child. Some of the reference arrows indicate (using a number on the line) that BASE provides more than one reference for the child. If you click the down arrowhead beside **BASE in CREO_BASE_MB.PRT** you will see four entries (an edge, two surfaces and an axis). As you move your mouse across each entry, they will highlight on the model and the reference arrows to the various child features will highlight in cyan. There is clearly a lot of information available here even for this simple model. Close the **Reference Viewer** window. We will return to it in a later lesson.

The relations between parents and children can become very complicated. It is crucial when starting a new model to carefully consider how these are going to be set up - which features should depend (and how) on which others, and which should be independent. Insufficient planning here (or, even worse, none at all!) will result in a lot of grief later when the model needs to be modified. This concludes our exploration of the model of a single part.

Now, of course, very few parts exist in isolation from one another. Most parts are used in assemblies that must perform some function. Let's move on to an assembly and spend some time exploring how it is set up.

Anatomy of an Assembly

We will now load an assembly that uses the base part we have just been examining. Using the Graphics toolbar, set up the screen as follows: shaded display (colors ON preferred) with datums off (they just clutter up the view!). You will need to have all the tutorial files installed as discussed earlier. Select the second tab (***Folder Browser***) in the Navigator window to show the Common Folders, then select the ***Working Directory***. The Browser window shows a listing of files in the chosen location. If you pick the ***Preview*** button at the lower right, you can see a preview of the object selected above (Figure 33). The usual

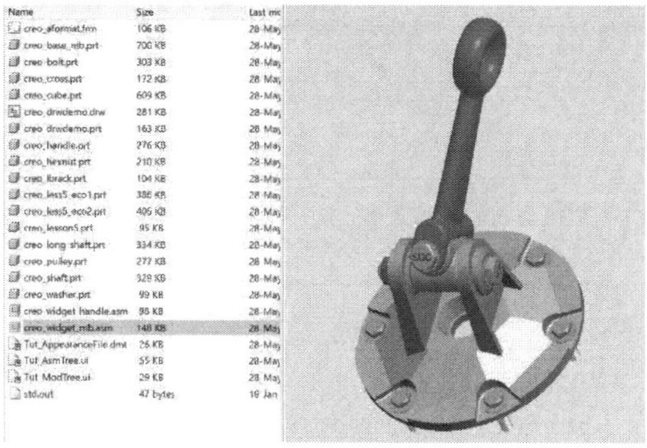

Figure 33 The Browser with preview window and file details turned on

dynamic view controls work within this window.

Before we proceed, select the ***Views* ➤ *Details*** option in the Browser window. Observe the size of the file **creo_widget_mb.asm**. This is the assembly file we will open in a minute. Compare its size to the part file **creo_base_mb.prt** that we have been examining. Although the assembly contains the base part, and many others, the assembly file is quite considerably smaller. Why? The reason is that the assembly file contains only the necessary information for *how* the parts are put together in the assembly. *The assembly file does not contain the parts themselves.* The most common cause of an error when retrieving an assembly is not having the appropriate part files available (where Creo Parametric can find them). This is a common mistake that many new users make when transferring work from one computer to another, thinking that all they need is the assembly file itself.

Double-click the file **creo_widget_mb.asm** in the Browser to open the file. Figure 34 shows the assembly model tree and the model itself. The groups and commands in the Model ribbon have changed a bit. Some commands specific to parts have been removed (like the **Engineering** and **Editing** groups) while ones for assemblies have been added (like the **Component** group).

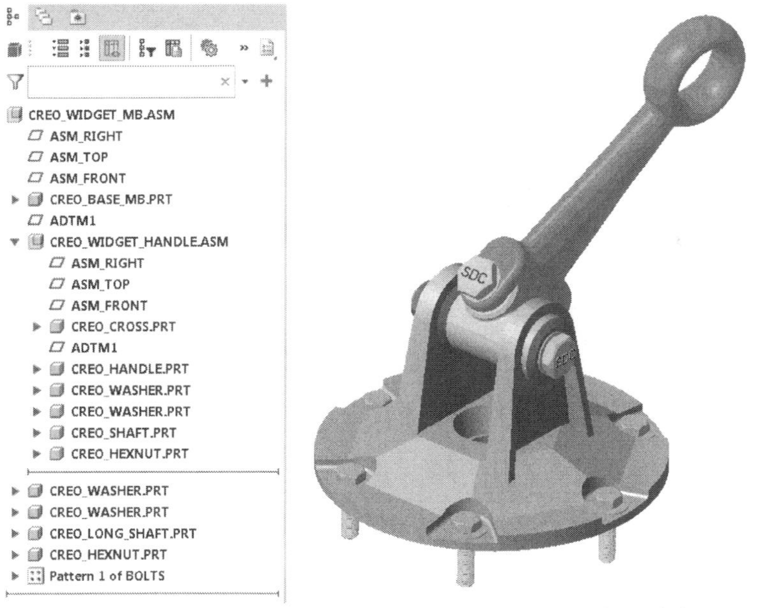

Figure 34 The assembly **creo_widget_mb.asm** and model tree

If the selection filter is set to ***Part***, moving the cursor back and forth across the assembly will highlight the components and bring up pop-up boxes that list the part names. The parts in Creo Parametric are called *components*. Note that, like a part, the assembly contains datum planes. These are usually designated with ASM- or A- to distinguish them from datums in individual parts. Unlike a part, the model tree for an assembly does not contain a Design Items entry or a design tree. It does, however, have a Mechanism Tree. See the Mechanism Tutorial from SDC or the online Help for more details.

Helpful Hint: Managing Multiple Windows

If you already had the base part in its own window, a new window has been created for the assembly, and the base part window is still available. The best way to switch windows is via the ***{View}: Windows*** command - all currently created windows are listed - or use the ***Windows*** button on the top left Quick Access toolbar.

You can see that the model tree is laid out in the same way as features in a part. Using the ***Settings*** button in the model tree, turn on ***Preselection Highlighting***. Passing the mouse over a component in the model tree will cause it to highlight in orange on the model. Note the icons in the model tree. The second component is itself an assembly - usually called a sub-assembly. Opening the model tree of the sub-assembly to the next level shows that it is made up of six other components. Sub-assemblies can be nested like this to as many levels as you want and are a good way to organize the structure of a complicated assembly especially if the sub-assembly occurs in several places. Different groups in a company can even work on different sub-assemblies at the same time, although file management does become an issue for very complex models.

Some parts (like CREO_WASHER) are used several times in the assembly (10 in fact), but only a single washer part file is on the hard disk. This shows that components can be re-used multiple times within an assembly. Common components, like fasteners, can be used in many different assemblies. This can obviously save a lot of file storage space. In fact, many companies will maintain a digital library of commonly used parts and assemblies. Components like springs can be declared to be flexible, so that only one part file is required but the spring can be used in multiple locations each with a different length, for example.

Notice at the bottom of the model tree is a pattern. The first element in the pattern (the pattern leader) is a group (BOLTS) of two components (CREO_WASHER and CREO_BOLT). Each of the other groups in the pattern contains these components. This is similar to the structure of a pattern of grouped features.

An Assembly BOM (Bill of Materials)

Select (in the **Investigate** group)

{Model}: Bill of Materials

In the BOM window, select ***OK***. This opens the Browser window with a BOM report for the assembly. This report

Summary of items for CREO_WIDGET_MB:

Quantity	Type	Name	Actions
1	Part	CREO_BASE_MB	
1	Part	CREO_CROSS	
1	Part	CREO_HANDLE	
10	Part	CREO_WASHER	
1	Part	CREO_SHAFT	
2	Part	CREO_HEXNUT	
1	Part	CREO_LONG_SHAFT	
6	Part	CREO_BOLT	

Figure 35 Bill of Materials summary table

contains three tables; the last one is shown in Figure 35. The buttons in the three right columns allow you to highlight the part in the model display, get model info, or open the individual component. If desired, you can use the Browser controls to save or print this BOM. In any case, the BOM data is automatically saved to a file (in this case, *creo_widget.bom*) in the current working directory whenever you run this command. ***Close*** the Browser with the button at the top right.

Modifying the Assembly

Returning to the model tree in the Navigator, let's add the extra columns in the model tree as we did for the single part. Select ***Model Tree Settings* ➤ *Settings* ➤ *Import Tree Settings*** and select the file **Tut_AsmTree.ui (Import for Active Model Only)**. The columns are added (Figure 36).

The model tree now also shows what are called *assembly features* in the model[21]. For example, the first three entries on the model tree of the assembly are datum planes. Datum features are very common in assemblies, since they provide references for placing and constraining components. To see an example of this, select feature #5 **ADTM1** in the assembly model tree (the datum plane does not have to be visible on the screen). In the pop-up menu select ***Edit Dimensions***. An angular dimension will appear in blue that

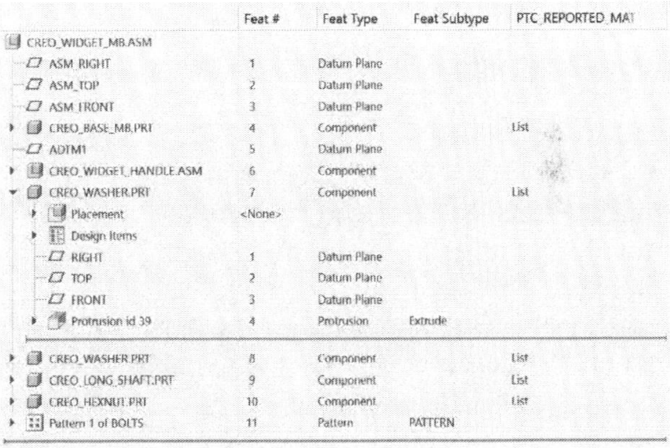

Figure 36 Assembly model tree

gives the rotation of this datum plane around the horizontal axis **A_2** in the base part. Set the RIGHT view to see this clearly. Change this angle by double-clicking on the number and typing in a new value, say ***90***, then press Enter. The value will stay blue until the model is regenerated. Regenerate <ctrl-g> it now.

The assembly will regenerate with the subassembly positioned at a new angle (Figure 37). The datum plane **ADTM1** in the top assembly serves as an orientation reference for the subassembly. When the datum plane moves, so does the subassembly. Repeat the ***Edit Dimensions*** command to put the

Figure 37 Modified assembly

[21] If not, in the model tree select ***Tree Filters*** and turn on the check box beside **Features**.

subassembly back in its original position (change the 90 back to **45**). Ctrl-g to regenerate it.

In the model tree, expand the sub-assembly CREO_WIDGET_HANDLE. It also contains an assembly datum ADTM1 which controls another angle. Change this to 30E and regenerate the model. Clearly, there is a problem with the model - at least two parts (the handle and the base) are interfering with each other. This interference is quite obvious. Later in these lessons we will find a Creo Parametric command that will identify interference automatically, often ones that you don't expect. For now, put the datum ADTM1 in the sub-assembly back in the original position (75E) before you continue. (HINT: Use the ***Undo*** command in the Quick Access toolbar or use ***Ctrl+Z***.)

In addition to assembly datums, you can create regular features like holes and cuts in the assembly that act on the components. A hole feature would allow you to ensure that holes through several components are aligned perfectly.

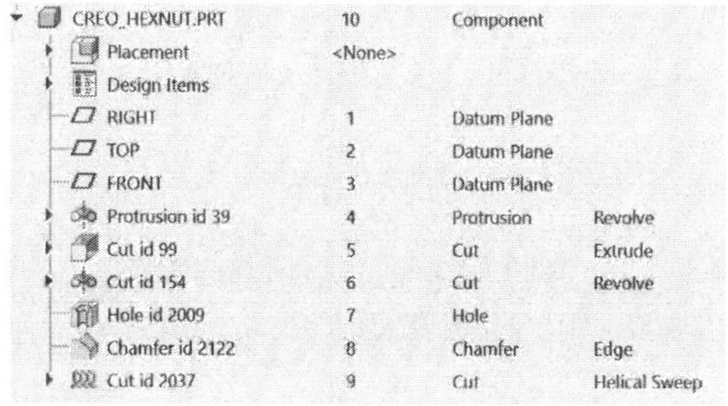

Figure 38 Model tree showing part features

Notice that each component listed in the model tree can be expanded to show its individual features: click on the ▼ sign to the left of the component icon on the model tree. For example, Figure 38 shows the expansion of component #10 CREO_HEXNUT. Selecting part features allows them to be modified the same as before. Try changing the thickness of the base part as we did before. What happens to the bolts and washers?

You can get the same sorts of information about components in the assembly (component info and parent/child relations) as you could for a part. These are usually launched, as before, with a right mouse click in the model tree. In the case of an assembly, the parent/child relations of components involve the datums, surfaces, and edges of components that are used to constrain each component in the assembly. We will look into this in depth in Lesson 8.

Also in a later lesson, we will see how we can create components in the assembly environment. This means that existing components can provide references for the creation of new components. Thus, you can create a new part to fit exactly to a previous part (for example, a shaft in a hole) by using existing geometry rather than manually transferring dimensions from one part to another.

There are just a couple more items we want to cover this lesson. Turn shading and color back on, if they aren't already. Close the Browser and Navigator windows if they are open. Put the assembly in the **Default Orientation** (ctrl-d).

Exploding an Assembly

Exploded views of assemblies are often helpful in showing how they are put together. If you have ever worked on a do-it-yourself project or kit, you have probably seen an exploded view in the plans. Getting this view in Creo Parametric is a snap. Select

{View} : Exploded View

which is in the **Model Display** group. The position of each component in the exploded assembly is determined by a default (Figure 39). However, the *explode position* of each component can be easily changed. These positions are stored with the assembly. It is possible to define several different sets of exploded positions within a single assembly file. These might show various steps in an assembly sequence. Note also that while the subassembly is exploded from the main assembly, the components in the handle sub-assembly (in this case) are not exploded from each other. This indicates that the *explode state* of various components can be individually set. It is also possible to add the explode lines to the view. We will spend some more time with layout of the exploded view in a later lesson. For now, unexplode the assembly by selecting the ***Exploded View*** button in the ribbon again.

Figure 39 The exploded assembly

The ***View Manager*** is a tool that allows you to set up defined view states of the assembly, including how much of the model is exploded (if any), showing or hiding components, changing the appearance of specific components (shaded or hidden line options), and other display attributes. It can be launched from the Graphics Toolbar. For further information, check out the online Help.

Opening Parts in an Assembly

When an assembly is being worked on, you will frequently want to open up an individual part by itself. This is easy to do. In the graphics window (Selection filter set to ***Part***), left click on the handle (part CREO_HANDLE). It will highlight in green. Select ***Open*** from the pop-up menu. The part will open in a new window by itself. Creo Parametric can have many separate parts and assemblies (and drawings) loaded into memory at once - these are called *in session* (check out the command ***{File}: Manage Session ➤ Object***

List) - each in a separate resizable window. To switch to another window, for example, back to the assembly, select (in the Quick Access toolbar at the top left)

> ***Windows ➤ CREO_WIDGET_MB.ASM***

This concludes our look at the structure and layout of a simple assembly until later in the tutorial. As always, you are encouraged to experiment further with the commands we have covered and see what you can discover on your own.

Obtaining Hard Copy

You may want to produce some hard copy documentation from time to time as you go through these lessons. There are several basic kinds of hard copy:

1. Plain text containing model, component, or feature information, feature lists, the model tree.
2. Formatted web documents.
3. Images of the graphics area of the screen.
4. Detailed engineering drawings.

For the first type, whenever you query the model for this type of information, Creo Parametric generally creates a printable text file in the working directory. Look for files with the extension *txt*, *inf*, *lst*, *err*, or *dat*. These can be printed with any simple text editor or word processor.

Web-formatted documents are available for any page viewed in the Browser window (such as the BOM report). Both "Save" and "Print" shortcut buttons are on the Browser toolbar.

Images of the graphics area ("screen shots") can be obtained easily if you have the appropriate printer attached to your system. Different printing methods may be required depending on whether the image is shaded or not. Sometimes the Windows Print Manager will take care of this for you. To obtain a screen shot of a line drawing (hidden line or wireframe image), issue the commands

> ***File ➤ Print ➤ Print***

In the **Print** window that opens up, in the Destination field select the appropriate output device. Windows users will usually select MS Print Manager. The default is to print exactly what is shown on the screen; select the Model tab and observe the default for a plot is **Based on Zoom**. It is not recommended that you do this with shaded images, particularly on some systems with shared printers.

The command

> ***File ➤ Print ➤ Quick Drawing***

allows you to easily produce a drawing containing your choice of paper size, views, scales, and display style. These can be predefined in drawing templates. You must be careful here about model orientation on the screen and scaling the desired views to fit the specified paper size.

For high quality shaded images, your best bet is to select

File ➤ Save as ➤ Save a Copy

and in the **Type** pull-down list, you can find a number of image formats (TIFF, JPG, EPS,...). The image file will be created in your working directory. These images can be very high resolution (up to 600 dpi and 24 bit color) but are also very large (think megabytes). There are special tools with Creo Parametric (see the ***Render*** tab) for very high-quality rendering of images suitable for glossy marketing purposes. These allow you to have multiple light sources, shadows, reflections, textures, background images, and so on. We will not have time to explore these functions here.

A quick and easy way of getting hard copy of a shaded image in Windows is to capture the screen contents to the clipboard. Use ***Alt+PrtScn*** to capture the active window. Then paste the clipboard image (a bitmap) into a graphics utility program or word processor. This image is stuck at the resolution and color depth of your screen, however.

Obtaining hard copy of engineering drawings will be covered later in these lessons. For these, there are extra considerations of sheet size, drawing scale, pen width, and so on.

Leaving Creo Parametric

When you want to quit Creo Parametric entirely, you can leave by using the ***Exit*** command in the **File** menu or the X at the top-right corner. If you accidentally do this, you can cancel the command.

Helpful Hint

Unlike many programs, Creo Parametric will not *automatically* save anything for you. This applies both during operation (like regular timed backups) and when you exit. If you leave the program without saving new work, it is basically gone! Anyone who says they have never lost work this way is probably lying! Try not to be like them.

Depending on how your system has been set up, Creo Parametric *may* prompt you to save your work when you exit[22]. This includes any parts, assemblies, and drawings.

---

[22] You can include the option *prompt_on_exit* in the configuration file accessible using ***File ➤Options➤Configuration Editor***. The default for this option is *No*.

Items *in session* (stored in memory) will be closed too. You will be prompted individually for each object in session. Reply with a ***Y*** or ***N*** to save each object in the current working directory. A middle click will accept whatever default is shown. If you are sure you have saved the most recent version of all objects in session, you don't need to do that again so press ***Q*** (for Quit).

This completes Lesson #1. You will no doubt be relieved to know that it is by far the longest in this series. As you can see, there was much fundamental material to deal with. You are strongly encouraged to experiment with any of the commands that have been presented in this lesson. Explore the other parts in the widget assembly, and experiment with the view controls. Like all sophisticated tools, the only way to become proficient with Creo Parametric is to use it a lot!

In the next lesson we will create our first part using simple basic features (protrusion, cut, hole) and spend some time learning about the Intent Manager in Sketcher. The important concept of *design intent* will be introduced, with examples.

Questions for Review

Here are some questions you should be able to answer at this time, or that may provoke some thought and/or further study:

1. What mouse buttons are used to pan, spin, and resize the object?
2. What is the purpose of the datum planes?
3. What are two ways to get on-line help?
4. How can you get a shaded image of the part? What about wire frame with no hidden lines?
5. How do you turn the datum plane visibility on and off?
6. How many ways can you think of to modify the value of a dimension?
7. How does a ribbon group indicate that it has more command buttons not currently visible?
8. Where does your system first look for stored part files? What is this called?
9. What is meant by *preselection*?
10. What is meant by the term *regeneration*?
11. What is the regeneration sequence? How can you determine it for a part?
12. How can you turn color display on and off?
13. How many ways can you get to the default view orientation? What are they?
14. How do you change the working directory?
15. What is the spin center?
16. What is the correct toolbar button to execute the following functions:
    a. toggle display of datum planes
    b. start Orient Mode
    c. regenerate

Appendix for further details on using the configuration file editor.

d. go to a previously named view
e. turn off the spin center
17. What is the function of the following buttons:

a)  b)  c)  d)  e) 

18. What are the three primary goals in creating models?
19. Describe all the shortcuts and functions associated with the middle mouse button.
20. What and where is the ***Command Search*** button?
21. Why do datum planes have two colors? What are they?
22. How large is a datum plane?
23. How do you toggle (turn on and off) the display of datum tags?
24. Where is the Quick Access toolbar?
25. What are the default buttons on the Graphics toolbar? How can you change the contents and location of the Graphics toolbar?
26. What is the meaning of the buttons  and where are they located on the screen?
27. What is the meaning of the following icons? Where do they appear?

a)  b)  c)  d)  e)  f) 

28. Where is the explode command for assemblies?
29. When editing dimensions, what is the meaning of the colors dark blue, green, and light blue for a displayed dimension value?
30. What is the difference in operation of the ***Hide*** command on a solid feature and a component in an assembly?
31. What happens if you preselect a dimension and then right click? This is useful if the screen gets really crowded.
32. What are the ways for opening a part file? (Hint: there at least four!)
33. What does "context sensitive" refer to, and when is this relevant?
34. A middle-click often means the same as ______?
35. What is meant by a "buried" feature? Is this good or bad?
36. What is the meaning of the red flag that appears in the message line?
37. Give two reasons why rounds are usually added to the model last.
38. What is the Model Player used for?
39. How many separate bodies can be contained within a single part?
40. Can bodies overlap?
41. Must a body be a single contiguous solid?
42. What features are NOT contained in a body?
43. What are the three Boolean operations that can be used on bodies?
44. Where do bodies show up on the model tree?
45. In your own words, define what is meant by a "body" in Creo.
46. How can you make an object transparent?
47. What is the purpose of a snapshot and how do you take one?
48. What is the design tree?
49. Where can you place the design tree on the screen?

Exercises

1. Try out the model player with the widget assembly. How does this differ from using it in part mode?
2. What are some downstream uses for a Creo Parametric model? Which ones are available to you in your local school or installation, and who uses them?
3. Open the part CREO_CUBE.PRT. Practice manipulating the display of the cube to show each numbered side facing the front of the screen, with the number the right way up. Do this as fast as you can. Hint: turn color and shading on (the cube is translucent, so you can see the numbers on the hidden faces). If your system cannot handle transparency, use hidden line display.

4. Obtain a printed hard copy of a hidden-line display of the CREO_BASE part.
5. Obtain a printed hard copy of the model tree for the CREO_BASE part showing the additional columns.
6. Obtain a hard copy of the Browser window showing the model information for the part CREO_BASE.
7. Obtain a hard copy of the Browser window showing the feature information for the BASE feature in the part CREO_BASE.
8. Obtain a printed hard copy of the assembly CREO_WIDGET.ASM (no hidden lines) showing the handle part perpendicular to the base.
9. Obtain a printed hard copy of a shaded image of the exploded view of the assembly CREO_WIDGET.ASM. Label this with the part names. Bonus marks if you can figure out how to do this with ***Annotations***.
10. Make an isometric sketch of a simple object (like a brick) located at the origin of the three datum planes (TOP, FRONT, RIGHT). Identify two ways to select references for each of the standard engineering drawing views (top, front, right, left, back, bottom) of the object.
11. In the **File Open** dialog window, find out what the ***Views*** option ***Thumbnails*** does.
12. Create a new ***Saved Orientation*** that will produce a perfect isometric view of the model showing top-front-right surfaces. (HINT: in the **Orientation Type** dropdown list, use the ***Dynamic Orientation*** option. Starting from a FRONT view, rotate -45E around a vertical axis, and +37.5E around a horizontal axis.)

Lesson 2

Creating a Simple Object (Part I)

Introduction to Sketcher

Synopsis

Creating a part; introduction to Sketcher; sketch constraints; creating datum curves, protrusions, cuts; sketch diagnostics; using the dashboard; saving a part; part templates.

Overview of this Lesson

The main objective of this lesson is to introduce you to the general procedures for creating sketched features. We will go at quite a slow pace, and the part will be quite simple (see Figure 1 on the next page), but the central ideas need to be elaborated and explored so that they are very clearly understood. Some of the material presented here is a repeat of the previous lesson - take this as an indication that it is important! Here's what we are going to cover:

1. Creating a Solid Part
2. Introduction to Sketcher
    - Sketcher menus
    - Intent Manager and Sketcher constraints
    - Sketcher Diagnostics
3. Creating a Sketch
4. Creating an Extruded Protrusion
    - Using the Dashboard
5. Creating an Extruded Cut
6. Saving the part
7. Using Part Templates

> ***Helpful Hint***
> It will be a good idea to browse ahead through each section to get a feel for the direction we are going, before you do the lesson in detail. There is a lot of material here which you probably won't be able to absorb with a single pass-through.

Start Creo Parametric as usual. If it is already up, close all windows (except the base window) and erase all objects in session using ***File ' Manage Session ' Erase Current*** and/or ***File ' Manage Session ' Erase Not Displayed***, also available in the ***{Home}*** ribbon. Close the Navigator and Browser windows.

Creating a Simple Part

In this lesson, we will create a simple block with a U-shaped central slot. By the end of the lesson your part should look like Figure 1 below. This doesn't seem like such a difficult part, but we are going to cover a few very important and fundamental concepts in some depth. Try not to go through this too fast, since the material is crucial to your understanding of how Creo works. We will be adding some additional features to this part in the next lesson.

We are going to turn off some of the default actions. This will require us to do some things manually and to give you a better understanding of what the many default actions are. Eventually you will come across situations where you don't want the defaults, and you'll need to know your way around these options.

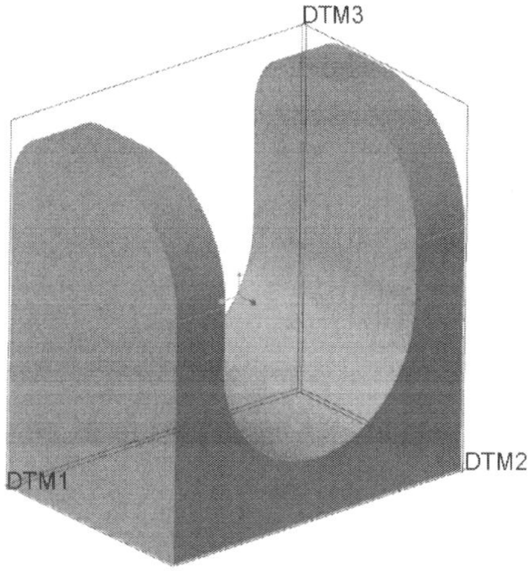

Figure 1 Part at the end of this lesson

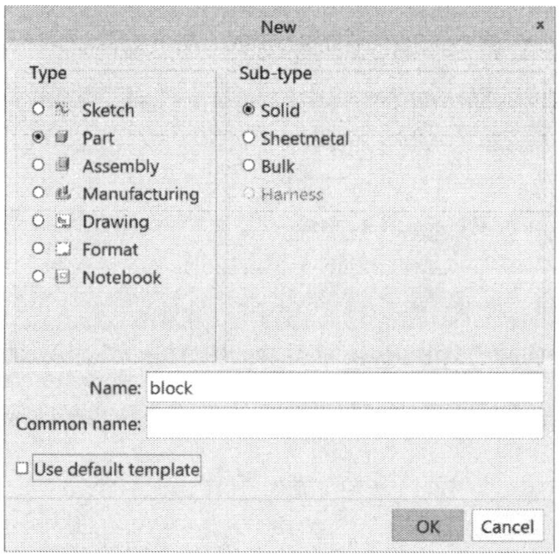

Figure 2 Creating a new part

Creating and Naming the Part

Click the ***{Home}: New*** button, or select ***File ' New***. A window will open (Figure 2) showing a list of different types and sub-types of objects to create (parts, assemblies, drawings, and so on). In this lesson we are going to make a single solid object called a *part*. Keep the default radio button settings

Part | Solid

IMPORTANT: Turn off (remove the check) the ***Use Default Template*** option at the bottom. We will discuss templates at the end of this lesson.

Many parts, assemblies, drawings, etc. can be loaded simultaneously in the current session. All objects are identified by unique names[1]. A default name for the new part is presented in the window, something like **[prt0001]**. It is almost always better to have a more descriptive name. So, double click (left mouse) on this text to highlight it and then type in

[block]

(without the square brackets) as your part name. The ***Common Name*** of the part is an option for specifying an even more descriptive name. For example, you might have a number of part files named using a part or catalog number such as "TG123_A29". This is not very descriptive, so you could enter a common name such as "small flat rubber washer". We will not use common names in this tutorial, so leave this blank and just press ***Enter*** or select ***OK***.

The **New File Options** dialog window opens. Since we elected (in the previous window) to *not* use the default template for this part, Creo is presenting a list of alternative templates defined for your system. As mentioned previously, we are going to avoid using defaults this time around. So, for now, as shown in Figure 3 at the right, select

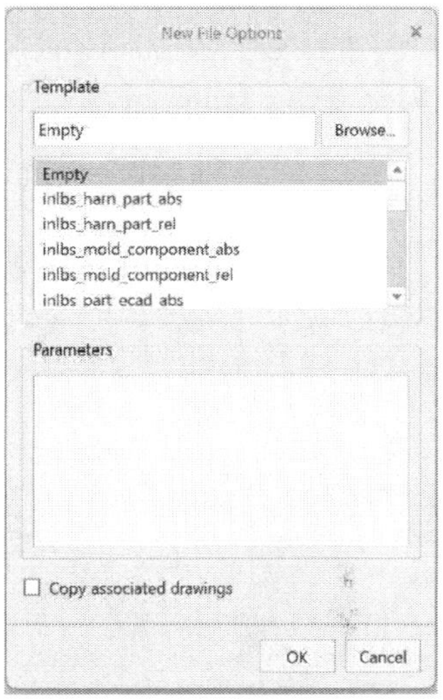

Figure 3 Options for new parts

Empty | ***OK*** (or middle click).

At this time, **BLOCK** should appear in the title area at the top of the graphics window and the part modeling ribbon has appeared.

Create Datum Planes

We will now create the first features of the part: three datum planes to locate it in space. It is not absolutely necessary to have datum planes, but it is a very good practice, particularly if you are going to make a complex part or assembly. Datum planes are created using the ***Plane*** command in the ***{Model}:Datum*** group. Note that this button looks quite similar to the buttons that control the display of datums and tags in the ***{View}:Show*** group of the ribbon. What's the difference?

Select the ***Plane*** button now. Since we currently have no features in the model, Creo rightly assumes that we want to create the three standard datum planes. Normally this command only creates a single datum plane at a time. In the ***{View}:Show*** group, make sure that display of all datums and tags is turned on.

---

[1] Creo can keep track of objects of different types with the same names. A part and a drawing can have the same name since they are different object types.

The datum planes represent three orthogonal planes to be used as references for features that will be created later. You can think of these planes as XY, YZ, XZ planes, although you generally aren't concerned with the X,Y,Z form or notation. Your screen should have the datum planes visible, as shown in Figure 4. Note that each plane has an attached tag that gives its name: **DTM1**, **DTM2**, and **DTM3**. This view may be somewhat hard to visualize, so Figure 5 shows how the datum planes would look if they were solid plates in the same orientation. An important point to note is, while the plates in Figure 5 are finite in size, the datum planes actually extend off to infinity. Finally, before we move on to the next topic, notice that the last feature created (in this case DTM3) is highlighted in green. This is a normal occurrence and means that the last feature created is always preselected for you as the "object" part of the object/action command sequence.

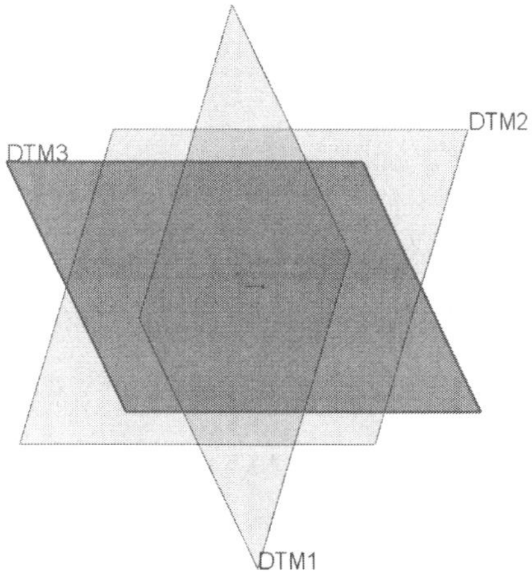

Figure 4 Default datum planes

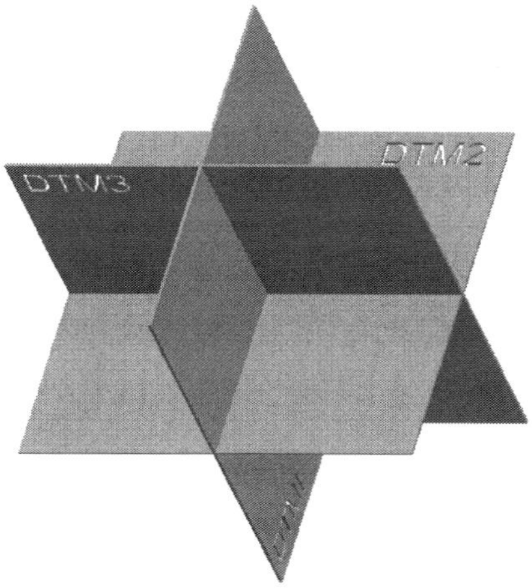

Figure 5 Datum planes as solid plates

Part Modeling Feature Overview

There are three command groups in the ***{Model}*** tab of the modeling ribbon that will create solid features. These are shown in Figure 6. Most will create solids; some can also create surfaces.

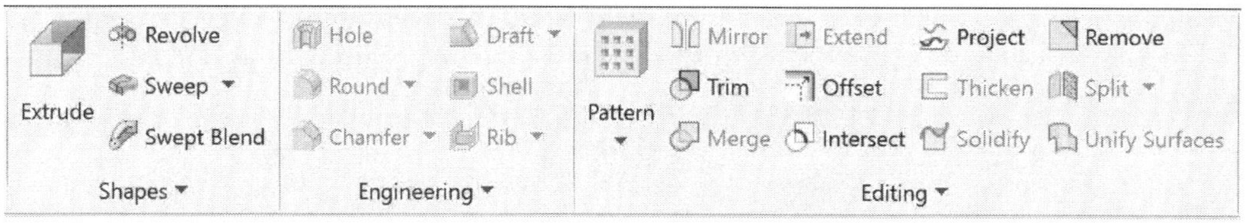

Figure 6 Ribbon groups for creating and editing solid features

Shapes (formerly known as "sketched features") - (extrusions, revolves, sweeps, ...) These features require the definition of a two-dimensional cross section which is then manipulated into the third dimension. All of these features can either add or remove material from the solid. These features will involve the use of an important tool called

Sketcher. This will be the focus of this lesson.

Engineering (formerly known as "placed features") - (holes, rounds, shells, ...) These are features that are created directly on existing solid geometry. Some always either add (rib) or remove (hole) material from the solid. Some (round, chamfer, draft) can do both, depending on the underlying geometry. Examples are placing a hole on an existing body, or creating a round on an existing edge of a part.

The third group of buttons, **Editing**, is used for modifying existing features. We will deal with some of these commands (***Mirror*** and ***Pattern***) later in the Tutorial.

In this lesson we will be using the ***Extrude*** command to create two types of shaped features (a protrusion and a cut). In the next lesson, we will use the ***Hole***, ***Round***, and ***Chamfer*** commands to create three engineering features. Before we continue, though, we must find out about an important tool - Sketcher.

Introducing Sketcher

Sketcher is arguably the most important tool for creating features in Creo Parametric[2]. It is therefore critical that you have a good understanding of how it works. We will take a few minutes here to describe its basic operation and will explore the Sketcher tools continually through the next few lessons. It will take you a bit of practice and experience to fully appreciate all that it can do.

Basically, Sketcher is a tool for creating two-dimensional figures. These can be either stand-alone features (*Sketched Curves*) or embedded elements that define the cross sectional shape of some solid features. The aspects of these figures that must be defined are location, shape, and size, roughly in that order. The sketching plane where we will create the 2D sketch is defined or selected first. Then, within Sketcher the location is further specified by selecting references to existing geometry. You will find the usual drawing tools for lines, arcs, circles, and so on, to create the shape. Finally, you can specify dimensions and/or alignments to control the size of the sketch and its relation to existing geometry.

Sketcher is really quite smart, that is, it will anticipate what you are going to do (usually correctly!) and do many things automatically. Occasionally, it does make a mistake in guessing what you want. So, learning how to use Sketcher effectively involves understanding exactly what it is doing for you (and why) and discovering ways that you can easily override this when necessary.

The brain of Sketcher is called the *Intent Manager*. We will be discussing the notion of *design intent* many times in this tutorial. In Sketcher, design intent is manifest not only in the shape of the sketch but also in how constraints and dimensions are applied to the

[2] Sketcher tools can also be used in drawing mode to create drawing entities.

sketch so that it is complete and conveys the important design goals for the feature. Completeness of a sketch implies that it contains just enough geometric specification so that it is uniquely determined. Too little information would mean that the sketch is under-specified; too much means that it is over-specified. One function of Intent Manager is to make sure that the sketch always contains just the right amount of information. Moreover, it tries to do this in ways that, most of the time, make sense. Much of the initial frustration involved in using Sketcher arises from not understanding (or even sometimes not realizing) the nature of the choices it is making for you[3] or knowing how easy it is to override these actions. When you are using Sketcher, Intent Manager must be treated like a partner - the more you understand how it works, the better the two of you will be able to get along!

The term *sketch* comes from the fact that you do not have to be particularly exact when you are "drawing" the shape, as shown in Figures 7 and 8. Sketcher (or rather Intent Manager) will interpret what you are drawing using a built-in set of rules. Thus, if you sketch a line that is approximately vertical, Sketcher assumes that you want it vertical. If you sketch two circles or arcs that have approximately the same radius, Sketcher assumes that's what you want. In cases like this, you will see the sketched entity "snap" to a particular orientation or size as Intent Manager fires one of its internal rules (this occurs on-the-fly while you are sketching).

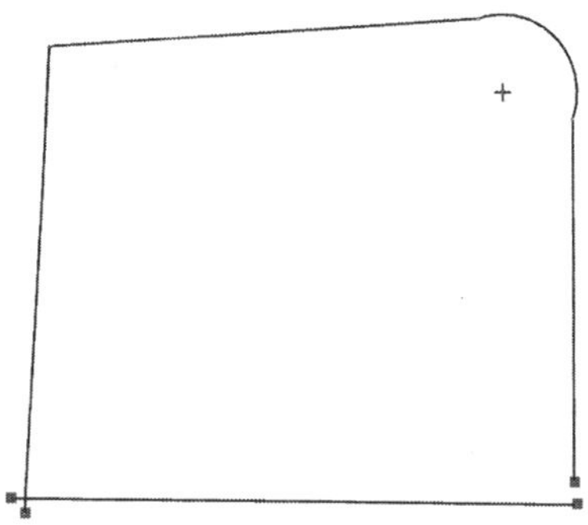

Figure 7 Geometry before processing by Intent Manager. Note misaligned vertices, non-horizontal and vertical edges, no tangent curves.

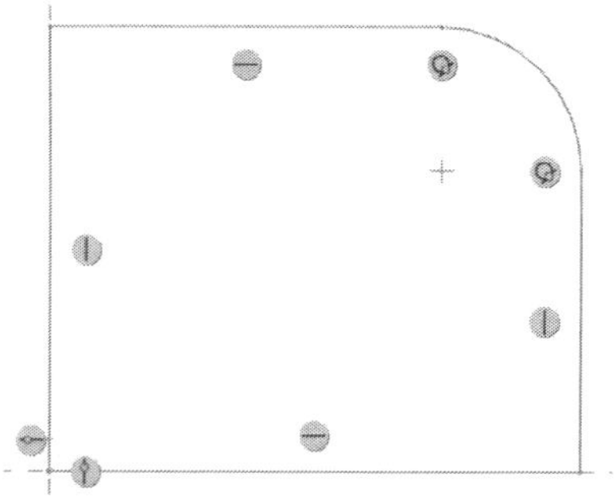

Figure 8 Geometry after processing by Intent Manager. Note aligned vertices, horizontal and vertical edges, tangent curves.

When Intent Manager fires one of its internal rules, you will be alerted by a symbol on the sketch that indicates the nature of the assumed condition. If you accept the condition, it becomes a *constraint* on the sketch. These constraints are summarized in Table 2-1.

---

[3] Intent Manager is able to "solve" sketches that are much more complicated than mere humans can do. To make sure it doesn't get ahead of you, keep your sketches simple.

Table 2-1 Implicit Constraints in Sketcher

Rule	Symbol	Description
Equal radius and diameter		If you sketch two or more arcs or circles with approximately the same radius, the system may assume that the radii are equal
Symmetry	 	Two vertices may be assumed to be symmetric about a centerline
Horizontal or vertical lines	 or 	Lines that are approximately horizontal or vertical may be considered to be exactly so
Parallel or perpendicular lines	 or 	Lines that are sketched approximately parallel or perpendicular may be considered to be exactly so
Tangency		Entities sketched approximately tangent to each other may be assumed to be tangent
Equal segment lengths		Lines of approximately the same length may be assumed to have the same length
Point entities lying on other entities or collinear with other entities	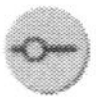	Point entities that lie near lines, arcs, or circles may be considered to be exactly on them. Points that are near the extension of a line may be assumed to lie on it.
Equal coordinates	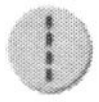 or 	Endpoints and centers of the arcs may be assumed to have the same X- or the same Y-coordinates
Midpoint of line		If you sketch close to the midpoint of a line the vertex will snap there

You should become familiar with these rules or constraints, and learn how to use them to your advantage. If you do not want a rule invoked, you must either

(a) use explicit dimensions or alignments, or
(b) exaggerate the geometry so that the rule will fail, or
(c) tell Sketcher to disable a specific instance of the constraint (Hint: RMB), or
(d) set up Sketcher to explicitly ignore constraints of a given type.

You will most often use option (a) by specifying your desired alignments and dimensions and letting Sketcher worry about whatever else it needs to solve the sketch. When geometry is driven by an explicit dimension, fewer internal rules will fire. Option (b) is slightly less common. An example is a line that must be 2E away from vertical - you would draw it some much larger angle (like 15E or so - this prevents the "vertical" rule from firing) and put an explicit dimension on the angle. Once the sketch has been completed with the exaggerated angle, you can modify the dimension value to the desired

2E. The third option is used on-the-fly while you are sketching to disable a constraint that Intent Manager indicates it wants to place. Finally, there are settings for Sketcher that explicitly turn off the rule checking (for all rules or selected ones only) during sketching. This is very rarely used.

An example of a sketch with geometric constraints is shown in Figure 9. Note how few dimensions are required to define this sketch. See if you can pick out the following constraints:

< vertical lines
< horizontal lines
< perpendicular lines
< four tangencies
< three sets of equal length lines
< equal radius and diameter
< vertical alignment

How do you suppose Sketcher is able to determine the radius of the rounded corners (fillets) at the top and bottom on the left edge? (Hint: this involves the solution of a system of equations.)

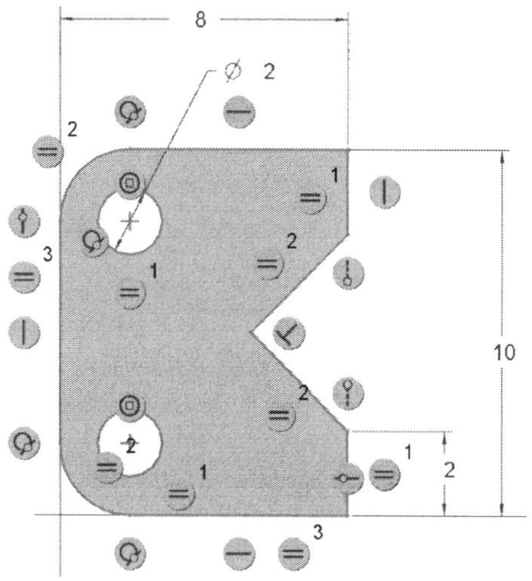

Figure 9 A sketch showing implicit constraints (constraint icons enlarged)

Sketcher has some very useful tools that make it quite easy to diagnose errors in sketches (such as duplicated edges) or improper sketches (such as open curves for features that require closed curves). We will investigate these tools by intentionally introducing some errors into our sketches.

Two Ways to use Sketcher to Create Shaped Features

In this lesson, we will use two methods to create a couple of shaped features. The creation of the sketch represents the majority of the work required in creating these features. The two methods will differ by when in the feature creation sequence you will invoke Sketcher. The two methods are shown schematically in Table 2 - 2.

In the first method, we invoke Sketcher first to create the cross sectional shape of the extrusion. This shape is defined in a *sketched curve* which becomes a stand-alone feature in the model. We then launch the extrude command, specifying the curve to define the cross section of the feature. A dependent copy of the sketch is brought into the extrude feature. The original sketch feature is automatically set to ***Hide***, although it can still be used as a reference by other features. In the second method, we do not create a separate curve but rather invoke Sketcher from inside the extrusion creation sequence. In this case, the sketch exists only inside the extrude definition and is not usable by other features. In the Table below, notice the difference in the appearance of the model tree following completion of the extruded feature.

Table 2-2 Steps to create a sketched feature

	Method #1	Method #2
Step 1	Create shape using Sketcher (sketch becomes stand alone feature)	Launch Extrude tool
Step 2	Launch Extrude tool	Create shape using Sketcher (sketch exists only inside the extrude feature)
Model Tree after completion of Extrude	RIGHT TOP FRONT Sketch 1 Extrude 1 Sketch 1	RIGHT TOP FRONT Extrude 1 Section 1

In terms of design intent, the first method would be used if the curve was used for more than one feature, for example an extrude and a subsequent revolve. The second (creating the sketch within the feature) is the traditional mode of operation, and would be the method of choice if the sketched shape was used only for that single feature.

Both features we will make in this lesson are extrusions: one will be a *protrusion* (which adds material) and the other is a *cut* (which removes material). Either of the two methods shown here can be used to create either protrusions or cuts; for either method, whether you add or remove material is determined by an option button, or the direction of the extrusion away from a solid surface (away produces a protrusion, into produces a cut).

There is a third way to use sketches to create complicated shaped features. This uses *Sketch Regions* formed by curves and lines from multiple separate sketches. This topic will be discussed later in these lessons. Sketches can also be used to create hole patterns.

Creating a Sketched Curve

The last feature we created in the Block model was the datum DTM3. If that is currently selected (highlighted in green), unselect it by clicking somewhere else in the graphics window. The behavior of Sketcher is different if we enter it with a plane already selected.

Now in the ***{Model}:Datum*** group select the ***Sketch*** tool Sketch.

Setting Sketch Orientation

The Sketch dialog window opens (Figure 10) where we must specify two references. The first is the **Sketch Plane**. Select the border of DTM3. It is now highlighted in the graphics window in cyan. This is the plane on which we will draw the sketch. Sketcher will automatically pick the second reference, probably DTM1, but this will depend on the orientation of the datum planes on your screen. It will automatically assign a **Sketch Orientation** - see Figure 10. This also depends on the current orientation of the datums[4].

A magenta arrow shows the direction of view onto the sketching plane. Spin the orientation with the middle mouse button to see the arrow. The arrow attached to the edge of DTM3 should be pointing back into the screen. The direction of view can be reversed by clicking directly on the magenta arrow or with the ***Flip*** button in the dialog window. Leave it pointing towards the back. In the dialog window, DTM1 is identified as the Sketch Orientation ***Reference***, with the ***Orientation*** set to ***Right***. What is all this about?

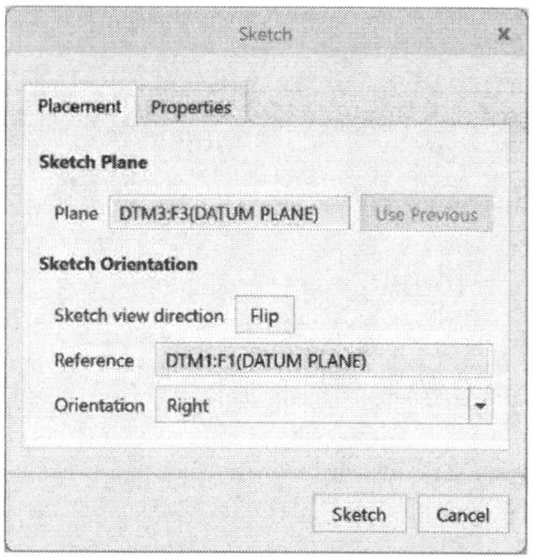

Figure 10 Defining the sketch plane and sketch orientation reference

The relation between the sketch plane and the sketch orientation reference generally causes confusion for new users, so pay attention!

The meaning of the sketch plane is pretty obvious - it is the plane on which we will draw the sketch - in this case DTM3. The best 2D view is always perpendicular to the sketch plane[5]. That is not enough by itself to define our view of the sketch since we can be looking at that plane from an infinite number of orientations (imagine the sketch plane rotating around an axis perpendicular to the screen). The ***Orientation*** option pull-down list in the dialog window (***Top***, ***Bottom***, ***Left***, ***Right***) refers to directions relative to the computer screen, as in "TOP edge of the screen" or "BOTTOM edge of the screen" and so on. We must combine this direction with a chosen reference plane (*which must be perpendicular to the sketch plane*) so that we get the desired orientation of view onto the sketching plane.

So that the sketch is the right way up, we can choose either DTM2 to face the Top of the screen, or (as was chosen automatically for us) DTM1 can face the Right of the screen. Note that both DTM1 and DTM2 are both perpendicular to the sketch plane, as required. The direction a plane or surface "faces" is determined by its normal vector which, for a datum plane, is perpendicular to the positive side. For a solid surface, the orientation is determined by the outward normal.

Read the last couple of paragraphs again, since new users are quite liable to end up drawing their sketches upside-down or (like the example below) sideways!

---

[4] If the dialog window does not look like Figure 10, cancel out of the window, press ***Ctrl+D*** to get to the default orientation, and launch the Sketch command again.

[5] It is possible to sketch in 3D (in fact that is the default) but all entities still lie only in the designated sketch plane.

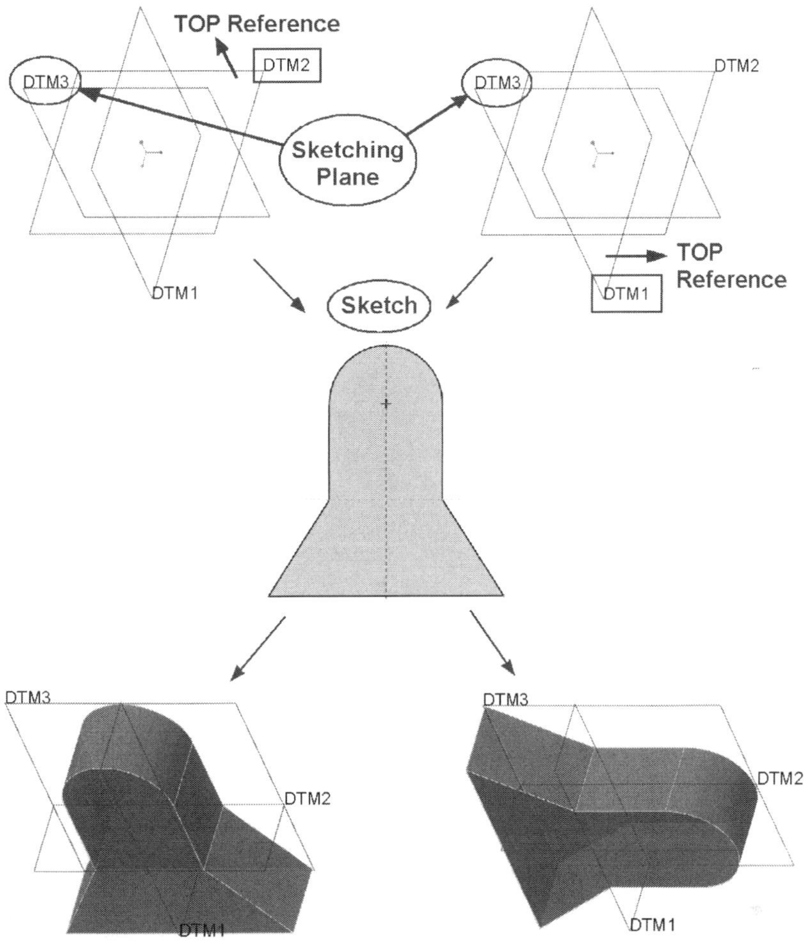

Figure 11 The importance of the sketching reference plane! The shaded images are both in the default orientation. The same sketching plane and sketch was used for both features - the difference in orientation of the solid is because of the choice of the sketching reference plane.

To illustrate the crucial importance of the reference plane, consider the images shown in Figure 11. These show two cases where the same sketching plane **DTM3** was used, the same sketched shape was drawn, the same reference orientation TOP was chosen, but where different datums were chosen as the sketching orientation reference. On the left, the TOP reference was **DTM2**. On the right, the TOP reference was **DTM1**. The identical sketch, shown in the center, was used for both cases (rounded end of sketch towards the top of the screen). However, notice the difference in the orientation of the part obtained in the final shaded images. Both of these models are displayed in the default orientation (check the datum planes). Clearly, choosing the sketching reference is important, particularly for the base feature (the first solid feature in the part).

Let's continue on with creating the curve. Make sure the **Sketch** dialog window is completed as in Figure 10. Select the ***Sketch*** button (or middle click).

Several changes occur on the screen. First, a new ribbon tab has appeared containing the sketching tools. Second, several new buttons have shown up on the Graphics toolbar. Take a moment to check these out with the pop-up tool tips. Finally, a couple of dashed cyan lines have appeared in the graphics area. What are these about?

To verify the meaning of the dashed lines, in the ***{Sketch}:Setup*** group select ***References*** (or select ***References*** in the RMB pop-up). This opens the **References** dialog window, Figure 12. In this window we can select any existing geometry to help Sketcher locate the new sketch relative to the part. In the present case, there isn't much to choose from, and two references have already been chosen for us - DTM1 and DTM2. The intersection of these datums with the sketch plane produces the two dashed lines. The number of references you choose is not limited (there may be several listed here) and a reference can be any existing geometric entity (plane, straight or curved edge, point, axis, projected edge, and more). You are also free to delete the ones chosen for you and/or add new ones. However, notice the **Sketch Status** at the bottom of this dialog. **Fully Placed** means enough references have been specified to allow Sketcher to locate your sketch. If there are not enough references, the status will be **Unsolved Sketch**. For now, do not proceed beyond this window unless you have a **Fully Placed** status indicated. If you have changed the references, use the ***Update the Sketch*** button at the lower right. Once the sketch is fully placed, select ***Close*** in the References window.

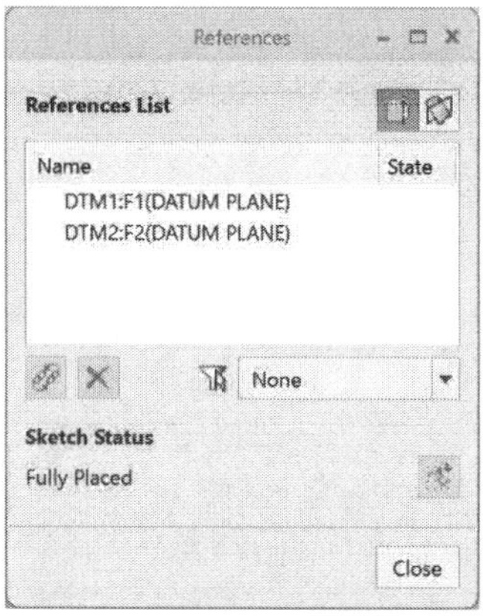

Figure 12 Choosing references in Sketcher

So that we are looking directly at the sketch plane, in the Graphics toolbar select the ***Sketch View*** button . The graphics window is shown in Figure 13. Note that you are looking edge-on to the datums DTM1 and DTM2. The positive side of datum DTM1 is facing the right edge of the screen, as specified back in Figure 10. Note that we could have obtained the same orientation by selecting DTM2 to face the top of the screen.

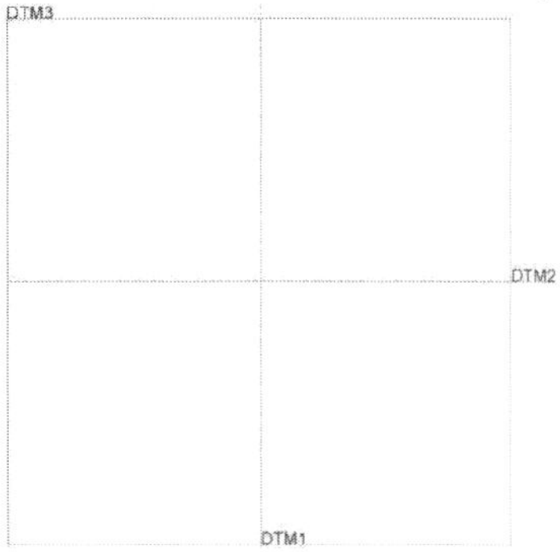

Figure 13 The drawing window

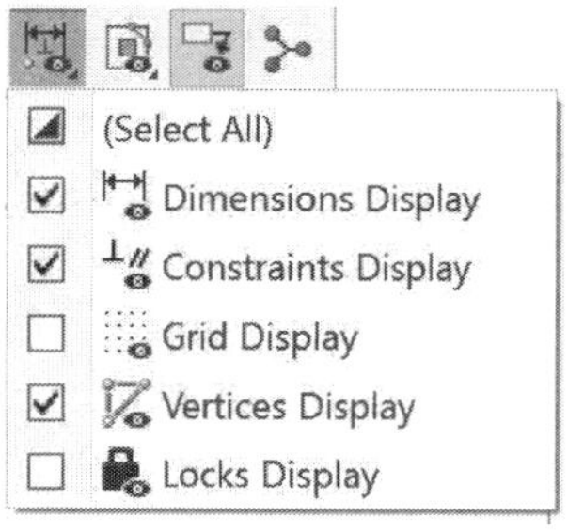

Figure 14 Filters for sketch entities

In the Graphics toolbar there is a new set of ***Sketcher Display Filters*** (Figure 14). Leave these options in their default value (***Dimensions***, ***Constraints***, ***Vertices***, all turned on). It is seldom (if ever) that you will need to turn on the grid in Sketcher. We will discuss dimension locks a bit later on. The major change on the screen is the Sketch ribbon. Let's take some time to explore it.

The Sketch Ribbon

The Sketch ribbon is shown in Figures 15 and 16. You should explore the groups for a while to familiarize yourself with the commands available. Some things you should note about the groups and commands are the following (starting from the left):

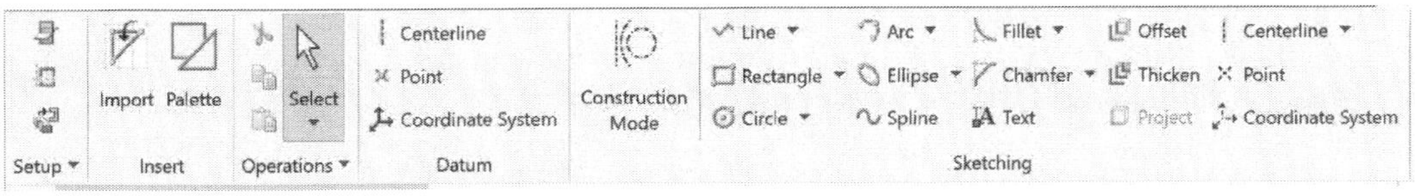

Figure 15 Left side of the Sketch ribbon (part 1)

Setup - allows you to reset the sketch plane and references, sketching options, and to reorient your sketch view (note this button is also on the Graphics toolbar).

Insert - import existing geometry or data from another source into the sketch. The ***Palette*** command lets you choose from several different families of 2D shapes that have already been created for you. We will use this command later in these lessons.

Operations – Windows based Cut (ctrl-x), Copy (ctrl-c) and Paste (ctrl-v) commands.

Datum - these buttons will create entities in the sketch that will appear and operate the same as datums in the model, but they are stored within the sketch. These were previously called geometry datums. They will create a datum element that is visible outside the sketch (and could be used as reference by another feature). For example, a ***Geometry Point*** creates a datum axis that will be perpendicular to the sketch plane.

Sketching - these will be your most-used commands. The ***Construction Mode*** button will toggle between real geometry and construction geometry. Real geometry will produce the feature, while construction geometry is only used to help define the sketch. An example would be using a construction circle to create an inscribed hexagon. You can also switch entities from construction to real (or vice versa) after they have been created. The rest of the sketching buttons do the actual geometry creation. Note that most buttons have flyouts to show variations of the command. For example, check out the rectangle button. We will spend a lot of time exploring some of these options.

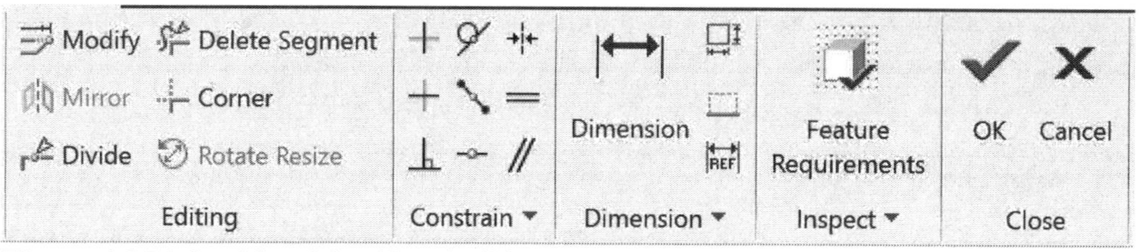

Figure 16 Right side of the Sketch ribbon (part 2).

Editing - contains commands for modifying geometry by changing dimensions using a special dialog window, trimming (and extending) entities, mirroring of entities, dynamic trim, and translating, rotating, and scaling a group of entities simultaneously.

Constrain - these are the constraints that Intent Manager will invoke automatically, but you can also explicitly assign these constraints to a sketch. See Table 2-3 on page 2-15.

Dimension - does pretty much what you would expect - allow you to create dimensions on the sketch. There are several varieties, including reference, ordinate, and perimeter.

Inspect - these contain tools to help you diagnose problems with your sketch. Some of these options are also available on the Graphics toolbar ***Inspection Filters*** dropdown. Under the pulldown you will see ***Shade Closed Loops*** and ***Highlight Open Ends*** active.

Close - lets you accept the sketch or quit without keeping or using the sketch. Note that the latter is irreversible (you cannot ***Undo*** it).

It is going to take a while to get comfortable with all the commands in the ribbon. Fortunately, there is a small subset that will be used repeatedly.

Helpful Hint

In the following, keep in mind that many functions in Sketcher are available in context sensitive pop-up menus attached to the RMB. For many sketches, you need never actually pick a ribbon button since almost everything is available in the graphics window using RMB commands only. Try it!

Table 2-3 Explicit Constraints in Sketcher

Constraint		
Vertical	Tangent	Symmetric
Horizontal	Mid-point	Equal
Perpendicular	Coincident	Parallel

Creating the Sketch

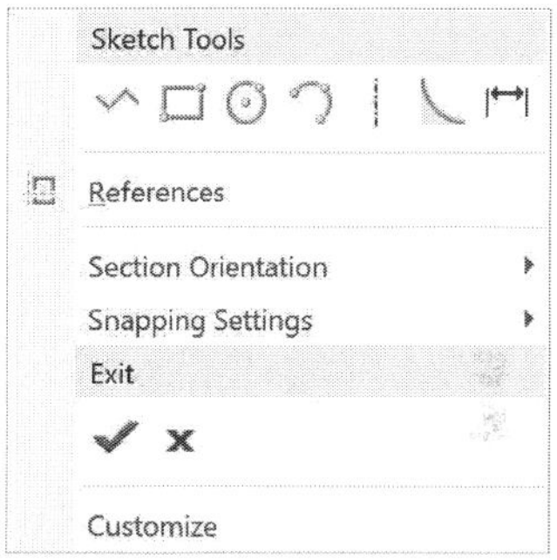

Figure 17 RMB pop-up menu in Sketcher

Select the ***Line Chain*** tool using one of the following three methods:

- using the ***Line Chain*** button in the Sketching group,

  OR (probably the easiest!)

- hold down the RMB and select ***Line Chain*** from the top row in the pop-up menu (Figure 17).

You will now see a small X with a dot at its center which will chase the cursor around the screen. Notice that the X will snap to the dashed references (which turn bright green) when the cursor is brought nearby, but not to the displayed edges of DTM3. While you are creating the sketch, watch for entities turning green that indicate Intent Manager is firing an internal rule to set up a constraint (Vertical, Horizontal, Equal Length). These symbols will come and go while you are sketching. The trick with Sketcher is to get Intent Manager to fire the rule you want, then click the left mouse button to accept the position of the vertex. Click the corners in the order shown in Figure 18[6]. After each click, you will see a straight line rubber-band from the previous position to the cursor position:

---

[6] The sketcher constraints will appear while you are sketching if the configuration option **show_sketcher_constr_dyn_edit** is set to **Yes**. The default is **No**.

1. left-click at the origin (intersection of **DTM1** and **DTM2**)
2. left-click above the origin on **DTM1** (watch for Vertical)
3. left-click horizontally to the right (watch for Horizontal and Equal Length - we do not want Equal Length)
4. left-click straight down on **DTM2** (watch for Vertical)
5. left-click back at the origin
6. middle-click anywhere on the screen to end line creation

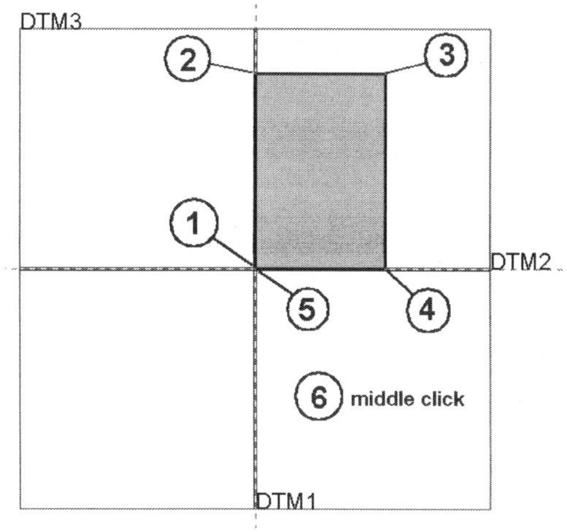

Figure 18 Drawing the Sketch

When you are finished with this sequence, you are still in **Line** creation mode (notice the X on screen and the ***Line Chain*** ribbon button). If you middle click again, you will leave that and return to ***Select*** mode - the same as if you picked on the ***Select*** button in the ribbon, but much faster.

Helpful Hint

From wherever you are in the Sketcher menu structure, a single **middle** mouse click will often abort the current command and return you to the toolbar with the ***Select*** command already chosen. Sometimes, you may have to click the middle button twice (the second time is to clear a selected object).

The sketched entities are shown in green. Since we have created a closed curve, the interior of the sketch is shaded[7] (because of the ***Shade Closed Loops*** button in the ribbon). Many features require sketches to contain only closed loops, so this is an easy way to verify that condition. At some point or other, you will create a sketch that you think is closed, but it will not shade. This usually means you have extraneous entities in the sketch (usually duplicated lines or edges) or the corners do not exactly close. We will see an example of duplicate lines in a few minutes.

Note that we didn't need to specify any drawing coordinates for the rectangle, nor, for that matter, are any coordinate values displayed anywhere on the screen. This is a significant departure from standard CAD programs. We also didn't need the grid (this is available if you want it, along with a snap function).

You can also sketch beyond the displayed edges of the datum planes - these actually extend off to infinity. The displayed extent of datum planes will (eventually) adjust to the currently displayed object(s).

---

[7] The color depends on a system color setting. See the Appendix.

Helpful Hints

If you make a mistake in drawing your shape, here are some ways to delete entities:
1. Pick the ***Select*** tool in the ribbon and left click on any entity you want to delete. Then either press the ***Delete*** key on the keyboard, or hold down the RMB and choose ***Delete***.
2. If there are several entities to delete, hold the CTRL key down while you left click on each entity. Then pick ***Delete*** as before.
3. You can left-click and drag to form a rectangle around a set of entities. Anything completely inside the rectangle is selected. Use ***Delete*** as before.
4. Notice the ***Undo*** and ***Redo*** buttons in the Quick Access Toolbar.

We will cover more advanced Sketcher commands for deleting and trimming lines a bit later.

After you have finished the sequence above, Sketcher will put two dimensions on the sketch - for the height and width of the rectangle. These will be in light blue, similar to those shown in Figure 19. For the first feature in a part, the numerical values of these dimensions are picked more-or-less at random (although they are in correct proportion to each other)[8]. For later features in the part, Sketcher will know the sketch size more accurately because it will have some existing geometry to set the scale.

If you select a dimension or constraint icon, the relevant entities will highlight in red - very handy!

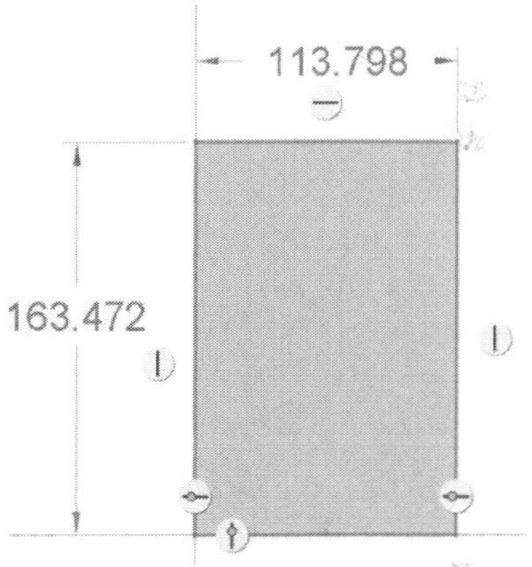

Figure 19 Initial sketch with weak dimensions

Weak vs Strong Dimensions

A dimension created by Sketcher (or actually Intent Manager) is called "weak" and is shown in light blue[9]. Strong dimensions (dark blue), on the other hand, are those that you create explicitly. You can make a strong dimension in any of three ways:

☞ modify the numeric value of a weak dimension

or ☞ create a dimension from scratch by identifying entities in the sketch and placing a new dimension on the sketch

or ☞ select a weak dimension and promote it to strong using the pop-up menu

---

[8] The default datum display with no other features present is actually ±250 units from where they cross.

[9] Incidentally, reference to entity colors assume the default system colors. Yours may be somewhat different due to system customization settings.

The special significance of weak and strong dimensions is critical. When Intent Manager is "solving" a sketch, it considers the sketch references, any implicit rules that have fired and any existing dimensions. If there is not enough information to define the drawing (it is *underconstrained*), Sketcher will create the necessary and sufficient missing dimensions. These are the weak dimensions. If Sketcher finds the drawing is *overconstrained* (too many dimensions or constraints) it will first try to solve the sketch by deleting one or more of the weak dimensions (the ones it made itself earlier). It will do this without asking you. This is one way for you to override Intent Manager - if you don't like the dimensioning scheme chosen by Sketcher, just create your own (automatically strong) dimensions. Sketcher will remove whichever of the weak dimensions are no longer needed to define the sketch. Sketcher assumes that any strong dimensions you have created shouldn't be messed with! However, if Sketcher still finds the drawing overconstrained, it will tell you what the redundant information is (either dimensions or constraints), and you can choose what you want deleted. Thus, although weak dimensions can be deleted without asking you, Sketcher will never delete a strong dimension without your explicit confirmation unless you are deleting one of the entities being dimensioned.

We want to modify the two weak dimensions on the rectangle in a couple of ways. First, we can make a cosmetic improvement by selecting the dimension text (the number) and performing a drag-and-drop to move it to a better location. Note in passing that preselection highlighting also works with dimensions and constraints. Do that now, so that the dimensions are located as in Figure 20 ("off the sketch")—note DTM display is off. Your font size and the number of significant digits (3 in this case) may be different.

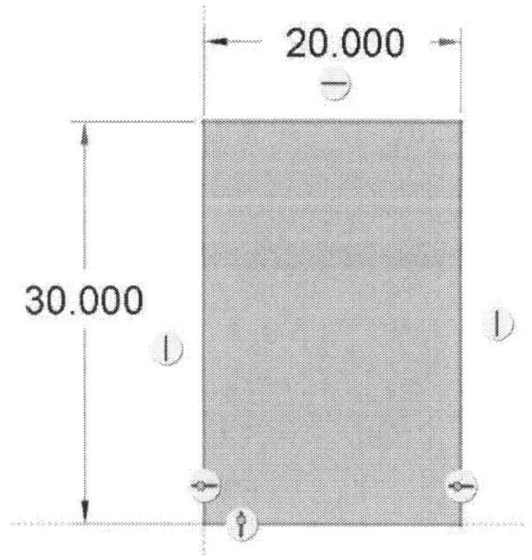

Figure 20 Modified sketch

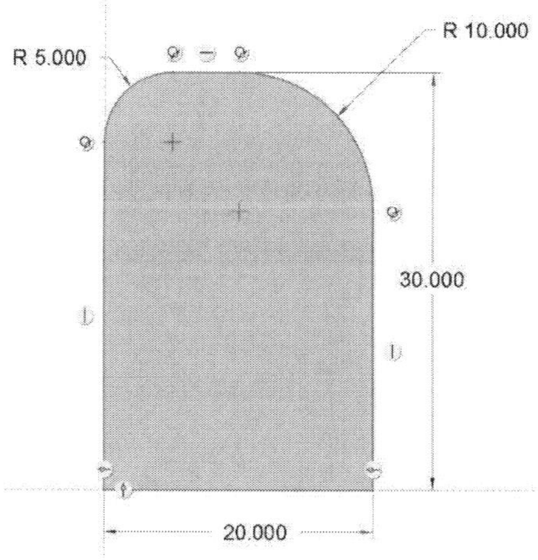

Figure 21 Completed sketch

Next, we can change the numeric values of the dimensions. It is explained later why it is not a good idea to do this now (since the shape of our sketch is not complete yet), but we will anyway to show why it is not a good idea! Double-click on the horizontal dimension. In the text entry box, enter the value **20**. When you hit ***Enter***, the sketch geometry will be updated with this new dimension which is now strong.

IMPORTANT: Since this is the first dimensioned feature in the part, Sketcher will automatically scale all the other weak dimensions in the sketch to maintain the sketch shape. This will only happen for the first dimensioned feature in a part. This avoids the cumbersome ***Select All*** ' ***Modify*** ' ***Lock Scale*** sequence that does the same job.

Change the vertical dimension to **30**. It will also now be strong. (Click anywhere on the graphics window to remove the green highlight and update the shading.) See Figure 20. Both dimensions are now in dark blue.

Notice that the indicated extent of the datum plane DTM3 adjusts to the sketch. You may want to ***Refit*** the sketch in the graphics window (or use zoom).

Now we'll add a couple of rounded corners, technically known as fillets, on the top corners of the sketch to help us "see" the orientation of the feature in 3D. Select the ***Arc*** command from the RMB pop-up menu ( ) and pick on the top and right lines in the sketch close to but not at the corner. A circular fillet is created to the pick point closest to the corner. Two tangent constraints are added, along with a weak dimension for the fillet radius. Do the same on the top and left lines. Middle click to return to Select mode.

IMPORTANT: Because our fillet command has removed two vertices on the top of the sketch, Intent Manager has removed our two strong dimensions (which used those vertices) and replaced them with weak ones (see the message window - you may have to scroll back a couple of lines). So really we wasted a couple of steps by previously modifying the values of those dimensions (and making them strong in the process) before actually completing the sketch shape. You can make them strong again by selecting them, and selecting ***Strong*** in the pop-up menu (or use ***Ctrl+T***). Modify the cosmetics and values of the fillet radius dimensions as shown in Figure 21.

Helpful Hint

When you are creating a Sketch, the most efficient order of operations is the following:

1. Make sure you have the **references** you want.
2. Create the **geometric shape** you want.
3. Make sure the **constraints** you want have appeared.
4. Add the **dimensions** that you want to control the sketch. Intent Manager will remove any weak ones no longer needed.
5. Modify the dimension **cosmetics** to take the dimensions off the sketch.
6. Change the dimension **values** to the ones you want. Be aware of decimal places since the displayed values may be rounded off (precision is a system setting).

If you do not do things in this order, you will likely have to backtrack (as we did above by respecifying the strong dimensions for the block).

Sketcher Diagnostic Functions

At this point, we have a completed a valid sketch. Select the ***Feature Requirements*** button in the Sketch tools menu at the top. The window that opens indicates that the sketch is okay (all requirements check out). Let's introduce a common error into the sketch to see what happens. Close the **Feature Requirements** window.

We will create a duplicate line on the sketch using ***Line Chain***. Start at the bottom left corner and end the line part way up the vertical edge towards the top, overlapping the existing line. When you finish the line and return to Select mode (remember to middle click twice), three things will happen. First, the sketch is no longer shaded, indicating that it is no longer a simple closed loop. Second, a big red square appears where the overlapping line ends. This indicates an open (that is, unconnected) end of the sketch (if the ***Highlight Open Ends*** button in the **Inspect** group in the ribbon is on). Third, a weak dimension will appear for the new line. Although the geometry of the sketch appears visually correct, if we tried to use it to create a solid feature it would fail[10]. Now pick the ***Highlight Overlaps*** button in the **Inspect** group. The offending vertical line will highlight in red (easier to see with the datum plane display turned off). At this point, you would have to do some detective work to figure out where the problem was. Select the short vertical line (note that preselection works here) and delete it. The red square disappears and the sketch is again shaded.

You should experiment with the Sketch Diagnostic tools periodically as we proceed through the tutorial.

This completes the creation of our first sketch. Select the ***OK*** in the ribbon, or in the RMB pop-up menu. On the top menu bar select **VIEW** and under ***Show*** make sure the Datum Plane display is on, and the Datum Plane Tag display (just below it) is on. This returns us to the regular graphics window with our new sketched curve shown in green (last feature created). You can spin the model around with the middle mouse button to see this curve from different viewpoints. When you are finished with this, return the model to approximately the default orientation - Figure 22. If you click anywhere else on the graphics window, the curve turns its default color of light blue. This is a good time to save your BLOCK (ctrl-s).

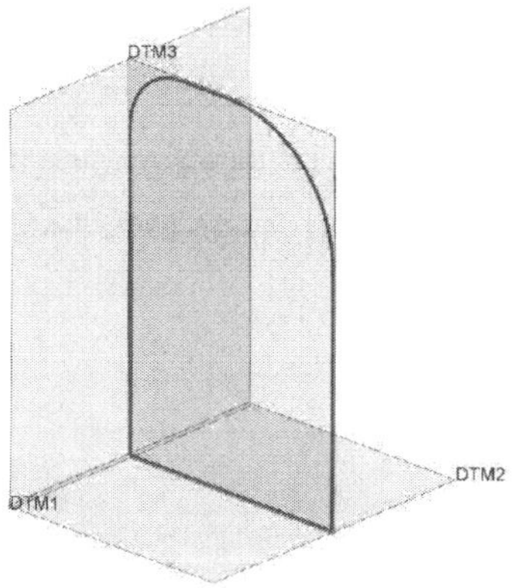

Figure 22 Sketched curve

[10] This sketch could be used to create an extruded surface feature, but with overlapping surfaces, probably not a good idea. For an extruded solid, the feature would fail with an error message about requiring a closed section only.

Creating a Solid Protrusion

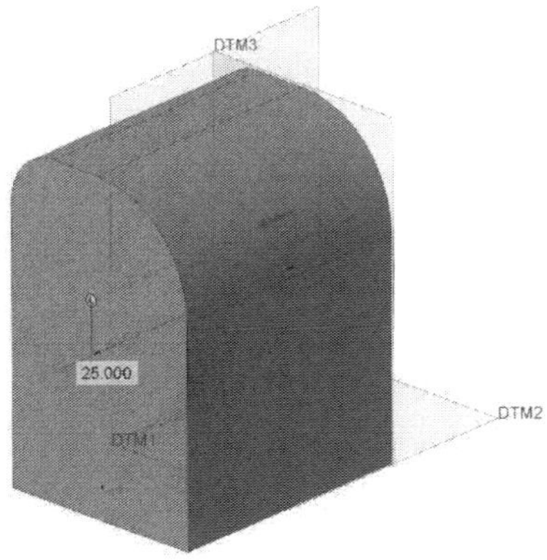

Figure 23 The protrusion preview

Most of the work to create this feature has been done already - creating the sketched curve that defines its shape. This curve should be highlighted in green. If you have been playing around with the model and the sketch is blue, just left click on it to select it again.

With the sketched curve highlighted, pick the ***Extrude*** button in the **Shapes** group of the ribbon or in the middle row of the pop-up menu.

What you will see now is an orange shaded image of the protrusion, Figure 23. On this shape, you will see a magenta arrow that indicates the extrusion direction, which by default comes off the positive side of the sketch. There is also a dashed line ending in a yellow circle. This is a drag handle. Click on this with the mouse and you can drag it to change the length of the extrusion. This length is also shown in a dimension symbol. You can even drag this extrusion out the back of the sketch to extrude in the opposite direction. This direct manipulation of the feature on the screen is called, in Creo Parametric vernacular, *Direct Modeling*. Bring the protrusion out the front and double click on the numeric dimension, and enter the value **30**.

Check out what happens if you hold down the RMB with nothing selected. What is in the pop-up menu? What happens if you select the yellow drag handle? It should turn blue. What is in the RMB menu now? We will explore these options later.

At the top of the graphics window is a new collection of tools. These comprise the *Dashboard* for the extrude feature. Many features are constructed with tools arranged using this style of interface element. It is worth spending some time exploring this one in detail, since you will probably be using it the most.

Helpful Hint

You may accidentally leave the dashboard with an inadvertent click of the middle mouse button. Remember that this is a short cut for ***Accept***. If that happens, with the protrusion highlighted, select the ***Edit Definition*** button at the top of the pop-up menu. This will bring you back to the dashboard. The ***Undo*** command, if executed immediately, will delete the feature.

The Extrude Dashboard

The dashboard collects all of the commands and options for feature creation in an easily navigated interface. Moreover, most optional settings have been set to default values which will work in the majority of cases. You can change options at any time and in any order, often by using a RMB pop-up menu.

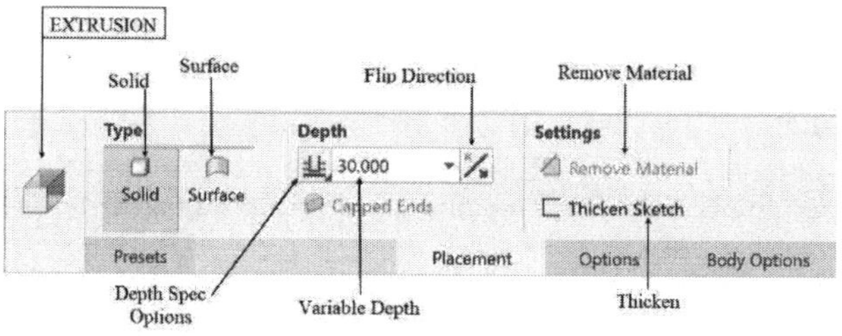

Figure 24 The *Extrude* Dashboard

The feature being created is identified by the ribbon tab at the top of the dashboard as well as the icon at the left end of the dashboard. The dashboard itself contains two areas. On the left (Figure 24) are commands, settings, and so on for the particular feature under construction. The ones showing are context sensitive so with a different feature (like a hole) you will see different commands. The following are for an extrude:

Solid and ***Surface*** buttons - these are an either/or toggle set. The default is to create a solid. If you pick ***Surface***, the sketch will be extruded as an infinitely thin surface (Figure 25). Return this to the ***Solid*** selection.

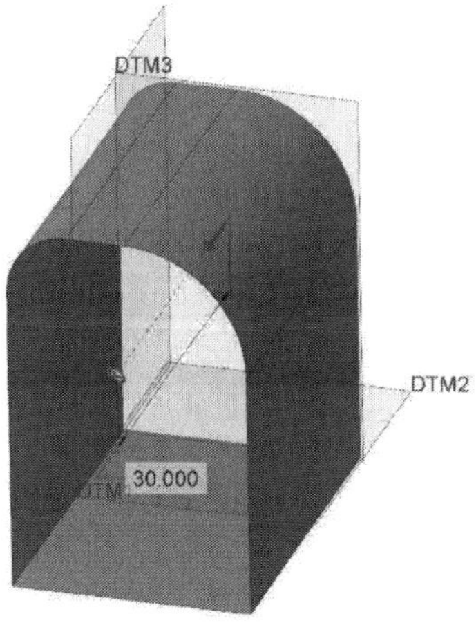

Figure 25 Extruded surface

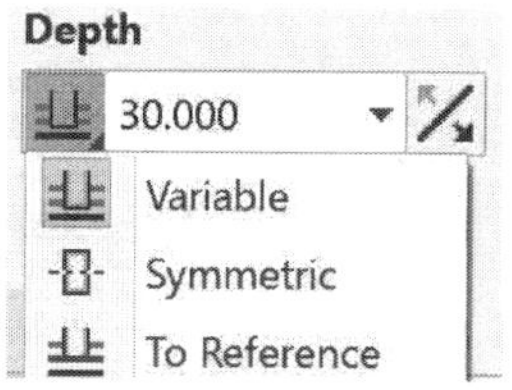

Figure 26 Depth Spec Options

Depth - the next button is a drop-down list of all the possibilities for specifying the depth of the extrusion. These are indicated in Figure 26. The default is a ***Variable*** depth (previously known as *Blind*), which means the extrusion is for a fixed distance. Other options may appear here as more part geometry appears (as in the cut which we will do next). The **Variable Depth** box contains the numeric value of the length of the extrusion. If the depth specification on the button to the left is not ***Variable***, this text input area is grayed out. Just to the right of the depth value is the ***Flip*** button for reversing the direction of the protrusion (the magenta arrow).

Remove Material - this allows you to change the meaning of the solid feature from a protrusion (which adds solid material) to a cut (which removes solid material). Since there is nothing to remove at this time, this command is grayed out - all we can do is add.

Thicken Sketch - press this to see the solid block replaced by a thin-walled extrusion (formerly called a *Thin Solid*). A new dimension appears in the graphics window and on the dashboard. This is the thickness of the solid wall. Try changing this thickness to something like 2.0. On which side of the sketched curve has this been added? Another ***Flip*** button has also appeared. Press this a couple of times - it controls which side of the sketch the material is added to. Actually, it is a three way switch since you can also add material equally on both sides of the sketch. Press the ***Thicken*** button again to return to a full solid protrusion.

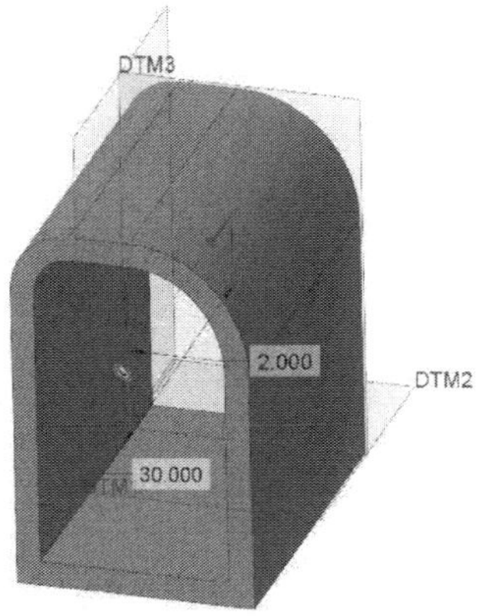

Figure 27 A ***Thick*** extruded solid

Just below this area of the dashboard are several drop-down panels which can be opened or closed by clicking on its label. The ***Placement*** panel is opened by default.

Placement - allows you to select, create, or modify the 2D section to be used for the feature. Since we preselected the sketched curve, it is now listed on this panel. If we had not preselected the curve, we could have chosen it now, or launched Sketcher from this panel to create a new sketch. This would involve selecting the sketching plane, sketcher reference, and so on. We will go this route in the next feature. If you wanted to change the sketch for the extrude, this is how you access it. The ***Unlink*** button is currently displayed on the **Placement** panel. This button appears if, like now, you have preselected a curve to serve as the sketch for the extrude. Thus, the extrude is linked to the previous curve feature; changes to the geometry or dimensions of the curve would drive changes in the shape of the extrude. The curve itself is a separate entry on the model tree. The purpose of unlinking is to break this (parent/child) connection to the original curve. If you were to select this command (**don't do this now**), the copy of the original curve stored in the extrude feature would become independent. In that case, a change to the original curve would not affect the extrude. The original curve could be modified, moved, or even deleted, and the extrude would still be able to regenerate. The use of external curves to drive feature geometry is an important aspect of an advanced modeling technique that makes use of *skeleton models*, especially for mechanism design.

Options - (click the box to open this panel now. The ***Placement*** panel will close.) contains more information about the depth specification for the feature. We will find out what is meant by "Side 2" a bit later. Check out the ***Add Taper*** option (drag on the yellow drag handle, return to 0 when done).

Body Options - this is where you can specify which body the feature should belong to (the default is the currently active body), or create a new body. Leave this set to Body 1. Almost all the parts in this Tutorial will only use a single body.

Properties - specify the name of the feature, currently EXTRUDE_1.

As mentioned, to close a drop-down panel, just click on its tab in the dashboard. The drop-down panels can be detached from the dashboard and placed as separate windows anywhere in the graphics area. See Figure 28. They will stay there (and open), including for extrudes you create later, until you attach them back to the dashboard. Do that by either dragging the panel back up or clicking the ***Attach*** button at the top right of the panel windows.

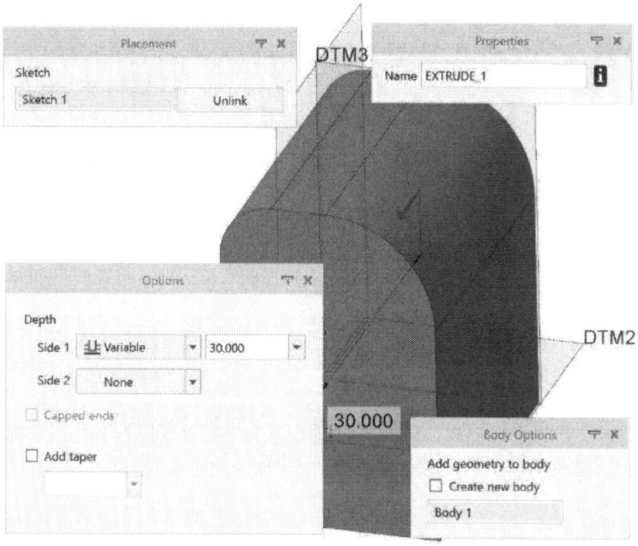

Figure 28 Detachable panels

As you explore the creation of new features in Creo Parametric you should investigate what is in each of these menus. They are context sensitive, so there is a lot of variety in what you will find, but all the dashboards basically operate the same way.

To the right on the dashboard are several common tools that appear for all features. See Figure 29. These function as follows:

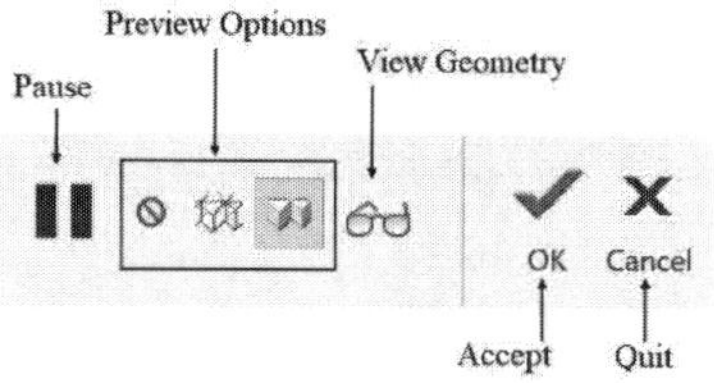

Figure 29 Common dashboard controls

Pause - allows you to temporarily suspend work on this feature so that you can, for example, create a missing reference like a datum plane, measure something in the model, etc. When you are finished with the side trip, press the symbol <that appears here to continue where you left off.

Preview Options - (default = ***Attached***) this is responsible for the shaded orange display of the feature under construction. Select the ***No Preview*** button - all you will see is the feature creation direction, drag handle, and depth dimension. Turn this back on.

View Geometry - this shows what the geometry will look like when the feature is fully integrated into the part. Not much happens with this first protrusion. Press again to return to preview.

OK and ***Cancel*** - do just what you expect!

Select ***OK*** (or middle click). The block now appears in the default neutral gray color, Figure 30, with its edges highlighted in green (last feature created).

We spent a lot of time discussing the creation of this feature so it maybe seemed like a lot of work. Let's create it again without all the discussion. Select the ***Undo*** button at the top - this will delete the protrusion. Now create it again:

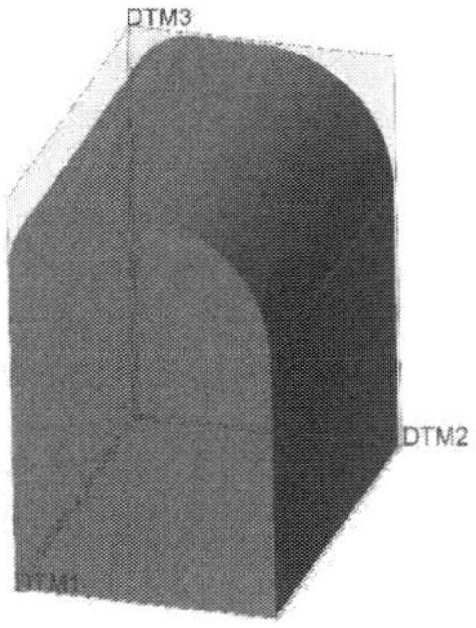

Figure 30 Completed protrusion

- ✓ highlight the sketched curve
- ✓ select the ***Extrude*** tool
- ✓ change the depth to **30**
- ✓ middle click

Because of all the defaults, this only takes four clicks!

Open the Navigator pane to see the model tree, Figure 31. The default datum planes are listed. Then comes the sketched curve **Sketch 1** (hidden) and the extrude **Extrude 1**. Open the listing for the extrude and you will see the internal sketch (also **Sketch 1**) that uses the hidden curve and is currently linked to it. If you have used the ***Unlink*** button in the **Placement** panel, the model tree will appear somewhat differently - the sketch stored inside the extrude is called S2D0001 and is independent of Sketch 1. The **Bodies** list under **Design Items** now indicates that it contains geometry (if you have the additional model tree columns as in Lesson 1). The extrude feature also indicates that it contributes to Body 1. Notice that datum features and sketches do not directly contribute to body definitions since they are not solid geometry. Close the Navigator.

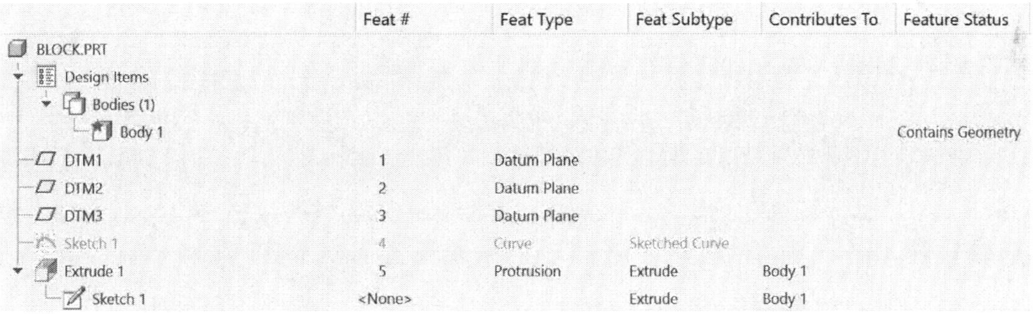

	Feat #	Feat Type	Feat Subtype	Contributes To	Feature Status
BLOCK.PRT					
Design Items					
Bodies (1)					
Body 1					Contains Geometry
DTM1	1	Datum Plane			
DTM2	2	Datum Plane			
DTM3	3	Datum Plane			
Sketch 1	4	Curve	Sketched Curve		
Extrude 1	5	Protrusion	Extrude	Body 1	
Sketch 1	<None>		Extrude	Body 1	

Figure 31 Model tree of *block.prt* with first protrusion created

We will now add another extruded feature - this time we will create a cut that removes material. Furthermore, instead of creating the sketch first, as we did for the solid protrusion, we will create the sketch within the feature itself. This will follow Method #2 of Table 2-2 back on page 2-9 of this lesson. **<u>This is actually the more common way to use Sketcher, especially if the sketch will be used just once.</u>** Before we do that, now is a good time to save the part (ctrl-s).

Saving the Part

It is a good idea when you are just getting started to save your model quite frequently, just in case something serious goes wrong. If you have to bail out of the program, you can always reload the most recently saved copy of the part and continue from there.

There are (as usual!) several ways to save the part:

- in the Quick Access toolbar, select the ***Save*** button, or
- in the pull-down menus select ***File*** ➤ ***Save***, or
- use the keyboard shortcut CTRL-S.

Make sure that the **Save Object** dialog is showing the desired working directory at the top. If not, select it in the Common Folders area in the Navigator. At the bottom of the dialog window, the name of the current active object (remember that you can have more than one object loaded into memory at a time) should already be in view. Accept the default model name **[block]** (this is the *active* part) by pressing the enter key or the middle mouse button. Creo will automatically put the part extension (*prt*) on the file. If you save the part a number of times, Creo will automatically number each new saved version (like block.prt.1, block.prt.2, block.prt.3, and so on[11]). Be aware of how much space you have available. It may be necessary to delete some of the previously saved versions; or you can copy them to an archive location. You can do both of these tasks from within Creo - we'll talk about that later.

> **IMPORTANT NOTE:**
> The ***Save*** command is also available when you are in Sketcher. Executing this command at that time will ***not*** save the part but will save the current sketched section with the file extension *sec*. This may be useful if the sketch is complicated and may be used again on a different part. Rather than recreate the sketch, it can be read in from the saved file (using ***Import***). In these lessons, none of the sketches are complicated enough to warrant saving them to disk.

Now we will proceed on to the next feature - an extruded cut. Make sure nothing is currently selected (highlighted) on the model (by clicking anywhere else in the window).

Creating an Extruded Cut

Following Method #2 of Table 2-2, start by launching the ***Extrude*** command from the ribbon. The extrude dashboard at the top of the screen opens. The ***Placement*** pull-down panel tab is red to indicate missing information there. Select ***Define*** or in the pop-up

[11] On computers running Windows you will need to turn off the option "Hide extensions for known file types" in the ***Folder Options*** > ***View*** settings in Windows Explorer to see the file revision numbers.

menu select ***Define Internal Sketch*** . The **Sketch** dialog window appears. This time, however, nothing has been preselected for us; we'll have to enter the data ourselves.

In the Sketch dialog window, notice the light green data entry field, called a *collector*, beside **Sketch Plane**. Whenever one or more references can be specified for something, they are placed into a collector. When a collector is active (ready to accept data) it is light green; otherwise, it will be white. If you want to make a collector active, just click it.

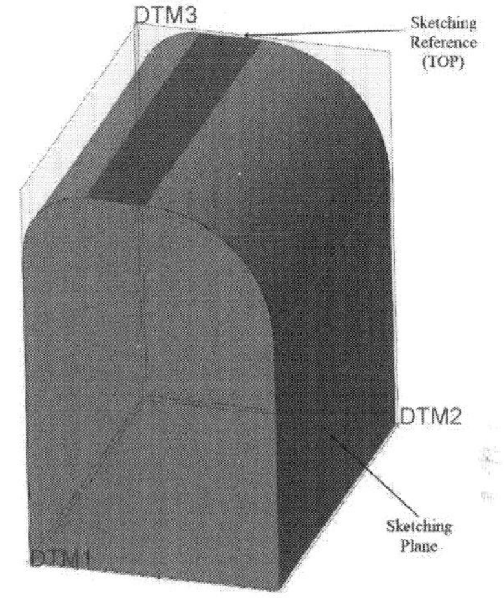

Figure 32 Setting up to sketch the cut

Since the Sketch Plane collector is highlighted, the dialog is waiting for you to select the sketching plane (see the message line at the bottom of the window). Pay attention to preselection here. You will not be able to pick an edge or a curved surface (both of these would be illegal). Pick on the right side surface of the block (see Figure 32). As soon as you pick the sketching plane (it highlights in cyan), a magenta arrow will appear showing the default direction of view relative to the surface. The ***Flip*** button can be used to reverse this direction, but leave it as it is. Creo makes a guess at a potential reference plane for you to use. This may depend on the current orientation of your view, and might result in a strange view orientation in sketcher (like sideways or even upside down). The sketch orientation reference serves another purpose that we will see in a minute, so we want to be a bit more careful and specific here. The collector for the sketch orientation reference surface is showing the default chosen. To override the default for the active collector, pick on the top planar surface (Figure 32) between the two tangent lines of the rounded corners; the surface will highlight in green. In the ***Orientation*** pull-down list, select ***Top*** so that the reference will face the top of the screen. We now have our sketch plane and reference set up, so select ***Sketch*** at the bottom of the dialog window. So that we are looking straight on to the sketching plane, select the ***Sketch View*** button in the Graphics toolbar[12].

In the Graphics toolbar, set Display Style to ***No Hidden*** We are now in Sketcher (Figure 33). Two references (dotted lines) have been chosen for us (the back and top surfaces of the object). We are going to create the U-shaped figure shown in Figure 34. Note that there is no sketched line across the top of the U - there is no inside or outside. Thus, it is technically called an *open* sketch (as opposed to a *closed* sketch for our previous feature). There are some restrictions on the use of open sketches which we will run across in a minute or two.

Use the RMB pop-up menu to select the ***Line Chain*** command. Start your sketch at vertex 1 in Figure 34 - the cursor will snap to the reference. Then drag the mouse down and pick vertex 2 (note the Vertical constraint), and middle click twice to end the ***Line*** command. Some weak dimensions will appear. Do nothing about them yet because, since

[12] A configuration option (*sketcher_starts_in_2D*) can make this automatic. This is also a handy command to add to the RMB pop-up menu using ***Customize***.

they are weak, they are liable to disappear anyway. If we make them strong, this will cause us extra work dealing with Intent Manager. The line is highlighted in green. Left click somewhere else to deselect the line.

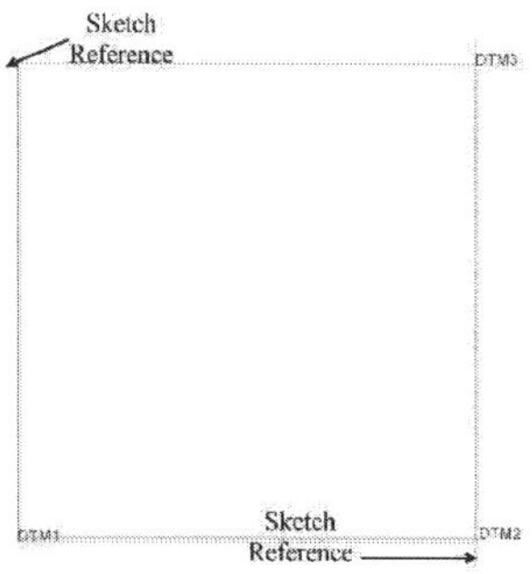

Figure 33 References for cut sketch

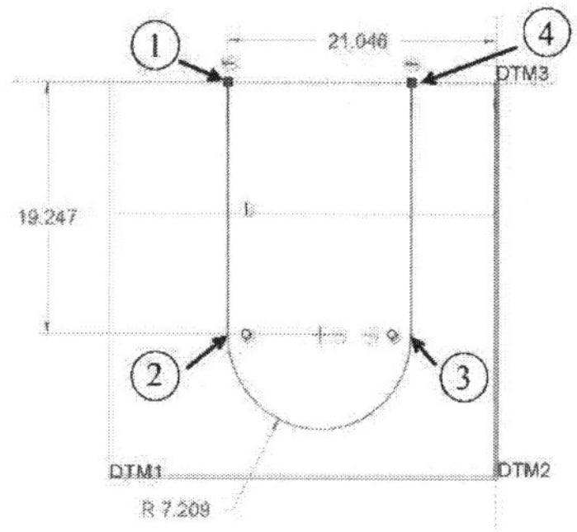

Figure 34 Sketch geometry

> **Helpful Hint**
> Wait until the shape of the sketch is finished before you start worrying about the dimensioning scheme or dimension values.

Use the pop-up menu again and select the ***3-Point Arc*** command. Pick on the end of the sketched line (vertex 2) and drag the mouse downwards in the direction of tangency. Once the arc has been established, drag the cursor over to the right (the arc will rubber-band while maintaining the tangency constraint) and click at vertex 3. (If you drag straight across to vertex 3 you will get a 3-point arc which is not automatically tangent at vertex 2.) You should see horizontal constraint symbols that indicate when vertex 3 is at the same height as the center of the arc. Left click to create the vertex 3. Use the RMB menu to pick ***Line Chain*** again.

Now left click at vertex 3 and draw a vertical line up to snap to the reference at vertex 4, making sure that you have a tangent constraint at vertex 3. Our sketch is complete. Use the middle mouse button to return to ***Select*** mode. Your dimension values may be different from those shown in Figure 35. Your dimensioning scheme may even be slightly different. It will be easier to see this if you go to hidden line display instead of shading.

All the dimensions should be weak. Drag them to a better location if necessary (off the part). Don't be afraid to resize your display so that you can see everything clearly. Compare the dimensioning scheme with the one in Figure 35. We want to have

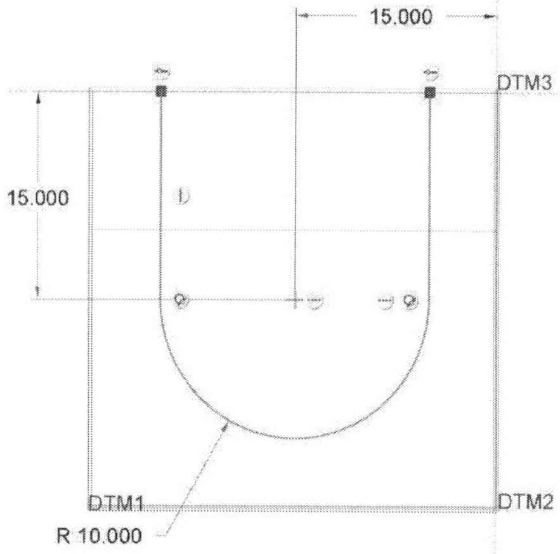

Figure 35 Final sketch for cut

a horizontal dimension of 15 from the reference at the back of the part to the center of the arc of the U. If you do not have that dimension, we'll have to add one manually. This will illustrate a case where we will override the Intent Manager.

To create your own dimension, select the ***Dimension*** command from the ribbon (or ***Dimension*** in the pop-up). Click on the vertex at the center of the arc (it will highlight) then click again on the dashed reference line at the right. Now middle click in the space above the part where you want the dimension text to appear. Set the value as shown in the figure. It's that easy! Note that this dimension shows immediately in dark blue since it is strong. One of the weak linear dimensions should be gone. Middle click to get back to ***Select***.

We now have the references, shape, implicit constraints and dimensioning scheme we want. So now we can worry about the dimension values. Get in the habit of dealing with your sketches in the following order (see the Helpful Hint on page 2-19):

Select References →
Draw Shape →
Assign Constraints →
Setup Dimension scheme →
Set Dimension values

Modify the values of the dimensions to match those in Figure 35.

Recall the hint above about open vs closed sketches. The sketch we have just produced will be able to produce the desired cut, but it is not the best we can do. The behavior of features constructed using open sketches is unpredictable. The danger lies in the possibility that something or someone may come along later to modify this part and add something to the top of the block earlier in the model tree. In that case, the cut feature would likely fail due to the uncertainty in what to do across the open ends of the sketch. So before we accept this sketch, let's explore the diagnostics a bit.

Helpful Hint

In general, try to keep your sketches closed - you will have fewer problems that way. Open sketches will be necessary only for special features or modeling requirements.

In the Graphics toolbar, select ***Inspection Tools Display Filters***  pulldown, and check ***Highlight Open Ends*** and ***Shade Closed Loops***. Since the sketch is currently open it will not shade; the two ends will show the red squares. Select the ***Feature Requirements*** button in the ribbon. Notice the warning - the system is telling us that although this is not a fatal error, it is not a good idea. Close this window and create a line across the top of the sketch to close it (use ***Line Chain*** in the RMB pop- up). Now the sketch will shade, the red squares disappear. The sketch is now complete, so click on the ***OK*** button in the toolbar[13]. If you are in No Hidden or Hidden Line display, return to Shading display.

The feature will now be previewed. A couple of new buttons have appeared on the dashboard. First, in the **Depth Spec** pull-down list, there are a few more options available (Figure 36). For this cut, we would like the sketch to be extruded through the entire part, so pick the ***Through All*** option. Note that the dimension for a blind extrusion disappears from the screen. To the right of this area, click the ***Flip*** button (or the magenta arrow) to make the extrusion go through the part. The ***Remove Material*** button needs to be selected - finally we indicate that we are making a cut. Now, there are two magenta arrows attached to the sketch. The one perpendicular to the plane of the sketch shows the direction of the extrusion. The other shows which side of the sketched line we want to remove material from. These should be set as shown in Figure 37.

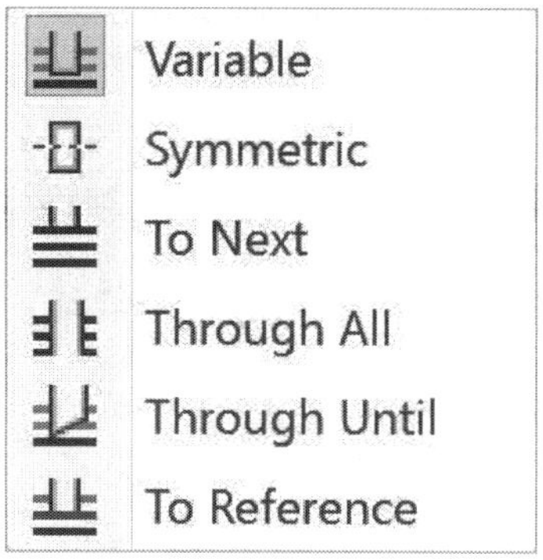

Figure 36 More Depth Spec options in the dashboard

Now select the ***Verify*** button in the right area of the dashboard. If you have the ***Remove Material*** button set wrong, Creo Parametric will create a protrusion instead of a cut[14]. Turn ***Verify*** off.

Another common error with cuts is having the material removal side set wrong (the second magenta arrow in Figure 37). If you do that for this part, you will end up with Figure 38. Make sure the material removal arrow points to the inside of the U. Plus, you should explore the **Placement**, **Options**, **Body Options**, and **Properties** menus on the dashboard before you leave.

We are finished creating this feature, so select the ***OK*** button at the right end of the dashboard. Open the **Bodies** list under **Design Items**. Select Body 1 and in the RMB pop-up select the ***Make Transparent*** command. As long as model colors are turned on, even the default neutral gray body will become partially transparent. You can set the transparency using the control in the graphics toolbar. The part should now look like Figure 39. The cut will be highlighted in green as usual, as the last feature created.

---

[13] This is one of the few times when a middle click does not mean ***Accept***, which is a good thing since inadvertent middle clicks happen often when you are in Sketcher.

[14] Even worse, if we had left the sketch open, the protrusion would fail. This sketch is ambiguous since when the sketch starts out from the sketch plane, the vertices at the ends are out in the open air; Creo Parametric would not know how to create the solid to attach it to the existing part. This problem does not occur with a cut as long as the open ends of the sketch stay outside or on the surface of the part. You might come back and explore this later.

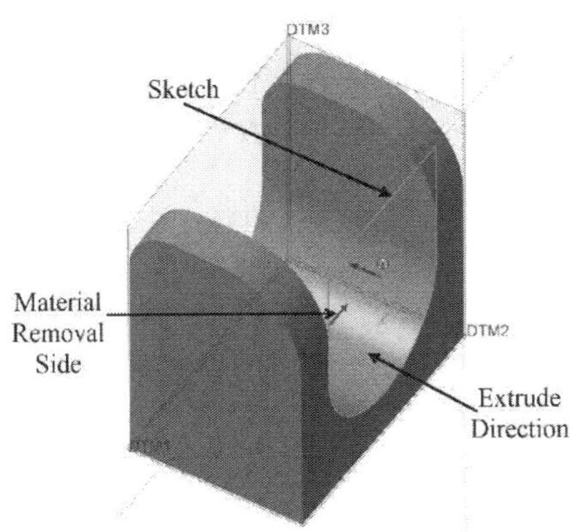

Figure 37 Defining cut attributes (direction and material removal side)

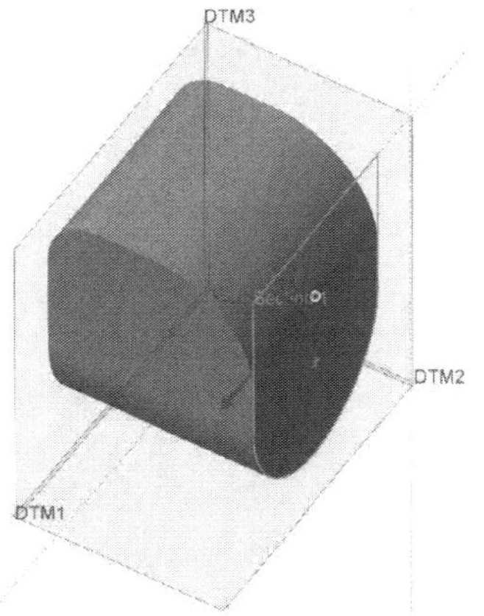

Figure 38 Removing from the wrong side of the sketch

Save the part (Ctrl-S). We will need it in this condition for the next chapter. Notice that there are no defined view orientations for the part, like FRONT, RIGHT, and so on. Since those are so commonly used, it will be handy if they could be created automatically. That is one of the reasons to use a part template. Read on!

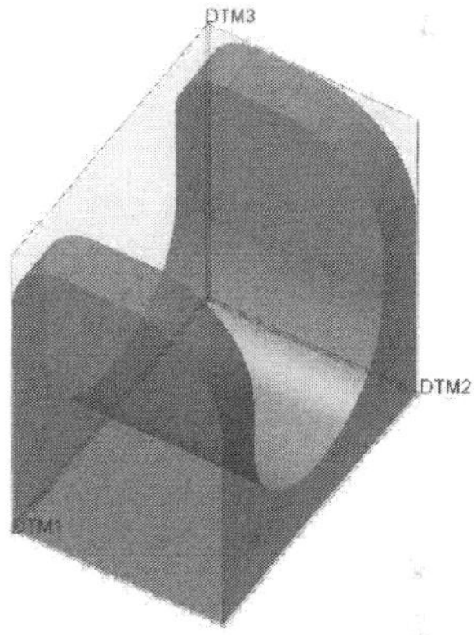

Figure 39 **Cut** feature completed

Using Part Templates

You will recall that, in creating the block part, the first thing we did was to create default datum planes. These are very standard features and aspects of part files, and it would be handy if this was done automatically. This is exactly the purpose of part templates.

A template is a previously created part file that contains the common features and aspects of almost all part files you will ever make. These include, among other things, default datum planes and named views. Creo has several templates available for parts, drawings, and assemblies. Your local installation may have more, and you can create your own templates as well. There are variations of the templates for each type of object. One important variation consists of the unit system used for the part (inches or millimeters). Templates also contain some common model parameters and layer definitions[15].

A template can be selected only when a new model is first created. Let's see how that works. Create a new part (note that you don't have to remove the block - Creo can have several parts "in session" at the same time) by selecting

[15] Model parameters and layers are discussed in the *Advanced Tutorial*.

File ➤ New

or using the "Create New Object" button in the Quick Access toolbar. The **New** dialog window opens. The ***Part*** | ***Solid*** options are selected by default. Enter a new name, like ***[exercise_1]***. Remove the check mark beside *Use default template* (we normally don't do this, but we should have a look at what is available) and then select ***OK***.

In the **New File Options** dialog window, the default template is shown at the top. It is likely "**Empty**" (unless your system has been set up differently). Scroll down to find "**inlbs_part_solid_abs**". This template is for solid parts with the units set to inch-pound-second. The "*_abs*" refers to a setting (absolute vs relative) for the internal accuracy of the computations in Creo. Absolute accuracy is recommended. It seems strange to have force and time units in a CAD geometry program. Actually, this is included so that the part units are known by downstream applications like Creo Simulate which perform finite element analysis (FEA) or mechanism dynamics calculations. These programs both require force, mass, and/or time in their operation and are very picky about units!

Note that there are templates available for metric units "**mmns_part_solid_abs**" and "**mmns_part_solid_rel**" (millimeter-Newton-second) for both absolute and relative accuracy versions. While we are mentioning units, be aware that if you make a wrong choice of units here, changing units or accuracy setting of a part is possible after it has been created (see ***File ➤ Prepare ➤ Model Properties***).

There are only two model parameters in the default template. *DESCRIPTION* is for an extended title for the part, like "UPPER PUMP HOUSING". This title can (eventually) be called up and placed automatically on a drawing of the part using, you guessed it, a drawing template. Similarly, the *MODELED_BY* parameter is available for you to record your name or initials as the originator of the part. Fill in these parameter fields ("Creo Exercise #1" and your initials) and select ***OK***.

The new part is created which automatically displays the default datums. They are even named for you (we will see how to name features in lesson 3): instead of DTM1, we have **RIGHT**. **TOP** replaces DTM2, and **FRONT** replaces DTM3. The part also contains a coordinate system, named views (look in the ***Saved Orientations*** List), and other data that we'll discover as we go through the lessons. The named views correspond to the standard engineering views. Thus, it is important to note that if you are planning on using a drawing template that uses these named views, your model orientation relative to the default datums is critical. The top-front-right views of the part are the ones that will be automatically placed on the drawing later. If your model is upside down or backwards in these named views, then your drawing will be too. This is embarrassing and not likely to win favor with your boss or instructor!

Now, having created this new part, you are all set up to do some of the exercises at the end of the lesson. Do as many of these as you can. Perhaps do some of them in different ways by experimenting with your sketch orientation, Sketcher commands, and so on.

This completes Lesson #2. You are strongly encouraged to experiment with any of the commands that have been presented in this lesson. Create new parts for your experiments since we will need the block part in its present form for the next lesson. In the next lesson we will add some more features to the block, discover the magic of relations, and spend some time learning about the utility functions available to give you information about the model.

Questions for Review

Here are some questions you should be able to answer at this time or that might require you to do some digging around the on-line Help area or just play with the software.

1. What is meant by a blind protrusion?
2. What is the purpose of the sketching reference?
3. How do you specify the name of a part?
4. Give as many of the Sketcher implicit rules as you can.
5. How do you save a part?
6. What is a template?
7. What is your system's default template? What units does it use?
8. Where does your system store your part files when they are saved?
9. What is meant by the *active* part?
10. How does Sketcher determine the radius of a fillet created on two lines?
11. What happens if you delete any of the constraints on a sketch?
12. In an extrude, what happens if you set the thickness of a thickened sketch greater than the radius of a filleted corner of the sketch?
13. What is meant by *Linking* to a sketch?
14. In Sketcher, what is the difference between light blue, dark blue, green, and red dimensions?
15. In Sketcher, how do you create an explicit dimension? Is this weak or strong?
16. In Sketcher, how do you indicate where you want the dimension text placed?
17. What are the commands available for diagnosing errors in sketches?
18. How do you turn on the Sketch grid?
19. What are three ways to create a strong dimension?
20. How do you create a weak dimension (trick question!)?
21. In Sketcher, how do you create a radius dimension? A diameter dimension?
22. In Sketcher, what dimension is created if you left click on an arc or circle, then on a planar reference, then middle click to place the dimension?
23. What kind of dimensions are created in Sketcher for the sloping line in the figure at the right if you pick the end vertices then you middle click in regions A, B, or C shown in the figure at the right?

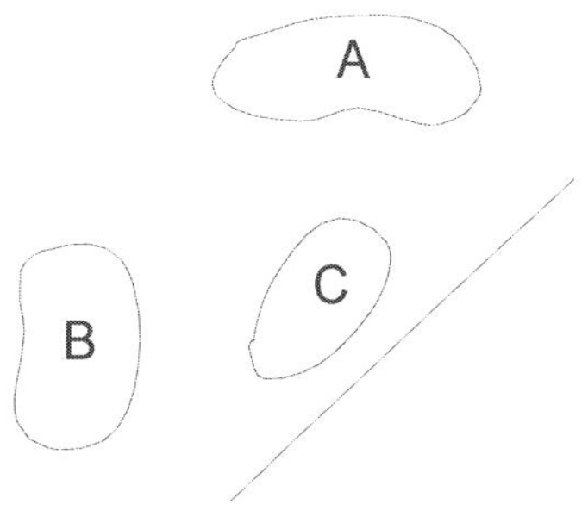

24. When would you normally create a sketch embedded within a feature, as opposed to a separately created curve?

25. What is the sketching reference plane used for?
26. What is the default direction for the normal to a datum plane? A solid surface?
27. What is the purpose of the *Pause* button on the feature creation dashboard? What about the *Preview* check box? How is *Preview* different from *Verify*?
28. In Sketcher, what is the difference between a geometry element and a construction element? What types of elements does this refer to?
29. What are the nine explicit constraints in Sketcher?
30. What are two ways of finding out if a sketch is open or closed?
31. How can you find overlapping entities in a sketch?
32. How do you specify whether you want to create an extruded solid or an extruded surface? What about an extruded cut?
33. In the extrude preview, what is on the RMB menu associated with the depth drag handle?
34. When/how do you attach a feature to a body?
35. What features will NOT contribute to a body?
36. What are the drop-down panels in the Extrude dashboard? What information do they contain? How do you attach/detach them?
37. What is a "collector"? What indicates that it is expecting input?
38. How can you customize the pop-up menus?

Exercises

Here are some simple shapes that you can make with simple extrusions. They should give you some practice using the Sketcher drawing tools and internal rules. Choose your own dimensions and pay attention to alignments and internal constraints. The objects should appear in roughly the same orientation in default view. Have a contest with a buddy to see who can create each object with the fewest number of dimensions. This is not necessarily a goal of good modeling but is a good exercise! Feel free to add additional features to these objects.

1

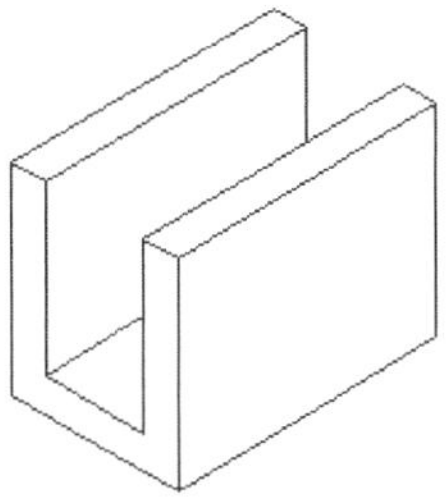

2

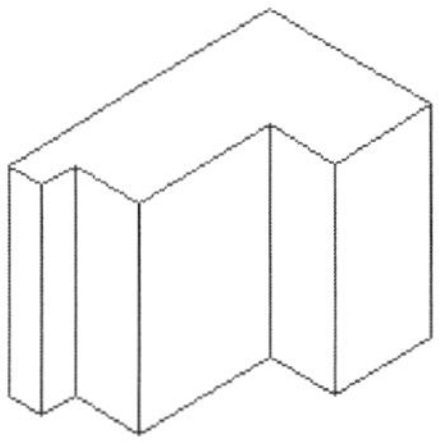

3

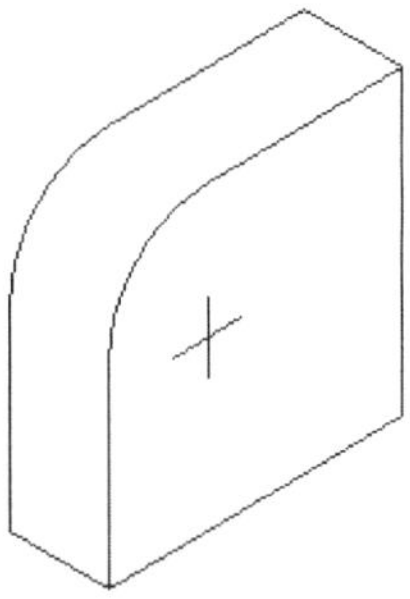

4

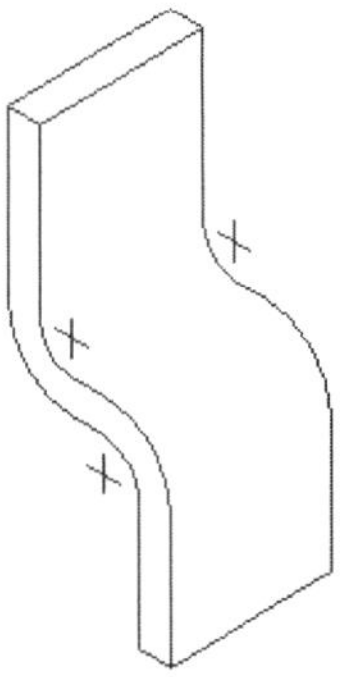

5

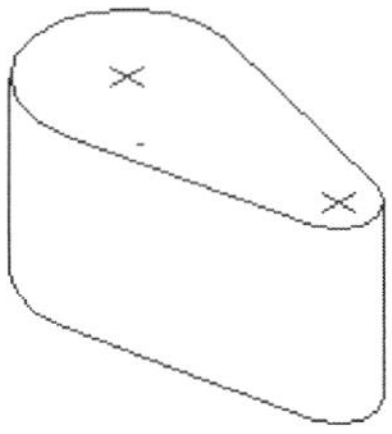

6

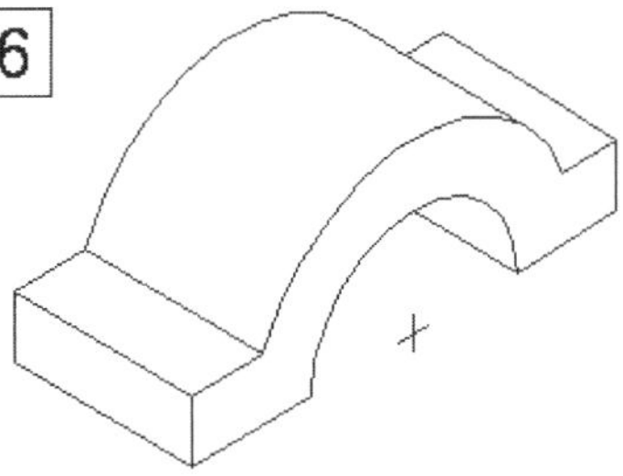

The following two parts “bushing” and “washer” will be needed later (Lesson #9 Assembly Fundamentals). Save them in a safe place.

7. Bushing

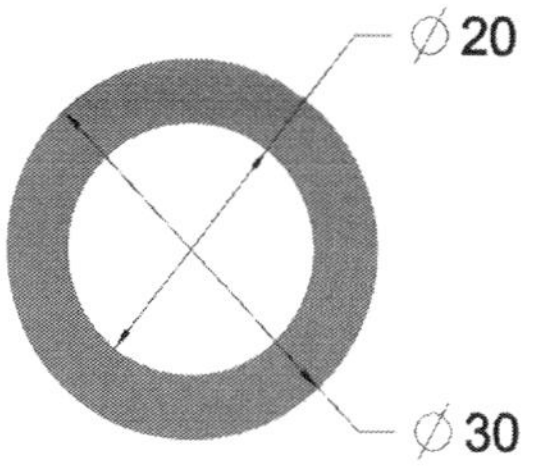

12.5 THICK

Bushing dimensions

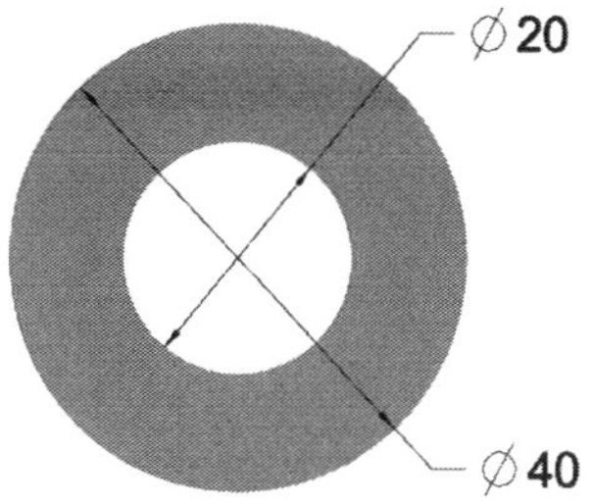

2.0 THICK

Washer dimensions

8. Washer

The easiest way to make this is to do a ***Save A Copy*** of the bushing part (giving a new name **washer**), ***File*** > ***Open*** the new part, then double-click on the protrusion and edit the dimensions.

The next parts are a bit more complicated, requiring two or more simple extruded features (protrusions or cuts). Think about these carefully before you try to make them.

9.

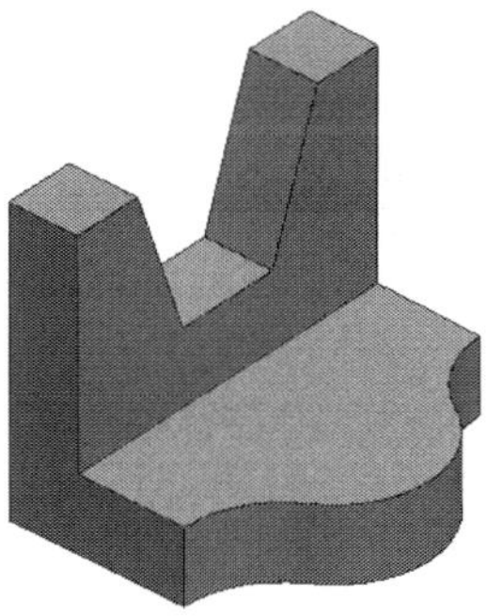

10.

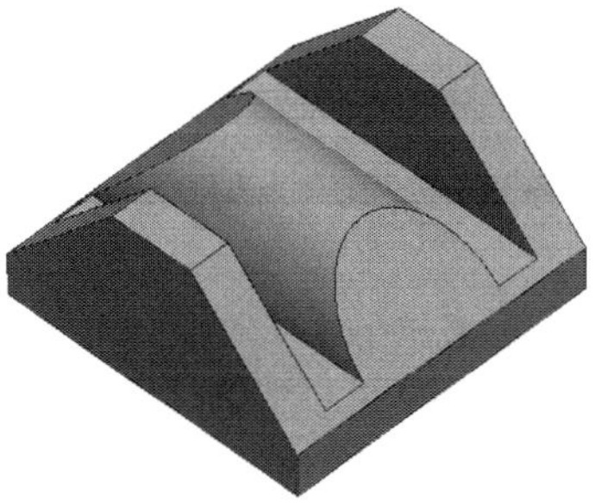

11.

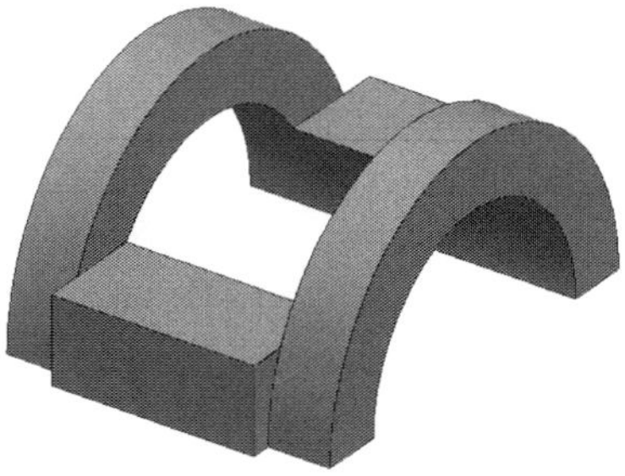

Lesson 3

Creating a Simple Object (Part II)

Synopsis

Engineering features (hole, round, chamfer) are added to the block created in Lesson #2; customizing the model tree; naming features; modifying dimensions; implementing *design intent* using relations; more Sketcher tools and options.

Overview of this Lesson

We will continue with the creation of the model you started in Lesson 2. We are going to add three new Engineering features to the block: a *hole*, a *chamfer*, and a *round*. These do not require Sketcher since their geometry is more-or-less predetermined. We only have to specify where they go - they are *placed* on the model, rather than sketched. Then, we will explore some more of the interface tools using the model tree. We will look at several ways to modify part dimensions, and then introduce the use of *relations* to adjust the geometry automatically. Finally, we will open up the sketch of the cut we made last lesson and look at how a feature's dimensioning scheme is used to implement *design intent*. Along the way we will come across some new tools and functions in Sketcher.

When we are finished with this lesson, the block part should look like Figure 1. Here are the major steps we will follow, which should be completed in order:

1. Retrieving a Part
2. Adding a Hole
3. Adding a Chamfer
4. Adding a Round
5. Customizing the Model Tree
6. Naming features
7. Modifying Dimensions
8. Adding Feature Relations
9. Implementing Design Intent in Sketcher

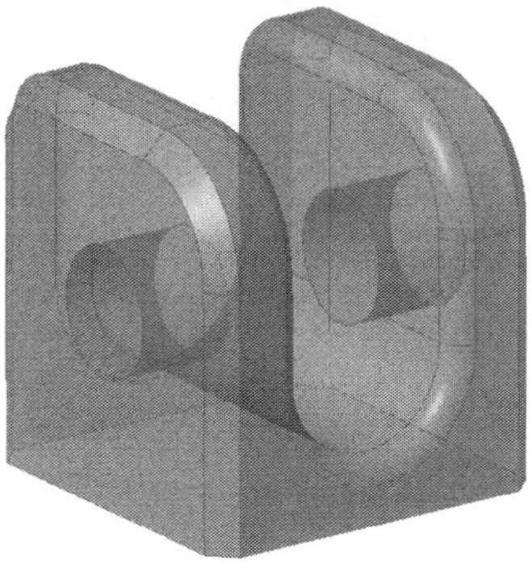

Figure 1 Final part geometry

We will be seeing a lot of new menus and dialog windows here. As usual, we will not discuss all options in detail although some important modeling and Creo Parametric concepts will be elaborated. You should quickly scan each new menu and pop-up to familiarize yourself with the location of the commands and options.

Retrieving a Part

If you haven't already, bring up Creo Parametric. If it is already running, make sure there are no parts in the current session (select ***Close*** in the Quick Access toolbar at the top, then select ***{Home}: Erase Not Displayed*** to remove any other parts in the session).

> **Helpful Hint**
>
> If you need to change the default directory, use the command:
>
> ***{Home}: Select Working Directory***
>
> and select the path to the desired directory for your *block.prt* file from the last lesson.

You can retrieve the block part we worked on last lesson using one of the following command sequences:

☞ in the pull-down menus, select ***File ➤ Open***

OR ☞ use the ***Open*** shortcut button in the Quick Access toolbar at the top

OR ☞ open the Folder Browser by clicking on ***Working Directory*** in the Navigator.

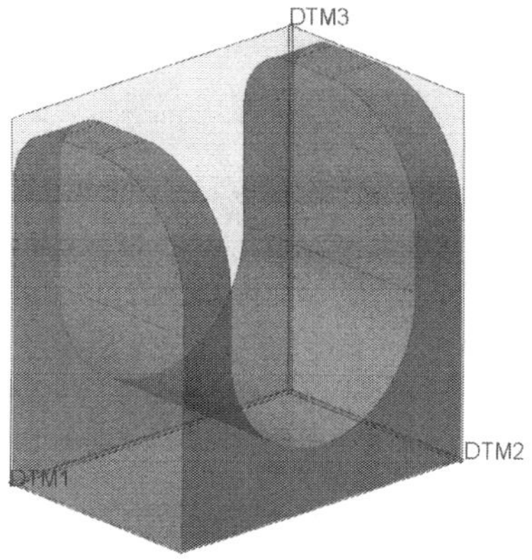

Figure 2 The **block** part at the end of the previous lesson

If you use either of the first two methods, in the dialog window that opens make sure you are looking at the desired working directory (shown at the top). In all cases there is a ***Preview*** button at the bottom. This will be useful when your directory starts to fill up with part files by making it easier to select the file you want (especially if you are not very careful in using descriptive file names!). Note that the dynamic view controls (spin, zoom, pan) work in the preview window, which can be resized by dragging on the horizontal sash. Also, it is easy to customize the displayed list of files by requesting only part files, assembly files, drawings, and so on. You can also change a setting that allows you to see all versions of a file (see the section *Creo Parametric Files Saved Automatically* at the end of this lesson) and even thumbnail images of the parts, assemblies, and drawings in the directory.

With **block.prt** highlighted in the file list select ***Open*** in the RMB pop-up (or double-click the name). Creo Parametric will bring the part into the session, as shown in Figure 2. Turn on the display of datum tags. Position the part in the default orientation (***Ctrl+D***). If they are currently being shown, turn off the visibility of the additional model tree columns using the button at the top of the model tree window.

Creating a Hole

The next feature we'll add to the block is the central hole. You will recall that the protrusion and cut were *Shaped* features requiring the use of Sketcher to define their shape. A hole is an *Engineering* feature whose shape is pretty much already defined, and all we have to do is specify its type, size and location on the model. Some other examples of Engineering features are rounds, chamfers, shells, pipes, and draft features. We will add some of those later this lesson. Although it sounds like a hole should be simpler than an extrusion, there are many variations of this feature. Some examples are shown in Figure 3.

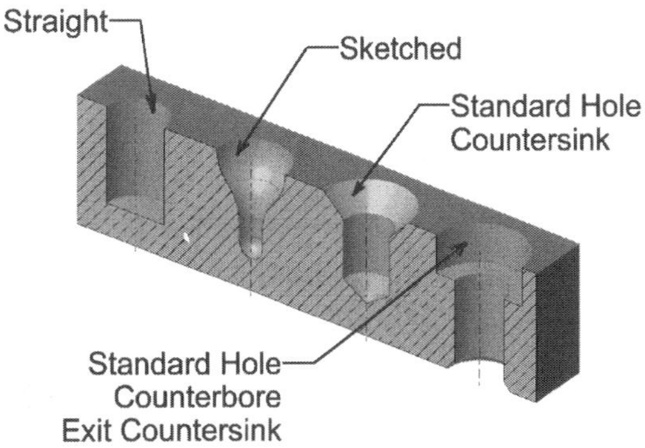

Figure 3 Example hole types

A *straight* hole is a simple cylindrical hole with a flat bottom, essentially what you get with an end mill. A *sketched* hole involves the use of Sketcher to define the hole cross sectional shape or profile. This shape is revolved through 360E to create the hole. This obviously gives considerable freedom in the hole geometry which is handy for holes with several steps or unusual curved profiles. The *standard* holes can be countersunk, counterbored, neither, or both! Notice the shape at the bottom of the holes. Standard hole sizes are built-in for common bolts and thread specifications, and can be either tapped or clearance holes. Standard holes can also be tapered. If you pick a common thread specification, this will automatically create a note (that can be included in a drawing, for example).

After its type and diameter, the next important variation in hole geometry is its depth. This is defined using one of the depth specifications shown in Figure 4. These are essentially the same as we saw previously for the cut feature. The ***Blind*** (similar to *Variable* for an extrusion) option drills the hole to a specified depth. ***To Next*** will create a hole until it passes through the next surface it encounters. A ***Through All*** hole, as you would expect, drills through everything. Finally, a ***To Reference*** (similar to ***Through Until***) hole goes up to a designated surface, edge, or point. If the hole is created "both sides" from the placement plane, then the hole depth can be defined either symmetrically or separately in each direction.

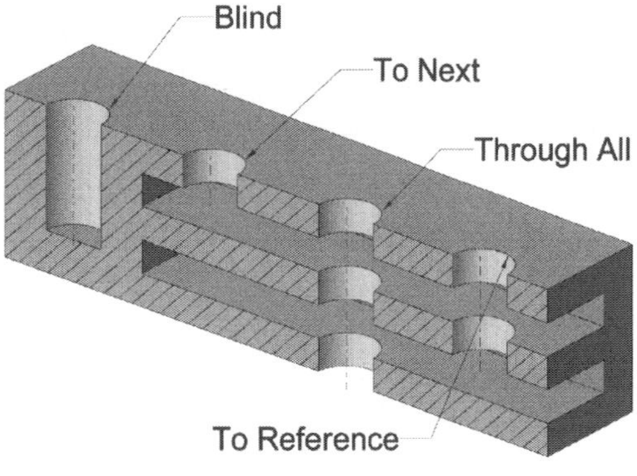

Figure 4 Some hole depth options

A final requirement for defining the hole is the method for specifying where the hole is to be created, that is, its placement surface and how it is located. As shown in Figure 5, this can be a *linear* dimensioning scheme (on the left) or a *radial* dimensioning scheme (on the right). Linear placement will position the hole using linear dimensions from selected references to its center point. The references are typically surfaces of the part or datum planes that are not necessarily orthogonal. Radial placement requires an axis, a radial distance from the axis, and an angular distance from a planar reference. Another common placement option (not shown) is *coaxial*, where the center of the hole is placed on an existing axis. A hole location is also uniquely determined by a previously defined datum point on a surface. Holes can also be placed on curved surfaces. Also, the hole orientation is not necessarily normal to the placement surface but can be determined (see the **Placement** panel in the dashboard) as parallel or perpendicular to another reference.

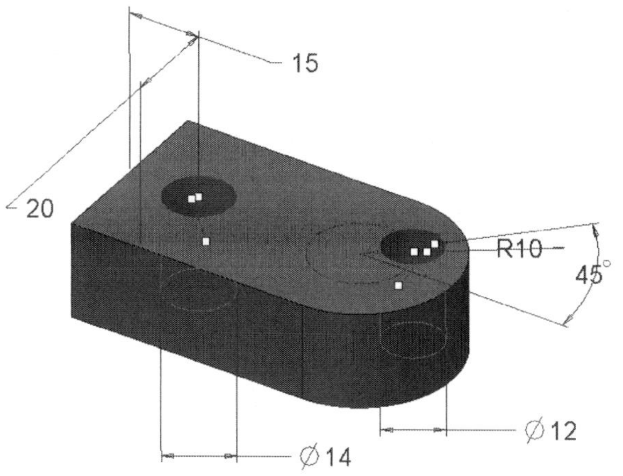

Figure 5 Hole dimensioning schemes: linear (left) and radial (right)

The ***Hole*** command is one of the more automated commands in Creo, and a lot happens here very quickly. Go through this slowly, and take time to explore some of the options and menus. As you select different options for the hole, the true size and shape will be previewed on the part.

If you create holes with threads, the actual thread will not be shown on the part. This would become computationally expensive if there were many holes. Instead, a cosmetic thread is created, which appears as a purple cylinder around the hole (easier to see in wireframe) showing the major thread diameter[1]. Full thread details are stored with the part, however, and can be placed on a drawing.

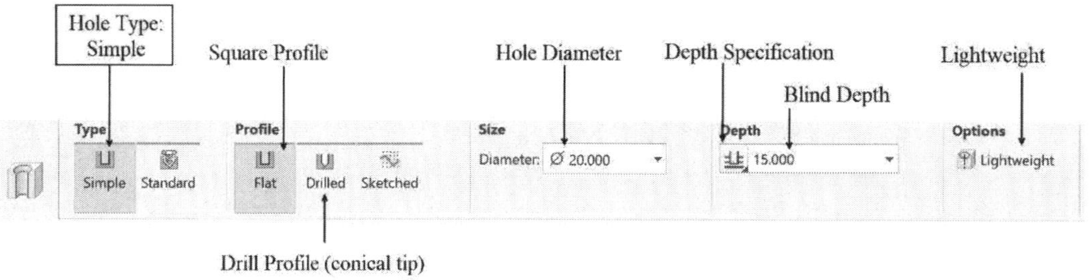

Figure 6 The ***Simple*** hole dashboard

[1] Cosmetic threads can also be placed on cylindrical solids to model bolts.

In the **Model** ribbon, select the ***Hole*** command in the **Engineering** features group. This will open the dashboard shown in Figure 6. The default settings are for a **Simple** hole. The ***Flat*** profile button is the default. The default ***Depth Spec*** is ***Blind***, that is, with a specified depth. The ***Drilled*** profile creates a conical bottom to the hole, as would occur if a normal drill bit was used. If you pick the drilled profile shape (try it!), some additional buttons will appear that specify how the depth dimension is to be interpreted (shoulder or tip). Toggle buttons allow creation of a countersink or counterbore on the hole. For the latter, opening the ***Shape*** panel[2] will allow you to specify dimensions. Also in the ***Shape*** panel, if you create a **Through All** hole, you have the option of specifying an exit countersink. The ***Lightweight*** option will only be relevant when you have a model with hundreds (or thousands) of holes[3].

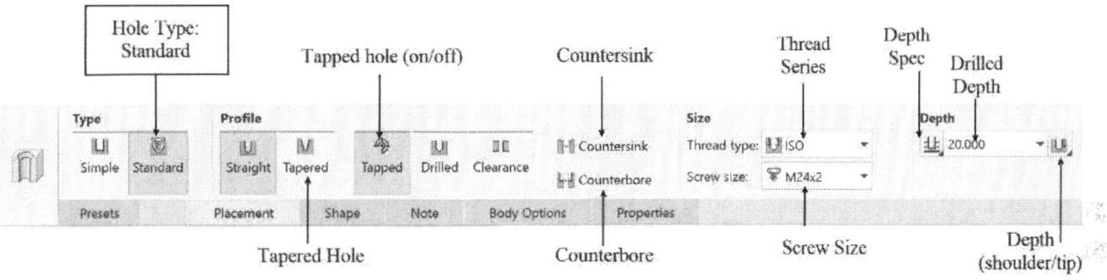

Figure 7 The ***Standard*** hole dashboard

The alternative to a Simple hole is a **Standard** hole (see Figure 7). This hole lets you specify a thread series (ISO, UNC, UNF, etc.), standard screw size, and a thread depth. Other options are for turning thread tapping on or off, and for creating a tapered thread. Tapered holes can also specify thread series (ISO, NPT, NPTF). The nominal hole diameter is set automatically when you pick the screw size. For a **Clearance** hole you can specify one of three fits (*Close*, *Medium*, and *Free*). It also lets you add a counterbore or countersink to the hole. All the hole information is placed in a note which becomes part of the feature definition. What is in the ***Note*** drop-down panel? Compare this to the ***Properties*** and ***Shape*** panels. In the latter, note the option for turning on and off the display of the thread surface. This is a cosmetic feature (a purple cylinder) that helps to differentiate between threaded holes and plain holes in complex models. This is a way of conserving memory, much like lightweight holes. Finally, the ***Body Options*** pull-down panel lets you add the hole to all existing bodies or a selected set.

---

[2] Do this while you are exploring the hole options.

[3] A lightweight hole contains all the hole data but is represented graphically by a single axis line and an orange circle on the placement plane and also has a special icon in the model tree. The purpose of lightweight features is to conserve memory and/or reduce regeneration time. This will only become relevant if you have a model with hundreds of holes. The ***Lightweight*** setting is a toggle that can be changed later, most easily using the RMB menu on the hole feature in the model tree.

For a Simple hole, another option for the hole shape is called a **Sketched** hole, Figure 8. Instead of a straight cylindrical hole, this lets you define the cross-sectional profile of the hole (on one side since it is axisymmetric) using Sketcher. This is useful for creating stepped holes. If the hole shape is complicated and used often, the sketch can be loaded from a separate file. An example of a sketched hole is discussed in Lesson #11.

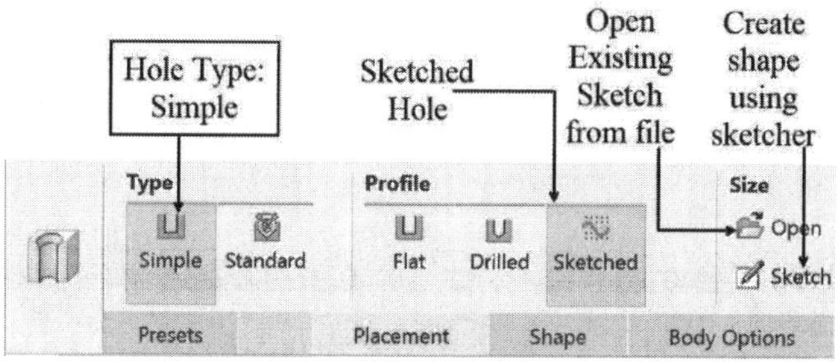

Figure 8 The ***Sketched*** hole dashboard

For now, all we want is a ***Simple***, ***Flat*** hole. This is the easiest hole imaginable to create that uses almost all default settings. If you have been exploring, make sure these are the settings in the dashboard.

Click on the front face of the block at the approximate location of the hole center (mid-way between the left and right faces, 1/2 of the way up from the bottom). This surface is called the *placement reference* for the hole[4]. You do not have to be very accurate with this since we will be setting exact dimensions next.

A hole preview will appear as shown in Figure 9. Notice that Creo Parametric has automatically figured out which way the hole should go. This feature preview has five drag handles whose functions are as follows (you may have to zoom in on the hole to distinguish between these):

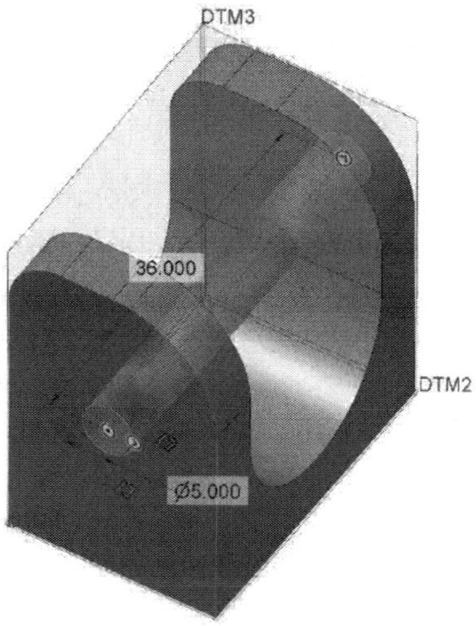

Figure 9 Default hole preview showing drag handles

- center location (yellow circle)
- diameter (yellow circle)
- distance to reference #1 (orange brackets)
- distance to reference #2 (orange brackets)
- depth (yellow circle)

You can modify the hole geometry and placement by selecting one of the handles and dragging them across the part. A selected drag handle also has its own RMB pop-up menu. If you accidentally click on a planar surface instead of a drag handle, the hole primary reference (placement surface) will change. Not to worry - just left click back on the front face of the block. Try moving the center of the hole and changing its diameter or depth. The default placement type is *linear*. The orange bracket drag handles are used to

[4] For a *coaxial* hole, the placement references are the surface where the hole will be located, and an axis from another feature.

select two edges, axes, planar surfaces or datum planes for linear dimensions to locate the hole center. We will use the right and top surfaces of the block for these references. All you have to do is drag each of the appropriate handles and drop it on either the right or top surface of the block. Each time you select a surface, be sure (using preselection) to drop the drag handle on the *surface* not the *edge*[5]. Once you have attached the handles to the references (they will change to yellow diamonds), double click on the dimension values to change the diameter to **10** and the linear distances to **10** from the right surface and **15** from the top (to center the hole on the front face). To change the depth from a blind hole to a **Through All** hole, use the ***Depth Spec*** button on the dashboard (or RMB on the depth drag handle). The hole preview will stop at the back surface of the part; the depth dimension disappears and the drag handle becomes a yellow diamond.

This completes the definition of the hole. See Figure 10. Before you accept the feature, select the ***Verify*** button to see the hole. Also, examine the contents of the **Placement** and **Shape** panels to see how other hole options may be specified. If you click the cursor in a reference field in these menus, the appropriate edge/surface on the model will highlight.

If the preview shows something wrong, you can correct any of the element definitions by selecting the appropriate area or data field in the dashboard and making your corrections. Assuming the hole is correct, accept the feature. Note that an axis line has been added automatically down the center of the hole.

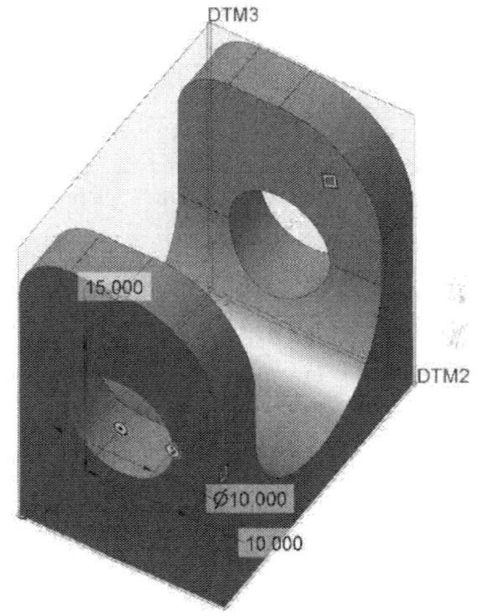

Figure 10 Hole definition completed

As you can see, there are dozens of possible hole types and combinations available with this feature. Some exploration of these is warranted, in particular the various **Depth Spec** options (for example: ***To Reference*** vs ***Through Until***). The **Shape** panel is particularly useful to help determine the parameters you need. There are a couple of exercises included at the end of the chapter for that practice.

Creating a Chamfer

A chamfer is another example of a placed feature - the only reference it needs is the edge (or edges) on which it will be placed.

---

[5] We prefer to use surfaces for references instead of edges because they are more robust. That is, they are less likely to be changed or removed later in a design revision.

We will use the object/action command style here. Using preselection (pick object - surface - edge with the left mouse button or use the selection filter), pick the two edges of the block highlighted in green in Figure 11. To pick multiple entities, after the first is selected you can hold down the CTRL key to include additional entities in a selection set. It doesn't matter which edge you pick first. You may have to click the RMB while holding down CTRL to cycle through possible entities at the pick point for the edge. Or use the ***Selection Filter***.

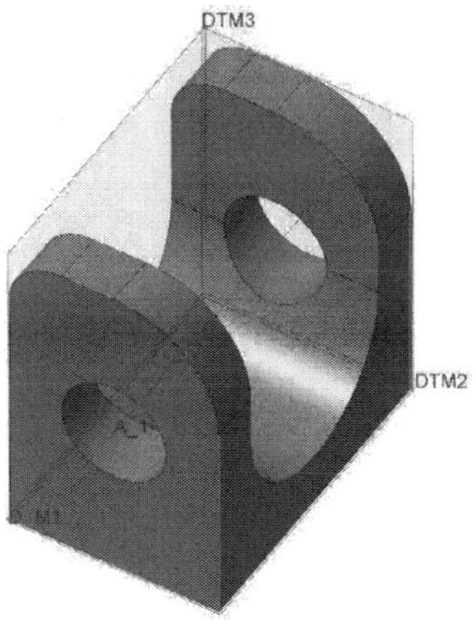

Figure 11 Picking two edges simultaneously for chamfer

With the edges highlighted, pick the ***Edge Chamfer*** tool in the Engineering group or pop-up menu.

A lot of things will happen on the screen at once. First, the **Chamfer** dashboard will open at the top of the screen (Figure 12). Also, the surface created by the chamfer will be shown in preview orange on the model. Two drag handles will be attached to these edges. See Figure 13.

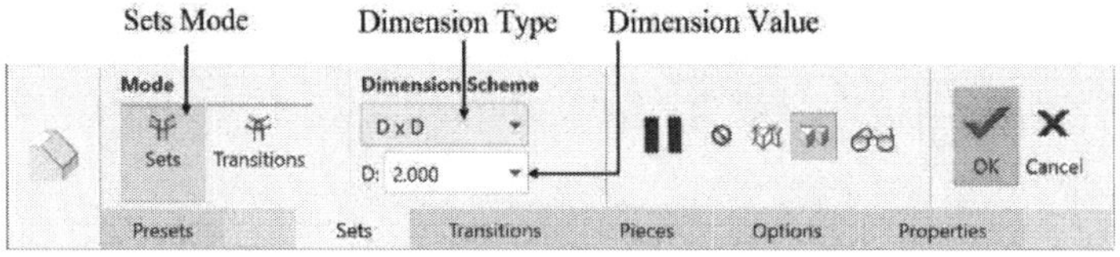

Figure 12 The ***Chamfer*** dashboard

Let's explore the dashboard a bit. The two buttons at the left end determine the dashboard mode of operation. In **Set Mode**, you determine the collection or set of edges that will all have the same chamfer properties. It is possible to create multiple chamfer sets within the same feature, each with different properties. The second button is used to enter **Transition Mode**. As the icon indicates, this mode is used to determine the geometry of chamfers that intersect or meet at corners. There are a number of ways that transition geometry can be set up[6]. None of these are necessary here since we do not have intersecting chamfers. The top **Dimension Scheme** drop-down list contains the options for setting the chamfer dimension type. The default (DxD) is an equal leg chamfer. The lower text field contains the value of the indicated dimension D. This value is also indicated on the screen next to the previewed chamfer.

You can change the chamfer dimension either by moving the drag handles, double clicking the value on the screen and typing in a new value, or by entering a new value in the dashboard. What happens if you make the chamfer too big (try sizes of 4, 5, and 6)? Set the final chamfer distance to **2**.

[6] Chamfer and round transitions are discussed at length in the *Advanced Tutorial*. We will not investigate them here.

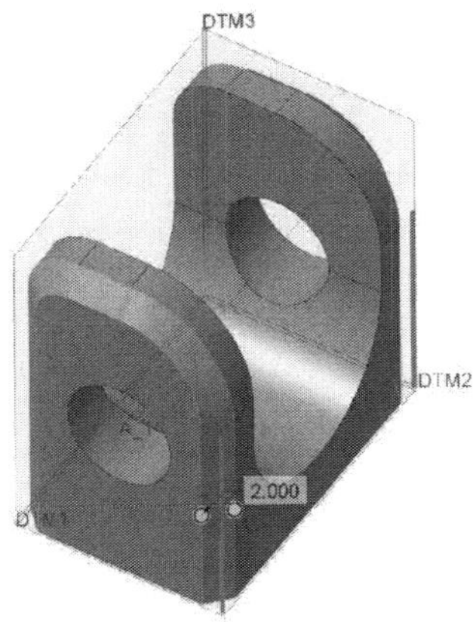

Figure 13 Creating the chamfers

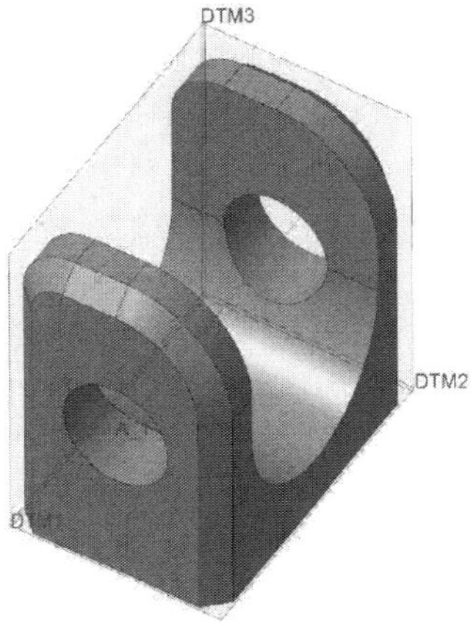

Figure 14 Chamfers complete

Before you accept the chamfer, have a look in the drop-down panels on the dashboard. The **Sets** panel shows our current set (*Set 1*) and the edges you selected in the References collector just below, with dimension type and value indicated. If we had multiple sets, they would be listed here. The **Transitions** panel is currently empty. Note that there is no option to assign a chamfer to a body.

We have basically used all the defaults to create this equal length chamfer. Note that Creo Parametric has placed the chamfer by following along the tangent edge chain starting from the single straight edges we first selected. To override this tangent-following behavior, use the ***Details*** button in the **Sets** menu (then select ***Rule-based***, and ***Partial Loop***). Use the ***Verify*** button on the dashboard to see how the final geometry will appear. Finally, ***Accept*** the feature (MMB). The block should now look like Figure 14.

Creating a Round

Rounds are very common features that are created in the same way as a chamfer. Rounds are normally considered *cosmetic features*, and are therefore added to the model quite late in the regeneration sequence. Because they are so easy to create and make the model look more realistic (in shaded mode or rendered images), newcomers tend to go overboard in adding rounds to their models. However, be warned that misuse of rounds can have several detrimental effects on the model, some serious.

The first is their appearance in drawings. If a model has many rounds, the display of tangent edges showing the round extents can clutter up the drawing considerably, as in the left side of Figure 15. On the other hand, if the display of tangent edges is turned off (which is standard practice), then important information about the part may sometimes be lost (e.g. the missing isometric view in the right side of Figure 15). A partial solution to this problem is to temporarily remove the rounds from the model when creating the drawing (which then shows the "virtual sharps" or defining edges for the rounds).

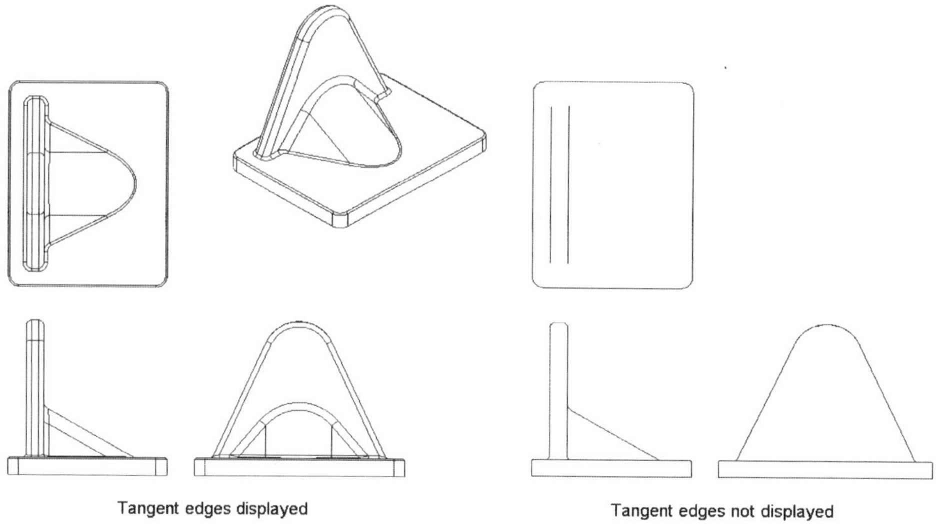

Figure 15 Problems caused by rounds in drawings

The second effect of rounds is felt if the model is used for Finite Element Analysis (FEA). In FEA, the presence of rounds leads to a substantial increase in the modeling effort, primarily in creating the number of elements required to represent geometry ("meshing"). This can drastically increase the model size and computational cost. Rounds do not usually have a large effect on the stress magnitudes in a part (unless they are at critical zones, especially fillets), so this increase in modeling effort is wasted (or may preclude the FEA modeling entirely). FEA models are often "de-featured" by eliminating unnecessary cosmetic features like rounds that occur in non-critical areas.

Finally, when a lot of rounds are present on a model, it is easy (but not good practice) to accidentally pick a tangent edge as a reference for a subsequent feature. Creating this type of parent/child relation will usually cause problems.

In these cases (drawings, FEA, or avoiding inadvertent references), it will be handy to have the rounds organized in the model near the end of the regeneration sequence, and in some sort of orderly presentation so that they can be easily and temporarily removed from the model (*suppressed*, see Lesson #5).

We have added all the key features to our block model, so now is a good time to add a final round feature. Using preselection, pick the edge shown in Figure 16. Now select ***Round*** in the ribbon at the top OR use the pop-up menu and pick ***Round***.

The **Round** dashboard appears as shown in Figure 17. The round surface is previewed on the model in orange with a couple of drag handles and a dimension (the round radius). The default round follows the tangent edge chain from the selected edge all the way around the model.

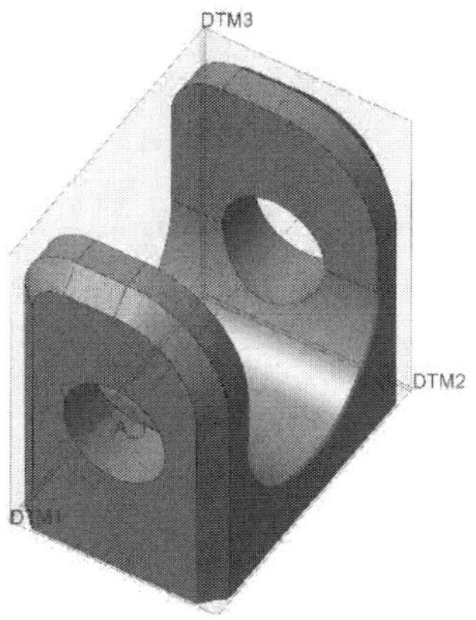

Figure 16 Selected edge for creating a round

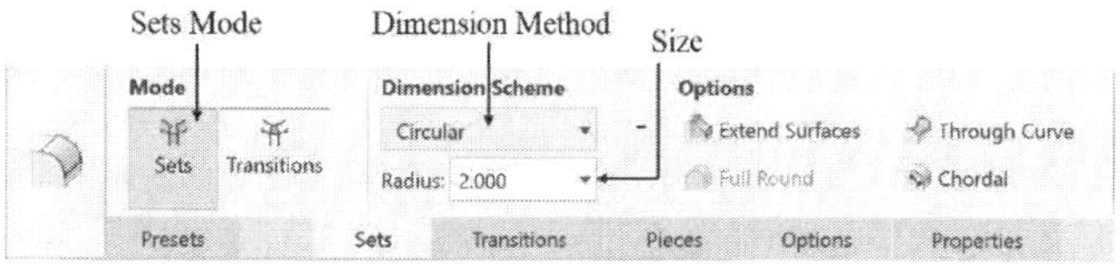

Figure 17 The ***Round*** dashboard

The dashboard looks and operates much the same as the chamfer dashboard. Selected edges to receive rounds are organized into edge sets (icon at left). A single round feature can contain several edge sets. All rounds within a given set have the same properties; different sets in the same feature can have different properties. Theoretically, all the rounds in a model could go into the same feature. You can use the transitions options to specify the type of geometry where rounded edges meet at corners.

Open the ***Sets*** drop-down panel to see the default settings for this edge/set. The round is *circular* (as opposed to *conic*, i.e. elliptical). The shape is a *rolling ball* round. Read the tool tips for descriptions of these settings. The references pane lists the edges in the present set. The ***Transitions*** panel is currently empty. As with ***Chamfer***, there is no option to assign the round to a body.

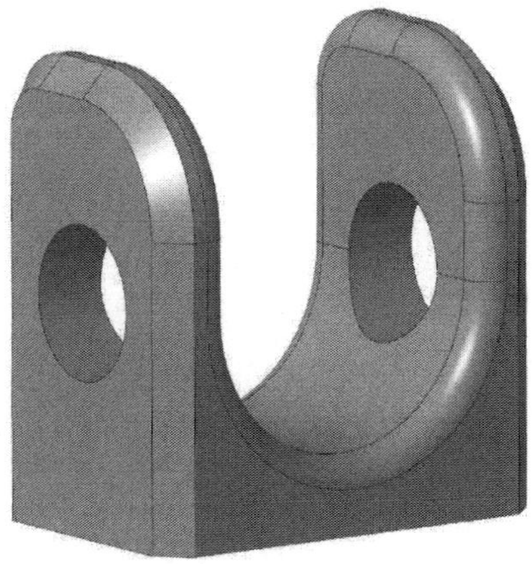

Figure 18 Completed round

Even for simple rounds (with no transitions), you will find that there are considerably more options than there were for chamfers. Some are indicated on the dashboard. A major variation available is the ability to create rounds whose radius changes along their length (*variable radius rounds*). The round extent can also be determined by other features (edges or datum curves). In this tutorial, we do not have time to explore all these options - we could spend an entire chapter just studying rounds[7]! Select the ***Chordal*** option button and observe the dimensioning scheme of the round on the model. Turn this option off and set the radius value to **2.0**. For our purposes, this round is fine, so go ahead and ***Accept*** the feature. The model should now look like Figure 18. Save it now, before we do any more explorations. And you might as well take a break before we continue.

[7] The *Advanced Tutorial* spends half a chapter on rounds, including variable radius rounds, surface rounds, and transitions, and still does not cover all the options.

Exploring the Model

Configuring the Model Tree

Open the Navigator window. The model tree should be displayed. If not, pick the left tab on the Navigator menu.

In the model tree, click the "►" icons beside the protrusion (EXTRUDE 1) and the cut (EXTRUDE 2). The tree now lists the sketches which were used with each of these sketched features. The other features do not have this because they were placed. The green line at the bottom of the tree is the *insertion point*. The next feature added to the model will be added to the tree here. We will find out in Lesson 5 how to move the insertion point around in the model, and what effect that has.

Review the features currently listed in the model tree. This is the regeneration sequence. Selecting a feature in the model tree will cause it to highlight on the model in the graphics window. To turn this highlight off, click anywhere on the graphics window.

If you followed instructions to the letter in the previous chapter, the model tree should have five additional columns (Feat #, Feat Type, Feat Subtype, etc.). If these are not being displayed, click the ***Show/Hide Columns*** button at the top - this is an on/off toggle. In the last chapter we created these by loading a model tree configuration file (*Tut_ModTree.ui*). These settings were saved automatically in a special file (*creo_parametric_customization.ui*) and loaded automatically at startup. If you did not do that, or want to change column settings (remove these or add new ones) here's how. First, let's reset the current settings. Select the ***Model Tree Settings*** button on the right, then ***Settings➤Reset Tree Settings in Mode***. This removes all the existing columns.

To create our own columns, select ***Tree Columns***. This opens up the **Model Tree Columns** dialog window (Figure 19). Have a look in the ***Type*** pull-down list. Keep the default option ***Info***. To add a column to the model tree just double click it in the pane on the left pane. Set up the columns shown in Figure 19. The ***Width*** control below the right pane will control the display of the selected column listed in the pane. Press the ***OK*** button. The new columns are added to the model tree in the Navigator window (Figure 20). You might like to drag the right edge out a bit to see the entire tree. The column width can also be modified by dragging on the column separator bars.

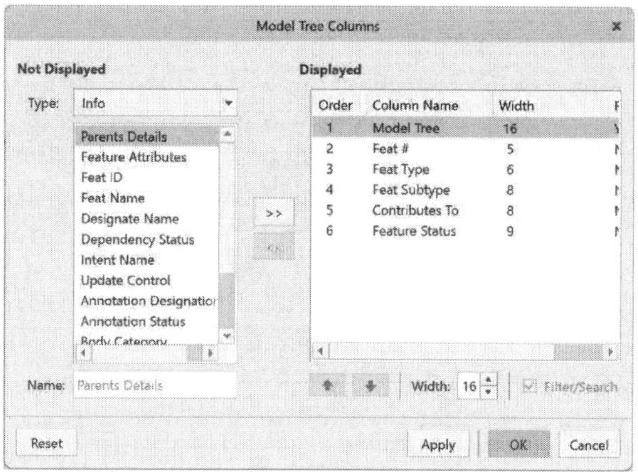

Figure 19 Configuring the Model Tree

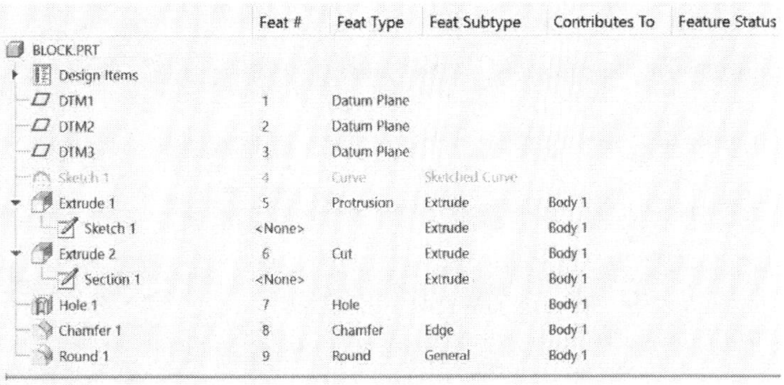

Figure 20 Model Tree with added columns

Select ***Tree Filters*** on the Navigator toolbar. This opens the dialog window shown in Figure 21. Make sure the checks beside **Placement Folders**, **Annotations** and **Suppressed Objects** are turned on. It will be helpful to have everything turned on so that you won't miss anything. Open the other tabs (*Cabling, Piping, NC*) to see what other information Creo Parametric can display in the model tree. Select the **Body/Quilt** options and check ***Consumed Bodies*** and ***Consumed Quilts***. This will show these under the Design Items list after some Boolean operations. Looks pretty complicated! Due to space and time constraints, we won't be formally dealing with any of these other items in this tutorial, but you may run across them during your own exploration of Creo. Select ***Apply*** (they won't change the model tree for this part but will be important later).

As mentioned above, model tree settings (columns and filters) are saved automatically in a "ui" file and (if present) loaded automatically from the startup directory. If there is a particular configuration of the model tree you want to use sometimes, use ***Model Tree Settings ➤ Settings*** and select ***Export UI Settings***. Select the directory where you want to save the file and enter an appropriate name (e.g., *my_tree_settings*). Go ahead and save this file now (the *ui* is automatically appended). You can retrieve it whenever you want with ***Settings ➤ Import Tree Settings***. You will be given the opportunity to restrict the new settings to the Active mode or all modes.

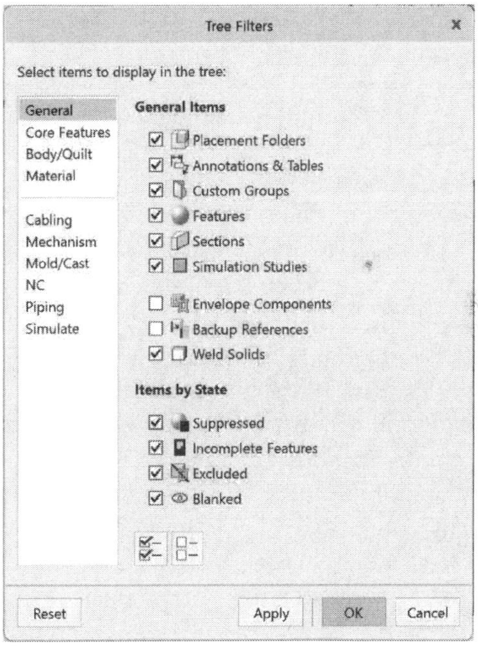

Figure 21 Setting items displayed in Model Tree

Naming Features

As was mentioned back in Lesson #1, it is a good idea to name the features in a model, or at least the major ones. This is important when dealing with large complex model trees. It is also indispensable for someone who must use a model created by someone else.

Naming the features is quite easy. Select the feature in the model tree - it highlights on the screen. If you click twice on the feature in the model tree or use ***Rename*** in the RMB pop-up, the text field opens to let you change the default name to something more meaningful. Do that now with the features in the model tree, using names similar to those in Figure 22. The name is not case sensitive (as you type it in) and will be converted to all upper case but cannot contain any spaces.

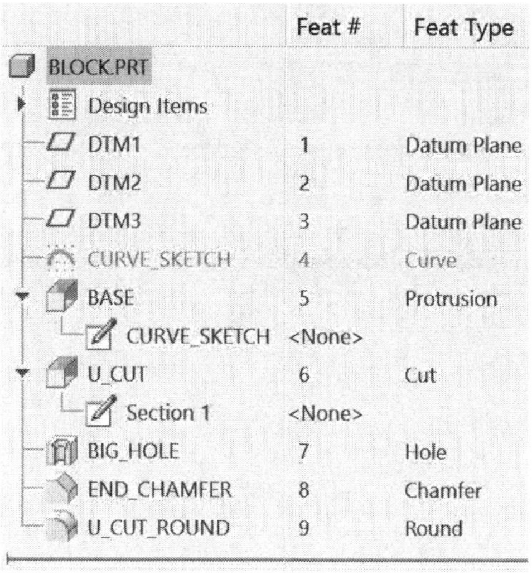

Figure 22 Named features

> **Helpful Hint**
> Just a reminder to periodically save your model. Now is a good time. Try using the keyboard command Ctrl-S. Check the message line below the graphics window.

Exploring Parent/Child Relations

As you recall from Lesson #1, when a new feature is created, any existing feature that it uses for reference is called a *parent* feature. The new feature is called a *child.* It is crucial to plan for and keep track of these parent/child relations. Any modification to a parent can potentially change one or more of its children. This may be precisely the desired model behavior and is the reason why we set up parent/child relations on purpose. However, with poor (or no) planning, changes to the parent may sometimes result in unintended (and often unwanted) changes in the children. In the extreme case, deleting a parent will normally result in deletion of all child features that reference it (and their children and so on...). In these cases, Creo Parametric will ask you to confirm the deletion. If you don't want to delete the child, you will have to change its references using techniques discussed in Lesson #5. Sometimes, if you are not careful about modifications to a parent, the child will be unable to regenerate. This is a symptom of a poor feature selection and/or referencing scheme. So it is important to be aware of what parent/child relations are present when new features are added. Be aware of the intent of your part geometry and build the model accordingly.

As you might expect, parent/child relations can become quite complicated when the model starts to accumulate features. This is a good reason to keep your models as simple as possible, and to think about your modeling strategy *before* you start creating anything! A parent can have many children, and a child can have several parents. Choosing (dare one say *designing?*) the best parent/child scheme for a part is a major difference of Creo Parametric from earlier types of CAD programs. A clearly thought-out and implemented parent/child scheme is crucial to having flexible and robust models. Poor planning of the model will almost guarantee big problems later on if the model must be changed in any way. Fortunately, there are a number of utility functions to help manage the parent/child relations in a model. These include changing the dimensioning scheme and/or replacing current references with new ones (called *rerouting*). In the worst case, reference elements of a feature can be *redefined*. We will be discussing these functions at length in Lesson #5. For now, you might keep as a general rule that, as in many things, simpler is better.

To reinforce these ideas, let's explore the parent/child relations of the cut. First, turn off the display of the additional columns in the model tree. Then select the cut in the model tree (or use preselection to select it in the graphics window). Now hold down the right mouse button and in the pop-up menu select

Information ➤ Reference Viewer

Turn off the **System** checkbox in the **Reference Types** list on the left, then close the filters pane on the left with the thin textured button on the vertical sash. What is left is a feature graph (Figure 23). The current object is in the middle, the parents at the left, and the children at the right. The body containing the current object is shown at the top left. The BASE feature is providing three references for the cut. Meanwhile, the cut is supplying a reference for the round feature.

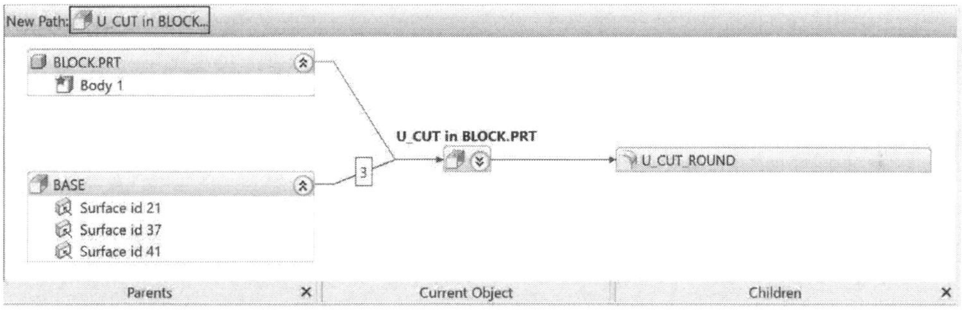

Figure 23 The **Reference Viewer** window for the cut

Expand the reference list beside BASE by clicking on the symbol. Each of the three surfaces is listed. Use Ctrl to select all three; they will highlight on the model as in Figure 24. When selected, using the RMB pop-up and selecting ***Info ➤ Reference Info***, an information window will open describing the nature of the reference. Look for the sketching plane and the section dimensioning references in the **Used As** area.

The feature *BASE* must have other children as well. Select it in the parent pane, click the RMB and select ***Set As Current***. The feature *BASE* appears in the Current Object area. It has several children, but it is interesting to note that the round does not appear on the child list although the chamfer does - both use surfaces of the BASE feature. We already know that the round is a child of the cut. Evidently, the parent/child relations explicitly reported by Creo Parametric are only one generation deep (no grandchildren or grandparents!).

Figure 24 Parent references for cut

Other than the body it contributes to, the *BASE* feature has one parent (the sketched curve) and several children. You recall we used this curve to specify the shape to be extruded to make that first protrusion. Use ***Set As Current*** to find out what the parents for the sketched curve are. The shape of the curve has been copied into the sketch attached (in the model tree) to *BASE*. If we were to ***Unlink*** the sketch in the first extrude, we would expect that the three datums would become listed parents of the extrude. Let's try that. Close the **Reference Viewer** window. You should become comfortable using this window since it will be important in deciphering model structure.

Select the BASE extrude in the model tree. In the pop-up menu, select ***Edit Definition*** in the **Edit Actions** list in the top row of the pop-up menu. We are now in the extrude dashboard and the model has been "rolled back" to its condition at the time the extrude was created. In the ***Placement*** panel click the ***Unlink*** button. A warning window will open to tell us what is about to happen - in this case exactly what we want. Select ***OK*** and then accept the extrusion. The geometry will not change, but there is a subtle change in the model tree - the sketch associated with the BASE feature is now Section 1 instead of CURVE_SKETCH. Furthermore, examining the BASE feature using the Reference Viewer shows the curve is no longer a parent and has been replaced by the datum planes, exactly as we anticipated. Close the reference viewer window.

Put the model back in default orientation.

Modifying Dimensions

There are numerous ways that you can change the shape of the model. You will do this most often by modifying its dimensions, as we saw briefly in Lesson #1. There are (at least) three ways of doing this. Let's experiment with the sketched curve. (We can mess around with this all we want without worrying about damaging the model, since in the last section we unlinked the curve from the BASE feature!) Select CURVE_SKETCH in the model tree and if necessary pick ***Show*** in the pop-up menu. Also, in the **{Model}:Operations** group, in the ***Regenerate*** pull-down menu turn off the option ***Auto Regenerate***.

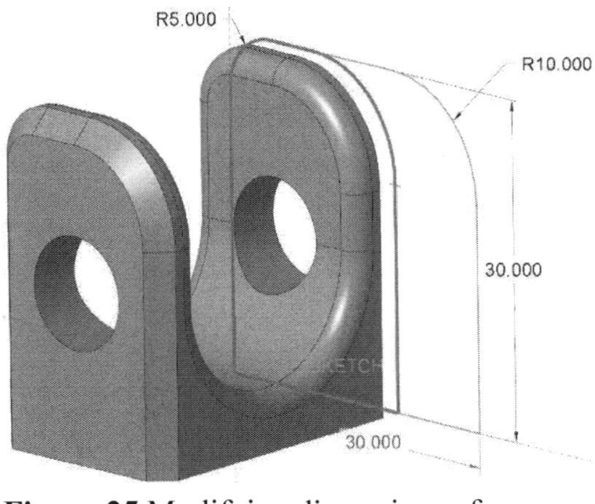

Figure 25 Modifying dimensions of curve

In the graphics window, double-click on the sketched curve. All the dimensions used in the sketch are shown. Put the mouse cursor over the line on the right edge - the cursor arrow loses its tail. Now dragging the mouse will cause the line to move. The movement of the line will be consistent with the existing dimensions and constraints on the sketch - tangency points must be maintained, vertical lines stay vertical, and so on. The only thing that changes is the dimension for the width of the sketch. Drop the line by releasing the left mouse button. What color is the affected dimension? What happens if you grab one of the corner fillets? What happens if you grab a vertex where a fillet meets an edge? If you double click directly on the dimension value, you can type in precisely the desired value. Do that now, using the values shown in Figure 25.

If you double click somewhere else on the screen, the dimensions disappear and the curve is shown in its original shape. Double click on the curve again. The shape with modified shape is still there with the new dimension.

New dimensions do not take effect until you ***Regenerate*** the model, using the button in the Quick Access toolbar OR selecting ***Regenerate*** in the ribbon OR pressing **Ctrl-G** on the keyboard. The new curve displays in blue. There has been no change in the protrusion (confirming that it is not a child of the curve).

Use ***Undo*** to reverse the dimension changes we made and return the curve to its original shape. Turn ***Auto Regenerate*** back on and do the previous modifications again to see the difference in behavior.

Two other ways of modifying dimensions involve selecting the feature (do that now with the sketched curve) in either the model tree or on the screen. In the pop-up menu, select ***Edit Dimensions*** . This will display all the dimensions as before. Double click on a dimension to change its value. What happens if you enter a negative number for the height or width dimensions? What is the value the next time you launch the ***Edit*** command?

We have found that we don't need the sketched curve in the model. Select it (either in the model tree or on the screen), hold down the right mouse button, and select ***Delete*** from the pop-up menu. You will have to ***Confirm*** this deletion. Note that ***Undo*** brings it back.

We would like to make the *BASE* feature wider. Select it (double click on the end surface in the graphics window) and change the dimension 20 to a new value of **30**. It should look like Figure 26. Notice that the other features have all adapted to this change in the base feature - the cut is longer (it was a *Through All* cut), and the chamfers and round have also lengthened. The hole has maintained its distance from the right reference surface of the block, just the way it was dimensioned.

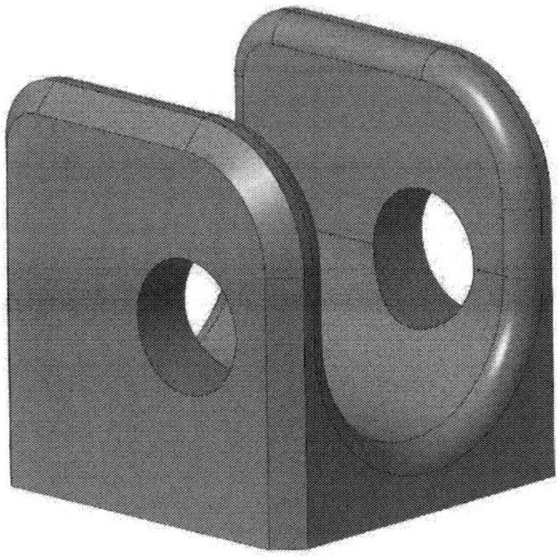

Figure 26 Modified *BASE* feature

What can go wrong here? Before we experiment, you should probably save the part. Now, double click on the *BASE* protrusion again and notice the size of the two rounded corners on top of its sketch. What happens if we try to make the sketch narrower than the sum of these two radii? We might expect trouble if we do that. Try it - say by entering a value of 12.5 for the width of the part. Clearly there is a problem! Zoom in on the top of the sketch for the protrusion and observe the shape. The *BASE* feature cannot regenerate with the sketch crossing through itself. A number of affected features are shown in red in the model tree. The first failed feature is bold. We won't explore error recovery techniques right now, so select ***Undo*** (or ***Ctrl+Z***).

You might try some other "silly" dimensions (for example, make the diameter of the hole bigger than the height of the block), to see what Creo Parametric will do - in particular, what messages does it give you? Try changing the location of the hole so that it is completely off the left side of the block. As long as datum axes are turned on, you can still find it using preselection. What happens if you decrease the height of the *BASE* feature to 20? In particular, what happens to the round? If you get into serious trouble here, just erase the part from the current session (***File ➤ Manage Session ➤ Erase Current***), and retrieve the previously-saved part (You did save it, right?) from disk (***File ➤ Open ➤ block.prt***). Before you proceed, return the dimensions to their original values and ***Regenerate***.

Can we set up the model to ensure that the hole always stays in the center, regardless of the width of the block? We'll find out in the next section.

Note that when the dimensions are changed, Creo Parametric will still maintain all the geometric constraints that you set up during feature creation. A simple example of this is alignment or snapping to references - the edges of the cut were aligned with the upper surface of the block, for example. If the block height is increased, the top edges of the cut sketch move with the top surface of the block. If a feature is *completely* defined by this type of constraint (i.e., all geometry is defined with alignments to references), then you will not be able to modify it directly by its dimensions since it has none! We saw this in a round feature in the *creo_base* part in Lesson #1. You will only be able to affect it via its parent(s). There is another way that we can define dependency between dimensions of different features that is equally powerful - these are feature relations.

Creating Feature Relations

A *Relation* is an explicit algebraic formula that allows a dimension to be automatically computed from other dimensions or parameters in the part (or in other parts in an assembly) or from a numeric formula[8]. This is an important way of implementing design intent. We will set up two simple relations to ensure that the hole in the block is always centered on its width, and mid-way between the top and bottom faces for any block size.

Turn the datum planes and axes off, close the model tree, and put the model in the default orientation. If you double-click on the hole, you will see its location dimensions: 10 from the right surface, 15 from the top surface. What we are going to do is create two formulas that will be used to compute these values based on the current dimensions of the block. There are a couple of ways to do this. One way is very quick, but you won't see the available options or commands for dealing with relations. So, we will do it the long way around first! In the **Model** ribbon, open the ***Model Intent*** overflow[9], (on the far right side of the **Model** ribbon) and select

d= **Relations**

The relations toolbar buttons are shown in Figure 27.

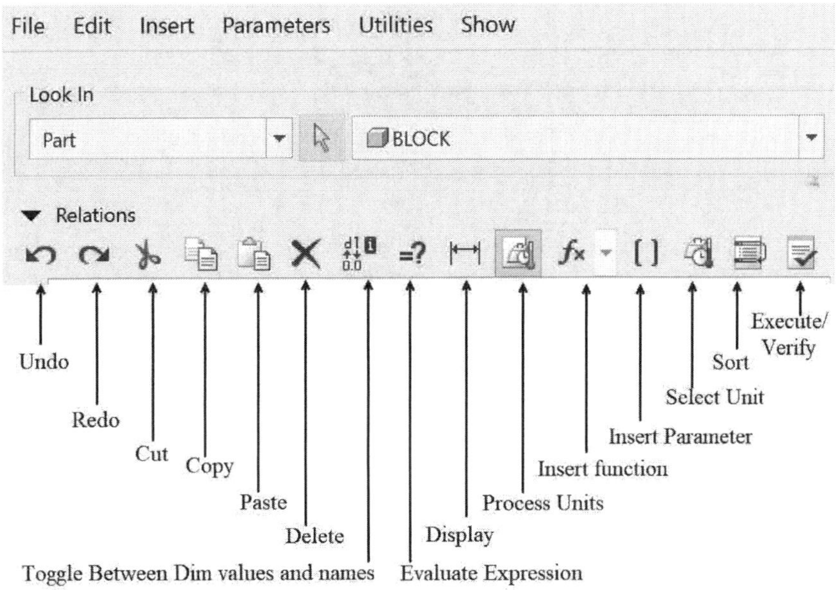

Figure 27 The **Relations** dialog window

Also, the dimension text on the hole showing the numerical values has been replaced with some symbols "dxx". See Figure 28. The "d" indicates a dimension; the number "xx" is a unique number defined by Creo, basically in the order that dimensions are added to the

---

[8] Relations also allow values to be assigned by external programs like MathCAD.

[9] The ***Relations*** command is also available in the **Tools** ribbon.

part. We will find out in a later lesson how you can create your own symbolic names for dimensions - this makes finding and using them a lot easier.

The ***Switch Dimensions*** button in the **Relation** window toolbar is a toggle that switches between symbolic and numeric display of dimensions. Try that now. Note what the dimension symbolic names are in your model – in particular notice that dimensions **d22** and **d23** in Figure 28 are used for locating the hole. Your symbolic names may be different.

We need to find out the symbolic dimensions for the width and height of the protrusion. Click on the front surface of the block. Its dimensions should now also show up on the screen. Take note of the symbolic names for the width and height of the block (dimensions **d7** and **d8** in Figure 28).

IMPORTANT: Your dimension labels might be numbered differently from these. Make a note of *your* labels!

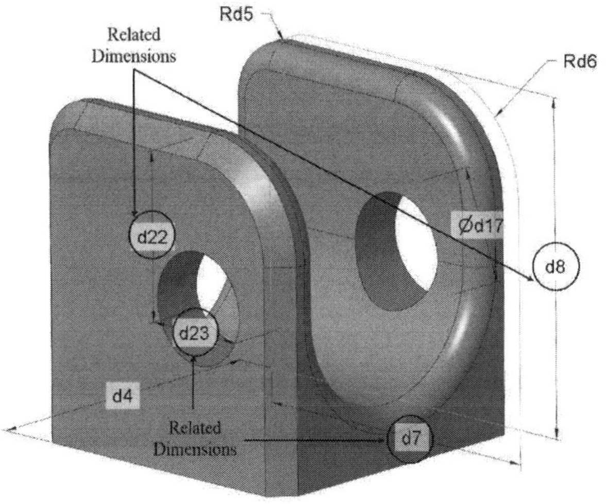

Figure 28 Symbolic dimensions of hole and protrusion

Helpful Hint

Make sure the ***Look In*** setting at the top of the **Relations** window is set to ***Part***. Otherwise, relations get attached to features and may be harder to find in the model later.

Now for the actual relations. In the text area in the **Relations** window, type in the following lines (see Figure 29) (**use your own dimension labels!**):

/* hole centered at half width
d23 = d7 / 2

The first line of the relation, starting with /*, is a comment line that describes the nature of the relation. This comment is not mandatory, but is a very good idea for clarity. You can put any text here that you like (multiple lines starting with /* are legal, too; blank lines are ignored). The next line defines the relation itself - the distance from the end face to the center of the hole is half the width of the block.

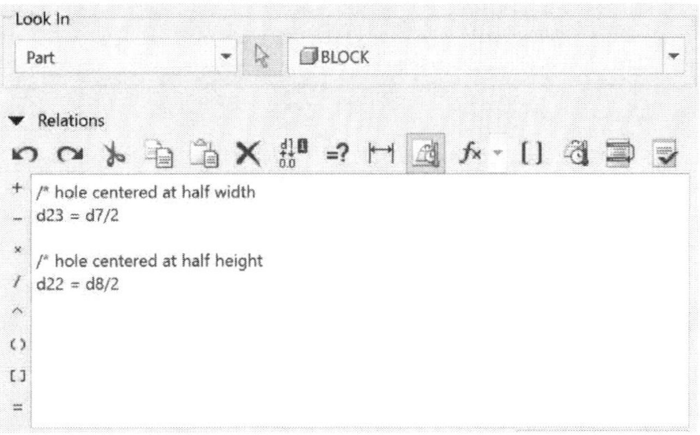

Figure 29 Relations for centering the hole

Here is another way to enter dimension symbols into the **Relations** window. Select ***Switch Dimensions*** so that they are showing up in the normal numerical form. Now type in the next line of text:

/* hole centered at half height

In the next line, instead of entering the dimension symbol text via the keyboard, pick directly on the hole dimension from the top of the block (**15**). The appropriate symbol (like **d22** in Figure 29) will appear in the Relations editor. Type in an equals sign (or pick it from the palette at the left) then pick on the height dimension for the block (**30**). The symbol (like **d8** in Figure 29) will appear. Complete the relation as shown in Figure 29. Notice that by using this method, we don't need to first look up the symbolic names.

Select the ***Execute/Verify*** button in the relations toolbar. This will confirm that the relations are correct (or at least that they contain no syntax errors). Before you leave the **Relations** dialog window, in the pull-down menu select

Show ➤ Info...

This opens a Browser page that lists all the relations in the part. It also reports the values computed by the relations. A lot of other information is presented that deals with mass properties of the part, which we will discuss later in these lessons. Close the Browser page, then accept the **Relation** dialog window with ***OK***.

Note that the relations haven't taken effect yet. Select the ***Regenerate*** button (or Ctrl-G). The hole should move to the center of the block. Double-click on the block and change the width and height dimensions to ***60*** and ***40***, respectively. The hole should be exactly centered on the block. Change the dimensions back to their original values (**30** and **30**) using ***Undo***.

Figure 30 Centered hole using relations

Try to change either of the dimensions that locate the hole - Creo Parametric won't let you! And it even tells you in the message area what relation is driving that dimension.

More about relations

Relations can take the following forms:

/* explicitly define a dimension
d4 = 4
/* explicitly define a parameter
length_of_block = 30
/* use a parameter
d6 = length_of_block
d16 = length_of_block / 2
/* set up a limiting value for a dimension
d14 > 2

Explicitly defined dimensions are just that - they create constant values for dimensions that cannot be overridden. The right hand side of a relation can contain almost any form of arithmetic expression (including functions like sin, cos, tan, ...). The inequality form can be used to monitor the geometry during regeneration of the part. If the inequality is violated, then Creo Parametric will catch the violation and show you a warning message.

These are the simplest form of relations - simple assignments. There are a number of built-in functions that can be used in relations (like locating entities in a model, logical branching, and so on). Relations can be used almost like a programming language. For example, relations can be used to solve systems of simultaneous equations involving part parameters, or be used to perform design calculations that yield dimensional values. Very high-level functionality can be obtained using a module of Creo Parametric called PROGRAM (see the *Advanced Tutorial*).

You recall that our block part has some default units (inches or mm) obtained from the part template. You can use mixed units in relations (for example adding inches to mm) as long as you set things up properly, and tell the relations manager you want it to monitor units. See the Help topic "About Units in Relations" for more discussion and examples.

All the relations for a part go into a special database that is consulted/executed when the part is regenerated. These relations are evaluated in a top-down manner, so that the order of relations is important (just like the order of feature creation). You can't have two relations that define the same dimension, and a relation is evaluated based on the current values on its right-hand side. If one of the right-hand side values is changed by a subsequent relation, then the relation using the previous value will be incorrect. Creo Parametric has a utility function that will let you reorder the relations to avoid this. When re-ordering, any comment lines above a relation will be moved with the relation when the database is reordered.

Considering Design Intent

The notion of *design intent* arises from the fact that there are always alternate ways of creating the model. Even for our simple cut, there are a number of possible dimensioning schemes that would all describe the same geometry. We must choose from these alternatives based on how we want the feature to relate to the rest of the part (or to itself) and how it will respond to changes elsewhere in the model. This is called *design intent.*

Our design intent can be implemented by a combination of feature selection and creation order, explicit constraints, parent/child relations, feature relations, and the dimensioning scheme. We have seen some of these methods. Let's go back and see how the dimensioning scheme of a feature can be used to implement design intent. Use preselection to select the U-shaped cut. When this is selected, use the pop-up menu to select ***Edit Definition***. This opens the extrusion dashboard. On the far left, pick the ***Placement*** panel, then the ***Edit*** button. This takes us to Sketcher with the sketch for the cut displayed (Figure 31). You can close the model tree. Change the orientation of the sketch using ***Sketch View***.

Design Intent Alternative #1

The dimensioning scheme we used before expresses a design intent as follows:

- center of cut is 15 from the back of part
- radius of cut is 10
- center of U-shape is 15 from top of part

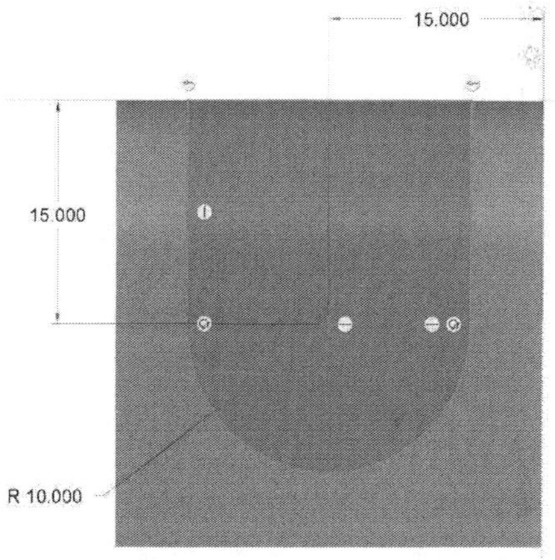

Figure 31 Sketch of cut with first design intent

Design Intent Alternative #2

Suppose we wanted a different design intent as follows (see Figure 33):

- center of cut is 15 from the back of part (same as before)
- radius of cut is 10 (same as before)
- clearance from bottom of cut to the bottom of the part is 5

We saw previously that decreasing the height of the block caused the existing cut to pass through the bottom of the part. Our new intent is that the bottom of the U should always be exactly 5 from the bottom of the block. We can easily set this up using a different dimensioning scheme in the sketch.

Select ***Dimension*** (either in the ribbon or in the RMB pop-up). Click once on the arc (not the center) and once on the bottom edge of the part. Move the cursor to the side and middle click at the location where you want the dimension text to be placed (see Figure 33).

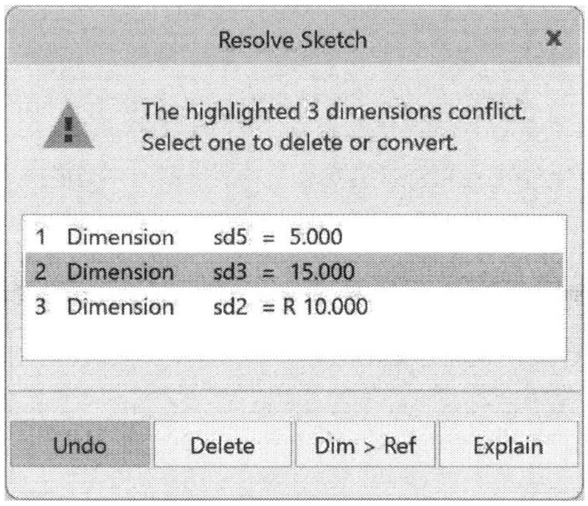

Figure 32 Resolve Sketch window showing conflicting dimensions

Since we are now over-dimensioning the sketch, the **Resolve Sketch** window will open (Figure 32). Recall that Intent Manager will not ask for confirmation if it wants to delete a weak dimension - it just does it. All the existing dimensions, however, are strong. Intent Manager won't delete any strong dimension or constraint without asking. The **Resolve Sketch** window lists the conflicting dimensions and these are highlighted in green on the screen. The dimension currently selected in this window will have a blue box around it. At least one of the three dimensions is no longer necessary. A dimension we can afford to lose is the 15 dimension. Select it and press the ***Delete*** button at the bottom[10]. The modified sketch with our new design intent implemented is shown in Figure 33. Accept the new sketch and the feature.

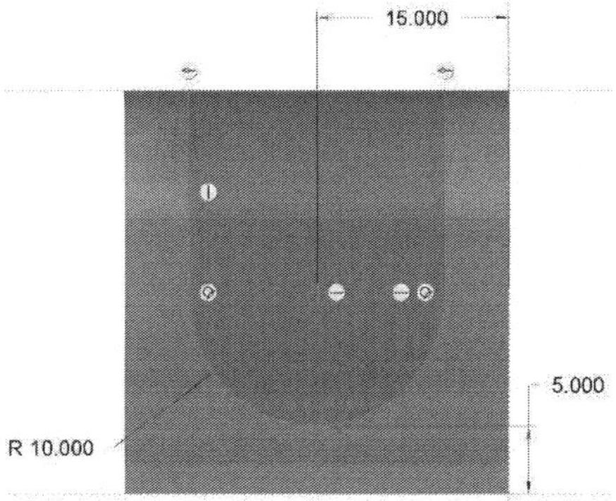

Figure 33 Sketch with second design intent

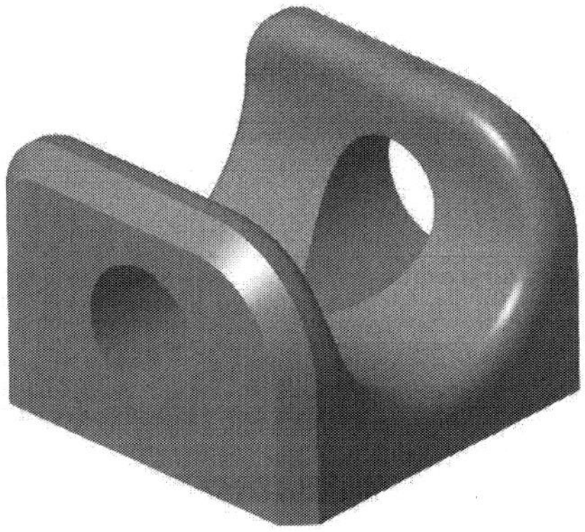

Figure 34 Design Intent #2

With this new design intent, change the height of the block to **20** and ***Regenerate***. If you recall, this is the value we used before that resulted in the part splitting into two pieces. That won't happen this time, due to our intent (Figure 34). Change the height back to **30**.

[10] You can also convert this to a *reference dimension*. Reference dimensions cannot be used to change the sketch geometry but just indicate values in the sketch. In Creo Parametric jargon, they are *driven* dimensions rather than *driving* dimensions.

Design Intent Alternative #3

Let's try one more variation on the design intent. Suppose we wanted the following:

- ensure a thickness of 5 between the vertical sides of the cut and the front and back surfaces of the block
- clearance from bottom of cut to the bottom of the part is 5 (same as previous)

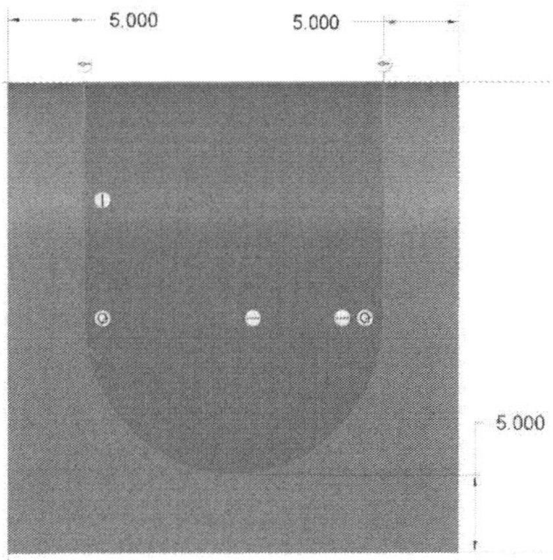

Figure 35 Sketch showing third design intent

Select the cut feature again, and use the right mouse button to select ***Edit Definition***. Re-enter the sketch (using the RMB pop-up, select ***Edit Internal Sketch***). To make sure we don't lose the dimension at the bottom, select it and in the pop-up menu select ***Toggle Lock***. The dimension will turn red. Now pick the ***Dimension*** command and add the dimensions shown in Figure 35. You will have to deal with the **Resolve Sketch** window again, and delete some of our previous dimensions. You will also have the opportunity to delete constraints here (which we don't want to do).

Don't leave this sketch just yet - there are more tools to investigate.

More Sketcher Tools

The *Modify* Command

We have seen how to modify an individual dimension by double-clicking on it. You can also change the sketch by grabbing a sketched entity and dragging it with the mouse. Here is yet another way.

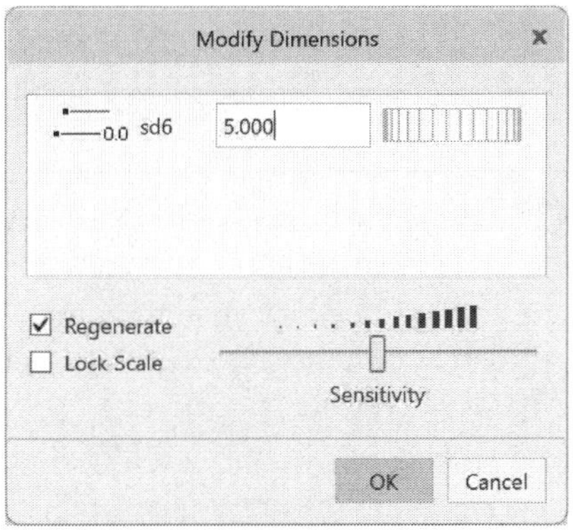

Figure 36 The **Modify** window

Click on the clearance dimension at the bottom of the cut. Now select ***Modify*** in the pop-up. The **Modify Dimensions** window appears (Figure 36). To the right of the dimension value is a thumbwheel. Drag this with the left mouse button. Experiment with the Sensitivity slider[11]. Return the dimension to its original value (5.**0**) by typing it into the data field and hitting ***Enter***.

---

[11] The sensitivity will be affected by a system setting for number of decimals.

Now Ctrl-click on the other two dimensions in the sketch so that all three appear in the **Modify Dimensions** dialog window. Check the box beside ***Lock Scale*** and drag the thumbwheel beside any of the dimensions. All dimensions change simultaneously, in the same proportion.

Helpful Hint

The ***Lock Scale*** option is automatically used when you are sketching the first dimensioned feature in a part. Recall that the numerical values created for the first feature are chosen at random since there is nothing else in the model to set the scale. If only one of the dimensions is changed, the shape of the sketch would likely be destroyed. If your sketch is more or less the right shape this is a great time-saver.

Remove the checks beside both ***Lock Scale*** and ***Regenerate***. The latter option will delay the simultaneous and/or immediate regeneration whenever a single dimension is changed. This is sometimes necessary when you want to change several dimension values at the same time, but don't want to regenerate until all new values are entered. This would avoid trying to regenerate to a geometry with some old and some new dimensions, which might be incompatible. When you are finished experimenting, return the dimensions to the original values and close the **Modify** window (or select the X symbol).

Sketcher Relations

We would like to ensure that the thickness of the part at the front and back are equal. Using Design Intent #3 from above, we have two dimensions that we want to make equal. We could do that using a relation defined at the part level as we did before. There is another way here that is quicker and more appropriate in this case.

While you are in sketcher, in the ribbon tabs at the top select

{Tools}: Switch Dimensions

The dimension labels on the screen will change to their symbolic values, with an "s" in front indicating they are sketch dimensions. Note the dimension at the top right (**sd7** in Figure 37).

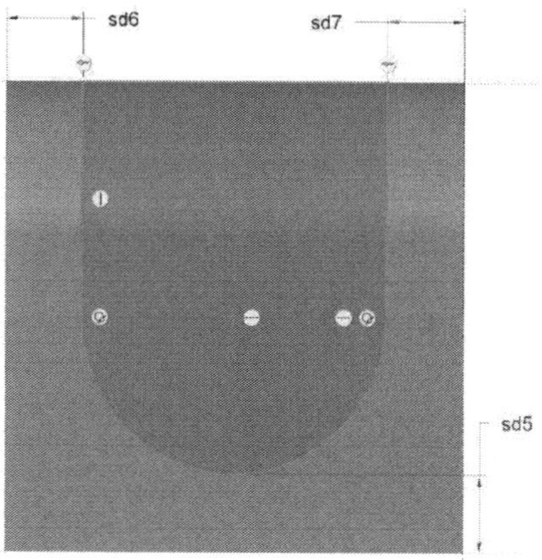

Figure 37 Sketch dimensions

To very quickly enter a relation, make sure you are in Select mode and double-click on the thickness dimension on the top left (**sd6** in Figure 37). Instead of typing in a numeric value, just type in the symbolic name of the other dimension on the top right, **sd7** in Figure 37. A message window asks if you want to add the relation

sd6 = sd7

Middle click to accept this. Select

{Tools}: Switch Dimensions

again to get the numerical display back.[12] Try changing the value of the top left dimension. You can't. Change the value of the other dimension to **8**, as in Figure 38. Both dimensions change, indicating that the relation has executed.

Accept the new sketch and the extrude feature.

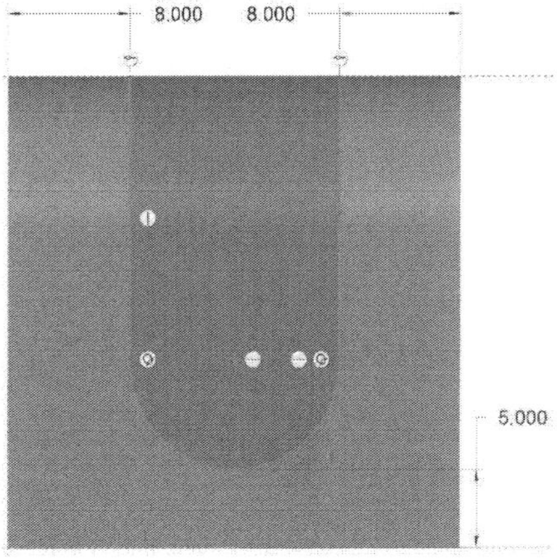

Figure 38 Final sketch for cut

Sketcher Preferences

There are quite a number of options for how you want Sketcher to behave. To investigate these settings, select (in the top pull-down menu)

File ➤ Options ➤ (Core) Sketcher

This brings up the ***Options*** window shown in Figure 39 on the next page. Come back to these and experiment with the settings later. Some useful options are to have the sketch open parallel to the screen, to change the default number of decimal places, and to automatically create references when you select background geometry. If you want to use these changes every time you run Creo, they can be stored in a configuration file (*config.pro*) in your start-up working directory using ***Export Configuration*** command at the bottom. See the appendix for more information. For now, select ***Cancel.***

Using *Redo*

Another useful command is ***Redo*** **.** Look for it on the very top left Quick Access Toolbar. Each time you select ***Undo***, you move backwards through any changes you have made in the part, one at a time. You can move forward again (with some restrictions) using ***Redo*** .

---

[12] By the way, notice that as you pass the mouse over each dimension its symbolic name appears in a small pop-up. This is quicker than ***Switch Dimensions***.

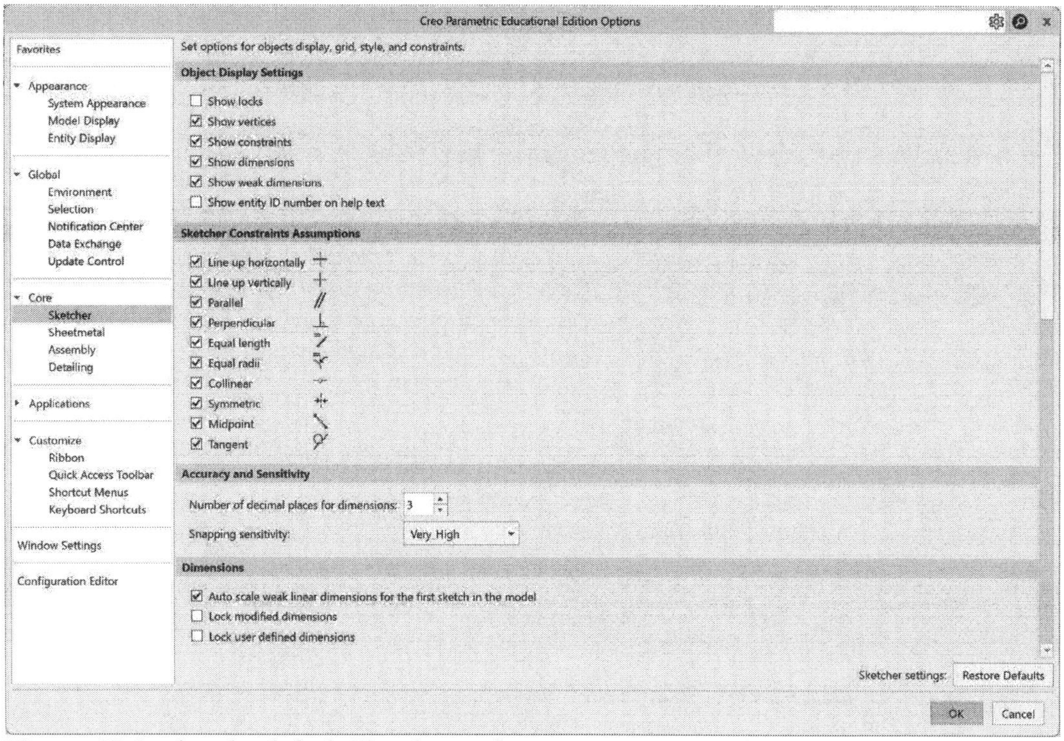

Figure 39 Setting the default Sketcher options

We are (at last!) at the end of this lesson - it's been a long one. The final part should look like Figure 40. Before you leave, make sure that you save the current part with

File ➤ Save

or use ***Ctrl+S***. You can now exit from the program.

Figure 40 Final part geometry

Creo Parametric Files Saved Automatically

Have a look at the files in your default disk space or Creo Parametric working directory. You should see files listed that include the following forms[13]:

---

[13] Make sure the Windows folder option "Hide extensions for known file types" is NOT checked.

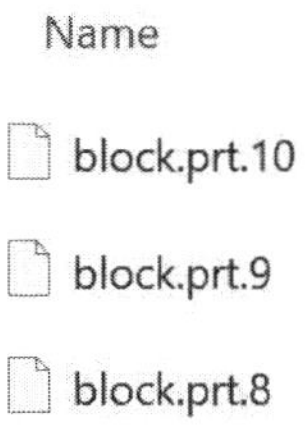

block.prt.10 block.prt.9 block.prt.8 . . . . block.prt.1

Each time you save a part (or drawing or assembly), a new file is created with an automatically increasing counter. As an example, in the above case, the block was saved 10 times. CREO does this so that you always have a back-up available if something goes very wrong. On the other hand, this can eat up your disk space very quickly since the part files can get pretty large. If you are sure you do not need the previous files, you can remove them (all but the highest-numbered part file, of course!). You should copy final part files to another storage location anyway for back-up.

Other files written to your disk space might include the following:

trail.txt.xx
This is a record of all keystrokes, commands, and mouse clicks you made during a session. For an advanced user, this may be useful to recover from catastrophic failures. Each new session you launch starts a new trail file, with an automatically incremented counter (xx). These files are often stored somewhere else on your system, away from your working directory.

feature.lst
The same list of features obtained using ***{Tools}:Feature List***

feature.inf
The data about features obtained using ***{Tools}: Model***

and other *.inf files.

Unless you have a good reason to keep these, remove them from your disk space as soon as you leave Creo Parametric (and not before!). Some programs are available for download from the internet (some free) that will automatically purge these files from your directories.

In the next lesson we will look at a number of new features, including revolved protrusions, mirrored copies, and more Sketcher tools that will extend our repertoire of part-creation techniques. In the meantime, here are some questions for you to think about. Some review material we have covered and others will require you to do some exploring on your own.

Questions for Review

1. What elements are required to define a simple hole?
2. What is meant by linear placement of a hole?
3. What is the difference between the terms "protrusion" and "extrusion"?
4. What are two methods of obtaining a model's feature list?
5. What commands are used to name features?
6. Suppose you have a very complex part with many features and you want to identify/locate (i.e. show graphically) a specific feature in the model. How would you do it?
7. When might you want to turn off *Regenerate* in the **Modify Dimensions** dialog window?
8. Once a feature has been created, how can you change its dimensions?
9. What do we call an equation that computes a dimensional value?
10. What is the difference between a standard profile and a standard hole?
11. What are the "junk" files that Creo Parametric creates in your disk space? How do you get rid of them?
12. How can you go back and edit a previously defined relation?
13. How can you find out the internal symbolic names for feature dimensions?
14. What happens to the relations if you delete a feature whose dimensions appear (a) on the right side, or (b) on the left side of a relation?
15. What is the difference between the **Feature Number** and the **Internal ID**?
16. What does it mean when the stem (tail) of the cursor arrow disappears?
17. What is the meaning of dark blue dimensions? What about the light blue, red, and green ones?
18. How can you strengthen a dimension? What does this mean?
19. Where did we see the ***Lock Scale*** option? What does it do?
20. How do you override the default dimensions placed by Intent Manager?
21. How do you change the location of the dimensions (on the screen) after they have been placed by Intent Manager?
22. What are the default shapes of rounds and chamfers?
23. What is the difference between linear and radial dimensioning schemes for holes?
24. What is the keyboard shortcut to ***Regenerate***?
25. Find out how to turn off the constraints presented by Intent Manager (not just turn off the display, but actually get rid of them).
26. What is the minimum number of Sketcher References needed by Intent Manager? The maximum number? What do these do?
27. Are the following sets of references sufficient or not for defining the Sketch (can the sketch be "fully placed")?
    a. a single vertical reference line
    b. a single point at the center of a circle
    c. a pair of parallel lines
28. In your own words, describe what is meant by "design intent." How was design intent implemented in the part created in this lesson?
29. Examine some simple everyday objects and describe how you might implement design intent in a computer model of the object. Make some freehand sketches to illustrate.

30. How many different variations of the RMB and pop-up menus can you find in Sketcher? When creating Holes?
31. How do you switch between numeric and symbolic dimension display?
32. How can you change the number of decimal places shown by Sketcher?
33. In the model tree settings menu, what does ***Reset Tree Settings*** do to the customization file that is stored and loaded automatically?
34. If there are several bodies in a part and you launch the ***Chamfer*** or ***Round*** commands, on which body does that new feature appear? What happens if two bodies share an edge or a tangent edge joins another body?

Exercises

Here (and on the next page) are some simple shapes you should be able to make using the features covered so far. When complete, the shapes should be approximately in the positions shown in default view. Before starting in on any new part, take a few minutes to plan your modeling strategy. For example, where should the datum planes be located? This will pay dividends in the ease with which you can model the part, and particularly with how you will be able to modify it afterwards.

❶

❷

❸

❹

❺

❻

❼ Hint:
This part can be made with just two features (a protrusion and a cut), although this requires fairly complicated sketches and so would not be good modeling practice.

These parts are a bit more complicated, and use more of the features covered in this lesson (holes, chamfers, rounds).

8.

9.

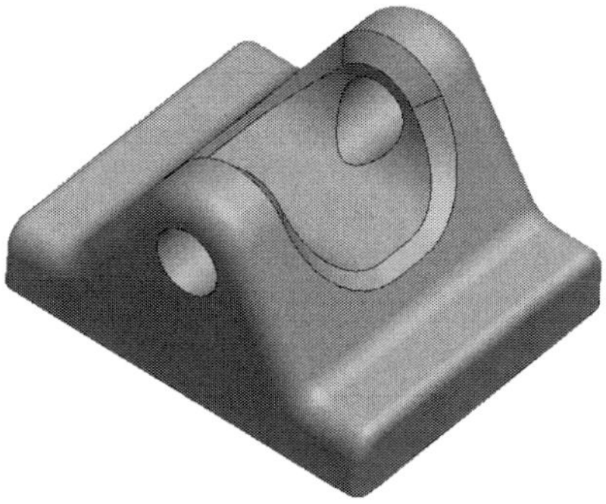

10.

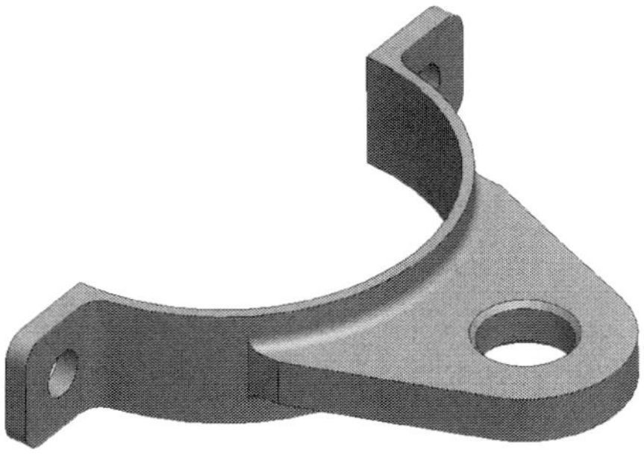

11.

12. Create the "hole sampler" block shown here in cross section, containing a variety of holes. Although these may be standard holes, cosmetic threads are not shown. Go ahead and create threads if desired.

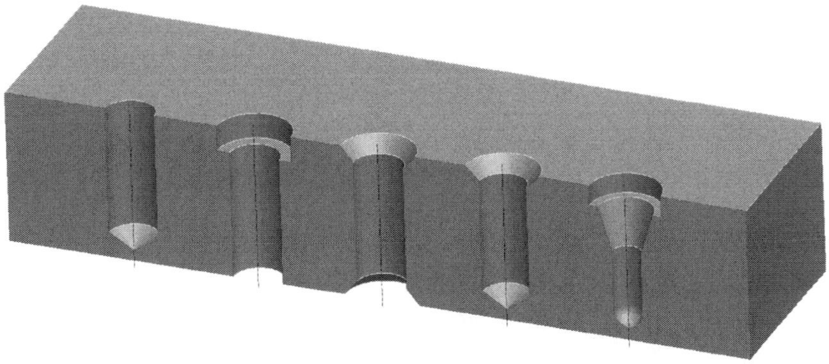

13. **The Base Plate**

Create a part called **bplate** according to the dimensions shown. Dimensions are in millimeters. The plate thickness is ***20*** mm. Note the two planes of symmetry. How can you exploit this? We will need this part when we get to Lesson #9. Save it in a safe place.

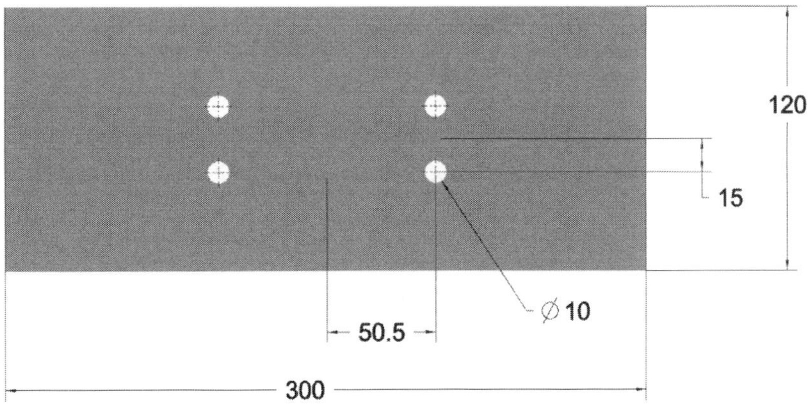

Project

Here is the first part in our assembly project. It uses only the features covered in the previous two lessons. All units are in millimeters. As usual, take a few minutes to plan your modeling strategy. For example, where should the datum planes be located? How should you orient the part in the Front-Top-Right system of datums (assuming you are using a template). Which is the base feature?

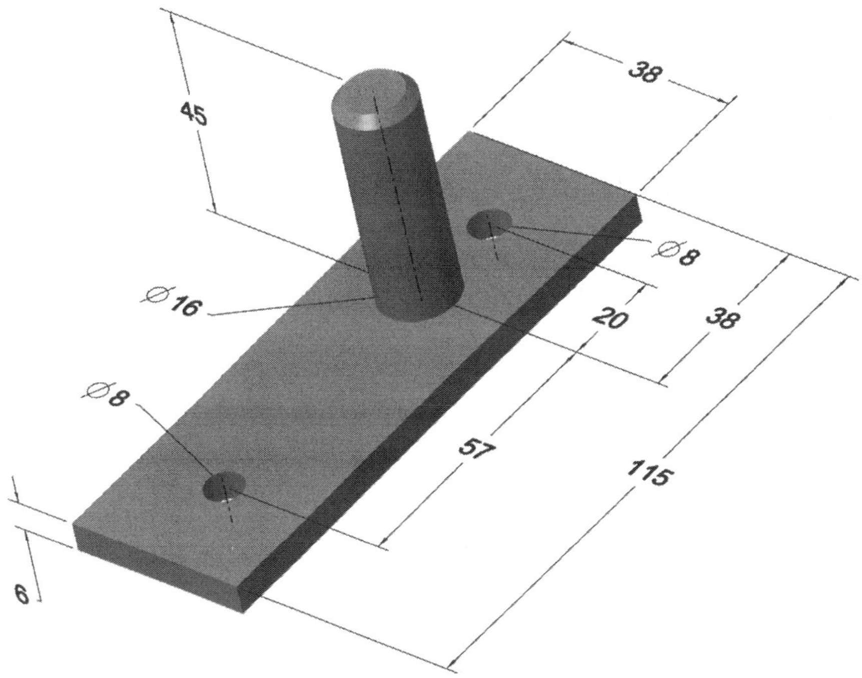

Lesson 4

Revolved Protrusions, Mirror Copies, Model Analysis

Synopsis

A new part is modeled using a number of different feature creation commands and options: both sides protrusions, an axisymmetric (revolved) protrusion, a cut, quick rounds, and chamfer edge sets. More Sketcher tools. Mirrored features. Error recovery. Model analysis functions.

Overview of this Lesson

This lesson will introduce you to an important feature geometry (a revolved protrusion), and give you some practice using features introduced in the first two lessons. Because Sketcher is such an important tool, we will spend some time exploring more tools and functions, and discussing how it can be used most effectively. The part modeling steps should be completed in order. Remember to scan through each section before starting to enter the commands - it is important to know what the goal is when you are going through the feature creation steps. If you can't finish the part in one session, remember to save it so that you can retrieve it later and carry on. The finished part should look like Figure 1. Here are the steps:

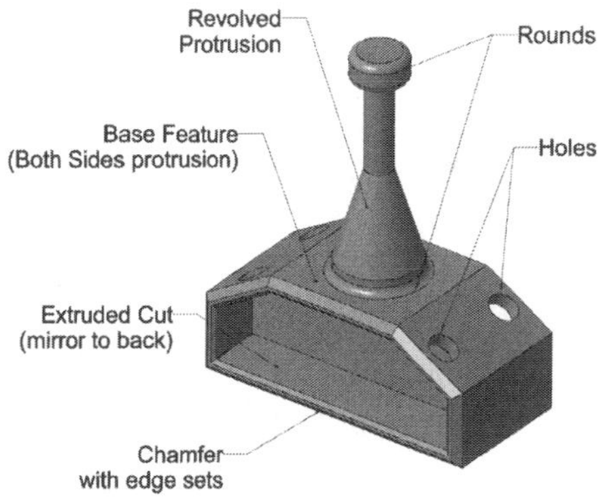

Figure 1 Finished part

1. Creating the base feature
2. Adding a revolved protrusion
3. Adding and mirroring a cut
4. Adding and mirroring holes
5. Adding rounds and chamfers
6. Model analysis tools
7. Exploring "What Can Go Wrong?"
8. Introduction to Multibody Modeling

IMPORTANT:

Be sure to complete the section 7. You will learn a lot about how Creo works, beyond which button to push.

As usual there are some Questions for Review at the end, some exercises, and another part for the project.

The instructions are going to be a bit more terse this lesson, especially for commands we have covered previously. You should be getting in the habit of scanning both the ribbon and pop-up menus. Don't forget to explore the RMB menus, too. By now, you should also be fairly comfortable with the dynamic view controls obtained with the mouse.

So, get started by launching Creo Parametric as usual. Close the Browser and Navigator windows. Study the object in Figure 1 carefully before proceeding. We will create the model first as a single body. In the last section, we will explore some of the commands used to create a model with multiple bodies. You never know when these might come in handy.

Create a part named ***guide_pin*** using the ***New*** button in the **{Home}** ribbon then set:

Part* | *Solid* | *uncheck ☐ Use default template ***| [guide_pin] | inlbs_part_solid_abs***

The default datum planes with datum tags RIGHT, TOP and FRONT should be displayed, if not, turn on the display of datum tags.

Creating the Base Feature

The rectangular block at the base of the part will be our first solid feature, also sometimes called the *base feature*. We will be creating the base feature so that the **FRONT** and **RIGHT** datum planes can be used for mirroring of features we will create later. This is an example of planning ahead. This one was easy - only "one move ahead." Like good chess players, good modelers are always looking many "moves" ahead.

> **Helpful Hint**
>
> Whenever you have symmetry in a part, it is a good idea to use the datum planes on the plane(s) of symmetry. That way, they will be available for mirroring and serving as references for symmetric features.

We will create the first feature as a *Symmetric* extruded protrusion: the sketch will be on FRONT and the protrusion will extend on both sides of the sketching plane. Rather than creating the sketch first and then launching the extrude command as we did for the block of lesson #2, we will create the sketch within the extrusion (an *internal* sketch). Thus, select the ***Extrude*** tool in the **Model** ribbon. In the **Placement** panel that opens automatically, select ***Define***. Alternatively, in the graphics window and ***Define Internal***

Sketch in the RMB pop-up menu. Now you need to select a sketch plane and sketch orientation reference plane. Choose **FRONT** as the sketch plane. The **RIGHT** datum is automatically chosen as the ***Right*** orientation reference. This is what we want, so just middle click to accept the dialog and enter Sketcher. The following will be easier if you are in 2D sketch mode (which is not the default). Select ***Sketch View*** in the Graphics Toolbar[1]. The sketching references (dashed cyan lines) have been selected automatically for you.

Before we start the sketch, recall the sequence we want to follow with Sketcher:

1. Make sure the **desired references** are selected; if you missed any on your way in to Sketcher, use the RMB pop-up menu to select ***References*** and add them. You can do this any time you are in Sketcher[2].
2. **Sketch the shape** using the chosen references for alignments, constraints, etc.
3. **Strengthen desired dimensions** created by Intent Manager. If this is the first geometric feature in the part, the ***Modify ➤ Lock Scale*** option will be automatically set for the first dimension you modify. Then modify any other dimensions you want to keep. Remember that this converts them to Strong dimensions (that Intent Manager will not delete unless you let it). Leave the rest for later.
4. **Add desired constraints** (equal length, equal radius, tangent, etc.) to implement your design intent.
5. **Change the dimension scheme** so that it implements your design intent.
6. **Modify the remaining dimension values** to those desired for the feature.

Steps 3 and 4 are interchangeable and you will often do both at the same time. The final sketch we want to create is shown in Figure 6; the previewed feature is shown in Figure 7. We'll get there in several steps, corresponding to the sequence just given.

Step 1 - Selecting References

The first step has been done for us - as we entered, Sketcher picked the only two references possible at this time. With these identified, you can turn off the datum planes as they will not be needed for a while. Some time you should try entering Sketcher with datum planes turned off to see what happens.

---

[1] So that this happens all the time, set an option using ***File ➤ Options ➤ Sketcher*** and check the box beside "Make the sketching plane parallel to the screen" or set the config option (see the Appendix) *sketcher_starts_in_2d* to **Yes**.

[2] With a config option (*sketcher_auto_create_references*) or a setting in ***File ➤ Options ➤ Sketcher***, you can automatically create a reference if you dimension to it or constrain an edge or vertex to it.

Step 2 - Sketch Geometry

Use the RMB pop-up menu to select the ***Line Chain*** drawing command. Letting the cursor snap to the references, you can create the entire sketch using a single polyline (left click, left click, ... seven times as shown in Figure 2). Middle click (twice) to leave ***Line*** mode and return to ***Select*** mode. Intent Manager will put some constraints and weak dimensions on the sketch in light blue. Don't worry if the constraints don't match the desired ones just yet. Also, since this is the first feature of the part, dimension values will be chosen based on a default setting for model size. Some dimensions and constraints will be the ones you want; others won't. Except for the first geometric feature, **DO NOT bother modifying these dimension values yet - this will result in wasted effort.** What we are interested in first is getting the shape and proportions of the sketch right, using the constraints we want.

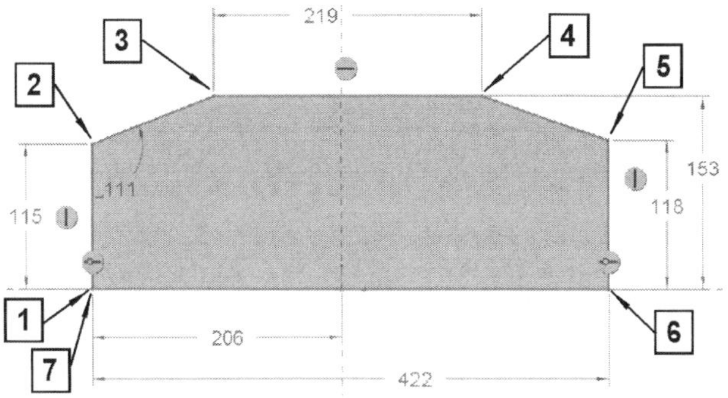

Figure 2 Creating the sketch for the base feature (constraints and dimensioning scheme not completed)

Step 3 - Strengthen Dimensions

Move the dimensions off the part (pick to highlight in green, then drag with the left mouse button) and observe the constraints and dimensions that have been created by Intent Manager. The final arrangement is shown in Figure 6. In this step we select dimensions we want to keep (by making them strong). Select one of the dimensions you want to keep (any one of the **219**, **422**, or **153** dimensions in Figure 2). When it is highlighted in green, enter the desired value **(10**, **20**, **or 8**, respectively). Since this is the first geometric feature, the ***Modify ➤ Lock Scale*** option will automatically be applied. The selected dimension will become strong (green if selected, dark blue if not). The other linear dimensions will also change in value but will remain weak. From here on, any dimension you modify will become strong without changing the value of any other. Continue doing this for any dimensions corresponding to those in Figure 6. If none of these exist, don't worry - we will be creating them soon.

Step 4 - Specify Constraints

Now we can implement any missing constraints that have not been deduced by Intent Manager. This follows after step 3 because as we add our new constraints, previous weak dimensions can be deleted by Intent Manager. The dimensions we made strong in the previous step are immune from this.

We can implement a left-to-right symmetry about the RIGHT datum (the vertical reference) as follows. Select the ***Symmetric*** constraint in the **Constraint** group in the ribbon (Figure 3). Read the message window, then select the vertical dashed reference (it turns green) and the two vertices on the end of the top horizontal edge. With the ***Symmetric*** constraint still active, do the same with the lower horizontal edge. When the constraints are created, some of the weak dimensions will be removed. Middle click to return to ***Select*** mode. ***Repaint*** your screen to look for the small symmetry constraint arrows.

Another way to create the symmetry constraint is to use the ***Centerline*** button in the ribbon Sketching group before you draw any physical edges, *OR* use the RMB pop-up menu and select ***Construction Centerline***. Sketch a centerline on the vertical reference; middle click when done. Now when you create physical edges if your sketch is close to being symmetric about this line, Sketcher may automatically apply the symmetry constraint.

If your current sketch is missing some of the other constraints shown in Figure 5, we can set them up now. The ***Constrain*** group in the ribbon contains the nine explicit constraint options shown in Figure 3. Examine these carefully, and add any constraints on your sketch so that it matches the desired figure. Note that as constraints are added to the sketch, the dimensioning scheme will probably change. Middle click to return to ***Select*** mode.

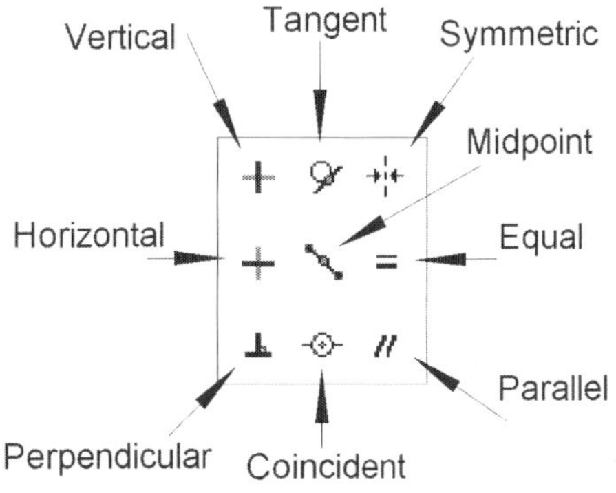

Figure 3 Explicit constraints on Sketcher toolbar flyout

As you are adding constraints, the weak dimensions will continue to change. When you spot one that you want to keep, strengthen it using the RMB pop-up menu.

A very quick way to add constraints is to select two entities, for example the vertical edges at each end (use CTRL to pick the second edge), then use the constraints in the pop-up menu shown in Figure 4 (bottom row). In this case choose ***Equal***. If required, add this to your sketch. The sketch constraints should now look like Figure 5.

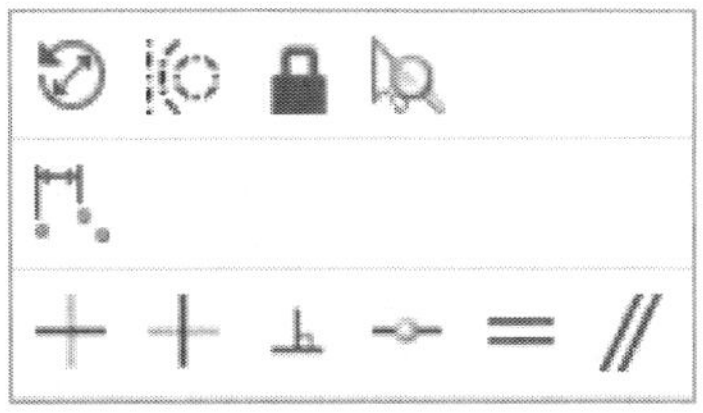

Figure 4 Pop-up menu in Sketcher with two lines selected

Step 5 - Finish the Dimensioning Scheme

So far, the shape and constraints are set the way we want, but the dimension scheme and values probably are not. Again, compare to the dimensioning scheme shown in Figure 5. Strengthen any dimensions you want to keep. If any dimensions are missing, create them explicitly. Pick the ***Dimension*** command from the dimension group or in the RMB pop-up. Recall that these will be strong dimensions and Intent Manager will remove redundant weak dimensions automatically.

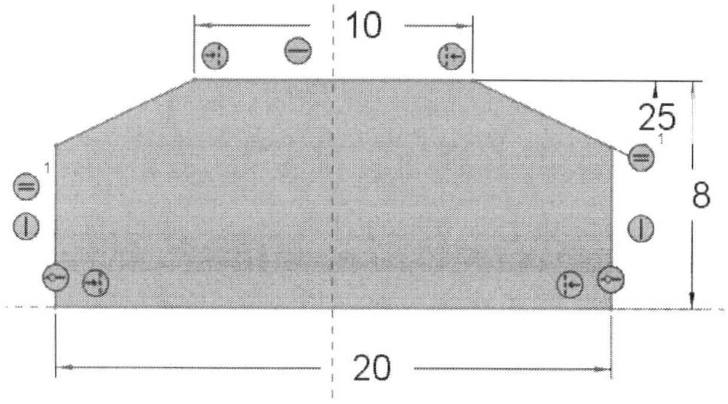

Figure 5 Sketch with desired constraints and dimension scheme

To dimension the angle, click on the two intersecting lines and middle click where you want to place the dimension text. Sketcher assumes that if two lines intersect you must want to dimension the angle between them. When you are finished adding dimensions, middle click to return to ***Select*** mode.

Step 6 - Modify Dimension Values

Finally, once you have the shape, constraints, and dimensioning scheme you want, you can modify the dimension values. As discussed above, the initial dimension values chosen by Sketcher for the base feature are fairly arbitrary. We have seen that ***Lock Scale*** is used automatically to maintain the shape of the first sketch. The sketch should now be approximately the right shape but perhaps with a few incorrect dimension values. Go ahead and correct them now by double clicking on them and entering the desired value. For sketches added after the first one, you can use Ctrl to select several dimensions simultaneously (or use a bounding box) then select ***Modify*** in the pop-up menu. Note that ***Lock Scale*** (in the ***Modify*** window) can be used whenever multiple dimensions are selected.

This should complete the sketch and it should look like Figure 6. So, select the ***OK*** button (the check mark ✓) in the ribbon or on the RMB pop-up.

You should review this Sketcher sequence again - using it properly can save a lot of frustration, especially taking advantage of the initial ***Modify* ➤ *Lock Scale*** function.

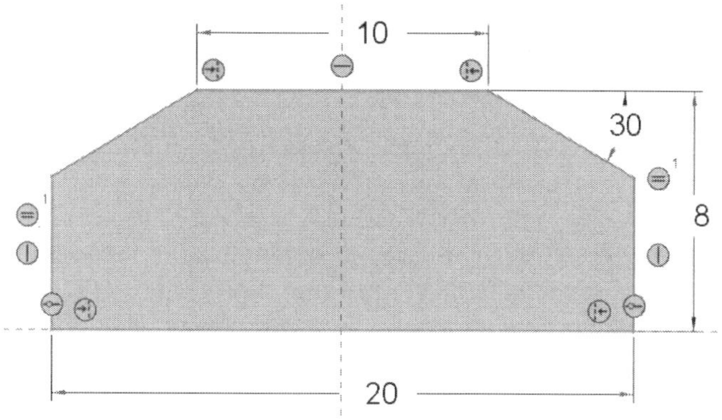

Figure 6 Final sketch for base feature

We now see a preview (in orange) of the protrusion. The default is a one-sided, variable depth protrusion (turn the datum planes back on to see where our sketch was on

FRONT). On the dashboard, open the **Depth Spec** list, and select the ***Symmetric*** option.[3] The dimension value specifies the total depth (symmetric about the sketch plane). Enter a value of **10** (either in the dashboard or on the dimension shown in the graphics window). The preview should look like Figure 7.

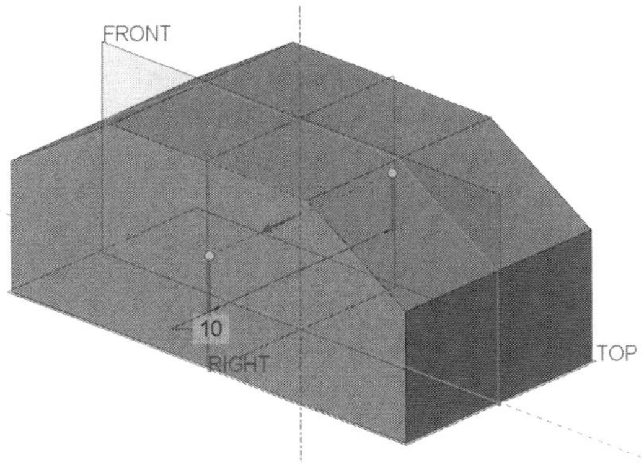

Figure 7 Base feature preview (*both sides, blind*)

You can now ***Verify*** the protrusion. Assuming everything is satisfactory, ***Accept*** the feature. Save the part using CTRL-S. Make sure it is being saved in the proper working directory, then middle click.

Creating a Revolved Protrusion

We'll now add the vertical axisymmetric shape onto the top of the base feature. In 3D solid modeling terms, this is a "revolved solid", created by taking a 2D sketch and rotating it around a specified axis. In Creo Parametric, we can use revolved features to create protrusions or cuts. The angle (or 'depth') of the rotation is adjustable using the same type of options as an extrusion (variable, up to surface, one or both sides, and so on). For this part we will do a one-sided, variable (360E) revolve. Only a half cross-sectional shape is required. See Figure 8. Depending on the feature shape and model geometry, the sketch can be either an open or closed curve (closed curves are safer). The axis of the revolve can either be included in the sketch of the section, or can be specified externally.

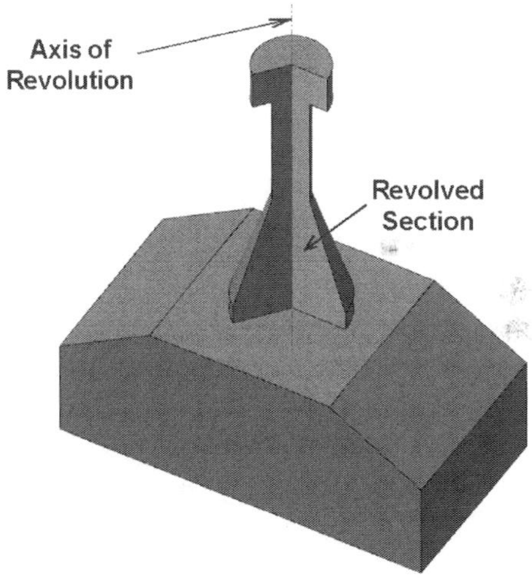

Figure 8 Section to be revolved to form protrusion

For this feature, we will again use an internal sketch so select the ***Revolve*** tool in the ribbon. The **Revolve** dashboard looks the same as the **Extrude** dashboard, and offers the same options (thin feature, remove material, depth spec, and so on). The ***Placement*** drop-down panel opens by default (observe the option to specify an external revolve axis). Select ***Define***. Select **FRONT** as the sketching plane. Set up the **RIGHT** datum plane as the **Right** sketch orientation reference, then middle click or select ***Sketch***.

[3] Remember you can also set this by selecting the depth drag handle, opening the RMB menu, and selecting ***Symmetric***.

Notice the two automatically chosen sketch references. The sketch we are going to create is shown in Figure 10. This is essentially the right half of the cross section of the revolved feature. We want the lower edge of the sketch to lie precisely on the top of the block. The easy way to do this is to make the top surface a reference. You may want to spin the object to see this surface. (If you do, reorient your view of the sketch back to the standard view.) In the RMB pop-up menu, select

References (or find the ***References*** command in the ribbon ***Setup*** group***)***.

Spin the model and left click to select the top surface now. This will create another horizontal reference in the sketch (see Figure 9). The one on the TOP datum can be deleted. The reference status should still be "Fully Placed." If it reads "Unsolved Sketch", just push the ***Update*** button on the right. Close the **References** window.

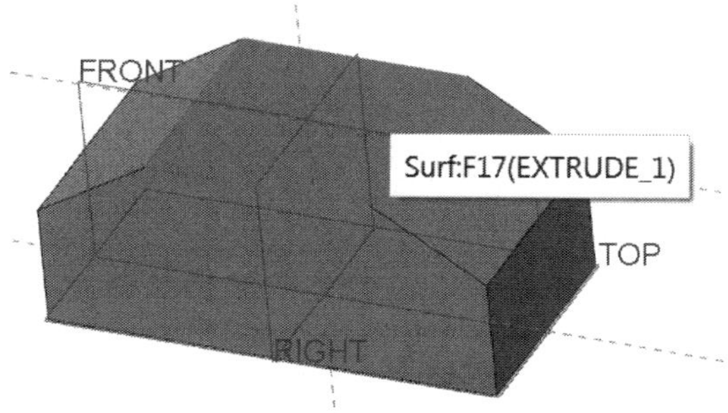

Figure 9 Adding the top surface of the base feature as sketching reference

Before attempting the sketch in Figure 10, first create a **Datum Centerline** <u>not</u> a *Sketching Centerline.* See step 2 below, then create the sketch.

Create sketch using Figure 10 below.

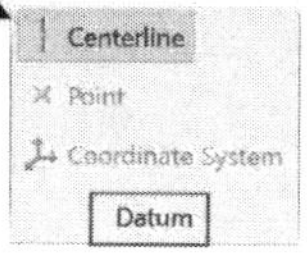

The sequence is:

- select the appropriate references (done that!)
- create vertical **Datum Centerline** as the *Axis of Revolution* (done that!)
- sketch the desired shape (doing that!)
- strengthen any dimensions you want to keep
- add explicit constraints
- add your own dimensions (start with smaller dimensions) to get the scheme you want
- modify dimension values to get desired size

Here are a few more tips for using Sketcher effectively:

1. While you are sketching lines, if a constraint appears, clicking the RMB will cycle through three possible states for the constraint: ***Lock / Disable / Enable***. Observe the change in the constraint symbol for locked and disabled constraints.
2. The *Axis of Revolution* must be specified on an internal sketch using a centerline. To place a centerline along the vertical reference in the **Datum** group select

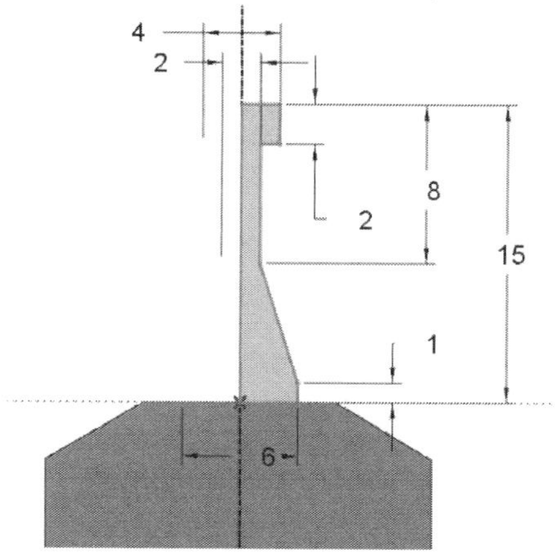

Figure 10 Sketch for revolved protrusion

Centerline. Alternatively, in the graphics window RMB pop-up menu, in the **Sketch Tools** row select ***Create Axis of Revolution*** (be careful not to select the sketching construction centerline button). Click on the vertical reference; the centerline will automatically snap to vertical when you click to create the second point on the reference. Notice some of the weak dimensions will change to show the diameter of the feature[4]. If you want to keep them, make them ***Strong*** (using the pop-up menu).

3. Revolved features are usually specified using diameters rather than radii. This is automatic if you use the previous axis of revolution command. If you want to create your own diameter dimension: left click on the sketched line or vertex, then on the centerline, again on the same line/vertex, then middle click to place the dimension text. This obviously only works if you have a centerline and are therefore using an internal axis for the revolve.

4. When you are modifying the dimension values one at a time, it is *beneficial to do the smaller dimensions first*. This ensures that the geometry will stay close to the desired shape throughout the changes. Notice we don't need to use ***Lock Scale*** here.

Helpful Hint

If you have several construction lines in the sketch (they all have the same line style as a centerline), which one becomes the axis of revolution? You must choose the one you want by selecting it then holding down the RMB to find the ***Designate Axis of Revolution*** command.

You might like to check out the ***Feature Requirements*** function at this time. If you have left the sketch open, there will be a warning here. To see the open ends, turn on the ***Highlight Open Ends*** button. In this case, the feature will regenerate, but to be more robust you should probably close the sketch. Do that now!

After finishing the sketch, select the ***OK*** button. The feature will preview in orange. If you have not designated an internal axis, the feature will not preview. The **Placement** panel will indicate that you are using an internal axis. Having an external reference (probably a datum axis, discussed in Lesson 6) for this might be handy if several features were going to be revolved around the same reference (if the reference moved, so would all the features). If you are going to use an internal centerline as the axis, you should get in the habit of always creating the revolve axis first when you enter Sketcher. It is then available for creating diameter dimensions. If you forgot to do this, just select the ***Define*** button in the **Placement** panel to re-enter Sketcher, or use the ***Edit Internal Sketch*** on the RMB pop-up.

---

[4] It is possible to use a ***Construction Centerline*** (in the pop-up menu) then in the RMB pop-up select ***Designate Axis of Revolution***. This is not a fatal error but will result in a warning, and the automatic conversion of the centerline to a ***Geometry Centerline***.

You should now be back in the Revolve dashboard. The default is a variable protrusion, which in this case means that the angle of the revolve is specified. It should be 360°. If you left the sketch as an open curve, Creo Parametric must be told which side of the curve is to be made solid.

All elements should now be defined. ***Verify*** the part (it should look like Figure 11) and select ***OK***. Note that if you used an internal centerline, an axis has been defined as part of the feature. Although it doesn't appear in the model tree, the axis can still be used as another feature's reference (for example, for a coaxial hole).

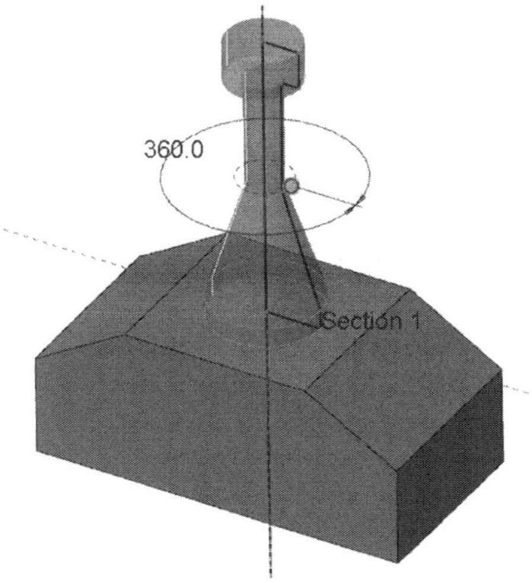

Figure 11 Preview of revolved protrusion

Adding and Mirroring a Cut

We'll now use an extruded cut feature to create a pocket on the front side of the base. We will then mirror it to the back side of the base. Our design intent here will be to leave a 1 unit thick wall around the pocket[5]. While we have created a cut feature before, we will introduce a new useful tool in Sketcher to create this geometry.

Select the ***Extrude*** button in the ribbon; its dashboard opens. Before we forget, click on the ***Remove Material*** button on the dashboard to produce the cut. Then, activate Sketcher using ***Placement* ➤ *Define*** or open the RMB menu and select ***Define Internal Sketch***. You should now be in the **Sketch** menu. Select the front surface of the block as the sketching plane, then select the top surface of the block as the TOP orientation reference. Middle click to enter Sketcher. When you arrive in Sketcher, references will have been already picked. We will now create our sketch using only a single dimension - the thickness of the wall around the pocket!

[5] Incidentally, if you wish to verify inch units, use ***File* ➤ *Prepare* ➤ *Model Properties***. We'll talk more about part units and how to change them in Lesson #8.

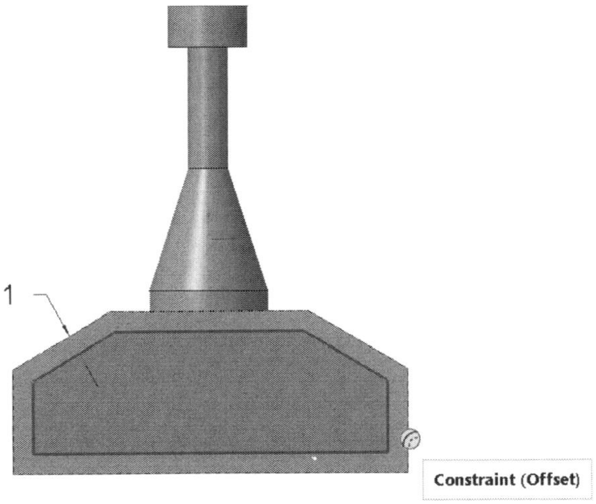

Figure 12 Sketch created using ***Offset***

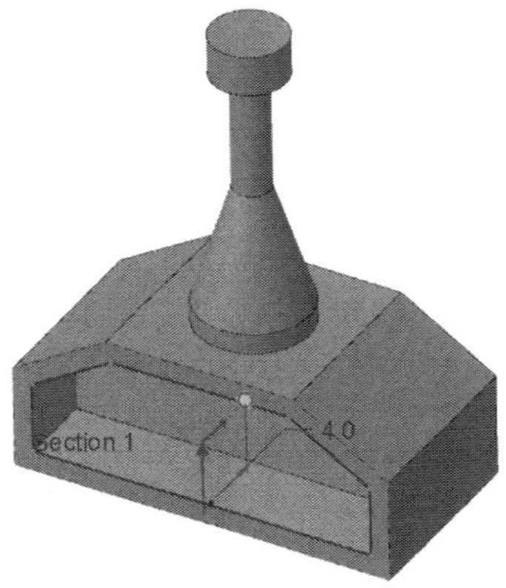

Figure 13 Pocket added using cut feature

In the **Sketching** group in the ribbon, select the ***Offset*** button Offset. Click on one of the edges of the front surface. It will highlight in green. Then, holding the Shift key, click on the front surface. That will define a loop containing the selected edge all the way around the surface. An arrow will appear on one of the green highlighted edges showing an offset direction. Accept the small **Loop Selection** window with a LMB click. If the sketch is outside the base feature, enter an offset value of ***-1***, otherwise enter ***1***. The sketch for our pocket is now complete, as in Figure 12. Look for a small constraint symbol that indicates the offset. Select the ***OK*** button.

The Creo Parametric default is a blind cut, with the material removal side on the interior of a closed sketch. All we have to do is set the depth. Enter a value of **4** and ***Accept*** the feature. The resulting pocket should look like Figure 13.

Select the cut feature, then in the pop-up select ***Edit Dimensions***. Notice that the offset distance *may* shows as "**1**" instead of "**-1**" as possibly given in the sketch. Once the direction was flipped with the minus sign, it was removed. This is a frequent occurrence in Creo.

Creating a Mirror Copy

Since the part is symmetrical, we can easily create the pocket on the back of the base by mirroring the first one. We only need a couple of mouse clicks to do this. For the following, in order to see a new function for locating features, turn off the display of the datum planes. The cut should already be highlighted in green as the last feature created. If not, just select it.

Select the ***Mirror*** button in the ribbon (**Editing** group). This opens the **Mirror** dashboard. See the message window. We want to mirror this pocket through the **FRONT** datum plane. To select this plane, we'll do something a little different. At the bottom of the screen beside the selection filter, select the ***Find*** button.

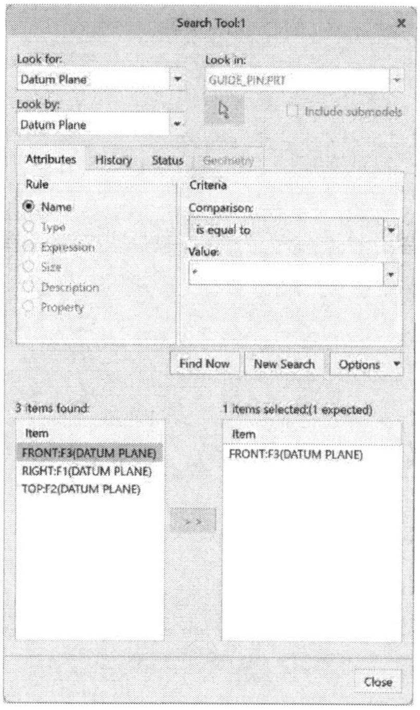

Figure 14 The Search tool

The **Search Tool** dialog window opens (Figure 14). Here you can select references by datum or surface, and by name, ID, or feature number, and many other variations. The ***Look for*** data field has already been selected for us. Set it to look for ***Datum Plane***. Press the ***Find Now*** button. All the datum planes in the model are now listed at the bottom. In this list, select **FRONT** (it shows on the model), then the >> button to move it to the selected window, and finally select ***Close*** (middle click).

The ***Find*** command is handy if, as in this case, the feature is not displayed or if the model becomes very complicated with many datum planes and/or features. It is also very helpful if the features have meaningful and unique names.

Before we accept the feature, open the ***Options*** panel in the dashboard. This contains a single toggle that determines whether the mirrored feature is dependent on the original. For example, if we changed the offset value of the original cut (currently 1.0) then we control whether the mirrored cut would also change (dependent) or not (independent).

Unfortunately, there is no ***Verify*** function here, so just accept the new feature (middle click a couple of times). The result is shown in Figure 15[6]. How does the mirrored feature appear in the model tree?

By the way, have you saved the part recently?

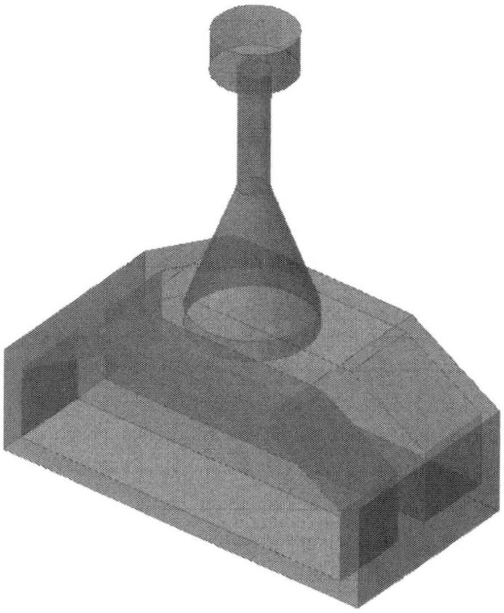

Figure 15 Part with mirrored pocket

[6] In the Model Tree, open **Bodies(1)** under **Design Items**, select Body 1 and in the pop-up, select the ***Make Transparent*** button.

Creating Holes

We already came across the hole feature in Lesson #3. We are going to add four holes as shown in Figure 1. We are going to do something a little different here with the depth specification, plus use the mirror command a couple of times after creating the first hole. Holes are placed features, so in the Engineering group select ***Hole***.

The **Hole** dashboard opens. The default is a ***Simple*** hole. For the primary reference (the placement plane), make sure the correct sloping surface of the base is preselected (highlighted), then click at approximately the position where we want the hole center to be. This is shown ("placement plane") in Figure 16. Use the orange bracket drag handles to select the **FRONT** datum plane (reference #1) and the upper edge of the end surface of the base (reference #2) for the linear references. This is one time where we must use an edge as a dimensioning reference, which we normally want to avoid. (Why?) The distance from each reference will be *3*. Set the hole diameter to ∅ **2.0**.

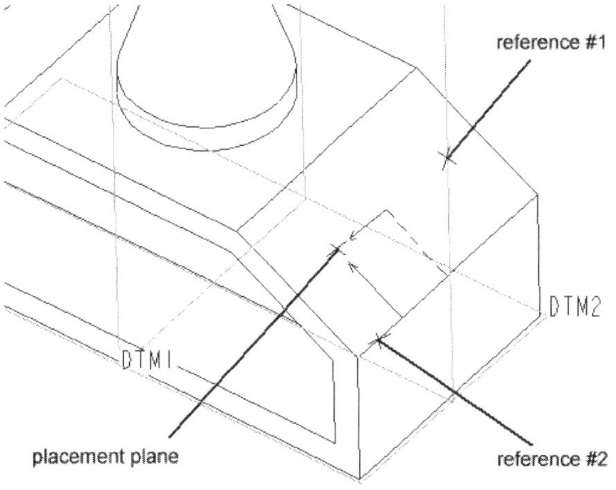

Figure 16 Placement plane and linear references

In the **Depth Spec** list, select ***To Next*** ("Drill up to next surface"). All the depth specification options are also available if you select the drag handle on the end of the hole and use the RMB pop-up menu. As might be expected, this creates the hole until it passes through the next surface it comes to, wherever that is. The only restriction on *To Next* is that the sketch or hole must be entirely within the terminating surface. That is, if only part of the sketch or hole intersects the surface, the feature will just keep going through! We will see some examples of the problems this might cause when we get to the last section of this lesson.

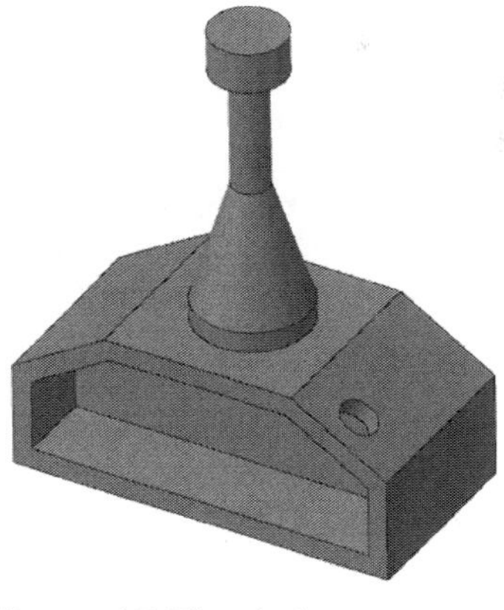

Figure 17 First hole

In this part, *To Next* means that the hole goes until it meets the surface of the pocket formed by the cut. A blind hole ending somewhere in space inside the pocket may have achieved the same geometry but would not be in keeping with our design intent, plus it would add an irrelevant dimension to the part database.

You can now ***Verify*** the hole. Assuming all is well, ***Accept*** the feature. See Figure 17. Notice that the hole feature automatically contains an axis.

We can use the ***Mirror*** command to make copies of the hole. For practice, turn off the datum planes and use the ***Find*** command again. First, mirror the single hole using FRONT as the mirror plane.

Repeat this process to mirror both holes to the left side of the part at the same time. Using the CTRL key, select both holes, then launch the ***Mirror*** command. The mirror plane is RIGHT. Accept the feature.

The part should now look like Figure 18.

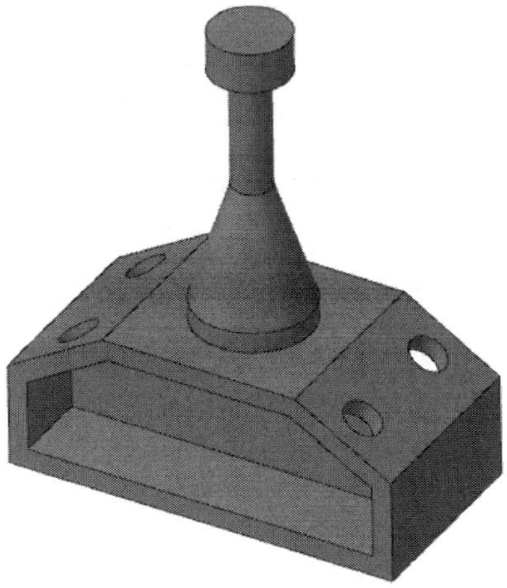

Figure 18 Holes added to base using ***Mirror***

Having Problems Mirroring?

If you have trouble creating mirrored features, it is likely that your underlying geometry is not perfectly symmetrical about the mirror plane. We should not have that problem here, because we used the symmetry constraints on our base sketch, and a both-sides blind protrusion, which is also automatically symmetric. If you ever do have problems, you may have to double check the geometry to ensure that its dimensions are exactly correct. We will investigate this potential problem later on in this lesson. Geometric conditions at the location of the mirrored featured must be "legal" for the creation of the feature. For example, if the left side of the block did not have the same slope as the right side at the location of the hole, we should expect problems trying to do the mirror operation from right to left if the hole is defined as perpendicular to the surface.

Creating Rounds

We will use a very handy short cut to add a couple of simple rounds to the top of the guide pin, and the edge where the shaft meets the base. Technically, these are called a round and a fillet, respectively. A round removes material from an edge, while a fillet fills in (adds) material.

For the first round, use preselection to pick the edge where the base of the revolved protrusion meets the block (Figure 19). Note that only half the circular edge needs to be chosen - the feature will follow the tangent edge all the way around. The selected edge is highlighted in green. Select ***Round*** in the pop-up menu. The display will show a preview of the fillet in orange. Set the radius to **0.5** by entering the value, or using the drag handles. To accept the fillet, just middle click. That's fast! (How many mouse clicks?)

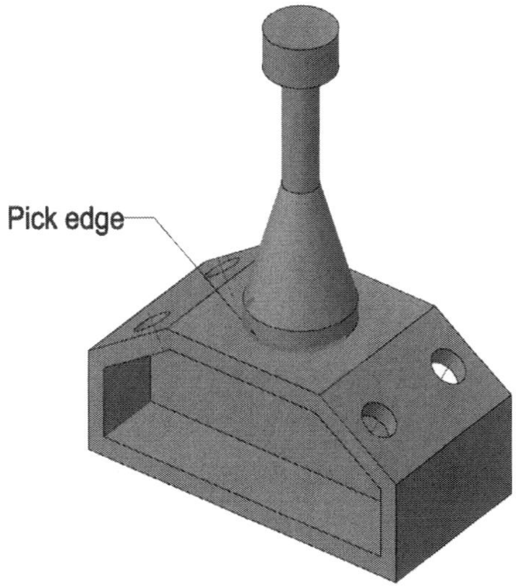

Figure 19 Creating fillet at base

Figure 20 Preview of fillet at base

Let's do the same for the edges at the top of the shaft. Use preselection once again to highlight a single edge. Then use CTRL-click to select the second edge (see Figure 21). Once again, select ***Round*** in the pop-up menu. Both edges will preview (Figure 22). Set the radius to 0.5.

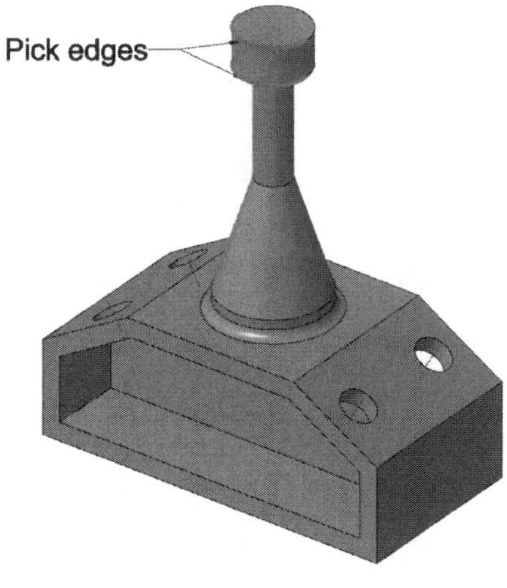

Figure 21 Selecting edges at top

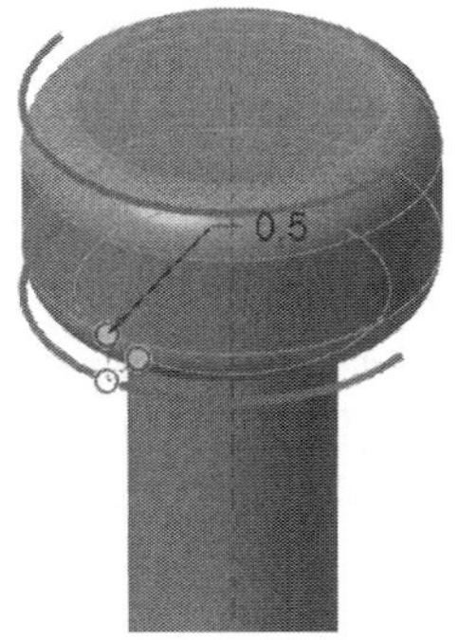

Figure 22 Preview of rounds at top

If you look in the **Sets** panel in the dashboard, you will see one set listed, containing two edges. All edges in the set have the same properties. Note that we could have put all rounds into the same set here. What would that imply about our design intent?

This round feature is OK for now, so just middle click. To toggle the display of the tangent edges formed by the rounds, select

File ➤ Options ➤* (Appearance) *Entity Display

Tangent edges display style (Dimmed, or ***Solid,*** or ***No Display, ...)***
Apply ➤ OK

You might experiment with the image (shading, hidden line, no hidden, etc.) to see what the rounds look like in different displays, in particular the appearance of the tangent edges.

Using Edge Sets with *Chamfer*

The last feature addition to this part is a chamfer all around the edge of the pocket and the parallel outside edge on the front and back of the base. We will include all these edges in a single chamfer feature by organizing them in two edge sets. All edges within each set have the same size. Edge sets are also used in the round feature. The main trick with edge sets is making sure a chosen edge is in the set you want it to be. Doing this is largely a matter of being careful when you are selecting the edges, and watching the screen carefully. The reason edge sets are useful is that the model tree can be simplified considerably by having multiple chamfered (or rounded) edges contained in the same feature. So, all chamfers or rounds of the same size can be modified simultaneously with a single dimension. Furthermore, when rounds or chamfers meet at corners, you can control the transitions between them. This can only be done for chamfers or rounds contained in the same feature.

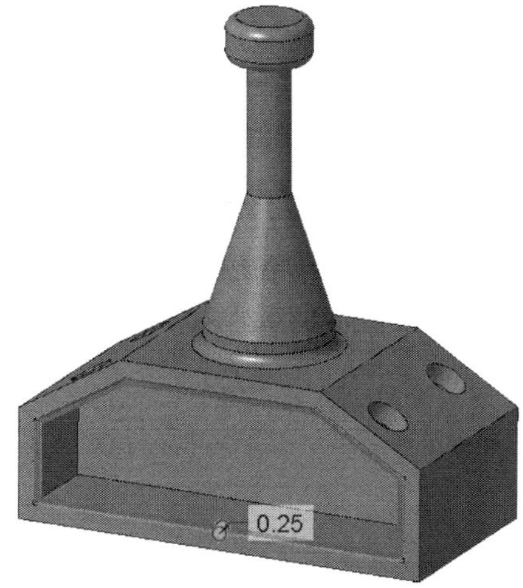

Figure 23 Chamfer edge set #1

You might find the following easier to do in hidden line or wireframe display. In shaded mode, some of the edges may not be really clear (or use ***Shading With Edges***). Start by selecting the ***Chamfer*** command in the Engineering group. The dashboard is now open. You can immediately start selecting edges. Hold down the CTRL key and pick the six edges shown in Figure 23. The edges will highlight in green and the chamfer will show in preview orange. When the last one is picked, adjust the size to **0.25** for this edge set. If you accidentally select a wrong edge, just pick it again to remove the selection.

Now, left click on an edge going around the outside of the base - see Figure 24. **As soon as you left click (without holding down CTRL), Creo Parametric assumes you are starting a new edge set.** The previous set is still shown in preview. Now holding down the CTRL key you can continue to pick edges for the second set. Set the dimension for this set to **0.5**.

Open up the **Sets** panel. The two sets are listed, with the edges of the highlighted set shown in the pane below. Selecting Set 1 or Set 2 will highlight the various edges on the model.

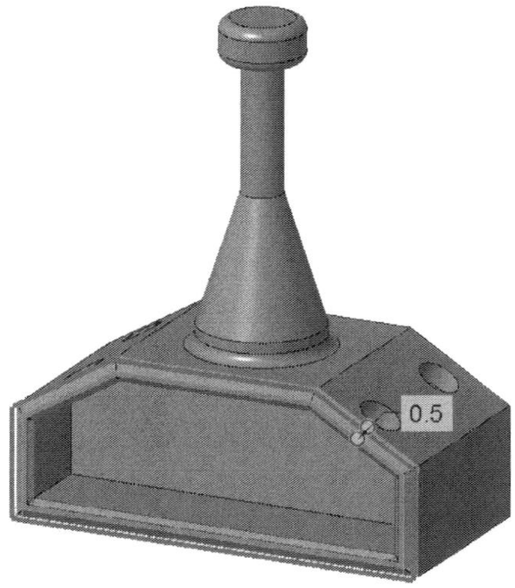

Figure 24 Chamfer edge set #2

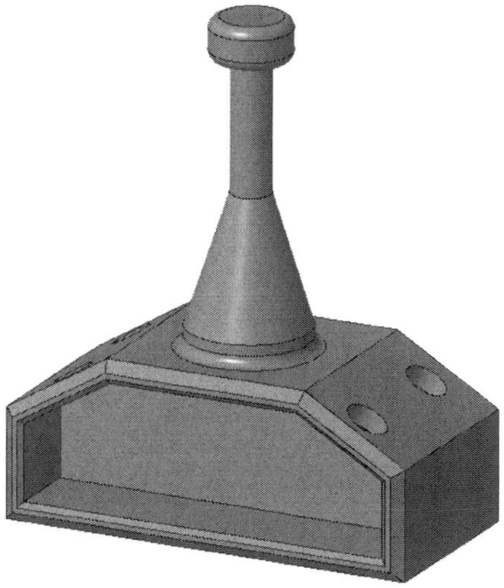

Figure 25 Chamfer completed

There are no Transitions here (the next dashboard panel button), since none of the edge sets intersect. Go ahead and ***Accept*** the chamfer. See Figure 25. Save the part.

Open the model tree and observe that the last feature is the chamfer containing all 12 edges although they are not listed. The previous two features were the rounds. Notice the listing of the mirrored holes and the mirrored pocket. It might be a good idea to come back later and rename all these Mirror features in the model tree to help distinguish them.

Try to mirror the chamfer to the edges of the pocket on the back face of the base using mirror plane **FRONT**. This seems like a reasonable kind of thing to do. However, if you pre-select the chamfer, the ***Mirror*** button in the ribbon is not available. We'll have to do something a bit different here.

To get the chamfer on the back pocket we have two options:

1. Delete the existing chamfer and create a new one containing edge sets with edges on both front and back surfaces, or
2. Redefine the existing chamfer by adding new edges to the feature (this involves commands discussed in Lesson #5), making sure that the edges we add go into the proper edge set.

For now, you might as well try the first of these two. With the new chamfer with edges on both sides, the part is completed. You should come back later and try the second option, perhaps also investigating the ***Details*** button on the **Sets** panel. This lets you select a chain or loop of connected edges, using a starting edge and a defining surface that contains all the edges of the loop (This uses the ***By Rule*** option). You can have multiple chains (for example at the front and the back of the guide pin) in the same edge set.

Saving the Part

Don't forget to save your part:

File ➤ Save

(or use CTRL-s) and if you have been saving regularly, get rid of previous copies of the part file by using

File ➤ Manage File ➤ Delete Old Versions ➤ Yes

Model Analysis Tools

Quite often, you need to find some information about a part - distances between points, surface area, center of gravity, moments of inertia. There are lots of tools available in Creo Parametric to query the model. Let's start with something simple. In the **Analysis** ribbon select:

Measure ➤ Summary

This opens a narrow window which can be expanded by selecting the plus sign at the right end. Also expand the **Setup** and **Results** areas, as in Figure 26. The icons across the top allow you to measure length of entities, distance between entities, angle, diameter, area, one-sided volume, and the transformation between coordinate systems. Select the second button (***Length***) then pick any edge on the model. Its length will appear in the window and on the screen. Select an edge of one of the holes (remember these are ø2). Is the reported length the full circumference or only half? What does this tell you about how Creo Parametric stores circular shapes?

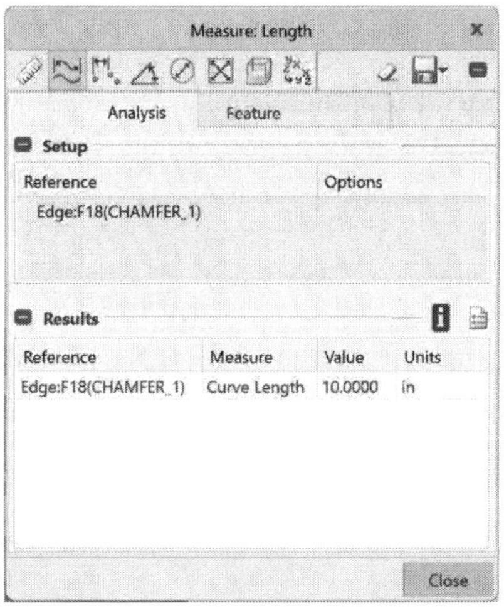

Figure 26 Obtaining a length measure

Try selecting ***Diameter*** then pick on the cylindrical portion near the top of the shaft. The diameter shows in the **Measure** window. Even more interesting is to pick on the conical part of the shaft. This reports the diameter at the pick point. What happens if you pick on the round at the base of the pin (this is curved in two directions)?

If you select ***Area***, you can select individual planar or curved surfaces (they highlight in green) or the entire model (click the part name in the model tree). See Figure 27. This might be useful to calculate paint quantities. You can also obtain projected area by specifying a projection plane or direction.

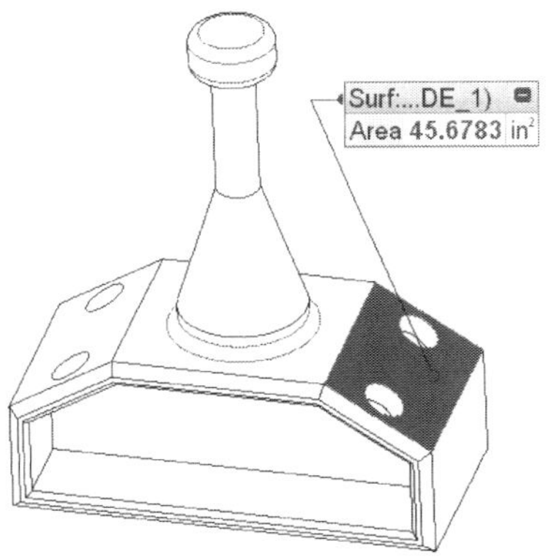

Figure 27 Area of surface

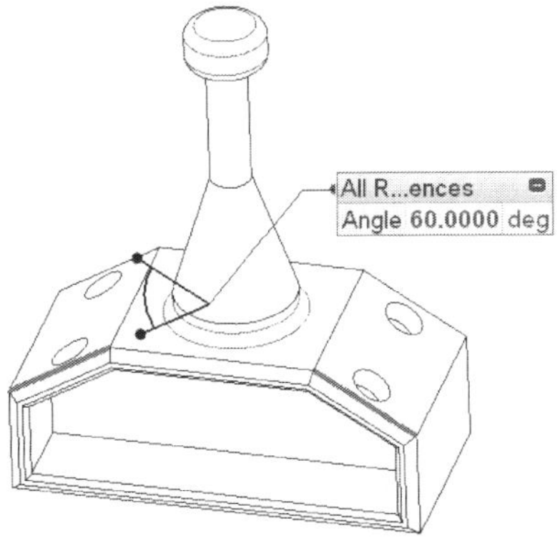

Figure 28 Angle between edges

By selecting ***Angle***, you can find the angle between the sloping edges at each end of the block (Figure 28) or between the normals to two solid surfaces. ***Distance*** lets you find the distance between any two entities. If one of the entities is an arc, the default will use the distance to its center. Figure 29 shows the measurement of the (shortest) distance between the tangent edge of the round and the chamfer on the front of the part. You will have to turn off the **Use as Center** option for the selected arc. Note that you can also measure the maximum distance, and you can also obtain projected distance. Close the **Measure** window.

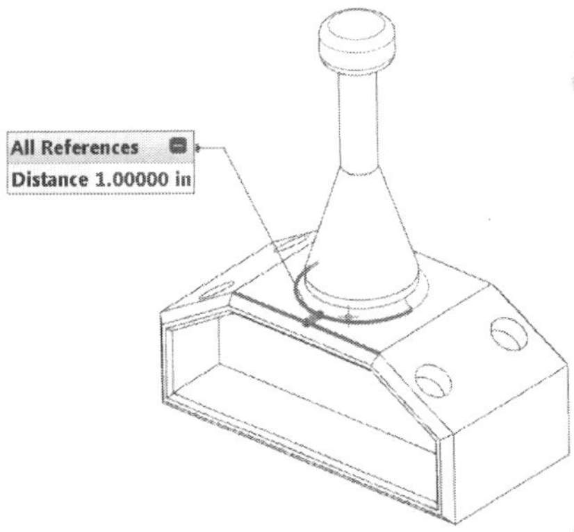

Figure 29 Distance between curves

Determining Mass Properties

We will now do a mass properties analysis of the part. In the process, we have to assign a material so that Creo knows the density. We could do that using ***File ➤ Prepare ➤ Model Properties*** but there is a more convenient way right here. In the **Analysis** ribbon select

Mass Properties

This opens the window shown in Figure 30. There is a warning about a missing material density - not surprising since we haven't defined one yet. Select ***Assign Material ➤ Other***. In the **Materials** window that opens, double-click on **Legacy-Materials**, then again on **al2014.mtl** to add it to the model (see the small pane at the bottom) and ***OK***. Back in the previous window, open the pull-down list at the bottom and select ***AL2014 ➤ OK***. You can have several materials defined in the model (for different bodies in a part or components in an assembly) and they would be listed here. Now press the ***Preview*** button at the bottom left.

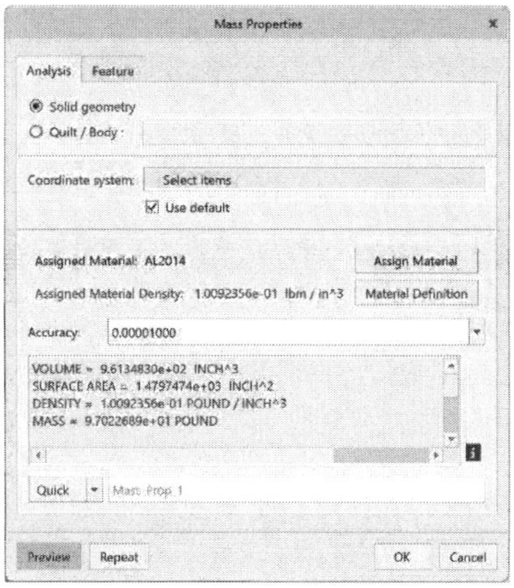

Figure 30 Model analysis showing mass properties

Figure 31 Part center of gravity and principal axes

The results (Figure 30) include the volume of the part, surface area, and mass, all expressed in current part units. A bit farther down in the results is the location of the part center of gravity (relative to a specified coordinate system). The location of the center of gravity is shown graphically on the object (Figure 31). The axes 1-2-3 refer to the principal axes of the solid. The results also include the mass moments of inertia about these axes, and the radius of gyration of the solid about each axis. This data is useful for dynamic analysis of the part. Close the **Mass Properties** window with ***OK***.

Other types of model analysis are cross section properties, one-sided volume (on one side of a selected plane), clearance between entities, determining if any edges are shorter than a specified length, or any thickness values are greater or less than specified values (important to molded plastic parts).

If you are going to perform one or more model analyses frequently, it is possible to create each analysis as a feature. This will then appear in the model tree, which can be set up (using ***Tree Columns***) to show the values of the calculated parameters which will be updated whenever the model is regenerated. These parameters can also be placed on notes in a drawing. This is discussed in the *Advanced Tutorial*.

Exploring the Model, or "What Can Go Wrong?"

Now comes the fun stuff! Here are some things you can try with this part. These explorations are very important, so DO NOT SKIP THIS SECTION!! We will review some of what we covered in Lessons #2 and #3. More importantly, some things we'll try here will show you how Creo Parametric responds to common modeling errors. Being comfortable with these methods to respond when an error occurs is an important aspect of your modeling proficiency. A common tendency among newcomers to Creo Parametric is to retreat from these errors and try to create the model in another (usually more familiar but less efficient) way. This will not expand your knowledge of modeling practice, or allow you to anticipate errors before they happen. Spend the time to learn this now, and it will save you much time later.

Here are the exercises:

1. We found out before how to name the features of a part. Do that now for the ***guide_pin***, using whatever names you like. Obtain model information using (in the RMB pop-up menu)

   Information ➤ Feature Information

   or

   Information ➤ Model Information

   What is the difference between these lists?
2. Make the following dimensional changes to various features of the model. (HINT: preselect the feature, use pop-up menu and select ***Edit Dimensions***). ***Regenerate*** the part after making each dimensional change. Observe what Creo Parametric does and see if you can explain why. You can usually recover from any errors that might occur by selecting ***Undo Changes***, or ***Quick Fix ➤ Delete***. If things really go wrong, you should be able to use ***File ➤ Manage Session ➤ Erase Current***, and retrieve your stored copy of the part file.

   < Change the radius of the round on the base of the revolved protrusion to the following values: (***0.75, 1.5, 3.0***). For each value, see if you can predict what Creo Parametric will do before you actually execute the regenerate command. Reset to the initial value (0.5) after these modifications.

   < Change the diameter of the first hole to the following values: (***1.0, 3.5, 4.0, 5.0***). Again, try to predict how Creo Parametric will handle these changes. Reset to the initial value (∅2.0) after these modifications. Try changing the diameter of one of the mirrored holes on the back of the part. When you click on this hole, where do the placement dimensions show up on the screen?

   < Change the location of the first hole from 3 to **1.5** away from the datum plane **FRONT**. Where does the hole now terminate? Why? Now change the same dimension to **5**. What happens and why? Reset to the initial value after these modifications.

   < Change the location of one of the holes from 3 to **(1.5, 1.0)** away from the edge reference on the end of the block. Where does the hole now terminate? Now change the same dimension to **(7.0, 8.0)**. What happens? Reset to the initial value after these modifications.

   < Change the height of the base block from 8.0 to ***6.0***, then to ***4.0***, then ***3.0***. Explain what happens and reset to the initial value after these modifications.

   < Change the depth of the base block (10.0) to (***9.0, 8.25, 8.0***). What happens each time? Reset to the initial value after these modifications (10.0).

   < Change the length of the base block to (***16.0, 12.0***). Shade the view. What happens each time? Reset to initial value after these modifications (20.0).

   < Change the diameter of the base of the revolved protrusion (6.0) to the following values: (***8.0, 9.0, 10.0***). What happens and why? Reset to the initial value after these modifications (6.0).

   < Change the radius of the rounds on the top of the revolved protrusion to the following: (***0.75, 1.5, 2.0***). What happens? Reset to the initial value after these

modifications.

< Change the edge offset dimension (1.0) for the pocket to the following: (***2.0, 3.5***). What happens? Reset to the initial value after these modifications.

< Change the depth dimension for the pocket to the following: (***4.5, 5.5***). Reset to the initial value (4.0) after these modifications.

3. Set up a relation so that the distance of the holes from the datum ***FRONT*** is such that the hole is always centered on the depth of the pocket. Add another relation that will give a warning if the web between the two pockets down the center of the part becomes less than 1.50 thick. Your relations will look something like this (your dimensions symbols will probably be different from these):

```
/* hole centered on pocket depth
d38 = (d5 - d14) / 2
/* narrow web warning - message is generated if false
(d5 - 2*d14) > 1.5
```

   Check these relations by changing the depth of the base feature from 10 to 20. Then change the depth of the pocket to 9.5. Follow the prompts in the message window. When the part is regenerated, open the **Relations** dialog window. Reset the values to remove the relation violation.
4. Examine the parent/child relations in the model. What are the parents of the pocket? What are the children of the pocket? Do the relations added in question 3 change the parent/child relations?
5. Delete the front pocket and all its children. Now, try to create it again. What happens to the holes? Since this new feature will be added after the holes, you might anticipate some changes in the model. This points out again the importance of feature creation order.
6. Explain why centering the base feature (the block) on the datums was a good idea.
7. Try to delete the revolved protrusion. What happens?
8. Try to delete one of the holes. What happens?
9. Select the first hole we made and in the pop-up select ***Edit Definition***. Open the **Placement** panel. At the bottom of the panel is an area marked **Hole Orientation** that contains a reference collector and a drop-down option list. Can you figure out what this placement function does? You may have to revisit the mirrored holes and use ***Reapply Mirror*** on the dashboard.

Introduction to Multibody Modeling

Although certainly not required for this model (in fact, it involves more modeling effort!), we will see how multibody modeling might be used for this part. Close the current model and start a new part called **guide_pin_MB**.

Create the base feature and add the pockets and holes as before. By default these will all be put into **Body 1**. Turn on or add the usual model tree columns. You can also select the ***Quilt/Body Evolution Tree***  ' ***Contributing Features***  buttons at the top of the model tree. See Figure 32.

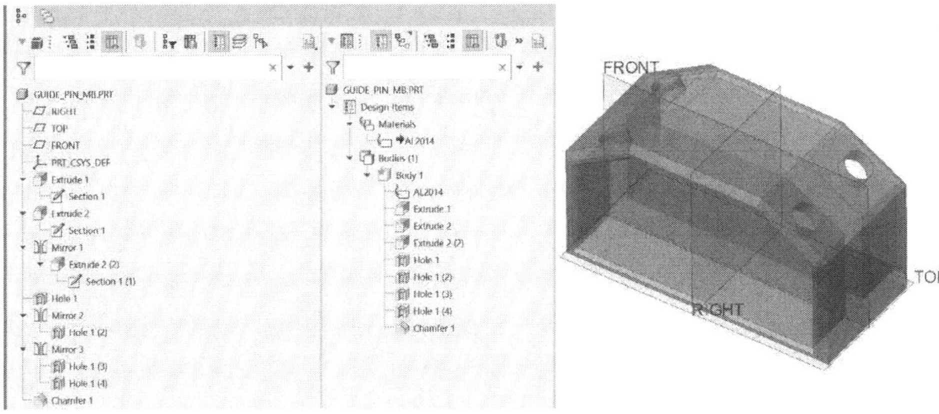

Figure 32 Body 1 created

We'll now create the revolved feature as a new body. Launch the ***Revolve*** command as usual and create the sketch shown in Figure 10 as before. When you accept the completed sketch, you will return to the ***Revolve*** dashboard.

Open the **Body Options** panel in the dashboard and check the option **Create new body** and accept the feature. A new body (**Body 2**) will be created - see the model tree. More importantly, this is now the **Active Body** as indicated by the small star on the body icon. This means the next feature created will automatically be added to it.

Go ahead and add the rounds. Can you put them all in the same feature? In this case, the answer is NO, because the lower round would involve Body 1. We'll have to deal with that later. The model and model tree now look like Figure 33.

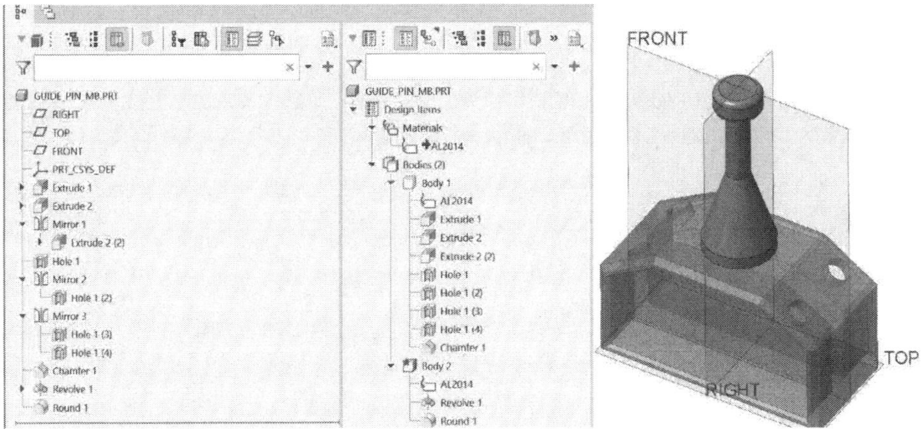

Figure 33 Body 2 created with revolve and round features

We can now do a Boolean operation to join the two bodies. We can either merge Body 2 to Body 1, or the other way around. The geometry would be the same either way but the logical structure of the model would be different. We will do the former.

Select the ***Boolean Operations*** ***Body Merge*** button in the **Body 1** group in the ribbon. The collector for **Body 1 to modify** is highlighted (cyan). Next, the **Modifying bodies** collector is active - pick on a surface of the revolve. The model will now highlight in orange. Notice the option to ***Keep Bodies***. We'll see what this does in a minute. Accept the Boolean operation with ***OK***.

Check out the model tree (Figure 34). A **Body Merge** feature has appeared at the bottom. If you have the **Body Evolution Tree** open, it will show Body 2 as a **Consumed Body** (notice the icon) and the active body is now Body 1.

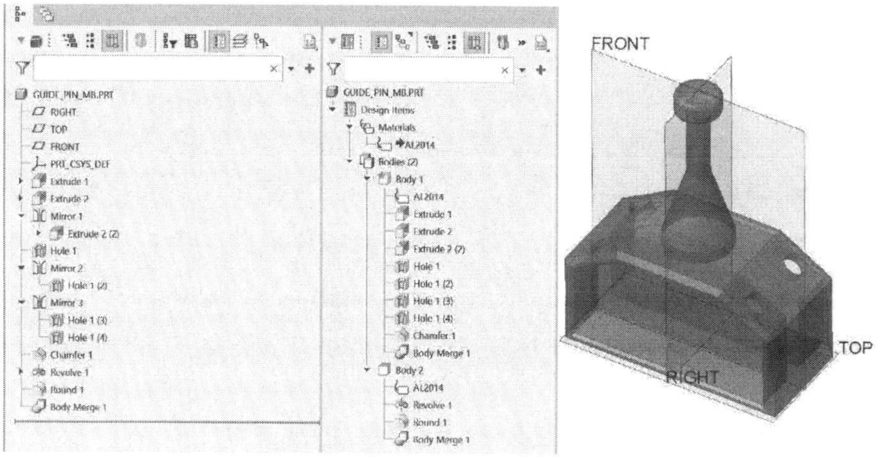

Figure 34 Body 2 merged with Body 1

If we had checked the option to **Keep Bodies**, the second body would have been merged with Body 1 but it would not be consumed. It would be available to do other operations - like making a copy or pattern (discussed in Lesson #6).

Go ahead and add the round at the edge where the revolve meets the base feature. The model is now complete. This was not such a complicated part and would not normally require creating a multibody model. But it was maybe worth the short side trip to see how it is done. Save the part.

In the next lesson we will discuss Creo Parametric utilities for dealing with features, including examining parent/child relations in detail, suppressing and resuming features, editing feature definitions, and changing the regeneration order. These are often necessary when creating a complex model, and to recover from modeling errors or poor model planning.

Questions for Review

1. When sketching with Intent Manager, why should you deal with and set up your constraints before setting up the dimensioning scheme? Why do you set the dimension values last?
2. What surfaces can be legally chosen as sketching planes?
3. In Sketcher, how do you easily create an arc tangent to a line at an endpoint?
4. What does the ***To Next*** depth specification do? What is a requirement for this?
5. In Sketcher, where are the ***Corner*** and ***Divide*** commands? What do they do?
6. What elements are required to create a **revolved protrusion**?
7. What is meant by a **linear** hole? What are the alternatives?
8. What is meant by a **dependent** copy?
9. What is the difference between a round and a fillet?
10. What types of chamfer are available?
11. When you are creating a mirrored copy can you:
    < select more than one feature to mirror at once?
    < select more than one mirror plane at the same time?
12. What happens when a chamfer meets a round at the corner of a part?
13. What happens when two rounds of different radii meet at a corner of a part?
14. The figure at the right shows a sketch of two four-sided polygons. What is the difference between these polygons? Notice the appearance of the vertical line on the far right.
15. Where do the placement dimensions of a mirrored feature appear under ***Edit Dimensions***?
16. What are the options for setting the depth of a blind, both-sides protrusion?

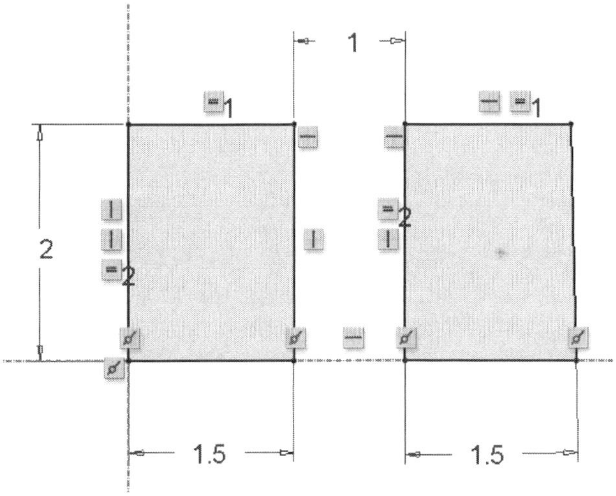

Figure for Question 14

17. Could the rounds we made on the top of the guide pin be created as part of the **revolved protrusion**? What advantages/disadvantages would there be in doing that?
18. How many of the **Measure** commands can you list?
19. What information is returned with the ***Mass Properties*** command?
20. How could you measure the complete circumference of a circular arc? The perimeter of an irregular plane surface?
21. How do you delete old versions of a part file?
22. Does the depth option ***To Next*** behave the same for holes and extruded cuts?
23. What happens if the **Depth Spec** of a revolved protrusion is set to ***Up To Surface*** and a surface parallel to the sketch is chosen?

Exercises

The first two parts here will be used later in these lessons (see Lesson 9 on assembly modeling). Make these now and keep them in a safe place until later!

1. The Axle

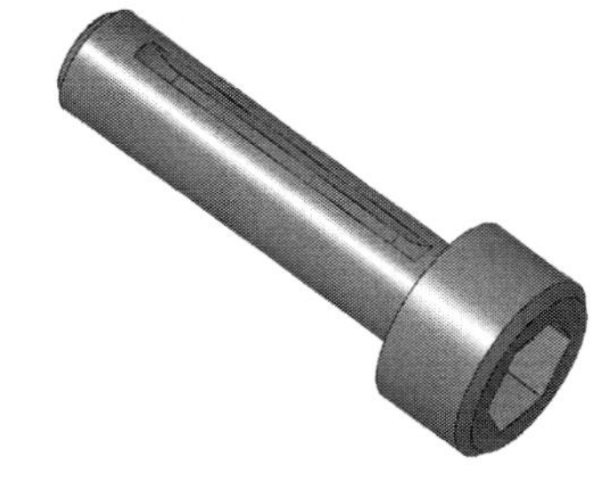

Create a part called **axle** as shown in the figure at the right. Use the dimensions shown below (we will change some of these later when we are in assembly mode). See the hint below for creating the hexagonal slot in the head. We will discover another way to do pre-defined shapes in Sketcher in a later lesson. Dimensions are in mm.

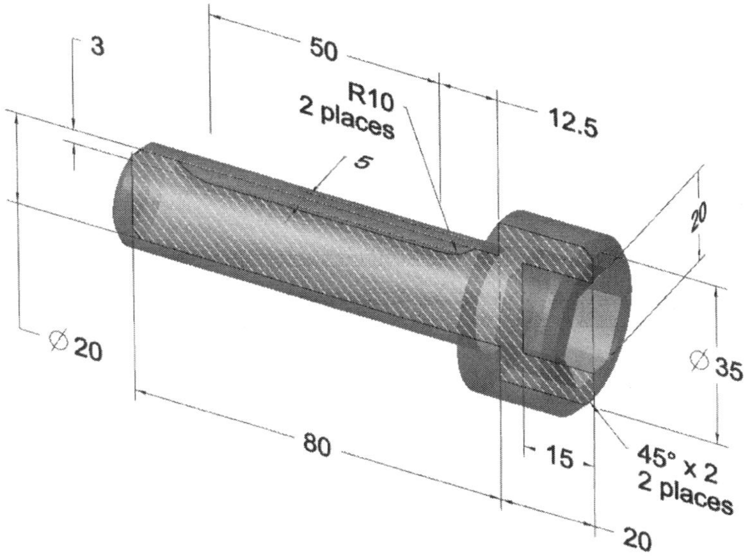

Helpful Hint

To make a hexagonal sketch a bit easier (if you don't want to use the sketcher palette tool), first create a circle, highlight it then right click and select ***Construction***. You can now sketch the six sides of the hexagon with vertices on the circle (observe the snaps that happen with Intent Manager). Add a dimension for the width across the flats. Then use the sketcher constraints to eliminate all the other dimensions.

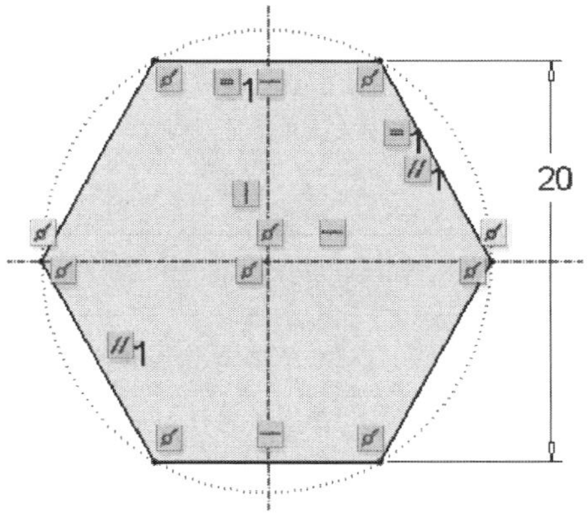

2. The Bolts

We will need several bolts in the final assembly. These will all come from the same part file **bolt** containing only a single bolt. Note that the threads have not been included for simplicity here. If you wanted to include the thread, you could use a helical cut or what is called a *cosmetic thread*. The dimensions of the bolt are shown below. See the hint above for creating the hexagonal sketch for the head. The beveled edge on the head is created with a revolved cut.

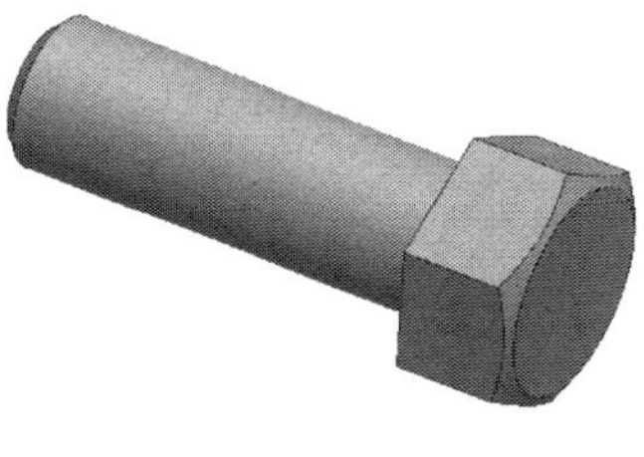

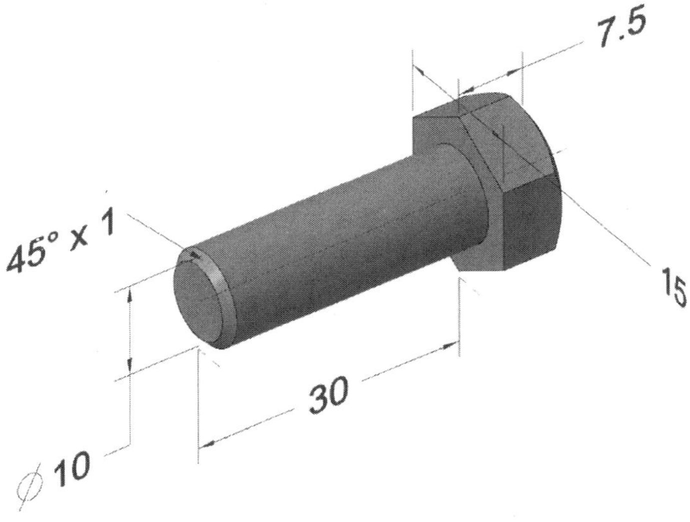

Bolt dimensions in mm

Here are some more simple parts to make that use the features introduced in this lesson.

3. A revolved feature sampler

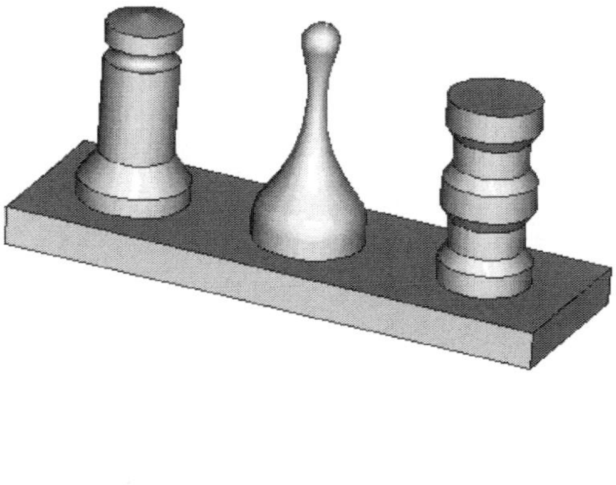

4. Mirrored pockets

5. Where should you put the default datum planes for this?

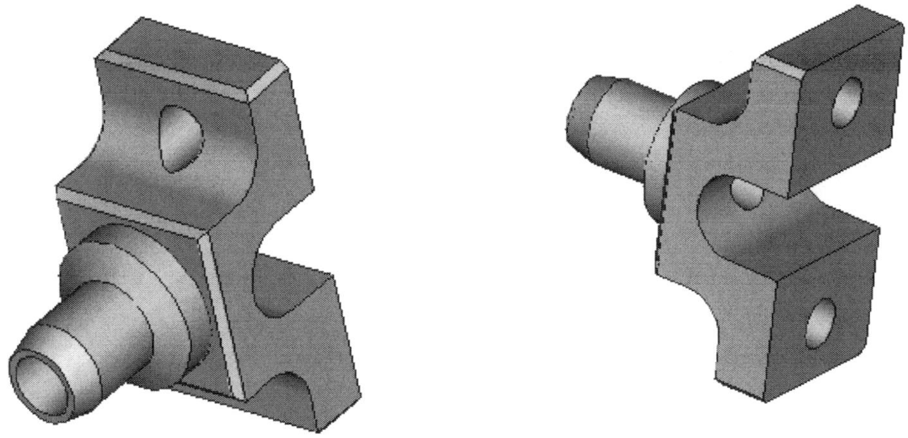

The next three parts will be a bit more challenging. HINT: at this stage, keep your features as simple as possible, and plan ahead!

6.

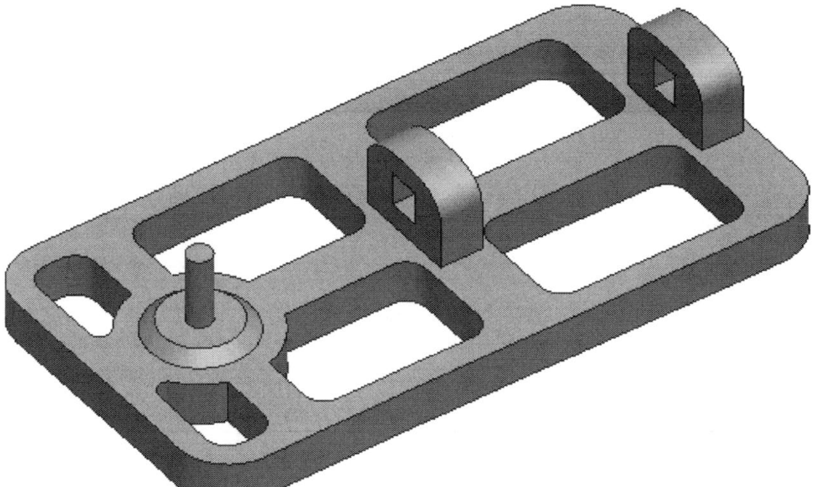

7.

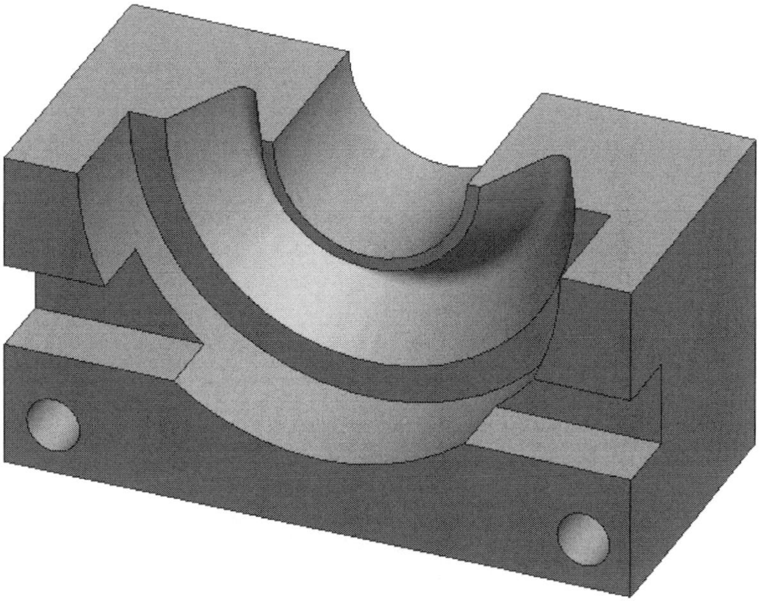

8.

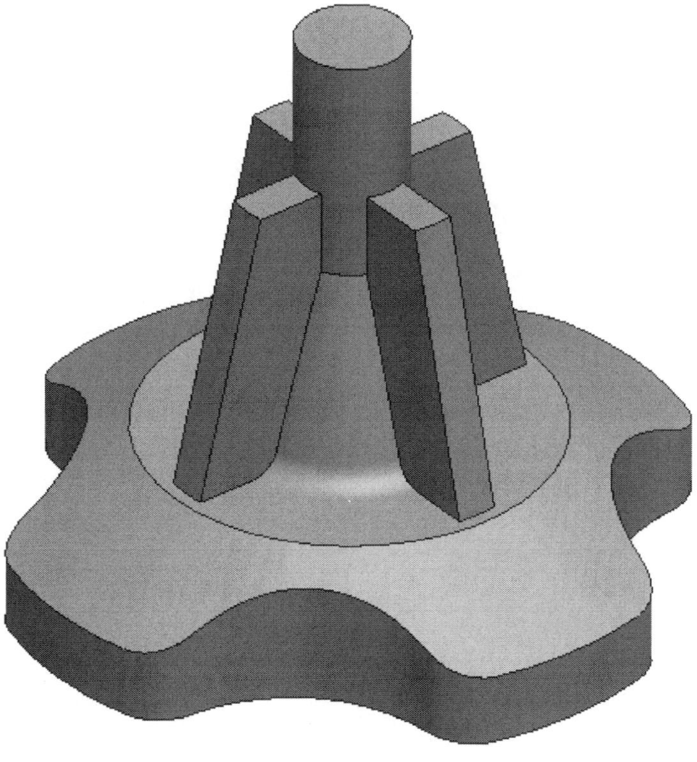

Project

Here is another part for the vise project, using features introduced in this lesson (a revolved protrusion, some mirrored cuts, and some rounds). All units are in millimeters.

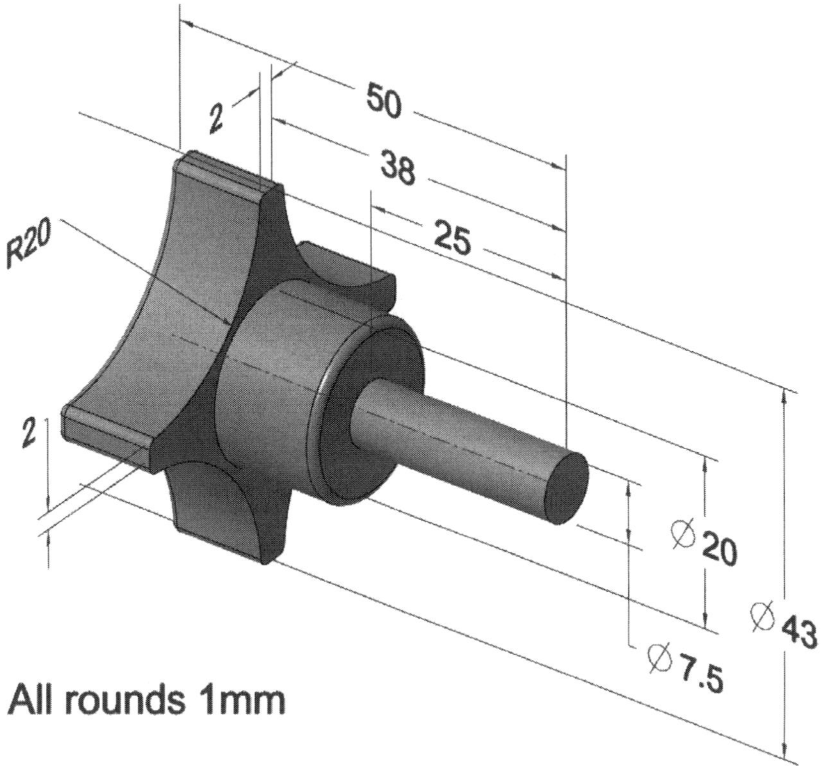

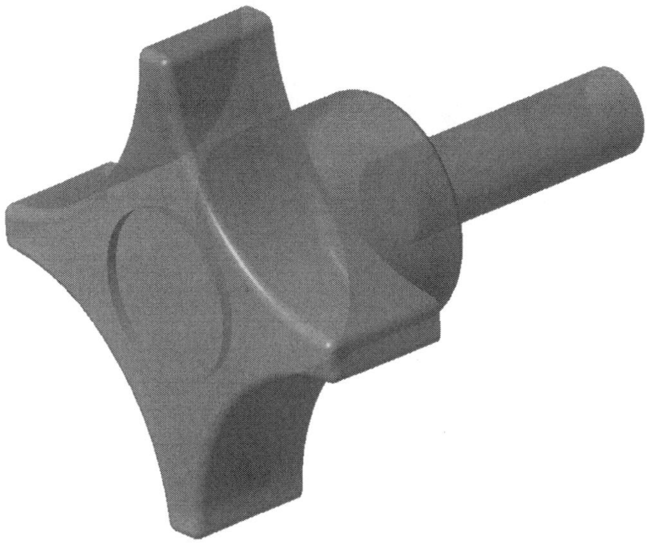

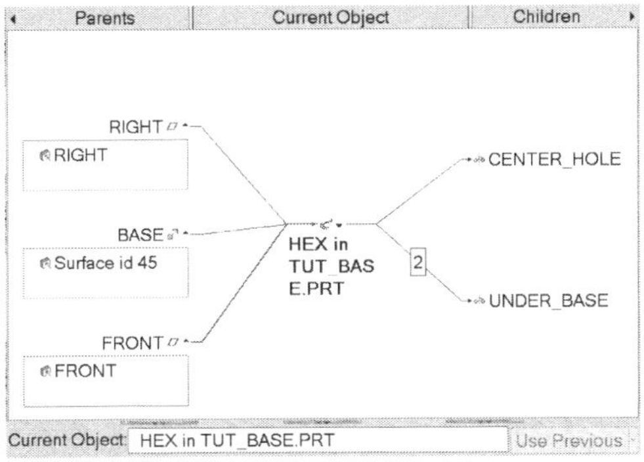

Lesson 5

Modeling Utilities

Synopsis

Utilities for exploring and editing the model: finding relationships between features, changing references, changing feature shapes, changing the order of feature regeneration, changing feature attributes, changing the insertion point, suppressing and resuming features.

Overview of this Lesson

In this lesson we are going to take a break from creating new features. Instead, we will discuss the equally important utilities for dealing with existing features. When you are creating complicated models, it is almost inevitable that you will have to change the geometry and/or structure of your model at some point. If your model becomes even moderately complex, you will need to know how to modify the model data structure and/or recover from poor model planning. This could be because you discover a better or more convenient way to lay out the features, or the design of the part changes so that your model no longer captures the design intent as accurately or cleanly as you would like. Sometimes, you just run into difficulty trying to modify the model (usually caused by the logical structure of the features) or have made errors in creating the model. This lesson will review the ways of obtaining information about parent/child relations, suppressing and resuming features, editing feature definitions and references (the commands formerly known as the 3 R's - ***Redefine***, ***Reroute***, and ***Reorder***). We will also introduce **Insert Mode** for adding new features to the model anywhere in the regeneration sequence. We have seen some of this before, so it will let you review that material.

The lesson is in four sections:

1. Obtaining Information about the Model
   - Regeneration Sequence
   - Obtaining a Feature List and Using the Model Tree
   - Getting Information about a Specific Feature

- Parent/Child Relations
2. Suppressing and Resuming Features
   - Single Features
   - Handling Features with Children
3. Modifying Feature Definitions - the 3 R's
   - Changing feature references with ***Edit References*** (formerly ***Reroute***)
   - Changing feature attributes with ***Edit Definition*** (formerly ***Redefine***)
   - Changing creation order with ***Reorder***
4. Insert Mode

As usual, there are Questions for Review, Exercises, and a Project part at the end of the lesson. Be sure to try the ECO (Engineering Change Order) exercises!

These utilities are most useful when dealing with complex parts with many features. To illustrate these commands we will look at their application to a very simple part that will be provided for you. This part has a number of modeling “errors” that must be fixed. With parts this simple, it might actually be easier to just create a new part and start over again (you may find it necessary to do that occasionally anyway). However, when parts get more complex, and contain many features, starting over will not be an option and these utilities will be indispensable.

In order to do this lesson, you will need a copy of the file *creo_lesson5.prt* that is available on the tutorial web site[1].

Once you have the part file, launch Creo Parametric, retrieve the part and continue on with the lesson. The part should look like Figure 1 in default orientation.

This model contains the default datum planes and four features, all in a single body. The base feature is a rectangular block. The other features are another solid protrusion and two cuts. The features are named as shown in Figure 1.

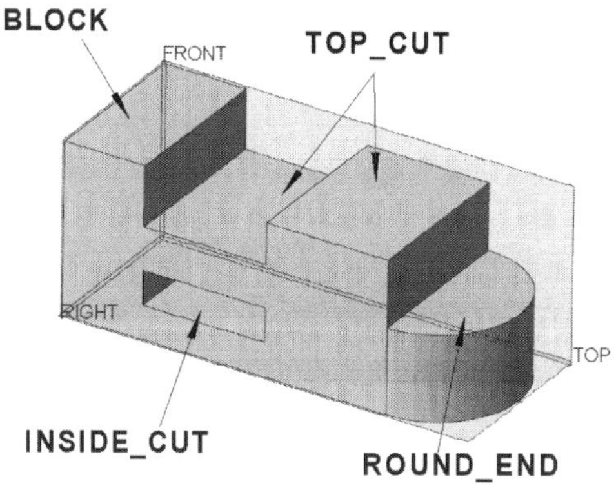

Figure 1 Initial part for Lesson #5

Obtaining Information about the Model

Once your model gets reasonably complex, or if you “inherit” a model from another source such as we are doing here, one of the important things to do is to have a clear idea of the structure of the model. Which features were created first? Which features depend on other features? How do the features reference each other? Answers to all these questions are easily available!

[1] http://www.sdcpublications.com/downloads/978-1-63057-837-4

The Regeneration Sequence

The order of feature creation during part regeneration is called the *regeneration sequence*. Features are regenerated in the order in which they appear in the part database[2]. (We will talk about changing the order of the regeneration sequence in a later section of this lesson.) To observe the regeneration sequence select the **Tools** tab in the ribbon, then:

> ***{Investigate}: Model Player***

Enter a **1** in the **Feat #** box to go to the first feature. Then press the ***Step Forward*** button to step you through the creation of the model one feature at a time. The model player window will tell you which feature is currently being created. As you progress through the sequence, the menu gives you a chance to get more information about the current feature, including its dimensions.

For example, when you get to feature #6, select ***Feat Info*** in the **Model Player** window. This opens a Browser window with a page that gives you lots of information about the feature. Look for the following: *feature number* (#6), the *internal feature ID* (52), the feature name (TOP_CUT), the IDs, names, and feature numbers of the parents and children of this feature, the *feature type* (an extruded cut), dimensions. One of the data panels (*Feature Element Data*) has scrollable text. Note that the depth of this feature is 10. This will be important later on. Also, note the difference between the feature number (the placement within the regeneration sequence) and the feature ID (Creo Parametric's internal bookkeeping). It will be possible to change the feature number (by reordering), but, once created, you can never change a feature's ID.

Close the Browser window and continue through the regeneration sequence until you have all seven features. Then select ***Finish***.

The Feature List

You can call up a table summary of all the features in the model by selecting:

> ***{Investigate}: Model Information***

This brings up the Browser page shown in Figure 2. The system of units for the model is shown at the top. Below this is a table that lists all its features. Information for each feature includes the feature number and ID in the first two columns, a name for the feature (defaults to feature type) and the type of feature. There are two action buttons for each feature which will highlight the feature on the model, or open up the feature information page we saw previously. If you have many features, it is a good idea to name

---

[2]This is called "history-based" modeling. Some modeling programs, like Creo Direct, do not have this mode of operation - they regenerate everything simultaneously. These are called "direct modelers."

them - in a part with possibly hundreds of features there is nothing worse than seeing a whole bunch of features all identified with just "Hole" or "Cut" in this table.

By the way, whenever you see a Browser window like this in Creo Parametric, you can easily print or save it by using the buttons at the top of the Browser window. This may be useful for design documentation. ***Close*** the Browser.

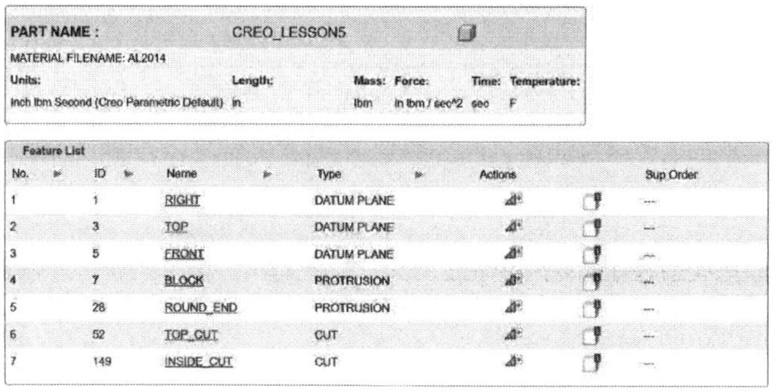

PART NAME : CREO_LESSON5
MATERIAL FILENAME: AL2014
Units: Inch lbm Second (Creo Parametric Default) Length: in Mass: lbm Force: in lbm / sec^2 Time: sec Temperature: F

Feature List

No.	ID	Name	Type	Actions	Sup Order
1	1	RIGHT	DATUM PLANE		---
2	3	TOP	DATUM PLANE		---
3	5	FRONT	DATUM PLANE		---
4	7	BLOCK	PROTRUSION		---
5	28	ROUND_END	PROTRUSION		---
6	52	TOP_CUT	CUT		---
7	149	INSIDE_CUT	CUT		---

Figure 2 The Browser page for **Model Info**

The Model Tree

The model tree was introduced earlier and you have seen it many times by now. If it is not currently displayed, open it. You should see the columns shown in Figure 3. If those are not visible, either load the model tree configuration file we used earlier (*Tut_ModTree.ui*) or use

Tree Columns

to add and format columns. The usual columns you will use are ***Feat #*** and ***Feat Type***, and ***Contributed Body***. Also, while we're here, select

Tree Filters

Figure 3 Model tree with added columns

This brings up a dialog window with a number of checkboxes for selecting items to be displayed in the model tree. For example, select **Core Features**, remove the check mark beside **Datum Plane**, then select ***Apply***. This might be useful if the part contains many datum planes which are cluttering up the view of the model tree feature structure. Turn the datum plane display back on. Under **General**, make sure the check boxes beside **Annotations & Tables** and **Suppressed** are on and exit the window with ***OK***. The model tree should now look like Figure 3.

Left click on any of the feature names shown in the left column of the model tree to see it highlighted in the model. (If the feature doesn't highlight, make sure that **Highlight Geometry** is checked using the ***Model Tree Settings*** button at the top right of the model tree.) This is an easy way to explore the structure of the database and the features in the model. Even better, under ***Model Tree Settings*** check the option for **Preselection Highlighting**. This highlights the features without having to select them first. But the model tree can do much more!

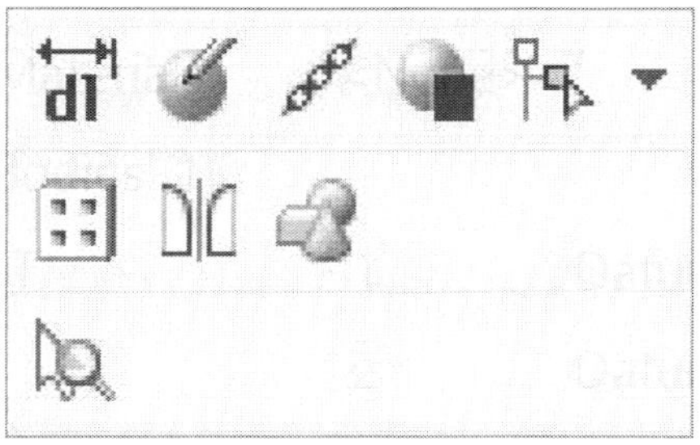

Figure 4 Feature pop-up menu

Select the BLOCK feature in the model tree. This pops-up the menu shown in Figure 4 (careful: it disappears if you move the mouse too far away!). In the top row are the commands for modifying features.

 ***Edit Dimensions***

 ***Edit Definition*** (aka ***Redefine***)

 ***Edit References*** (aka ***Reroute***)

 ***Suppress***

We will focus on three of these in this lesson since we have seen the ***Edit Dimensions*** command before. We will spend most of our time with ***Edit Definition*** and ***Edit References***, and also introduce ***Suppress*** (and its counterpart ***Resume***).

Parent/Child Relations

Using the commands given above, you can find out the regeneration sequence and internal ID numbers of parent and child features. There are several commands for exploring the parent/child relations in the model in considerably more detail. Select feature #5 in the model tree (ROUND_END) or preselect in the graphics window. Hold down the RMB and select:

Information ➤ Reference Viewer

The **Reference Viewer** window opens. In the filter pane on the left, turn off the check box beside **System**, then close the filter pane and expand the graphic area to see the feature names. Click the down arrow beside BLOCK to display the four surface references, as shown in Figure 5.

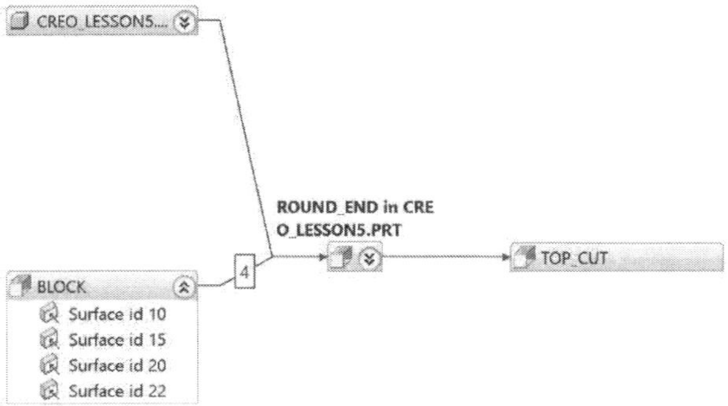

Figure 5 Graphic display in **Reference Viewer**

Mouse over on each of the four surfaces listed under BLOCK. As each is selected, the reference surface will highlight on the model. If you open the RMB pop-up for each surface or select the line joining BLOCK to ROUND_END (with the 4 in a box) and select ***Info ➤ Reference Info***, an information window will open describing the nature of the references (sketching plane or dimension reference), among other things.

On the children side, we see that the feature TOP_CUT is a child of the rounded end protrusion. What is the nature of this reference? Position the mouse cursor on the line joining the features - a pop-up information box will appear giving some information about the relation (Surface id 34 of ROUND_END is the horizontal sketcher reference). Explore the line joining the feature a bit more. Select the line (turns from aqua blue to red). Hold down the RMB and in the pop-up menu select ***Display Full Path***. This brings up the **Full Path Display** window shown in Figure 6. This shows something like the model tree structure, and the precise relation of the feature to its parent. ***Close*** this window.

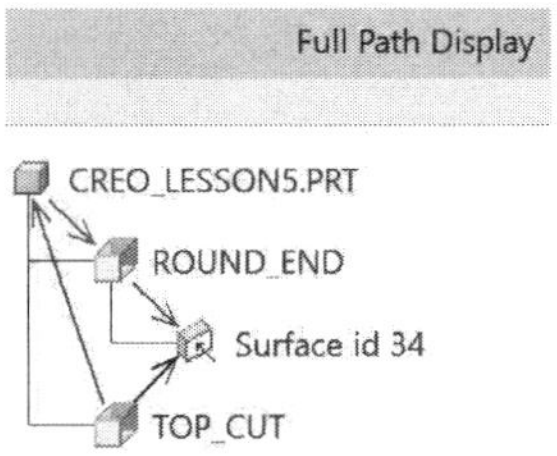

Figure 6 Graphical view of parent/child relations

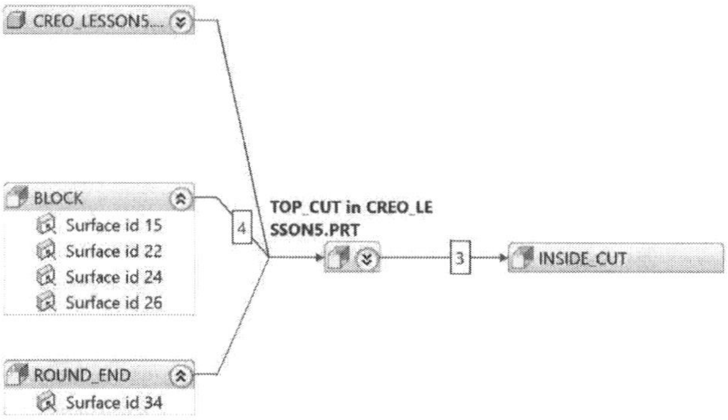

Figure 7 The Reference Viewer window for the feature TOP-CUT

Select TOP_CUT again in the **Reference Viewer** window and in the RMB pop-up select ***Set as Current***. The window now shows TOP_CUT as the current object, and lists its parents and children. Expand these lists (Figure 7). Notice the surface listed under ROUND_END in the parents area (Surface id 34). Select this surface and it is highlighted on the model. The reference information message (hold the cursor over the connecting line) again tells us that this was used as the horizontal sketcher reference for the cut feature #6 (TOP_CUT). This will be important to us later. The other four parent surfaces of TOP_CUT come from the BLOCK feature as follows:

1. (Surface id 15) the front of the block - sketching plane
2. (Surface id 22) right end of block - dimensioning reference
3. (Surface id 24) top of block - dimensioning reference used for aligning/dimensioning the cut
4. (Surface id 26) left surface of block - dimensioning reference

If you repeat this process for the inside cut (use ***Set Current*** and expand the parent features), you should see the following references:

1. the front of the block - sketching plane
2. the right horizontal surface of the top cut - horizontal reference plane
3. left vertical surface of the top cut - alignment/dimension reference
4. right vertical surface of the top cut - alignment/dimension reference
5. the Top datum plane - dimension reference

Now that we have explored the model a bit, you should have a good idea of how it was set up. Before we go on to ways that we can modify the model, let's have a look at a useful utility for dealing with features. Select ***Close*** in the **Reference Viewer** window.

Suppressing and Resuming Features

When you are working with a very complex model, it will often happen that many of the model features are irrelevant to what you are currently doing. Or, you want to avoid accidentally picking on some features as a reference for a new one. There is a command available that will temporarily remove one or more features from the regeneration sequence (and hence the model display). This is called *suppressing* the feature(s). It is important to note that this does not mean deleting the feature(s), it just means that they are skipped over when Creo Parametric regenerates the model. This will speed up the regeneration process thus saving you time.

When a feature is suppressed, it generally means that all its children will be suppressed as well. To bring the feature back, you can *resume* it. Let's see how suppress and resume work.

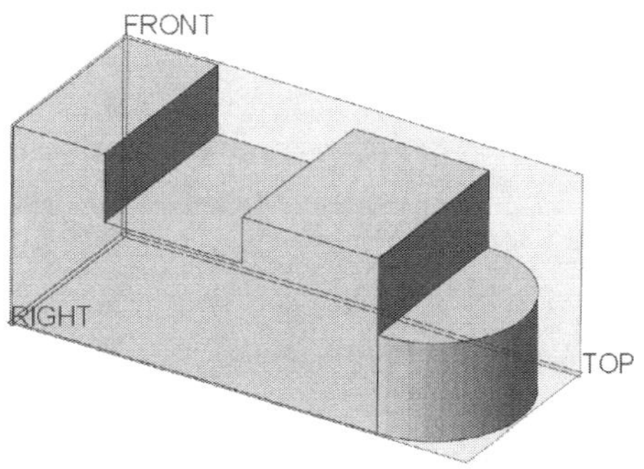

Figure 8 Part regenerated with cut suppressed

Preselect the feature INSIDE_CUT (or select it in the model tree). In the pop-up menu select

Suppress

Confirm the operation with ***OK***. The part will regenerate without the cut as shown in Figure 8. In the model tree (make sure that ***Tree Filters ➤ General(Suppressed)*** is checked), notice the small black square beside the name of the suppressed feature. This is called a "glyph" - there are a dozen or so of these symbols used to indicate a special status for a feature. See the online help for more information.

You will note that the suppressed feature no longer has a feature number (but it still has an ID). To get the feature back into the geometry, issue the commands (in the **Model** ribbon, **Operations** group overflow)

Resume ➤ Resume Last Set

or, even quicker, select ***Resume*** in the pop-up menu for the suppressed feature.

Now, try to suppress the TOP_CUT. Select it in the model tree and in the pop-up select ***Suppress***. A warning window appears. Move it out of the way to see the model. The TOP_CUT is highlighted in green, the INSIDE_CUT is highlighted in blue - it is a child of the TOP_CUT. You will have to decide what to do with it - the default is to suppress all the children. Select ***Edit Details***. This opens the **Children Handling** dialog window. This window allows you to identify the children and set options for how each child should be handled. The default action is to suppress all children with their parents. For now, select this with ***OK*** to suppress both cuts together. You should see the part as shown in Figure 9.

Check the display in the model tree. Both features have the small black square indicating their suppressed status. Try to resume the INSIDE_CUT by itself. Select it in the model tree then

Resume

using the pop-up menu. Both the selected feature and its parent (the other cut) are resumed - you can't resume a child without also resuming its parent(s).

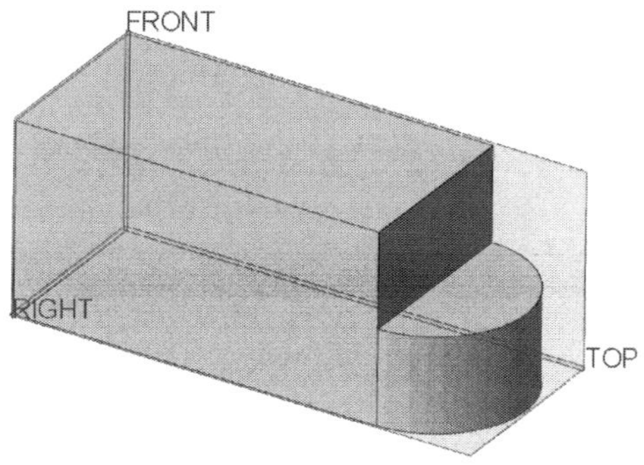

Figure 9 Part with both cuts suppressed

Using suppress and resume can make your life easier by eliminating unnecessary detail in a model when you don't need it. For example, if your part is a large valve, you don't need all the bolt holes in the flange if you are working on some other unrelated features of the valve. If you are setting up a model for Finite Element Modeling (FEM) for stress analysis, for example, you would usually suppress all fine detail in the model (chamfers, rounds, etc.) in order to simplify it. Suppressing features also prevents you from inadvertently creating references to features that you don't want (like two axes that may coincide, but may be separated later). Finally, suppressing unneeded features will also speed up the regeneration of the part.

Features that are suppressed are still included in the part data base, and will be saved with the part (with their suppressed status) when you save your model to a disk file.

In the **Model** ribbon **Operations** group, check out the other optional forms of the ***Suppress*** command:

> ***Suppress to End of Model*** - does just what you would expect
> ***Suppress Unrelated Items*** - suppress all but the selected feature (and its parents)

Helpful Hint

A trick used by advanced users who are dealing with very complicated parts is to suppress a large number of features before storing a part file. This reduces the file size, sometimes significantly. This can be useful when sending the part by email. When the file is opened by the new user, the features can be resumed. Be aware that this may be contrary to company policy (see next hint!).

When we get to drawings and assemblies in the last lessons, remember that suppressed features are carried over into these objects as well. That is, a suppressed feature will stay suppressed when you add its part to an assembly, or display the part in a drawing. Suppressed features in a part may even prevent the assembly from regenerating since some important references may be missing (although there are tools available in assembly mode to deal with this all-too-common occurrence).

Helpful Hint
When you inherit a part made by someone else, always check for suppressed features when you first open it. You may be (unpleasantly) surprised at what you find! Remember that the model tree default is to NOT display suppressed features.

Suppressing versus Hiding

Check out the pop-up menu for any of the datum planes. In addition to ***Suppress*** (in the top row), you will see another command - ***Hide*** (in the bottom row). Select that now. The datum plane disappears from the graphics window. All the datums in this part have children, so clearly we have not suppressed the datum, only removed it from the display. Observe that the datum entry in the model tree is now gray. This indicates its hidden status. To turn on its display again, select ***Show*** using the same pop-up menu. Two variations of the ***Show*** command are ***Show Only*** and ***Show All Except***. See the pop-up menu button tool tips for descriptions of these commands.

In a part, ***Hide*** and ***Show*** work for non-solid objects (datum planes, curves, points) and complete bodies. You cannot ***Hide*** a solid feature by itself, but you can hide a body. However, in an assembly you can ***Hide*** a component and still use it as a reference for other components. If a component in an assembly is suppressed, it cannot be used to provide references.

Modifying Feature Definitions

In previous lessons, we have used the ***Edit Dimensions*** command to change dimension values. We need some tools to let us modify the basic structure of the model. So, now we will look at ways to modify the parent/child relations in the part, and to modify the geometric shape of some features.

Suppose we want to take the original *creo_lesson5.prt* and modify it to form the part shown in Figure 10. This involves the following changes (some of these are not visible in the figure):

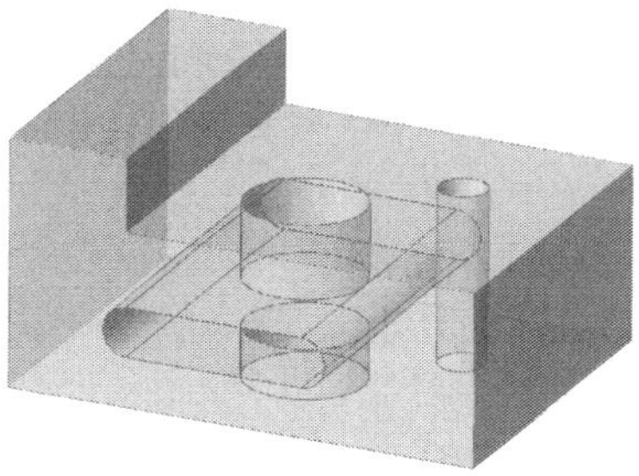

Figure 10 Final modified part

1. delete the rounded end
2. change the shape of the inner cut
3. change the dimensioning scheme of the inner cut
4. change the references of the inner cut
5. change the shape of the cut on the top surface
6. change the feature references of the top cut
7. increase the depth of the part
8. change the depth attribute of the top cut
9. add a couple of vertical holes

Some of these changes will require modifications to the parent/child relations that were used when the part was created. This will also result in a cleaner model.

If you haven't gone through Section 1 of this lesson on obtaining model information, now is a good time to do so, since a good understanding of the existing parent/child relations is essential for what follows. To see what we are up against, try to delete the rounded end of the part (the first thing on our "to do" list) by preselecting it, holding down the RMB and selecting

Delete

You will be notified that the feature has children (shown in blue) and asked what you want to do with them. In the warning window, select ***Edit Details***. This opens the **Children Handling** window we saw before. The default action is to delete the children. In this window, select the child TOP_CUT. ***Repaint*** to remove color highlights. To find out how it is related to the parent, hold down the right mouse button and select

Show References

You can now step through the references used to create TOP_CUT. The first reference is the sketching plane reference. See the message line below the graphics window. As you step through these with ***Next*** in the **SHOW REF** menu at the right, they will highlight in blue on the model. The next is the horizontal sketching reference for TOP_CUT. The surface used was the upper surface of the rounded end - this is the parent/child connection that has interfered with our plan to delete the rounded end. We could change that reference now, but we'll deal with that possibility later. We could also delete the child along with the parent. We would then have to decide what to do with the children of the children (that is, the inside cut) and so on! Keep selecting ***Next*** in the **SHOW REF** menu to step through the rest of the references. When you have gone through them all, select

Done/Return

in the **SHOW REF** menu. You are back to the **Children Handling** window. In the RMB pop-up menu for TOP_CUT, there are a couple of commands (***Edit References***, and ***Redefine***) which we will discuss shortly, but launching them in a different way.

In the **Children Handling** window, select the INSIDE_CUT and, in the RMB pop-up menu, select

Show References

If you step through these, you will see the sketching surface, the sketching reference surfaces, and a couple of dimensioning references.

In the **Children Handling** window, the Status column options for the two children are ***Delete*** or ***Suspend***. The former will remove them from the model immediately (along with the parent). ***Suspend*** will keep them in the model, but the next time the model is regenerated, these features will fail regeneration and require special processing (like recreating or reassigning the necessary references that have been lost) to keep them in the model.

Select ***Cancel*** a couple of times to back out of the deletion command that we launched previously. Clearly, deleting the rounded end is not going to be as easy as it first looked. It is very bad modeling practice to fix this problem by creating a new cut feature to remove the round end. That type of band-aid solution will come back to bite you later. We'll deal with our desired changes one at a time, and not necessarily in the order given above. For example, before we can delete the rounded end, we have to do something about its child references. Some careful thought and planning is necessary here. When you get proficient with Creo Parametric, you will be able to manage these changes more efficiently. Our main tools to use here are ***Edit Definition, Edit References***, and ***Reorder***. The first two commands were previously called ***Redefine*** and ***Reroute***, respectively, and those names still appear occasionally in Creo Parametric.

① Changing the shape of a sketch (*Edit Definition*)

The first thing we'll do is change the shape of the inner cut from its current rectangular shape to one with rounded ends. This requires a change in the sketch geometry of the feature. We'll take the opportunity to change the dimensioning scheme as well.

The ***Edit Definition*** (aka ***Redefine***) command allows you to change almost everything about a feature except its major type (you can't change an extrude into a revolve). Preselect the INSIDE_CUT (on the screen or in the model tree) and in the pop-up menu select ***Edit Definition*** (or press ***Ctrl+E***).

The **Extrude** feature dashboard will open, exactly as it would appear as the feature was being created. The feature is shown in preview orange. The ***Remove Material*** button is selected and the **Depth Spec** is set to ***Through All***. In the dashboard, select the **Placement** tab, then ***Edit*** (or use the RMB in the graphics window and select ***Edit Internal Sketch***). You will automatically be taken into ***Sketcher*** where we can proceed to modify the sketched shape of the cut. The desired final shape is shown in Figure 11.

First, delete the vertical sketched lines at each end: highlight both lines, open the RMB menu and select

> ***Delete***

Now add two circular arcs: use the RMB pop-up menu to select

> ***3 Point Arc***

and sketch the arcs at each end. Apply explicit tangent constraints as necessary. Now change (if necessary) the dimensioning scheme to the one shown in Figure 11. This may invoke the **Resolve Sketch** dialog to deal with any undesired existing constraints or strong dimensions once you add the new ones.

Note that the ends of the straight part of the slot are still aligned with the vertical faces of the top cut. We will deal with those later. Accept the sketch. In the dashboard, select ***Verify*** if desired, then ***OK***. If all went well, you should get the message

"Feature redefined successfully."

in the message area.

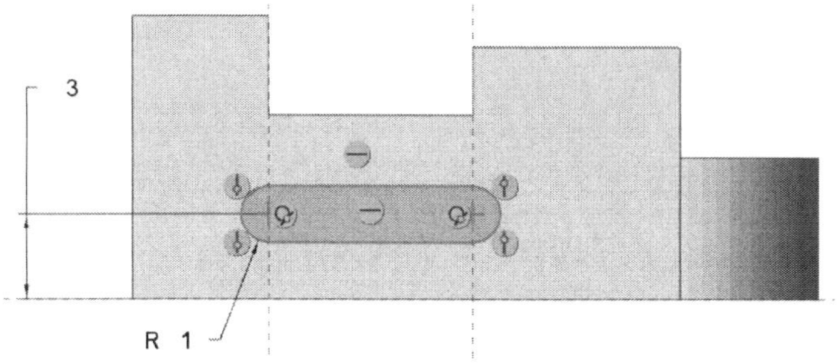

Figure 11 New sketch for the inner cut

② Changing a Feature Reference (*Edit References*)

Recall that the horizontal sketching reference for the inner cut was on the right side of the top cut, and we are planning on changing the shape of the top cut to remove that surface. We will have to change the reference for the inner cut to something else. This is done using the ***Edit References*** (aka ***Reroute***) command.

Preselect the inner cut, and use the pop-up menu to select ***Edit References*** which opens the window shown in Figure 12. This shows the body and references used in creating this feature. Select each in turn - they will highlight on the model. If you can't remember the nature of each reference, in the RMB pop-up select ***Reference Viewer***. In the viewer the chosen reference (Current object) is connected to the INSIDE_CUT with a line. Hold the mouse cursor over this line to see the nature of the reference. You should find the following:

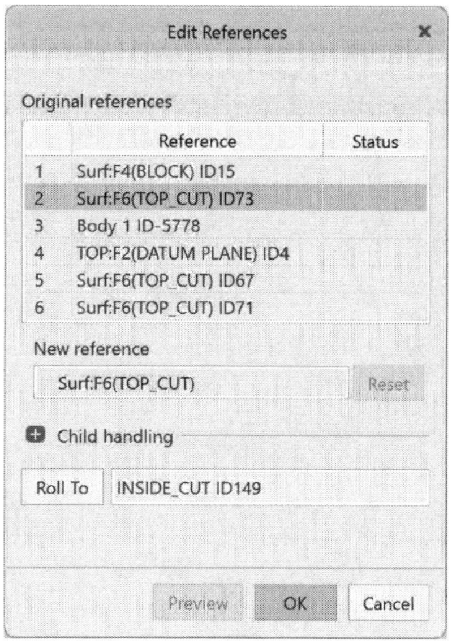

Figure 12 The Edit References window

1. Surface used as sketching plane
2. Surface used as horizontal sketch reference
3. Geometry reference
4. Surface used as dimensioning reference
5. Surface used as dimensioning reference

Select the second reference (which uses the surface of Feat #6, the top cut) - this is the reference we want to get rid of, since that surface will soon be removed when the shape of the top cut is changed. Click on the **New reference** collector, then pick a suitable replacement reference. In this case we could use the top surface of BLOCK but a better choice would be the TOP datum plane. Do that now. Notice the **Status** column in the **Edit References** window, and check for warning messages at the bottom of the screen as you move the mouse across the model. Some warnings may appear regarding surfaces being not colinear or parallel to the original reference. Obviously, we need to select a parallel reference. The non-colinear warning doesn't affect us now, but could if we were changing a dimensioning reference. That is all we need to do for now, so select ***OK***.

Select the inside cut and check with ***Information ➤ Reference Viewer*** to confirm that the horizontal surface of the top cut is no longer listed as a parent. There should still be a couple of references to the top cut, though. These are alignment constraints in the sketch of the inside cut. We'll still have to change at least one of these if we are going to modify the top cut as planned. Close the **Reference Viewer** window.

③ Changing the Sketcher Constraints (*Edit Definition*)

The ends of the straight part of the inner cut are still aligned with the vertical faces of the top cut. These alignments are still at work in the sketch. See Figure 11. To change these alignments, we need to redefine the sketch. A quick way to get directly into Sketcher is to open the INSIDE_CUT feature listing in the model tree, select the feature sketch **Section 1**, and in the pop-up select ***Edit Definition***. This has the added advantage of bypassing the dashboard both on your way in to the sketch and on your way out, saving a few mouse clicks.

Turn off the datum plane display. We want to do something with the sketch references so in the RMB pop-up menu select

References

Click on the left edge of the part. This should add an entry in the **References** window. Now, in the **References** window, select the other listed surface references (these will both be to feature #6, the top cut) and select the ***Delete*** button. The other two vertical references (shown in Figure 11) should disappear. Now select ***Update***. The sketch's **Reference Status** is still **Fully Placed**. Close the references window.

Change the dimensioning scheme to the one shown in Figure 13. Make sure the arcs are tangent to the horizontal lines. Intent Manager will do some of this for you automatically, but you may have to strengthen dimensions and/or add explicit constraints. Look carefully for any unwanted weak dimensions. Close out Sketcher to accept the redefined sketch. Notice we do not return to the extrude dashboard.

To make sure that there is now no relation between the top cut and the inside cut, preselect the TOP_CUT and, using the RMB pop-up, select

Information ➤ Reference Viewer

The inner cut should no longer be listed as a child!

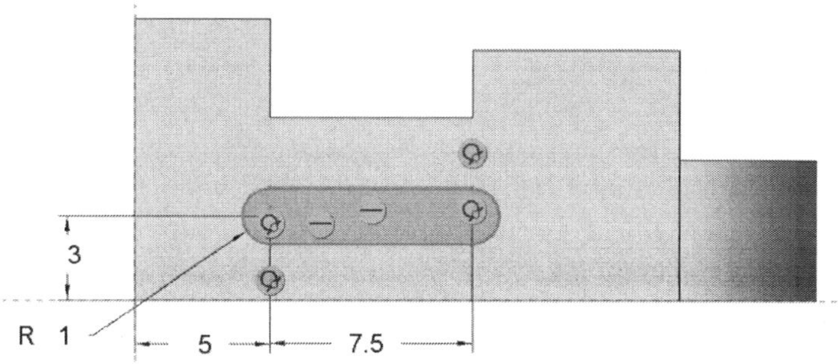

Figure 13 New dimensioning and constraint scheme for the inside cut sketch

④ Changing a Feature Reference (*Edit References*)

Recall that the rounded end is a parent of the top cut by supplying the horizontal sketching reference. We need to break this connection before we can delete the rounded end (which is on our "to do" list). This calls for another reroute operation. Pick on the top cut and select ***Edit References*** in the pop-up. At the bottom of the window click the ***Roll To*** button. "Rolling back" means temporarily returning to the part status when the top cut was created. In a complicated part, this would suppress all features created after the one we are interested in. (Note that this is automatic with ***Edit Definition***.) This is a good idea, since then it will not be possible to (accidentally) select a new reference that is "younger" than the cut (i.e. created after it). Unless you can think of a really good reason not to, you should **ALWAYS ROLL BACK THE PART!**

For the second reference in the list (surface of the rounded end, feature #5), select a new horizontal reference like the top of the block or the TOP datum (pick this in the model tree if datum display is turned off). This is all we have to reroute, so select ***OK***. Select the rounded end then in the RMB pop-up

Information* ➤ *Reference Viewer

observe that it now has no children. You can now go ahead and pick the ROUND_END feature in the model tree and using the RMB pop-up select

Delete

to remove the feature from the model.

⑤ Changing Feature Attributes (*Edit Definition*)

We want to change the shape of the top cut to get rid of the step. We will also change its depth attribute. To see why this is necessary, select the BLOCK feature in the model tree, then ***Edit Dimensions***. (Or just double click on the feature in the graphics window.) Change the depth from 10 to ***15***. If ***Auto-Regenerate*** is not turned on, you will have to ***Regenerate*** the part yourself. As you recall, the top cut had a depth of 10, so it doesn't go all the way through the new block as shown in Figure 14.

Let's change both the shape and depth of the top cut at the same time. Select it in the model tree and in the pop-up and select

Edit Definition

In the dashboard, change the **Depth Spec** option (or in the **Options** panel) from *Variable* to ***Through All***. This fixes one problem. Now, to change the shape of the feature, use the RMB menu command ***Edit Internal Sketch***.

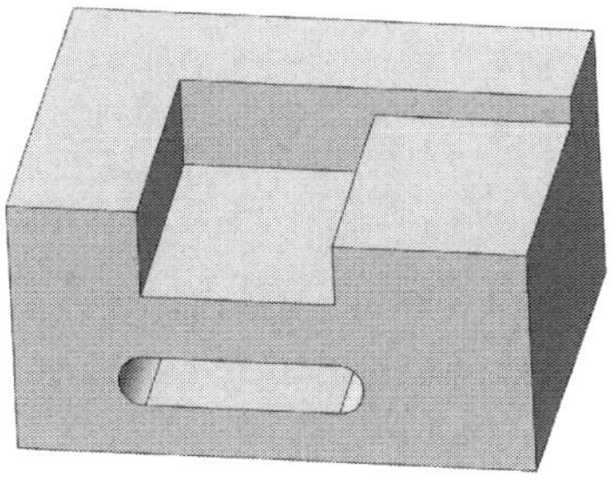

Figure 14 Block width increased to 15

Using the Sketcher tools, change the shape of the cut to remove the step, as shown in Figure 15. With the Intent Manager, you should be able to do this very quickly. Here are a couple of Sketcher tools to make this easier.

In the **Editing** group of the **Sketch** ribbon, select the Delete Segment tool (also known as ***Dynamic Trim***). As the icon implies, hold down the left mouse button and drag the mouse cursor through the two lines you want to get rid of. See Figure 15. Middle click to get back to Select mode.

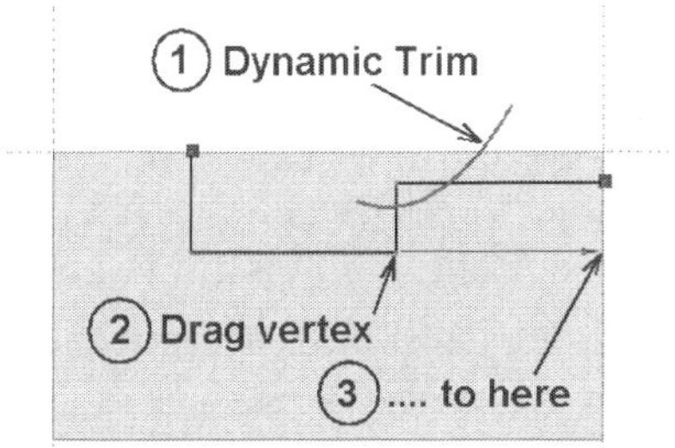

Figure 15 A couple of Sketcher tools

Now you can drag the right end of the horizontal line over to the right vertical reference of the sketch. To make sure the vertex sticks to this reference pick the ***Coincident*** constraint. Click on the vertex, then (with CTRL) on the dashed reference line. You might as well close the sketch while you are here. You should now have the sketch shown in Figure 16. Adjust your dimensions to match the figure.

When the sketch is complete, return to the feature dashboard. ***Verify*** the part, and if it looks all right, select ***OK***. The modified part is shown in Figure 17.

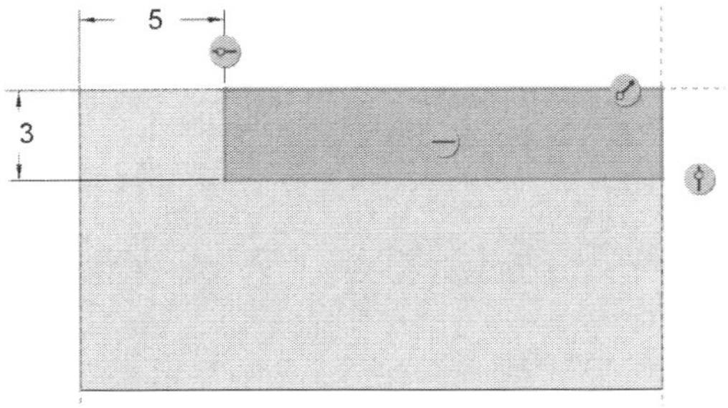

Figure 16 New sketch for the cut

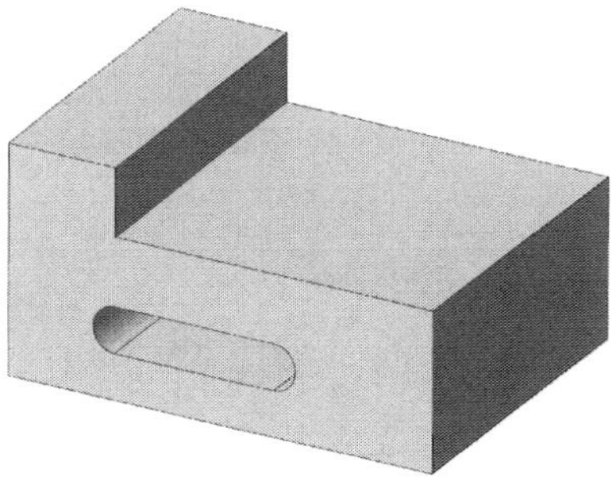

Figure 17 Part with redefined cut

⑥ Changing the Regeneration Sequence using *Reorder*

It is sometimes convenient or necessary to change the order of the features in the regeneration sequence. For example, an advanced technique involves grouping features in the regeneration sequence so that they are adjacent. Then the group can be patterned or copied (see Lesson #7). The major restrictions on reordering features are:

- ♦ a child feature can never be placed before its parent(s)
- ♦ a parent feature can never be placed after any of its children

The reasons for these restrictions should be pretty self-evident. Fortunately, Creo Parametric is able to keep track of the parent/child relations and can tell you what the legal reordering positions are. For this part we will just reorder the two cuts so that the inner cut comes before the top cut. Note that it is possible to reorder many features at once. To see how it works, start in the **Operations** group overflow to select

Reorder

and click on the inside cut. This cut (feature #7) was originally a child of the top cut (#6), but that relation was modified above. Thus, we should be able to create the cuts in any order, after the block (#4). In this simple part, there is only one legal possibility, that is, reorder the selected cut immediately after the block. In the **Feature Reorder** window (see Figure 18), select the ***After*** option, click in the **Target feature** collector to activate it, click on the BLOCK feature, and finally ***OK***.

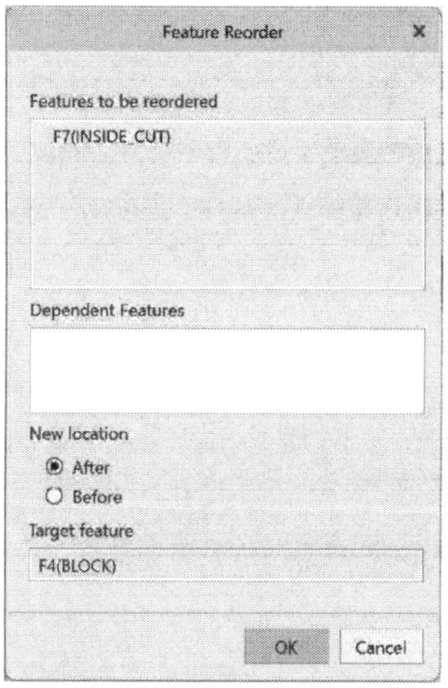

Figure 18 The Feature Reorder window

Note that the feature numbers of the cut and slot (shown in the model tree) have now changed (but the internal IDs are still the same).

Creo Parametric has made the ***Reorder*** command quite a bit easier by allowing you to drag and drop features in the model tree. Try that now by reordering the top cut back to where it was: click on the feature in the model tree and slowly drag the cursor upwards. The mouse icon will normally change slightly as you move back up the regeneration sequence to show you where legal reordered positions are. In this part, of course, there is only one valid position. You might try out this mode of reordering sometime when you get a more complicated part. You can reorder features both upwards and downward in the model tree.

⑦ Changing the Insertion Point

New features are typically added at the end of the regeneration sequence (notice the green line in the model tree). Sometimes it is necessary to create a new feature whose order you want to be earlier in the sequence. You could do this by creating it and then using the reorder command, being careful that you don't set up parent references to features after the targeted reorder position. Also, you would have to be careful not to create any new features that could interfere with existing features (like cutting off a reference surface). There is an easier way[3]!

You may have noticed in the RMB pop-up menu in the model a command ***Insert Here***. If you select a feature then select this command, the insertion point will move to be immediately after the selected feature. Try that with the BLOCK feature. What happens to features below the insertion point? To move the arrow back down after the final cut, you can go into the **Operations** group and use

Resume ➤ Resume All

You have probably guessed that you can also activate insert mode by dragging the ***Insertion Point*** line back up in the model tree. Click on the green line and drag it to just after the block feature. Create two *Through All* circular holes in the part as shown in the figure (the diameters are 2 and 5; placement is approximately as shown).

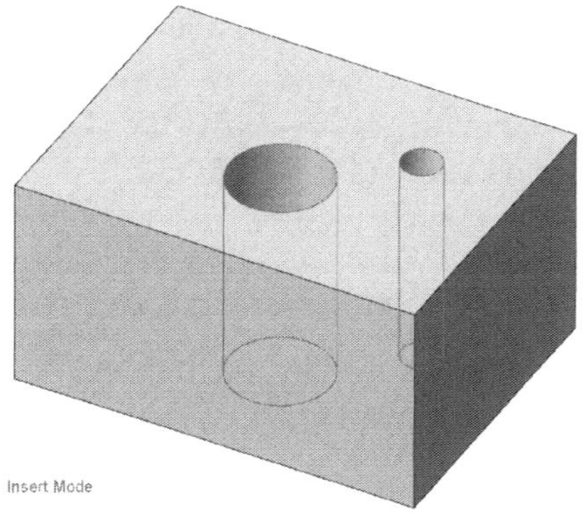

Figure 19 Two holes inserted after block feature

To return the insertion point line to the bottom of the model tree, just drag it down.

Check the model tree to see that the two holes have been added to the model after the block and before the cuts.

You can move the insertion point around pretty much anywhere in the model tree. There is one place it won't go - can you find it? All the features after the insertion point are automatically suppressed. If you resume any of these features, the insertion point will advance to the last selected feature; only the references for selected features are resumed.

Helpful Hint

A handy way to step through a model (much like using the model player) is to move the insertion point to where you want to start stepping through the model, then just drag the insertion point down through the model tree as many features as you want. If you use the pop-up ***Resume*** command on the feature below the insertion point it will also move down. You can select several contiguous features and resume them all at once - the insertion point will automatically go to the end of the list of resumed features. As you move through the model's regeneration sequence, you can query the model, add features, change references, and so on - much more than you can do with the model player.

---

[3] This used to be called entering "Insert Mode" but that terminology has been dropped.

Conclusion

The modeling utilities described in this lesson are indispensable when dealing with complex parts. You will invariably come across situations where you need to redefine, reroute, or reorder features. The information utilities are useful for digging out the existing parent/child relations and discovering how features are referenced by other features. The more practice you get with these tools (try the ECO exercises at the end of this lesson), the better you will be able to manage your models. As a side benefit, having a better understanding of how Creo Parametric organizes features will cause you to do more careful planning prior to creating the model, with fewer corrections to be made later. This will save you a lot of time!

In the next lesson, we will investigate the use of datum planes and axes, including creating temporary datums called "make datums". We'll also discover yet more tools and commands in Sketcher.

Questions for Review

1. How can you find out the order in which features were created? What is this called?
2. How can you find which are the parent features of a given feature?
3. How can you find the references used to create a feature?
4. How can you find any or all other features that use a given feature as a reference?
5. What is the difference between the Feature # and the internal ID?
6. What is the command to exclude a feature temporarily from the model?
7. What happens to the parents of a suppressed feature? To the children?
8. Is it possible, via a convoluted chain of parent/child relations, for a feature to reference itself? Can you ***Reroute*** (i.e. ***Edit References***) a feature's reference to another feature that occurs below it on the model tree?
9. What happens to suppressed features when the model is saved and you leave Creo Parametric?
10. If you are given a part file that you have never seen before, how can you determine if it contains any suppressed features? What about hidden features?
11. In Sketcher, how many variations of the right mouse pop-up menu can you find? In what modes are these active?
12. How can you restore previously suppressed features?
13. How can you change the sketch references when you are in Sketcher?
14. How many features can you suppress at once?
15. Can you resume a parent without resuming its children?
16. Is there any aspect of a feature that cannot be modified using ***Edit Definition***?
17. What is the difference between ***Edit Dimensions***, ***Edit References***, ***Edit Definition***? Which is the most general of these commands? (Which gives you the most options?)
18. What is meant by "rolling back the part"?
19. How can you remove unwanted alignments in a sketch?
20. How can you change the sketch orientation reference?
21. Which features can be hidden?
22. Can you hide a feature with children?
23. What symbol in the model tree indicates suppressed features?
24. What are the two fundamental rules of reordering?
25. Are there any restrictions on the location of the insertion point?
26. What happens if the insertion point is not at the end of the regeneration sequence when you save a part and then later retrieve it?
27. How do you easily move the insertion point from wherever it is to the bottom of the model tree?
28. What are the following buttons used for in Sketcher?

a)  b)  c)  d) 

Exercises

Here are some simple parts to model using the features we have covered up to here. Before you start creating these, think about where you will place them relative to the datum planes, what type and order you should select for the features, and how you should set up parent/child references and dimensioning schemes.

1.

2.

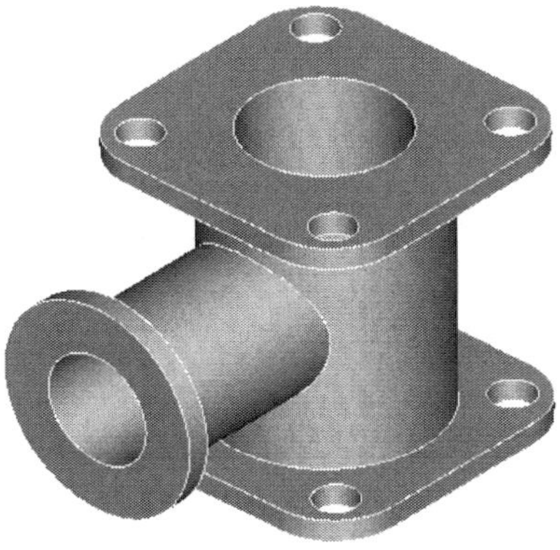

3.

4.

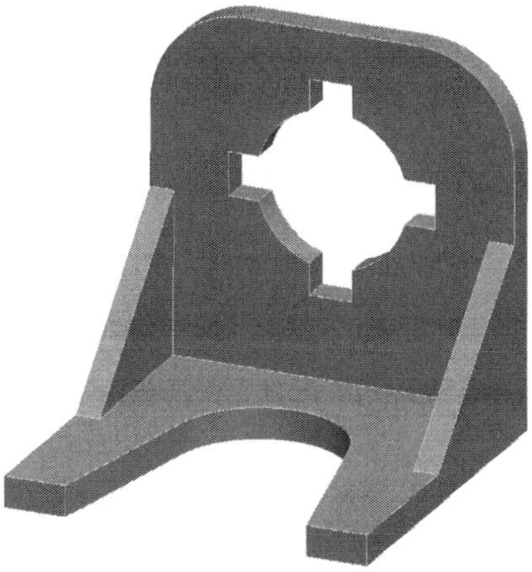

Exercises #5 and #6

"Engineering Change Orders" (or ECOs) are a fact of life. These can occur at any time during a product design, manufacture, procurement, assembly, or useful life. ECOs that occur late in the product development cycle cost a lot of time and money (or lead to costly "recalls"). The ability to modify parts created in Creo Parametric is critical, both to deal with ECOs and to modify existing parts for new design purposes.

In the following two exercises you are provided with a part and a list of ECOs for each part. The two parts can be downloaded from the web site (*creo_less5_eco1.prt* and *creo_less5_eco2.prt*). Make the listed changes to each part. You will probably not be able to do the changes in the order given so study them all before you get started, then plan a strategy. It may be possible to do more than one change at a time. You should attempt to keep all references as simple as possible (use explicit alignments instead of 0.0 dimensions; use Sketcher constraints rather than relations).

5. ECO Exercise Part #1

Part #1 : Original model

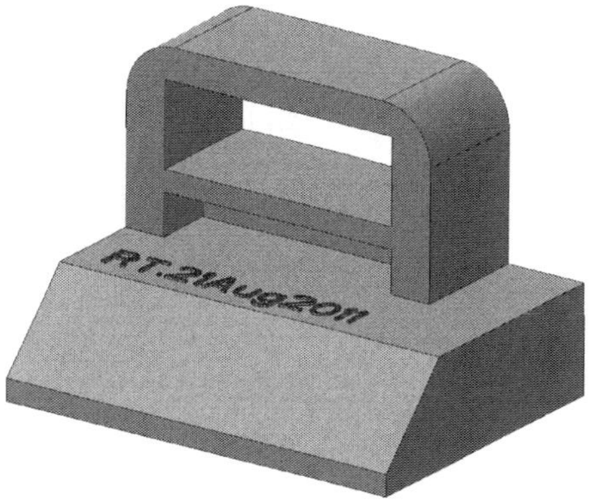

Part #1 : Final model after ECOs

- ❑ Increase the thickness of the base to 50mm.
- ❑ Make the top rectangular slot always go all the way through the upright, while the lower slot will always go exactly halfway through for any thickness of the upright.
- ❑ Delete the 45E chamfers on the front left and right corners.
- ❑ Add a 40mm chamfer along the top front horizontal edge.
- ❑ Increase the thickness of the vertical upright to 60mm and set it up so that its height above the top surface of the base is always 85mm.
- ❑ Make the bottom of the lower rectangular slot always flush with the top surface of the base plate.
- ❑ Make the two rectangular slots the same width, located and driven by the width of the upper slot.
- ❑ Change the parameter "MODELED_BY" to contain your initials and the current date (see ***Tools* ➤ *Parameters***).

When you are finished with the ECOs, test the robustness of your model by regenerating the part with the dimensions shown. No other modifications should be necessary except the dimension values:

height of base =	70
thickness of upright =	20
width of upper slot =	80

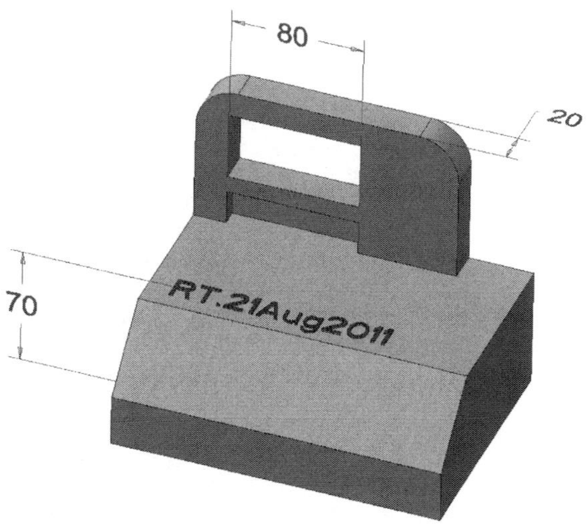

6. ECO Exercise Part #2

Part #2 : Original model

Part #2 : Final model after ECOs

- ❑ Increase the thickness of the base to 200mm.
- ❑ Set up the central rectangular cavity so that it is always 50mm from the front and back surface of the part.
- ❑ Delete the rounds on the top left and right corners.
- ❑ Add 60mm chamfers along the top left and right corners.
- ❑ Increase the depth (front to back) of the base from 250 to 300mm.
- ❑ Make the bottom surface of the rectangular cavity exactly halfway up the base block, for any size of base block.
- ❑ Change the outside radius of the cylindrical loop to 100mm, and ensure that the thickness of the cylindrical loop is always 10mm. The inner hole should be coaxial with the outer surface at all times.
- ❑ Change the parameter "MODELED_BY" to contain your initials and the current date (see ***Tools* ➤ *Parameters***).

When you are finished with the ECOs, test the robustness of your model by regenerating the part with the dimensions shown. No other modifications should be necessary except the dimension values:

height of base = 100
outer radius of hoop = 60
width of base = 340

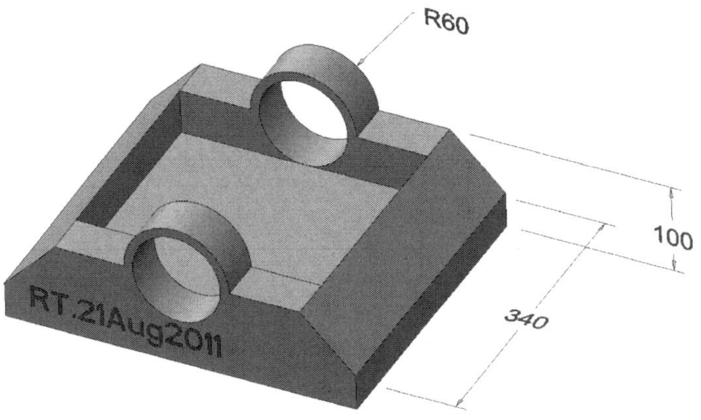

Project

Here are three small parts for the project. All dimensions in millimeters. For the acorn nut, you might like to investigate alternate Sketcher environments (see ***Sketch➤Options* | *Parameters*** when you are in Sketcher), including a polar grid, and the use of centerlines as construction aides (straight lines and/or circles). Note that the hexagon on the nut requires only one dimension (10 mm) to give its size.

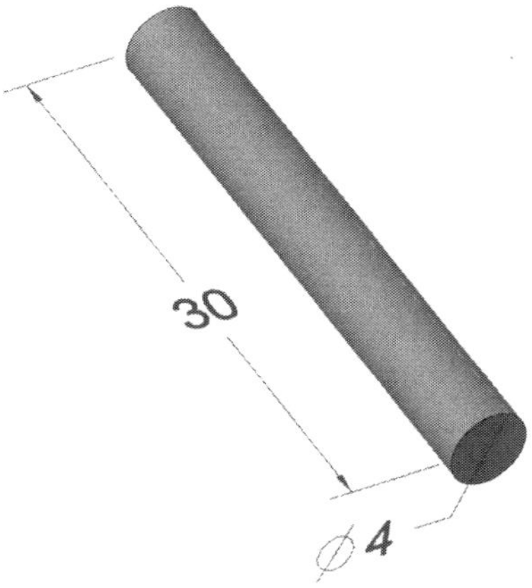

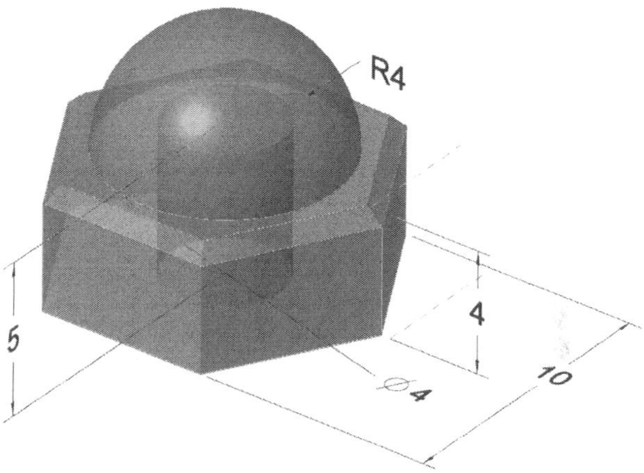

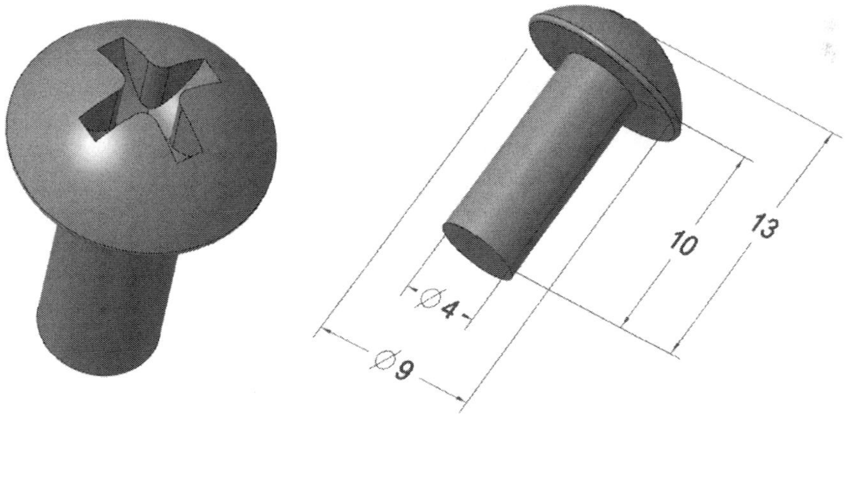

This page left blank.

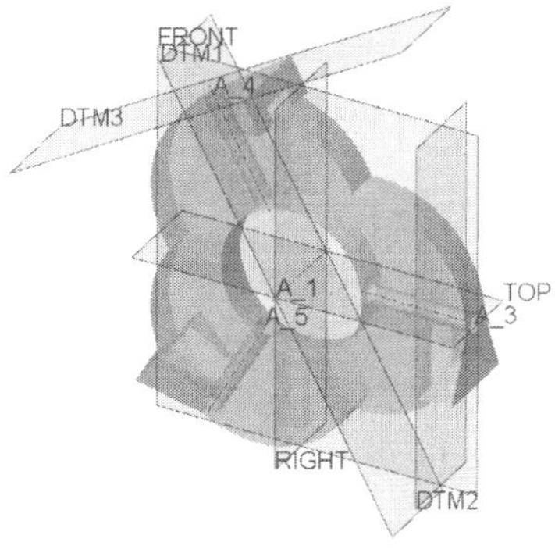

Lesson 6

Datum Planes and Sketcher Tools

Synopsis

The mysteries of datum planes and 'make datums' are revealed! What are they, how are they created? How are they used to implement design intent? More tools in Sketcher are introduced. Using *Sketch Regions*.

Overview of this Lesson

In this lesson we are going to look at some new techniques in Sketcher for creating sections, including relations and references to existing geometry. Our primary objective, though, is to look at the commands used to set up and use datum planes. Some of these datum planes, like the default ones (**RIGHT**, **TOP**, and **FRONT**), will become references for many features, or will appear similarly on the model tree. Others, called *make datums*, are typically used only for a single feature and are created "on-the-fly" when needed. These will appear on the model tree embedded in the feature for which they were created and automatically hidden in the graphics window. Along the way, we will discuss some model design issues and explore some options in feature creation that we have not seen before.

The part we are going to create is shown in the figure at the right. The part consists of a three-lobed disk with a central hole. Three identically-shaped triangular teeth are spaced at 120 degrees around the circumference. Each tooth includes a central radial hole that aligns with the central axis of the disk. Although there is no indication of it in the figure, each of these tooth/hole shapes will be created differently using different datum plane setup procedures. We will see what effect this has on the model at

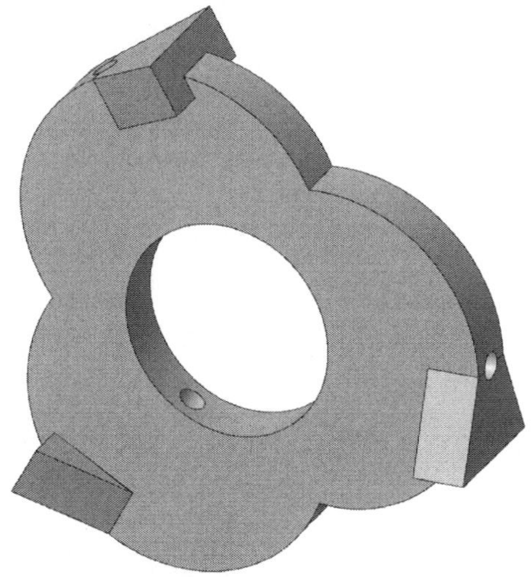

Figure 1 Final part - three tooth cutter

the end of the lesson[1].

Here is what is planned for this lesson:

1. Overview of datum planes and axes
2. Creating a datum plane and datum axis
3. Create the disk with hole, introducing some new Sketcher commands
4. First tooth - using an **Offset** constraint
5. Second tooth - using **Normal** and **Tangent** constraints
6. Third tooth - using make datums
7. Effects on the model
8. Things to consider about design intent

As usual there are some Questions for Review, Exercises, and a Project part at the end of the lesson.

Overview of Datum Planes and Axes

Datum features (planes, axes, curves, points and coordinate systems) are used to provide references for other features, like sketching planes, dimensioning references, view references, assembly references, and so on. Datum features are not physical (solid) parts of the model but are used to aid in model creation. Datum planes and axes extend off to infinity. By default, Creo Parametric will show visible edges of a datum plane or the datum axis line so that they encompass the part being displayed. It is possible to scale a datum plane or axis differently so that, for example, it will extend only over a single feature or surface of a complex part. This would be done to reduce screen clutter. When we use the word "datum" by itself, we usually mean a datum plane.

Let's consider how a datum plane can be constructed. In order to locate the position and orientation of the datum, you will choose from a number of constraint options. These work alone or in combination to fully constrain the plane in space. The major options for datum planes are:

Through
  the datum passes through an existing surface, axis, edge, vertex, or cylinder axis

Normal
  the datum is perpendicular to a surface, axis, or other datum

Parallel
  the datum is parallel to another surface or plane

Offset (linear)

[1]A better way to create this part would be to create a body consisting of a lobe, tooth, and radial hole. The body would then be duplicated around the axis using a ***Pattern***. The three bodies would then be ***Merged*** into the final shape.

the datum is parallel to another surface or plane and a specified distance away

Offset (rotation)

the datum is at a specified angle from another plane or surface

Tangent

the datum is tangent to a curved surface or edge

Some of these constraints are sufficient by themselves to define a new datum plane (for example, the **Offset(linear)** option). Other constraints must be used in combinations in order to fully constrain the new datum. Some constraints can be applied more than once for the same datum. An example is using **Through** with three existing vertices or points (must not be co-linear). When you are constructing a new datum, Creo will automatically pick an appropriate option (based on the entity selected), show you the results of the current constraints using a preview, and tell you when the constraints are sufficient.

Construction of a datum axis is similar, with the following constraint options:

Through

the axis is through a selected vertex, edge, or plane. May require additional constraints.

Normal

the axis is located using linear dimensions and is normal to a selected plane

Tangent

the axis is tangent to the selected reference at a specified point

Center

the axis is normal to and through the center of a selected planar circular edge or curve

We won't have time to explore all the variations of these options in this lesson. The general procedure is pretty similar for all options, however. With the preview capability it is quite easy to figure out what to do after you have seen the procedure a few times.

Let's see how this all works. Start Creo Parametric in the usual way, and clear the session of any other parts. Start a new part called ***cutter*** using ***inlbs_part_solid_abs***. The default datum planes are created for you as the first features in the part. Turn on the display of all datum tags. You can delete the datum coordinate system feature for this part since we won't need it, and it just clutters up our view (or just turn off its display with ***Hide***).

After the default datums are created, new datum planes and axes are created using buttons in the **{Datum}** group of the **Model** tab. When you get into Sketcher, there will be another location for these tools as well.

Creating a Datum Plane and Datum Axis

First, we will define a datum axis that will be the central axis of the cutter. This will be at the intersection of the existing datums **RIGHT** and **TOP**. Using the CTRL key, select both these datum planes. They should both be highlighted in green. Now, select the ***Axis*** button in the ribbon (also available in the pop-up in the graphics window). This creates the datum axis A_1, as shown in Figure 2. This is the appropriate axis for Creo Parametric to make if you pick two intersecting plane surfaces, so it skips over the **Datum Axis** dialog window. Let's have a look at that window.

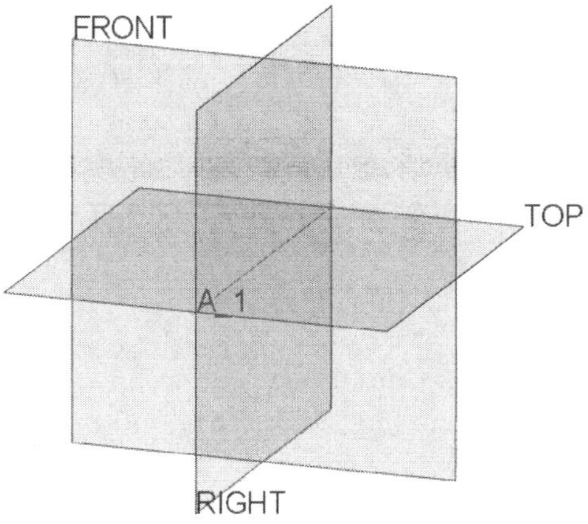

Figure 2 Datum axis A_1 created

With A_1 highlighted in green, use the RMB pop-up menu and select ***Edit Definition***. This opens the **Datum Axis** dialog shown in Figure 3. We see the two datums, RIGHT and TOP, each with the constraint ***Through***. This fully constrains the axis. Note that you can rename the axis feature using the **Properties** tab. Also, the **Display** tab lets you change the displayed axis length. Come back later to investigate these options. For now, close the window with ***OK***.

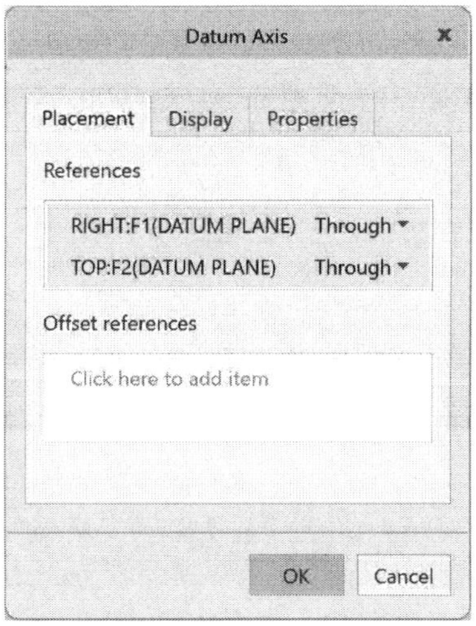

Figure 3 Datum Axis dialog window

In the procedure we just did, we selected the axis references first, then launched the datum creation command. This is an example of object/action execution. Alternatively, we could have launched the ***Axis*** tool first to open the dialog window, then picked on each of the existing datum planes and set the associated constraints shown in Figure 3 to ***Through***. This would be an action/object procedure. You can use whichever method you are most comfortable with - they have the same final effect on the model. Once you get more experience with the commands (that is, understand the defaults), you will probably find object/action to be more efficient.

Our next task is to create a new datum plane that passes through the axis A_1 we just made, and is at a specified angle to the TOP datum plane. We will use this as a reference in a couple of features later on in the part.

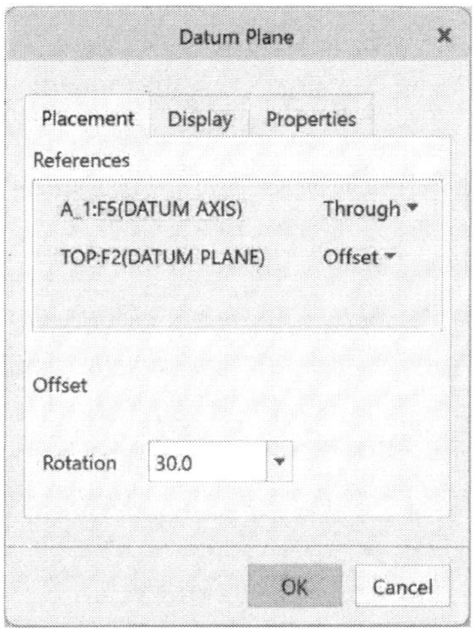

Figure 4 Datum Plane dialog window

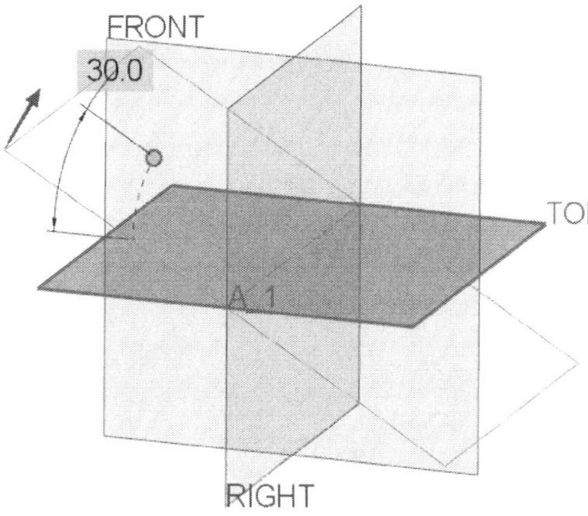

Figure 5 Datum plane with ***Through*** and ***Offset(rotation)*** constraints

We will use action/object for this. With nothing highlighted in the graphics window, in the ribbon select ***{Datum}: Plane***. The **Datum Plane** dialog window (see Figure 4) opens at the right. We need to specify the references for the new datum. If the axis was preselected it will already be listed. Otherwise, pick on the axis A_1. This is added to the reference collector in the dialog window with a ***Through*** constraint. Click on the listed constraint to see the alternatives. Leave it set to ***Through***. Now, holding down the CTRL key, select the TOP datum. This is now listed as well, with an ***Offset*** constraint. The ***Offset(rotation)*** constraint is the only one that makes sense with the existing ***Through*** constraint. The datum is previewed on the graphics window (Figure 5), with a drag handle to control the value of the offset angle. This value is also shown in the dialog window. Set the value to ***30***. If the plane jumps to the other side, enter ***–30***, because a negative value rotates it the other way. On one corner of the new plane is a magenta arrow. This defines the positive normal direction for the plane. To switch this direction, click directly on the arrow or use the ***Flip*** command in the **Display** tab of the dialog window. It is <u>not</u> necessary to ***Flip*** the arrow now.

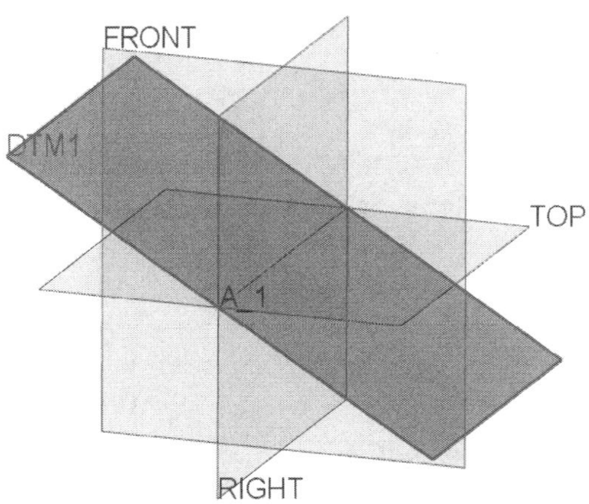

Figure 6 Datum plane DTM1 created at 30 degrees from TOP

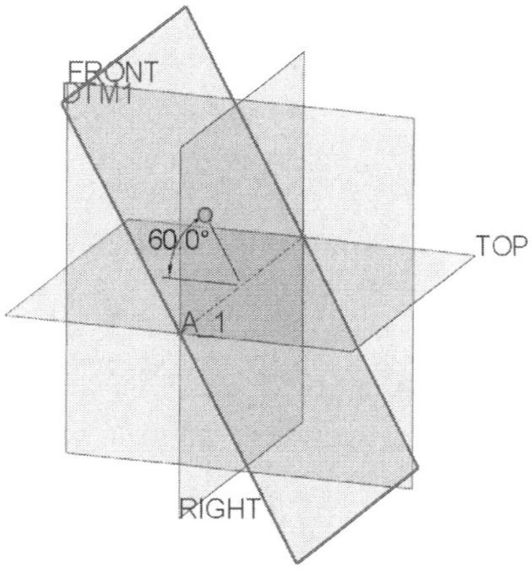

Figure 7 Datum plane DTM1 changed to 60 degrees from TOP

In the **Properties** tab of the dialog window you can change the name of the feature. Leave this as is and select ***OK***. The new datum plane DTM1 will be added to the model (Figure 6). If you open the model tree, you will see features A_1 and DTM1 there. If you double-click on the edge of DTM1, you will see the offset angle dimension. Change that to **60** (Figure 7), and ***Regenerate*** the part. This angle dimension is how we control the orientation of the datum plane.

These two examples have illustrated the general procedure for creating datum planes and axes. You might like to come back and experiment with these. There are some short cuts available that can save you some time (like preselecting features before launching the ***Datum*** command). To use the shortcuts, you need to have a very good grasp of how Creo Parametric will utilize defaults and the references you give it. You should also spend some time exploring the various constraint options for each chosen feature - these are available in a pull-down list beside each feature in the dialog window. Possibly because datums do not result in solid geometry, new users tend to find them a little tricky to deal with (especially make datums) and, as a result, often do not make very effective use of them. Remember that we are creating a model, not just a solid body, and datums are often crucial elements of the model structure.

We will leave datums for a bit now, so that we can create the base feature of the cutter. We will be creating several more datums as we go through this lesson.

Creating the *Cutter* Base Feature

Our base feature is a solid protrusion that will look like the figure shown at the right. We are going to go through the sketching procedure slowly here to illustrate a few new tools and techniques.

Our plan of attack is to sketch this shape on the FRONT datum plane. Since the part is symmetric front to back, we will make this a ***Symmetric*** protrusion. It also makes sense to center the feature where the datums TOP and RIGHT meet. The reason for DTM1 will be clear when we get into the sketch - it provides a reference for locating the geometry.

The sketch basically consists of three circular arcs. We'll call these the first, second, and third arcs, starting at the right and going counterclockwise.

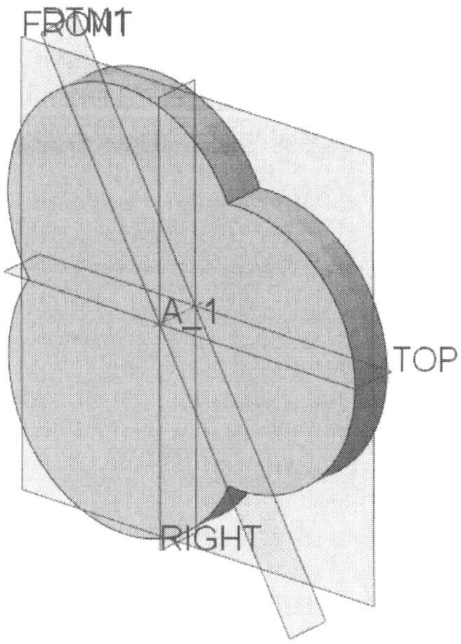

Figure 8 Base feature - a ***Both Sides*** protrusion

Select the ***Extrude*** tool. Its dashboard opens up.

Change the **Depth Spec** to ***Symmetric*** (also known as ***Both Sides***), and set its value to *2*.

Note that we can do this even before we have created the sketch. Now launch Sketcher using ***Placement*** > ***Define***. Click on the FRONT datum as our sketching plane. The RIGHT datum becomes our orientation reference facing the right side of the screen. Middle click to enter Sketcher.

In Sketcher, two references have been chosen for us. We want to add DTM1 (which is visible on edge) as a reference. In the ribbon, select

{Setup}: References

(or use the RMB pop-up and select ***References***) then pick DTM1 and then ***Close*** the **References** window.

The center of each of the three arcs is the same distance away from the axis A_1. We can implement this intent in Sketcher by creating a construction circle. Use the RMB pop-up menu to select ***Circle***, and draw the circle shown in Figure 9. Middle click and set the diameter to ***8***. Click on the circle so that it highlights in green and in the pop-up menu select ***Construction*** . This changes the line style to dotted (and removes the shading of the sketch). This curve can now be used as a sketching reference and will not contribute to solid geometry of the feature. You can toggle a construction line back to a physical edge using the RMB pop-up command ***Solid***.

Helpful Hint

If you are going to create a lot of construction geometry, put Sketcher into ***Construction Mode*** using the button in the ribbon. For only a few construction elements (or to change your mind), use regular Sketcher tools and the pop-up to toggle between Construction and Solid lines.

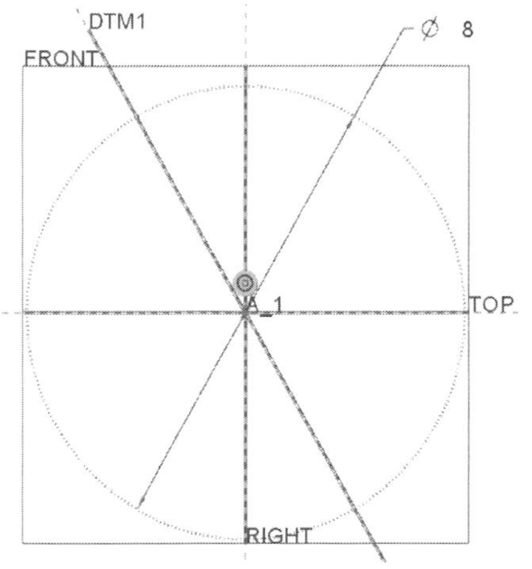

Figure 9 Construction circle

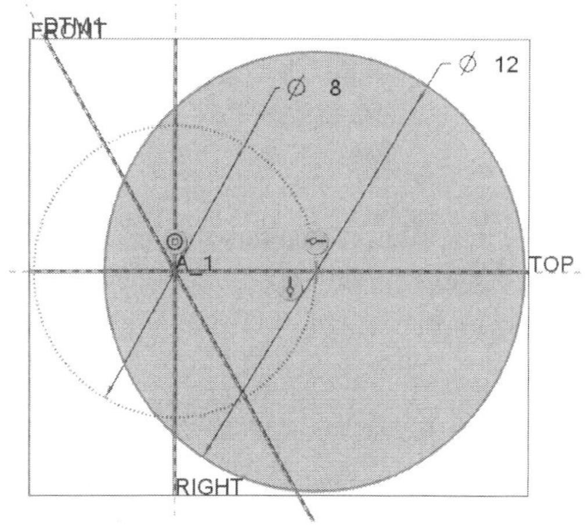

Figure 10 Sketching the first circle

Now (use RMB pop-up) select the ***Circle*** command again and draw the first circle, shown in Figure 10. Its center can be snapped to the intersection of the construction circle and the reference on TOP. Set the new circle diameter to **12**.

Select the ***Circle*** command again to draw the second circle. Its center can be snapped to the intersection of the construction circle and the reference on DTM1. Drag out the circle until the "equal radius" (with the same radius as the larger circle) constraint snaps in. Complete the circle - see Figure 11. You might turn the datum plane display off now, since the screen is getting a bit cluttered.

Select the ***Circle*** command for a third time. The center we want is on the construction circle and directly below the center of the second circle. The cursor should snap to this position. Watch for the icon that indicates vertical alignment. Drag out the circle until the "equal" constraint snaps. The sketch should look like Figure 12.

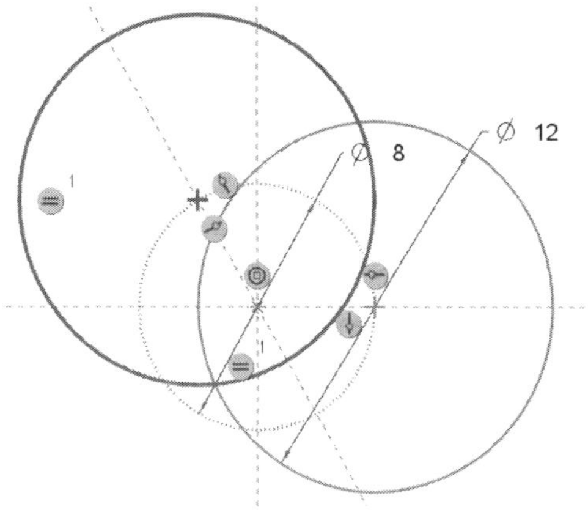

Figure 11 Sketching the second circle

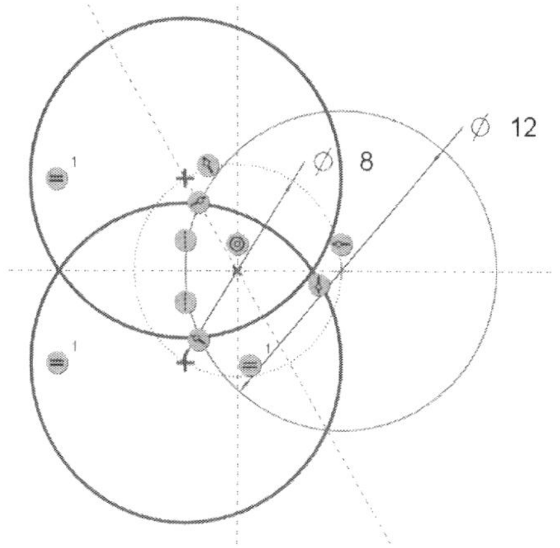

Figure 12 Sketching the third circle

Now we want to trim away all the geometry inside the three arcs[2]. Here are a couple of tools to do that. First, there is a nifty tool in the **Editing** group that looks like this . As the icon implies, all you have to do is swipe the mouse pointer across the edge segments you want to remove - this is ***Delete Segment***, also known as ***Dynamic Trim***. The edge will be trimmed back at both ends to the nearest intersection point or vertex. Try it! See Figure 13. Note that the dynamic trim command does not affect the construction lines. When using this tool, one thing you will have to watch out for is the presence of very small line fragments left behind after trimming. You can usually spot these either by the blue dots on the vertices or by dimensions that seem to go nowhere. Better yet, turn on the ***Highlight Open Ends*** and ***Shade Closed Loops*** functions in the top toolbar. To get rid of all these fragments at once, you may have to resort to another trick for deleting entities. Make sure ***Select*** is picked in the Sketcher menu. Then left click and drag out a rectangle that encloses all the offending lines. They should highlight

---

[2] An alternate and slightly quicker way of using this sketch, using a *Sketch Region*, is described on page 19 of this lesson. Check it out later!

in green. If you want to remove something from this selection set, use the CTRL key when you pick the item to toggle its selection status. For example, in this sketch we do not want to delete the construction circle, so remove it from the selection set. Also, beware of deleting constraints that you want to keep - remove them from the selection set as well. Hold down the right mouse button and select ***Delete***. The selected entities are all gone!

> **Helpful Hint**
> If things really didn't go the way you wanted, there is always the ***Undo*** button in the Quick Access toolbar to bail you out!

You may have accidentally deleted some Sketcher constraints (like vertical alignment of the centers of the second and third circle). Intent Manager is able to generate other constraints to keep the sketch solved. If these are not the ones you want, use the Constraints tool to explicitly create the ones you do want. Note that with two sketched entities selected, the RMB pop-up menu contains the relevant constraints possible for those entities. This is another example of object/action, this time dealing with the setting up of sketch constraints. The completed sketch should look like Figure 14. It should only require two dimensions (one diameter dimension has changed to a radius).

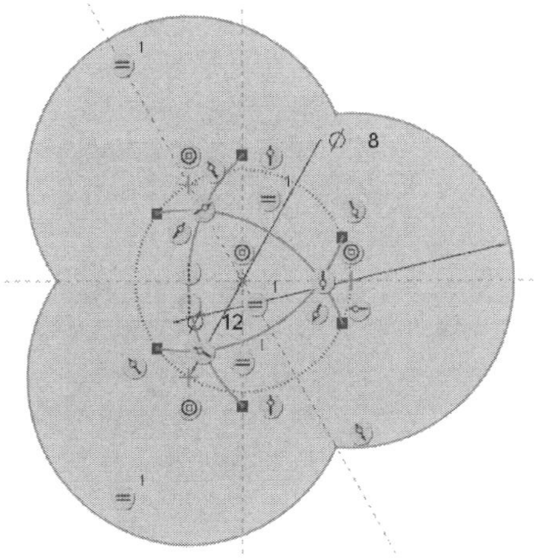

Figure 13 Some lines removed using ***Delete Segment***

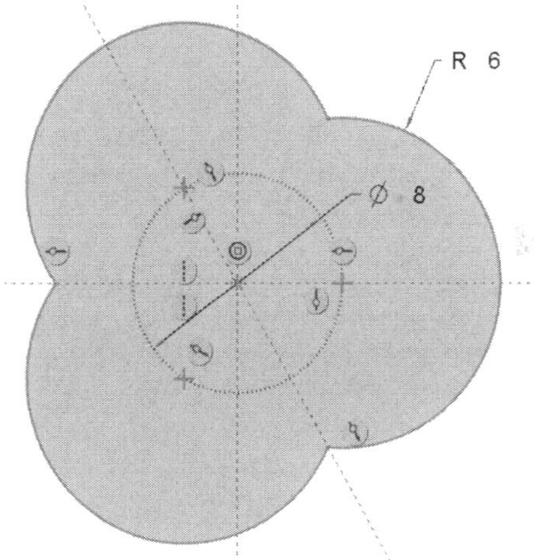

Figure 14 Completed sketch

If you have accidentally deleted an alignment constraint, you will likely see one or more "0.0" dimensions on the sketch. To see what these refer to, change the dimension value to something small, like 1.0. This will modify or shift the sketch slightly to show where the desired alignment has been lost. Restore the desired constraint(s) using the ***Coincident*** constraint. Make sure you return the dimensions to the values shown in Figure 14.

To test the flexibility of this sketch (and if the constraints are doing what we want), try changing either of the dimensions. If the sketch is robust, it should be able to regenerate correctly for a wide range of dimensional values.

Helpful Hint

To quickly explore the flexibility of a sketch, select all the dimensions and use the RMB pop-up command ***Modify***. Then use the thumbwheels to change the values of the dimensions, in various combinations. If a particular combination of dimensions is not possible, the sketch will not update and there will be some information in the message window. This is much faster than typing in new values for each dimension.

Accept the sketch. If you missed it before, set the depth to **2**. In the extrude dashboard, ***Verify*** the feature. The solid should be symmetric about FRONT (check the right side view). If everything looks like Figure 8 above, ***Accept*** the feature.

Creating a Both Sides, Coaxial Hole

We'll create the large center hole using some new options in the hole dialog window. Preselect the axis A_1. If you have trouble picking this, try setting the **Filter** at the bottom right to ***Datums*** or select the axis in the model tree or click with the right mouse button and select ***Pick From List***. Now, with A_1 highlighted, select the ***Hole*** command. A one-sided blind hole is now previewed. Change the diameter to ***8.0***. Check the **Placement** panel (Figure 15). Since we entered the command with A_1 preselected, Creo Parametric assumes that we want a ***Coaxial*** hole. It still needs to know what surface the hole should be defined on. The **Placement** collector should be highlighted (light green). With the CTRL key pressed, select the FRONT datum on the screen or in the model tree. The hole is still one-sided and blind (coming out the back). In the **Shape** panel, change the depth spec to ***Through All***. In the same panel, change the Side 2 depth spec also to ***Through All***. The preview now shows the hole coming in both directions off FRONT, with no depth dimension.

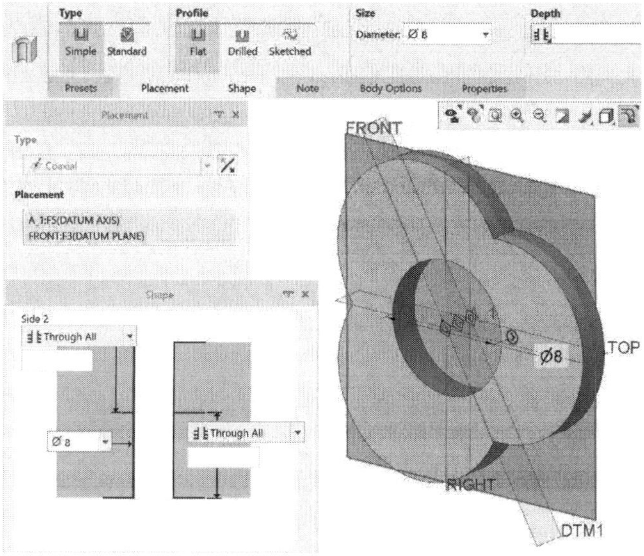

Figure 15 Creating a ***Coaxial*** hole (Note detached panels on dashboard)

That completes the hole, so click ***OK*** (or MMB). Have you saved the part yet? Now is a good time - make sure it goes in the desired directory.

First Tooth - Offset Datum

The first tooth will be the one at the right (3 o'clock position). Look ahead to Figure 18. The design intent for this tooth is that the inner side of the tooth will be a specified distance away from the disk axis. We will create a datum plane at the desired distance that we can use as a sketching plane. The tooth will be extruded outward (for a fixed distance) to the outer edge of the disk. Then we will place a hole, also on the new datum plane, using the both sides option to go radially inward and outward.

Start by selecting the ***Plane*** button in the ribbon. Then pick the RIGHT datum plane. This reference will be listed in the dialog window on the right, with the default constraint ***Offset***. Note that this is a translation. On the green preview of the new datum, you will see a single drag handle ⊙. Drag this out to the right. The dimension shows the offset distance from the reference. The offset dimension is also given in the dialog window.

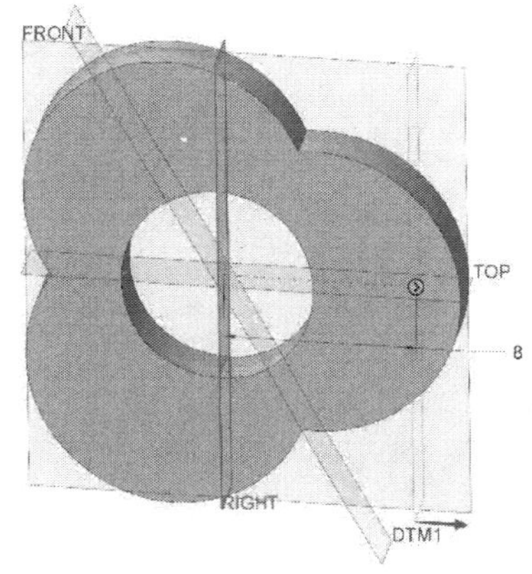

Figure 16 Creating an ***Offset(linear)*** datum

If you wanted to go to the other side of **RIGHT**, you could enter a negative offset or drag the handle to the opposite side. For now, enter a value of ***8***. Accept the new datum with ***OK***. It will be called **DTM2** and is highlighted in green.

Now we can create the tooth. Select the ***Extrude*** command. In the dashboard, select ***Placement*** > ***Define***. In the **Sketch** dialog window, select the new datum DTM2 as our sketch plane. (This would be automatic if DTM2 was highlighted when you launched the ***Extrude*** tool). If you are close to default orientation, the TOP datum will likely be automatically selected for the top orientation reference; if not, set that up now.

In Sketcher, the two references TOP and FRONT have been chosen for us already. We want to add a couple more to this list. In the ribbon select ***{Setup}: References*** (or RMB pop-up) then pick on the front and back surfaces of the base feature. We prefer to use surfaces for references instead of edges. You can delete the reference FRONT.

Make the sketch shown in Figure 17. Depending on your system settings, you may have to turn off the RMB pop-up menu option **Round Dimension Value** to see the decimal places. Notice that the top line in the sketch aligns with the horizontal reference and observe the dimensioning scheme. This sketch implements a design intent where the width of the tooth is determined by the overhang beyond the side of the disk. Can you think of different ways of using references and dimensions to create different design intents for this sketch?

Let's add a relation to make sure the two overhangs are the same. We'll do this the fast way presented in a previous lesson. Put the mouse cursor over the horizontal dimension on the right and note the dimension symbol label, *sdx*. Now double-click on the overhang dimension on the left. Enter the dimension symbol for the right overhang distance. You will be asked to confirm adding this relation to the sketch. As usual, when you have created relations you should test them to make sure they are working properly. Try to change the value on the left - you can't. Try changing the one on the right using the thumbwheel in the **Modify** window - they should both change. Return the value to the one shown in Figure 17.

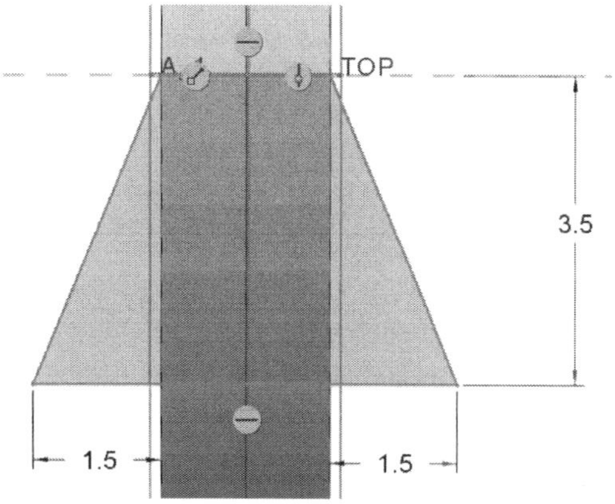

Figure 17 Sketch of first tooth

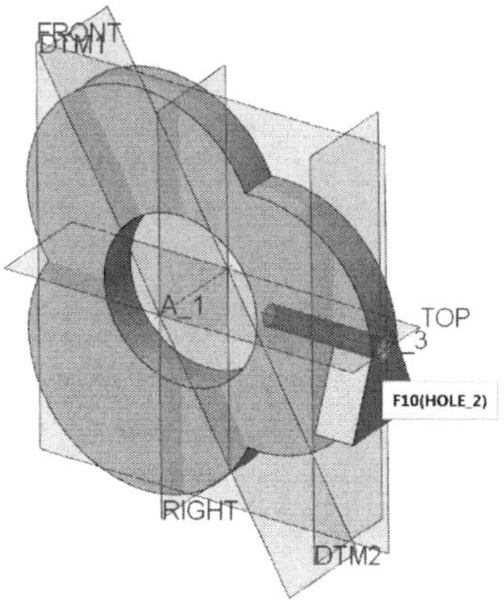

Figure 18 First tooth and hole complete

When you have a completed sketch, leave Sketcher, select a ***Variable*** depth of ***2***, and accept the feature. Turn the display of the datums back on.

Create the small hole using the new datum plane DTM2 as a placement surface. This is a straight, linear hole. Once you have selected DTM2 as the placement plane, open the Placement panel, click in the collector for **Offset References**, then just select the TOP and FRONT datum planes (with CTRL). Then change the Offset References from ***Offset*** to ***Align***. The hole diameter is ***1.0***. In the **Shape** panel, set both the Side 1 and Side 2 depths to ***To Next***.

Note that in one direction, a *Through All* depth would have gone completely through the other side of the disk, which we don't want. *To Next* extends the hole until it passes through the next part surface. The yellow hole preview may not show the hole depth correctly, so use the ***Verify*** button on the right of the dashboard. ***Accept*** the hole.

The tooth/hole combination should now be complete and look like Figure 18. Notice that an axis for the hole is automatically created. This does not appear as a datum axis in the model tree, but its display can be turned off along with all the other datum axes.

Second Tooth - Normal and Tangent Datum

The second tooth is the one at the top left of the part (on arc #2). The intent demonstrated here is to have the planar outer surface of the tooth tangent to the arc of the disk and to extrude the tooth inwards towards the center of the disk. So, we will create a datum to give us a flat sketching surface at the outer edge and tangent to the disk. We can make use of our existing datum **DTM1** which passes through the center of the disk and the second arc.

With nothing highlighted, select the ***Plane*** button in the ribbon. Select the curved surface of the cutter on the side of the second arc (you may have to set the selection filter to Surface). It highlights in green. A preview datum will show up in orange. The default is a **Through** constraint, going through the central axis of this surface. Hold down the CTRL key and click on **DTM1**. The preview datum is now normal to DTM1 and still through the center of the surface. In the **Datum Plane** dialog window, go to the curved surface reference and click on the ***Through*** constraint. In the pull-down list, select ***Tangent***. The previewed datum now moves to be tangent to the cutter and normal to DTM1 - exactly what we want. Notice the direction of the magenta arrow (should be away from the center of the cutter). Accept the new datum by selecting ***OK***. The new datum is called DTM3. See Figure 19.

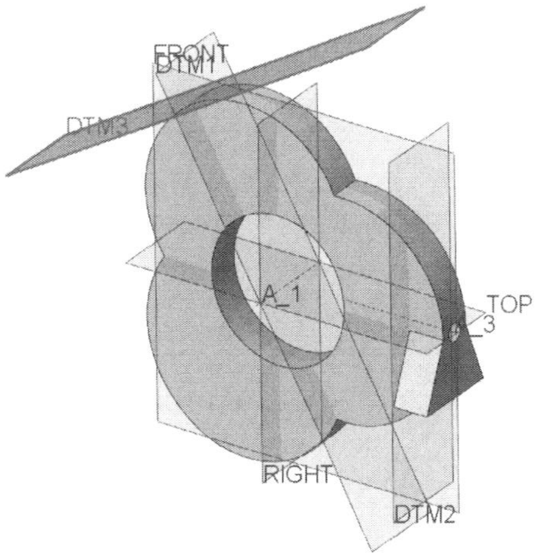

Figure 19 Tangent datum plane for second tooth

Next we'll create a one-sided solid protrusion on the new datum plane. Select the ***Extrude*** tool. In the RMB pop-up, select ***Define Internal Sketch***. Pick DTM3 as our sketching plane. The magenta arrow shows the direction of view onto the sketch plane. Flip this so that it points away from the center of the cutter. The sketch orientation reference is DTM1 and it should face the Top of the sketch. Now select ***Sketch***.

Helpful Hint

A default solid protrusion is created **towards** you off the sketch (coming out of the screen). A default solid cut is created **away** from you (into the screen).

The cutter will re-orient (if not, select ***Sketch View***). You might like to give the part a small spin to make sure you understand its orientation. You might find it easier to sketch when the display is set to wire-frame or hidden line. We're going to create the sketch shown in Figure 20.

Pick the following five sketching references: DTM1, both sides of the disk, and the two outer edges of the first tooth. Now create the sketch shown in Figure 20. This sketch only needs one new dimension because the lines and vertices snap and/or align to the various references. When the sketch is complete, leave Sketcher and choose a ***Variable*** depth specification and enter the value **2**.

Verify that the tooth is the correct geometry and accept the feature with ***OK***.

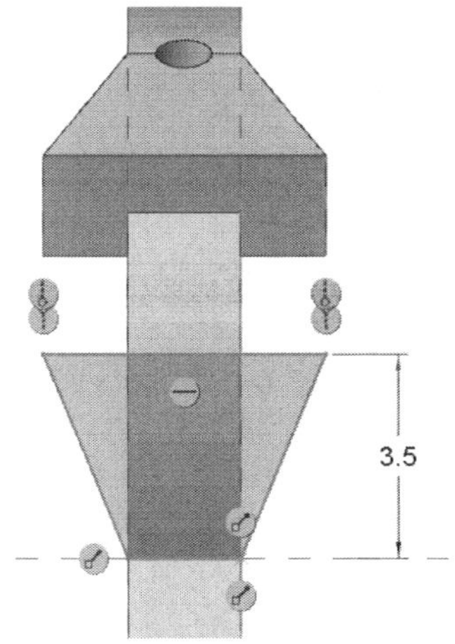

Figure 20 Sketch for second tooth

Create another ***Straight Linear*** hole using DTM3 as the primary reference (placement plane). Use the datums **FRONT** and **DTM1** for the linear placement references and ***Align*** to each. The hole has a diameter of ***1***. Specify a ***Blind*** depth and enter a value of ***8.0***. Once again, if any of the dimensions are hard to pick on the screen, you can set these in the dashboard or the slide-up panels. The completed tooth looks like Figure 21.

IMPORTANT NOTE:

Although this results in exactly the same solid geometry as the first tooth, notice our change in design intent. This tooth is to go a specific depth into the disk measured inwards from the circumference rather than outwards from the center. In this way, the tooth will be tangential to the disk regardless of the disk's size. Similarly, the hole's depth is a fixed value into the disk. At the present time, the hole goes through the surface of the inner hole. We will examine the effects of this later.

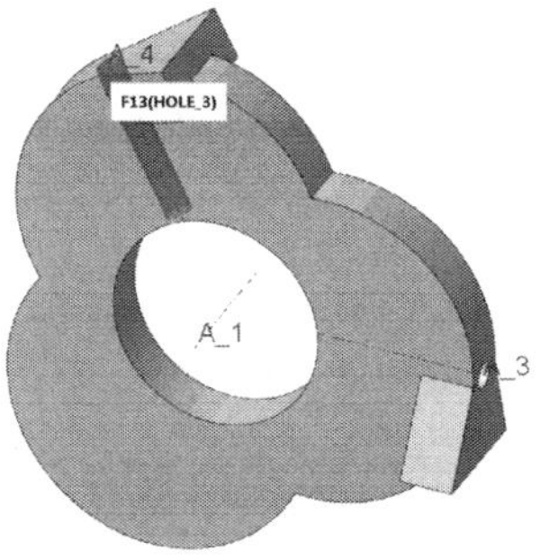

Figure 21 Second tooth completed

Third Tooth - Using Make Datums

The model is getting pretty cluttered up with datum planes which is making it more difficult to pick things out on the screen. One way to deal with this, of course, is to just turn off their display. This gets rid of them all, which cleans up the display but may make selecting them more difficult (you could always use ***Find*** to find them or use the model tree). A more selective way of controlling their display is to ***Hide*** them. Do that now with datum DTM1 - select the datum and in the pop-up menu select ***Hide***. The datum disappears (but not its children!). Do the same with DTM2 and DTM3 (use CTRL-click to select them both at the same time). The default datums should still be visible. Open the model tree and observe the entries for these datums have been grayed out.

All the datum planes we have created up to now have taken their expected place on the model tree, and could be used as parents for many other features. If a datum is only going to be used once to create another feature, it seems wasteful to create one that will be a stand-alone permanent feature on the model tree. Furthermore, we would likely want to ***Hide*** it to get it out of our way just as we did with the previous three datums. The solution used by Creo Parametric for both these problems is a *make datum*. This is a datum that is created just when needed ("on-the-fly"), and then is automatically hidden when the feature using it is accepted. Make datums are sometimes called "datums-on-the-fly" for precisely this reason. The official terminology for make datums in Creo Parametric is "asynchronous datums" which is a bit of a mouthful. We will continue to use the old terminology. One other new facet of make datums is that they are listed on the model tree, but in a special way which we will soon discover. There are a number of additional aspects to asynchronous datums that are treated in the online help (among them the ability to drag a datum into a feature, making it "embedded").

The rules and methods for constraining a make datum are the same as if it was a regular one. What determines whether a plane is considered a make datum is *when* it is made. All our previous datums were created *before* we launched the commands that used them as references. For example, for the first tooth we created DTM2 first, then picked Extrude, then identified DTM2 as the sketching plane. For a make datum, this sequence would be changed: pick the extrude command first, then when we are asked to identify a sketching or reference plane, make the datum "on-the-fly". This is sort of a "just-in-time" delivery notion.

We are going to do other things in a slightly different order for the third tooth, by creating the hole first. However, a hole requires a planar surface for its placement plane. We don't have such a plane at the desired angle. So, we will create the hole using a make datum to act as the placement plane.

Proceed normally to start the hole creation - that is, select the ***Hole*** toolbar button. You are asked (see the message area) to select a placement plane - but there isn't one in a suitable orientation. Here is where we will make the datum on-the-fly. At the right end of the Hole dashboard (you may have to select the <symbol at the extreme right end of the dashboard), select the Datum overflow and pick the ***Plane*** toolbar button. The Hole dashboard is now grayed out (technically speaking, it has been *paused*), and the **Datum Plane** dialog window appears. We need to specify the constraint references for the new datum. Select the reference A_1 of the cutter. This is entered in the **Datum Plane** window with the default constraint ***Through*** (just what we want). Now CTRL-click on the TOP datum. It is added to the collector in the **Datum Plane** window with the ***Offset*** constraint with

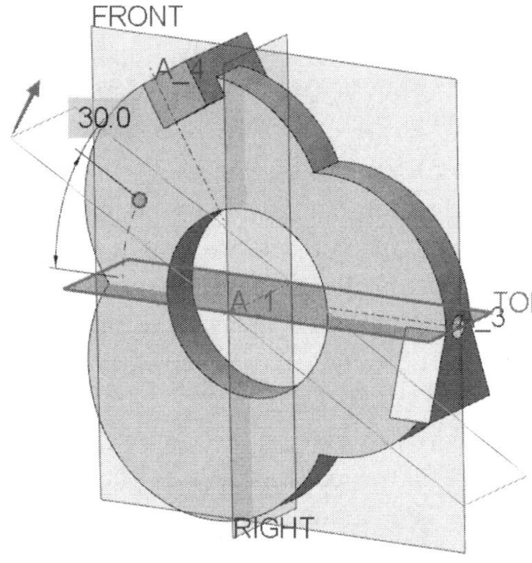

Figure 22 Creating the Make Datum

some rotation angle assumed. Change this angle to 30 degrees from the TOP datum, as shown in Figure 22 (you may have to use a negative angle). When this feature is finished, select ***OK***.

We can continue on with our hole creation. Select the ***Resume*** button ▶ in the dashboard (the only button active) to return to the hole creation. A previewed hole will appear on the new datum plane (which is called DTM4). Set the diameter to ***1.0*** and the **Depth Spec** to ***Through All***. Now drag the orange bracket handles to FRONT (you can drag to anywhere along the displayed edge) and A_1. In the **Placement** panel, change ***Offset*** to ***Align*** for both references. The hole preview is shown in Figure 23.

When the hole is accepted, there is no sign of the make datum we just created, although the hole does have an axis. Open the model tree. The other datums are all there (some are hidden), but the make datum DTM4 is not. However, the last feature on the model tree must be the hole we just created. Open the hole feature by clicking on the "▶" sign, and there is our hidden DTM4 along with the hole that used it. If you select it in the model tree, it highlights on the model. Select DTM4 in the model tree and select ***Edit Dimensions***. You will see the angle dimension associated with this make datum. You will also see this if you edit the hole itself.

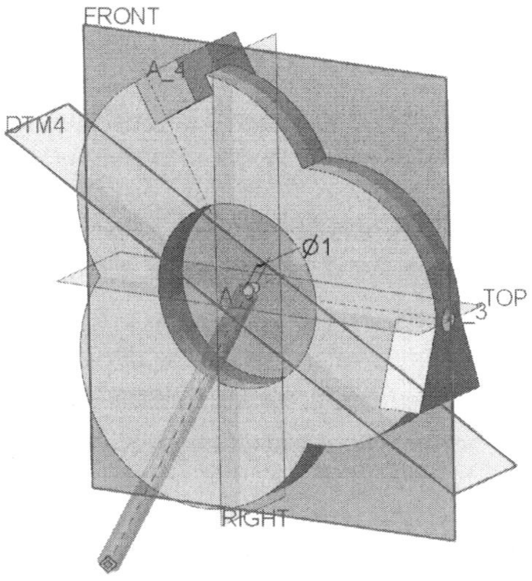

Figure 23 Creating the hole on DTM4

Now create the last tooth. This will also be a sketched protrusion on a make datum that is perpendicular to the axis of the hole (that's why we made it first) and tangent to the cutter surface. Select ***Extrude ➤ Placement ➤ Define***.

The **Sketch** window is waiting for us to specify the sketching plane. In the Datum overflow at the right select the ***Plane*** button. Select the axis of the hole we just made, and (using CTRL-click) the surface of the cutter where the hole comes out. In the **Datum Plane** dialog window, set the constraints for these references in the pull-down lists to ***Normal*** (for the axis) and ***Tangent*** (for the surface). The preview should show a datum plane at the correct location and orientation. See Figure 24. Accept this datum with ***OK*** and return to the **Sketch** window.

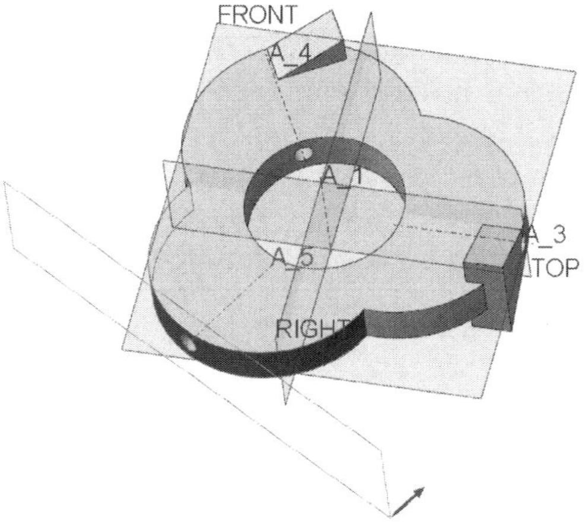

Figure 24 Make datum for the third tooth

Set the view direction away from the center of the cutter using ***Flip*** or clicking on the magenta view direction arrow on the edge of the sketching plane. Use the **FRONT** datum as the ***Left*** orientation reference plane for the sketch. Now select ***Sketch***. Once again, check your view orientation relative to the part. Pick the existing tooth edges as references, plus the front and back of the cutter body. Since these are all parallel, you may still be only **Partially Placed**. For the final sketch reference, pick axis A_1. Sketch the tooth as shown in Figure 25. Note that, in order not to fill in half the hole through the tooth, we must sketch around the circumference of the hole.

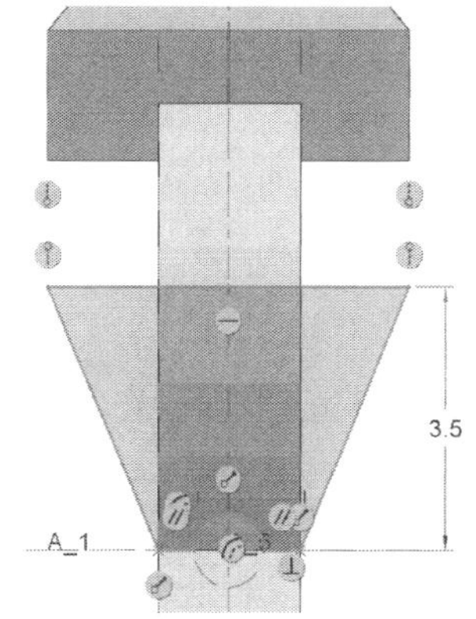

Figure 25 Sketch of third tooth

The ***Project*** button Project in the **Sketching** group is handy for this. Select this command and pick on the curved edge of the hole (use hidden line display to see this). You may have to pick edges on each side of the circle.
To create the geometry and close the tab, select the green arrow on the selection box.

Use ***Delete Segment*** to get rid of unwanted edges. Make sure you have a closed sketch with no open ends. When you are finished with the sketch, select a ***Variable*** depth of ***2***. We have now finished constructing the part, which should look like Figures 26 (***Appearance*** adjusted to show transparency) and 27.

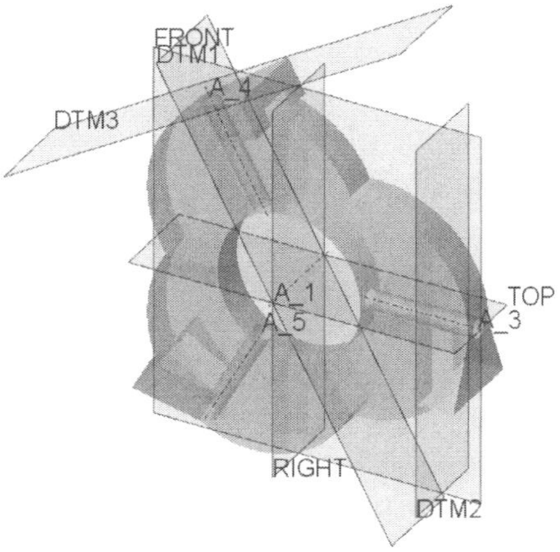

Figure 26 Finished part showing datums

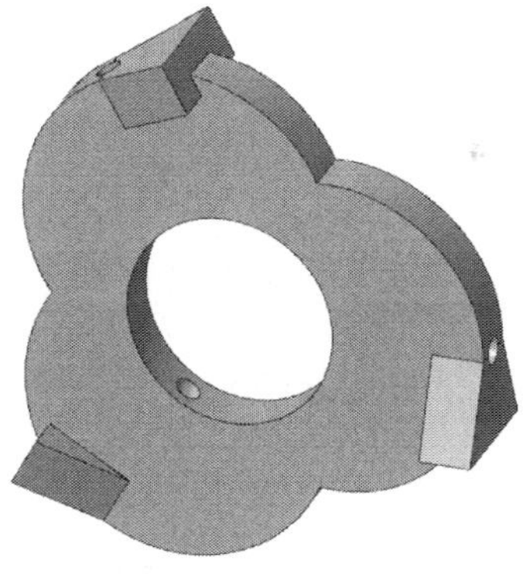

Figure 27 Completed part

Open the model tree and check how the last tooth is represented. In particular, where is the datum we used for the sketching plane and how is it shown? Save the part!

Exploring the Model

We have created three geometrically identical teeth using three different modeling strategies. Let's see what happens when we start to play with the dimensions of the features. Try the following and see if you can explain what is going on. Before you try any of this, save the part so that you can recover from any future disasters! In each case, change the geometry back to the original using ***Undo*** before making a new modification.

1. Change the radius dimension of the first circle of the disk (currently 6.0) to values of **5.0** and **8.0**. What happens to each of the tooth/hole features? Why?
2. Change the diameter dimension of the construction circle (currently 8.0) to values of **4.0** and **12.0**. What happens? Why?
3. Change the diameter of the large coaxial hole to 0.5. What happens? (This is easier to see in hidden line.) Why?
4. What happens if you try to ***delete*** the datum **DTM1**? What about **DTM2**? Don't actually delete the datums.
5. Turn all the datums off. Use the ***Find*** command to find **DTM1**. What happens if you try to change the angle of the datum **DTM1** to 45°?
6. Examine the other parent/child relationships in the model. It is possible that, rather than being related only through the width alignment, some of the tooth/holes refer to other features in ways that were not intended. A possible reason for this is when you were selecting references, the picks were made to axes or edges of previously created features rather than the datum planes or surfaces. How can you be more selective in choosing references?
7. Can you change the offset of **DTM2**? What happens if you use ***6.0*** or ***12.0***?
8. Can you change the diameter of the hole going through the third tooth? Where does the dimension appear for this? What happens if you set this diameter to ***1.5*** and ***2.0***? Why?
9. Can you change the angle of the Make Datum used to create the third hole? What happens if you change this angle to 60°?
10. Can you change the depth of the second and third teeth easily?
11. Suppress the central hole. What happens to the small radial holes? How far through does the first one go? Does anything happen to the third tooth? Why?
12. How many independent dimensions are there in this model? What is the minimum number that should be required? Set up the model so that only these dimensions can be modified.
13. Of the three methods used to create the teeth, which one would you say is the "best"? Keep in mind our three modeling objectives (simple, robust, flexible).

Considering *Design Intent*

You should be able to see once again that capturing the design intent is an important part of feature-based modeling and the model creation strategy. In this lesson we have added an important new consideration to this strategy - datum planes. Design intent involves consideration of the following:

- What is the design function of the feature?
- How does this influence the modeling strategy?
- How does the design function of a feature relate to other features?
- Which features should be unrelated in the part?
- How can you set up references and dimensioning schemes so that the

parent/child relations reflect the above?

- How can you create the model so that it is driven by as few dimensions as possible? Will this necessarily always be desirable?
- When should you use relations internally in the part to drive the geometry automatically, depending on the critical design dimensions?

Design changes are inevitable. Therefore, you should try to design the model so that it will be easy to make the kinds of changes you expect in as direct a manner as possible. This is hard to do if you know only a few methods to create new features since your choices will be limited. You can often create the correct geometry, but it may be very difficult to modify or change later. Furthermore, it is often difficult to foresee exactly how you might want the model to change later. One thing is for sure, if you just slap-dash your features together, sooner or later you will run into a serious modeling problem that can become a nightmare for making design changes.

Using *Sketch Regions*

Sketch Regions give you a shortcut in selecting geometry for shaped features (extrudes and revolves) that is composed of areas bounded by closed loops formed by multiple intersecting curves. The region can be from a single sketch or multiple sketches (on the same plane). Sketch regions eliminate the need for some of the trimming or projecting curves when creating new sketches.

The creation of a Sketch Region is launched by setting the **Selection Filter** at the bottom right of the graphics window to ***Sketch Region*** (shortcut **SHIFT+S**).

We will demonstrate this by constructing the base feature of the cutter (Figure 8 plus the central hole) in one extrude operation. Start a new part [***cutter2***] using the default template. ***Delete*** (or ***Hide***) the default coordinate system. Create the datum axis **A_1** and datum plane **DTM1** as we did before, to match the model shown in Figure 7.

Select the FRONT datum plane and launch the ***Sketch*** tool. Select DTM1 as an additional reference and create the sketch shown in Figure 28. Notice that this does not use a construction circle and we do not need to trim any of the lines in the sketch. Accept the feature and observe it in the model tree.

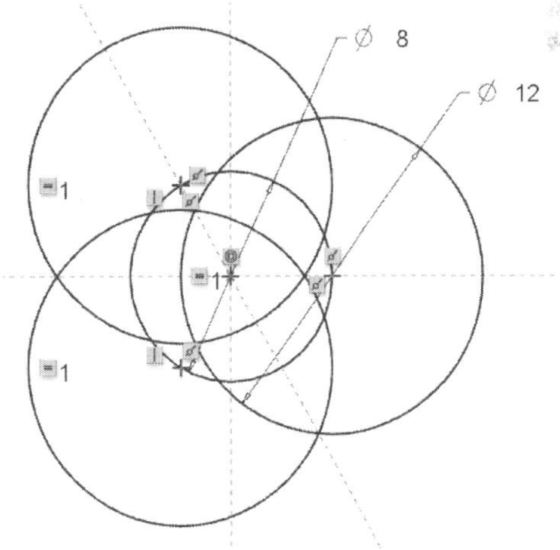

Figure 28 Sketch created to define geometry of Sketch Region

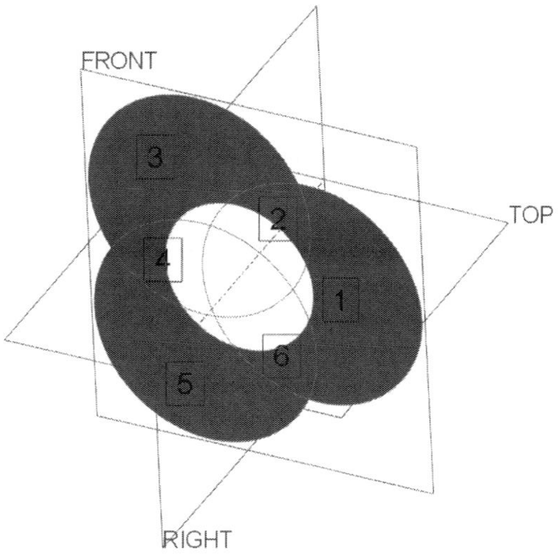

Figure 29 Sketch Region selected

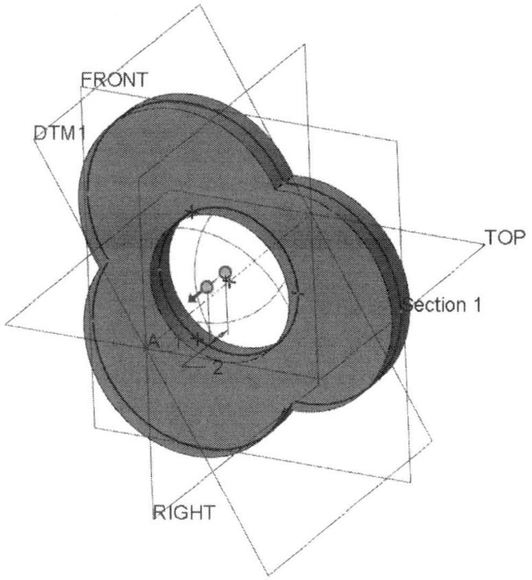

Figure 30 Preview of extrude using the Sketch Region

Now set the **Selection Filter** (lower right) to ***Sketch Region*** (shortcut SHIFT+S). Holding down the CTRL key, click on the six outer areas of the sketch - see Figure 29. As each is selected, it will highlight in green.

As the last area is picked (or using the RMB), select ***Extrude*** in the pop-up menu. This opens the standard dashboard and the feature is previewed in orange. Change the depth to ***Symmetric*** with a blind (total) depth of **2.0**. See Figure 30. Accept the feature. Pretty simple!

Here are some things to check out regarding the use of sketch regions:

1. What does the model tree look like for this combination of sketch + sketch region + extrude? Does the sketch region appear explicitly?
2. How can you edit the geometry of the resulting protrusion (there are three driving dimensions - where are they)?
3. Can you ***Hide*** or ***Suppress*** the original sketch with impunity?
4. Can you ***Unlink*** the section for the extrude from the original sketch?
5. What modifications can you make (using ***Edit Definition <Edit Internal Sketch***) to the section sketch in the extrusion?

Creo users have gotten along without *Sketch Regions* for years, but in the right circumstances they can save you a lot of time. So it is worthwhile to spend some time to experiment with them a bit. Keep your eyes open to the possibility of using them!

In the next lesson we will look at commands for creating patterned features (linear and radial patterns) and several ways of making feature copies. There will be more discussion of feature groups. We'll also see some new Sketcher tricks and a new type of protrusion (***Thin***).

Questions for Review

Several of these questions will require you to do some exploring of the program on your own.

1. What are the constraint types for creating datum planes? How are these different for make datums?
2. What are the constraint types for creating datum axes?
3. What combinations of constraints will lead to a completely constrained datum plane? Draw some freehand sketches to illustrate these.
4. What references are required to create a coaxial hole?
5. For a **Symmetric** solid protrusion, do you specify the depth in each direction, or the total depth? What about for a **Symmetric** cut?
6. What is the easiest way to create a datum plane parallel to a previous one at a specified distance away? What is this called?
7. Suppose you want to create a datum plane at an angle to another datum. You want to use the **Through** placement option, but there is currently no part edge or axis to use as a reference. How can you create the desired datum?
8. Does the order of selection of **Through** and **Offset(Rotation)** matter?
9. What is the difference between **Through All**, **To Next**, and **Up to Surf** when specifying an extruded feature's depth? For the last two, what happens if the extruded sketch does not completely intersect the specified surface?
10. Compare the advantages and disadvantages of using regular datums and make datums.
11. If you want to ***Edit*** a feature created using a make datum as a sketching plane, where does the sketch show up?
12. Can you use the ***Edit Definition*** function on a hidden feature (like a make datum)?
13. When the model starts to get cluttered up with surfaces, edges, datums, and axes, how can you make sure that you are making an alignment to the desired entity?
14. Can the dimensions of a make datum (offset distances or angles) be controlled using relations?
15. What is the difference between the symbols "**dx**" and "**sdx**"?
16. Are other feature dimensions available for use in Sketcher relations?
17. Find a simple mechanical part and try to "reverse engineer" the design intent. How would you implement this in Creo Parametric?
18. Where and how do make datums show up in the model tree?
19. Can you use a make datum as a reference for another make datum? That is, can make datums be *nested*?
20. Is there an axis equivalent to a make datum (sort of a "make axis") that behaves in the same way as a make datum? That is: you make it on-the-fly, it is automatically

included in the feature group, and its display is automatically hidden.

21. Does the ***Delete Segment*** tool affect construction lines?
22. How can you change the display extent of a datum plane or axis?
23. Can the ***Find*** command be used to locate Make Datums?
24. How can you quickly toggle between the ***Sketch Region*** and ***Geometry*** settings in the selection filter?
25. What happens if you preselect a plane and a perpendicular axis, then launch the ***Hole*** command?
26. For the first extrude in the *cutter*, what happens if you set **Side 1** depth to 6 and **Side 2** depth to -2 (i.e. a negative value)?

Exercises

Here are some objects for you to make. Don't worry about exact dimensions, but datums and make datums (and maybe Sketch Regions) will come in handy for these!

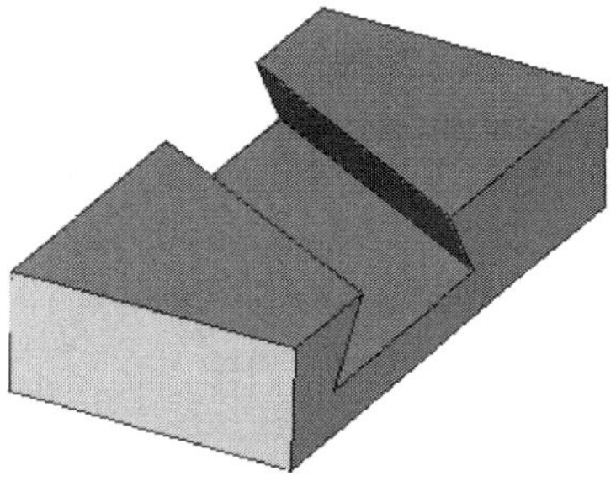

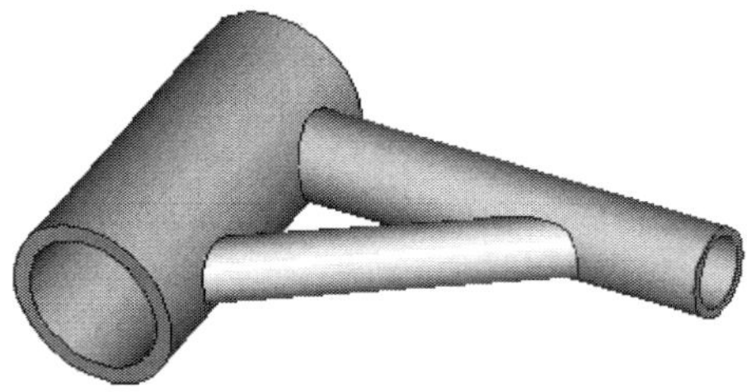

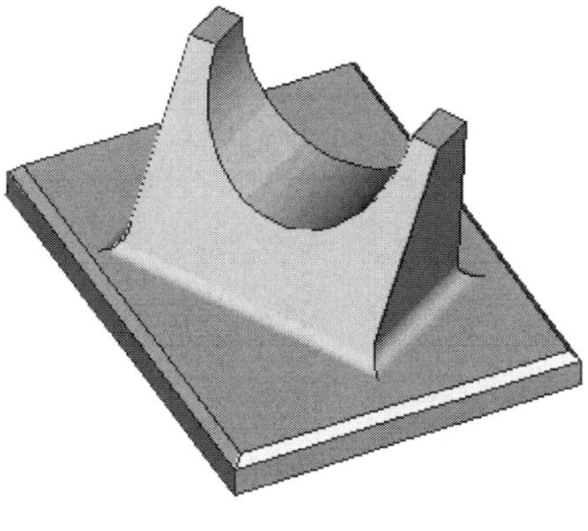

Project

Here is another part for the project. More pictures with dimensions are shown on the next page (all dimensions in millimeters). As usual, study the geometry carefully, and plan your modeling strategy before starting to create anything! Feature creation order is important to make this part as simple as possible.

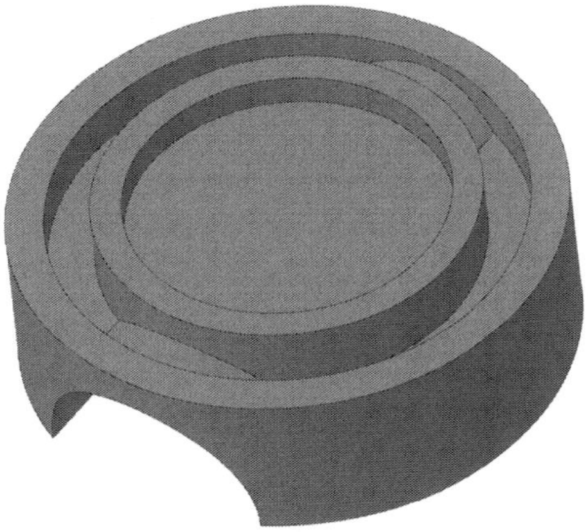

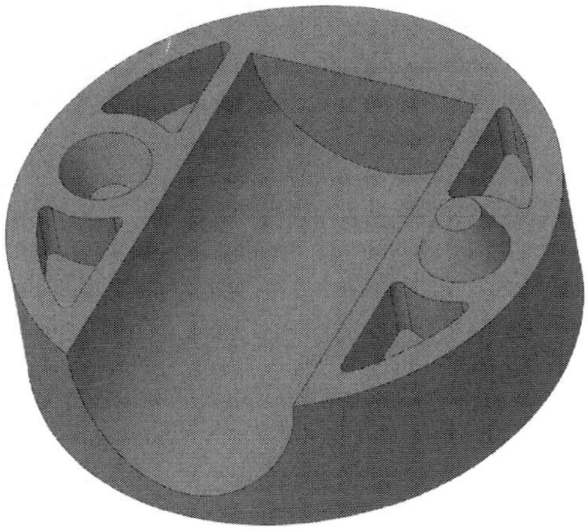

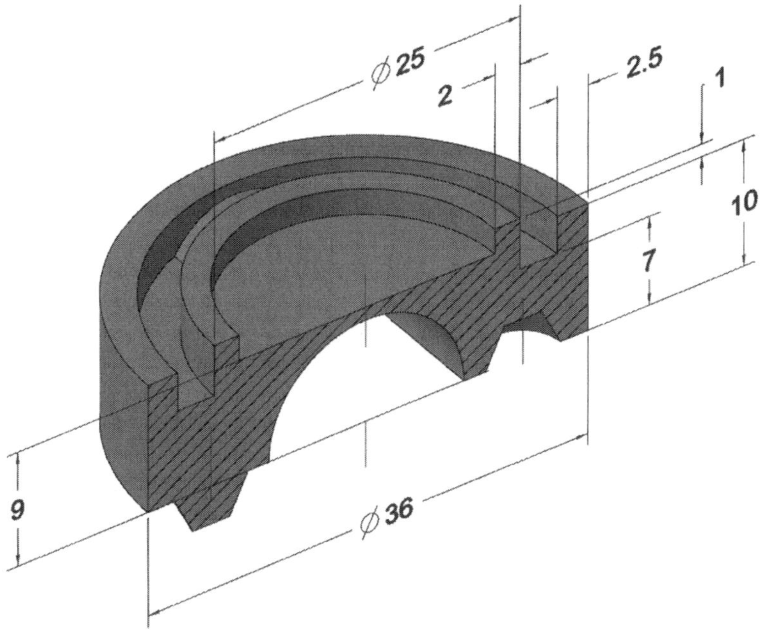

⌀25
2
2.5
1
10
7
9
⌀36

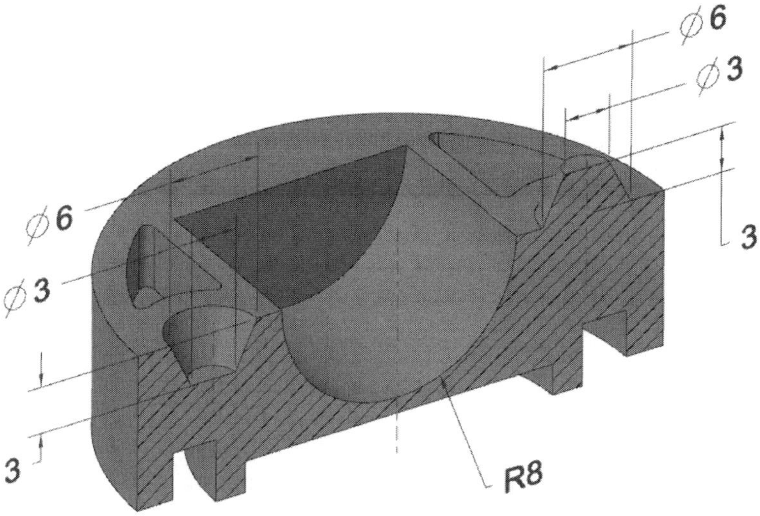

⌀6
⌀3
3
⌀6
⌀3
3
R8

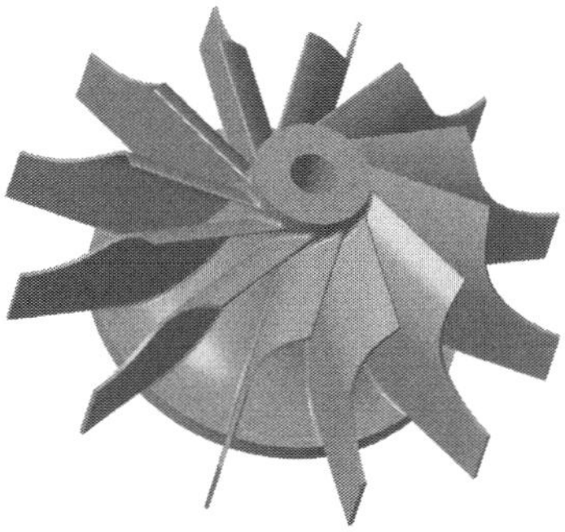

Lesson 7

Patterns and Copies

Synopsis

Naming dimension symbols. Dimension patterns in one and two directions. Creating a standard counterbored hole. Radial patterns of shaped and engineering features. Group patterns. Using the Copy command with Paste and Paste Special. Creating dependent and independent copies. Copies using translation, rotation. The Sketcher palette of defined shapes.

Overview of this Lesson

Models often contain repetitive instances or copies of the same geometric form. In Creo Parametric terms, these are called *patterns*. The name suggests a regular, geometrically repeated placement of features. We will find that patterns can do a lot more than this, including changing the shape of a feature as it is patterned. There are numerous options available when creating patterns. In this lesson we will look at the simpler and more commonly used pattern types.

When only a single duplicated feature is desired, it may make more sense to copy it rather than create a pattern. The copy command also allows more freedom in selecting references for the duplicated feature and making it partly or completely independent of the original. Copies can also be made by translation or rotation (or a combination of these) independently of the references used to create the original. Features can be copied to a totally different part.

Patterns and copies work on individual features, features arranged in groups, and (with some restrictions) on bodies. We'll look into the formation of groups a bit later in the lesson, but we will have to leave patterns of bodies for another time (they work much the same as individual features). To demonstrate the use of patterns and copies, we will be creating several different parts. The parts are totally independent of each other, so you can jump ahead to any one of these:

1. Patterned Features
   - ▶ simple uni-directional patterns

- bi-directional patterns
- radial pattern of holes with relations
- patterns of grouped features
- radial patterns of shaped features

2. Copied Features
   - using the *Paste* command
   - using the *Paste Special* command
   - creating translated and rotated copies with *Paste Special*
3. Design Considerations
   - some things to think about when designing with complex features

The use of named dimension symbols is helpful when dealing with patterns. We will see how to do that first. We will also use a Sketcher tool called the **Palette** that provides pre-defined shapes.

Patterned Features

Patterns are created by making duplicates of an existing feature, group, or body - called the *pattern leader*. There are eight kinds of patterns in Creo Parametric:

- dimension (not available with bodies)
- direction
- axis
- fill
- table
- reference
- point
- curve

In this lesson we will look at dimension patterns only[1]. This is the most commonly used pattern, and there are enough variations of these to keep us busy for a while. The simplest dimension pattern is created by incrementing a single dimension that locates the pattern leader on the part. Each increment of the pattern dimension produces a new *instance* of the feature at the incremented location. Patterns are even more powerful than just creating multiple instances: it is possible to form the pattern in two directions simultaneously and to change the geometry parametrically of each instance in the pattern. While the dimension(s) that locates the feature is(are) incremented, other dimensions of the pattern leader can be incremented so that the instances change size and/or shape. It is even possible to change size and shape of an instance without changing its location (see the exercises!). All instances in the pattern can be modified simultaneously, if set up to do so. You can see there are a lot of possibilities here.

[1] See the *Advanced Tutorial* for lessons on pattern tables, direction, axis, and fill patterns, reference patterns, pattern relations, *Identical*, *Varying*, and *General* patterns.

Naming Dimension Symbols

Create a new solid part called *pattern1* using the ***inlbs_part_solid_abs*** template. Create a base feature using an extruded protrusion. The Sketch plane is the TOP datum. As shown in Figure 1, the part is a **12 X 20 X 2** rectangular solid. While you are in Sketcher, check out the variations of the ***Rectangle*** command in the ribbon.

Now create an extruded protrusion near the front left corner of the base feature. This is a cylindrical protrusion of diameter **2** and height **1**. See Figure 2 for the dimensions and be sure to use the identical dimensioning scheme. This protrusion will be our pattern leader for the next several exercises.

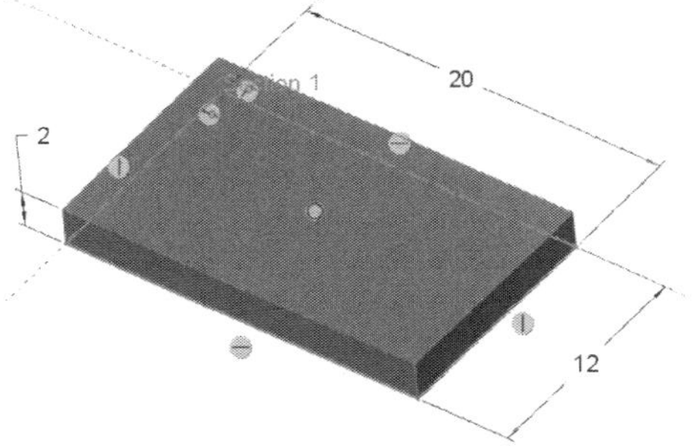

Figure 1 Base feature for part *pattern1*

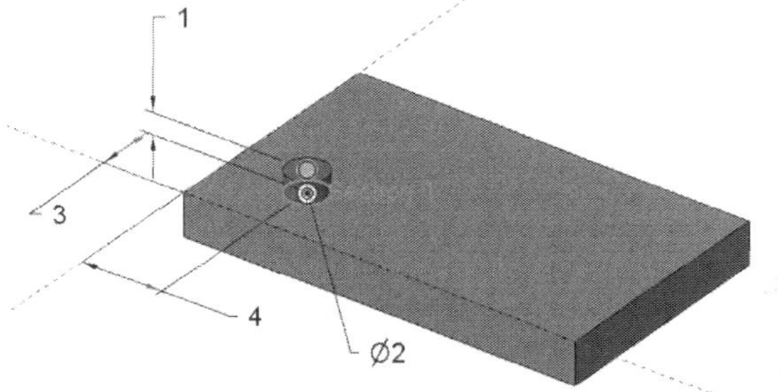

Figure 2 Dimensions of the pattern leader

In preparation for the pattern exercises, it will be helpful (though not necessary) to change the symbolic names for the dimensions shown in Figure 2. To see the current names, double click on the protrusion (Hint: set the preselection filter to ***Feature***) and select in either the **Model** ribbon or the **Tools** ribbon

{Model Intent}: Switch Dimensions

The symbolic names "***dx***" will be displayed. Let's make those symbols a bit more meaningful. Preselect the horizontal dimension (4 in Figure 2). When it highlights in green, the **Dimension** dashboard appears. Towards the left end is an area (**Value**) that lets you change the symbolic name and/or value. Enter a new name here [***L_X***] then middle click or press ***Enter***. Notice on the part that the "***dx***" symbol has changed to "***L_X***".

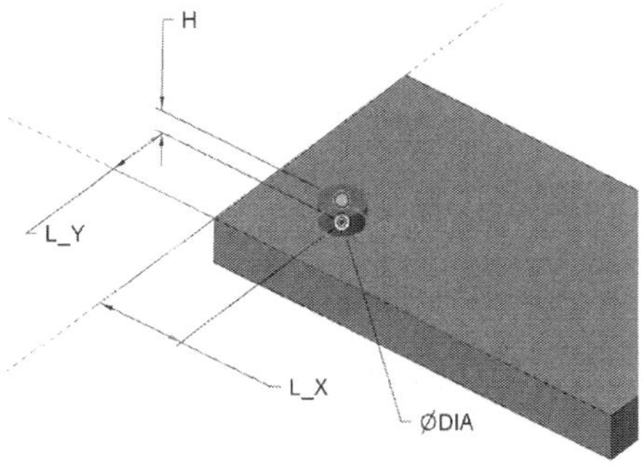

Figure 3 Renamed dimension symbols

Change the depth dimension (3 in Figure 2) text to "***L_Y***". Change the protrusion diameter (2 in Figure 2) text to "***DIA***". Finally, change the protrusion height dimension (1 in Figure 2) to "***H***". The protrusion dimensions should now be named as in Figure 3. Check out the ***Move*** command in the RMB pop-up for a highlighted dimension to rearrange the layout on the screen.

You can use almost any alphanumeric symbol for a dimension (note that upper case "X" and "Y", for example, are reserved symbols). Keep the name short and don't use spaces or punctuation. If a renamed symbol is used in a relation, Creo Parametric will automatically update the relation to use the new symbol.

Creating a Uni-directional Pattern

Turn off the datum planes and preselect the cylindrical protrusion. The ***Pattern*** tool  is available in the **Editing** group or the pop-up menu in either the graphics window or the model tree. Choose one of these options to launch the ***Pattern*** command.

You will now see the dimensions associated with the protrusion[2] and the **Pattern** dashboard opens. The pull-down list at the left end of the dashboard contains the eight types of patterns. Leave this set to the default ***Dimension***. In the following examples we will see what is meant by **1st Direction** and **2nd Direction**. In the dashboard, move your cursor slowly across the four text areas in **Settings** and read the pop-up tips. Two of these text areas are for entering the number of members (instances) in each pattern direction. The other text areas tell you which dimensions are being incremented in each direction. The item collector for **1st Direction** is active with a default of 2 instances.

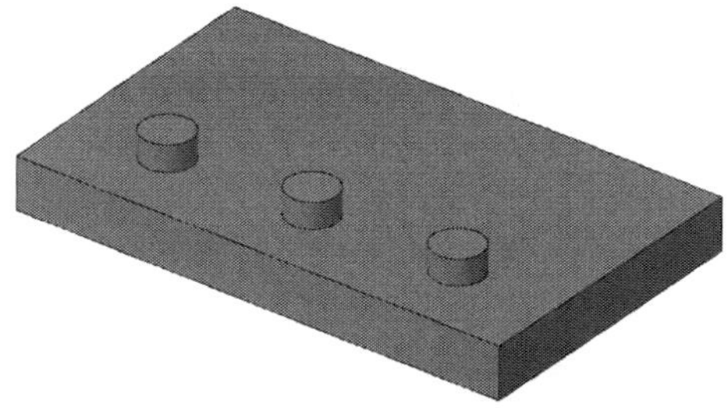

Figure 4 Simple pattern #1

Pattern #1 - Click on the horizontal dimension (4) for the protrusion. A text box appears which allows you to enter the increment to be used with this dimension. Enter a value of ***6***. This means the next instance will be 6 units over to the right from the pattern leader (that is, the 4 dimension is incremented by 6). The position is indicated by a small yellow circle. The one after that will be another 6 units over, and so on. In the dashboard, find the box for the **Number of Members** to be created in the first pattern direction. Enter ***3*** in this box. We have now provided enough information (dimension to be incremented, increment size, number of instances) to create our first simple pattern. The location of the three instances are indicated by the small yellow circles. ***Accept*** the pattern definition (shortcut: middle click). The part should look like Figure 4 (without the aqua blue color protrusions). All three protrusions are highlighted in green as the last feature(s) created.

If you double-click on the second or third protrusion in this pattern, you will see all the dimensions associated with that instance. In particular, you can see the increment 6, and the number of instances. Change the increment from 6 to ***4***, and the number of extrudes from 3 to ***4***. If ***Auto-Regenerate*** is turned on these changes will occur immediately, otherwise select ***Regenerate***. Change the increment back to ***6***.

---

[2] If the dimensions are still in symbolic form, select ***Switch Dimensions*** on the Tools ribbon then return to the Pattern ribbon using the tab.

For this feature, you can run the pattern off the right end of the part (in which case, observe the warning in the message window). This may not always be possible, especially if the feature being patterned was created with an open sketch. Running the pattern off the part can be a subtle error to catch if you are creating a pattern of cuts or holes (so pay attention to the yellow circles and the message window).

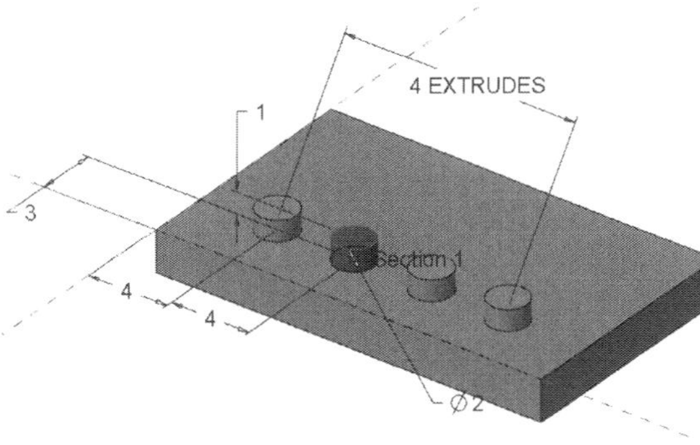

Figure 5 Pattern #1 dimensions

Open the model tree and expand the pattern. You will see each feature instance listed.

Pattern #2 - Let's play with this simple pattern some more. With nothing preselected on the model (no green highlights), in the model tree select the pattern and in the pop-up select ***Edit Definition***. This re-opens the pattern dashboard and shows all the dimensions for the pattern leader in the graphics window.

In the **Dimensions** panel the dimension "***L_X***" is indicated in direction 1. Select this and use the RMB pop-up, select ***Remove***. In its place, click on the dimension 3 for the protrusion - this is the dimension "***L_Y***". Enter an increment of ***6***. In the dashboard, specify ***2*** instances. ***Accept*** the pattern. See Figure 6.

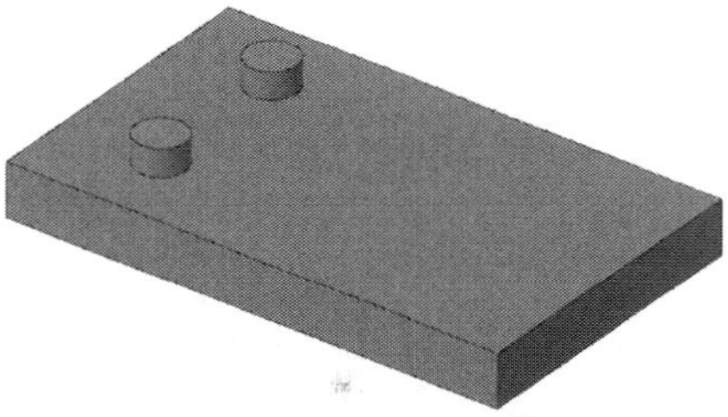

Figure 6 Simple pattern #2

So, the Direction 1 does not always have to be in the same direction on the part. In fact, the 1st and 2nd pattern directions are not physical directions at all but more logical ones. The 'directions' are a way of organizing which dimensions will be incremented together. It may even be that the feature being patterned does not move at all but just changes shape and size (see the pyramid exercise at the end of the lesson).

Can a pattern go only in the "***X***" and "***Y***" directions simultaneously? Let's find out.

Pattern #3 - Highlight the pattern in the model tree and using the pop-up, select ***Edit Definition***. Open the **Dimensions** panel. The "***L_Y***" dimension is listed in Direction 1. Click in the collector for Direction 1 and then CTRL-click on the horizontal dimension to add it. Set the increment to ***6***. The panel should look like Figure 7. ***Accept*** the pattern (Figure 8). So, we are not restricted to incrementing a single dimension in a pattern "direction". For that matter, we can increment in more than one physical direction at once. Let's explore this a bit more.

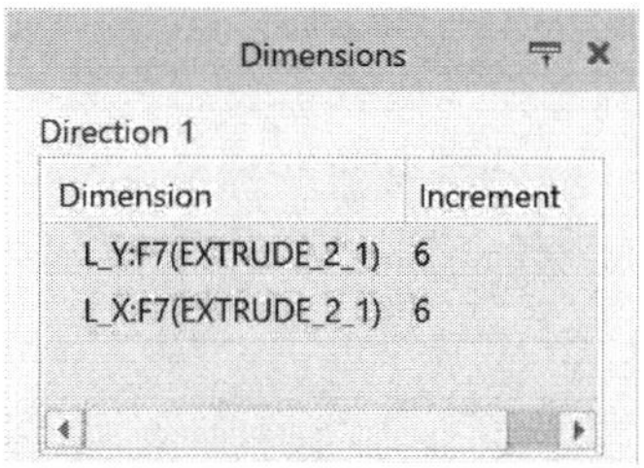

Figure 7 Pattern dimensions for pattern #3

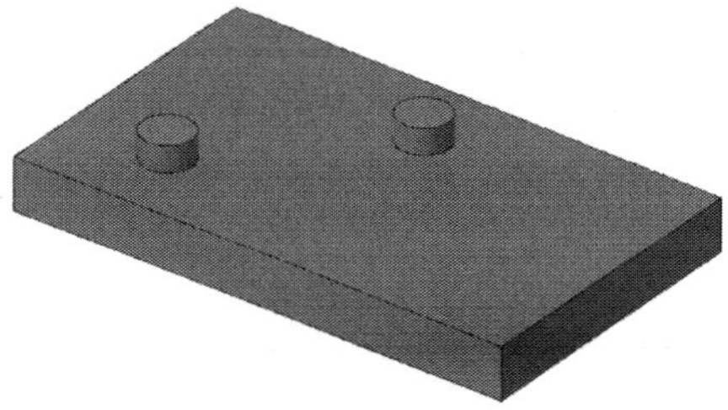

Figure 8 Simple pattern #3

Pattern #4 - Once again, select the pattern in the model tree and ***Edit Definition***. Open the **Dimensions** panel. ***Remove*** the "***L_Y***" dimension using RMB. Holding down the CTRL key, click on the protrusion diameter dimension in the graphics window. Set an increment of ***1***. Hold down the CTRL key and click on the height dimension. Set the height increment to ***3***. Change the number of instances to ***3***. The panel should now look like Figure 9. Accept the pattern and the model should look like Figure 10. The instances are created as before, but this time their diameter and height also change. Furthermore, it doesn't matter what order you specify the pattern dimensions shown in Figure 9. We could have picked the diameter dimension first, for example. Clearly, you can do more with patterns than just make duplicates!

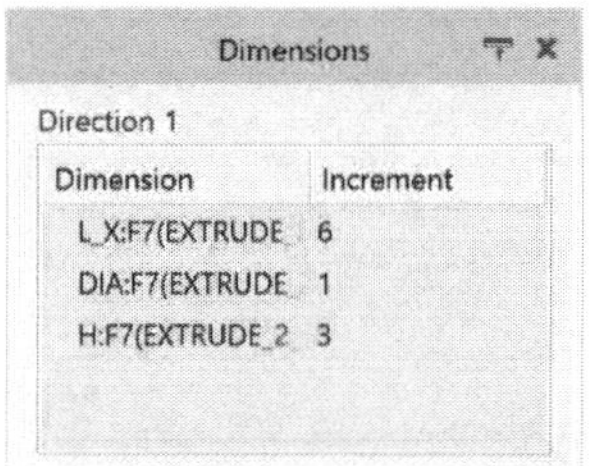

Figure 9 Pattern dimensions for pattern #4

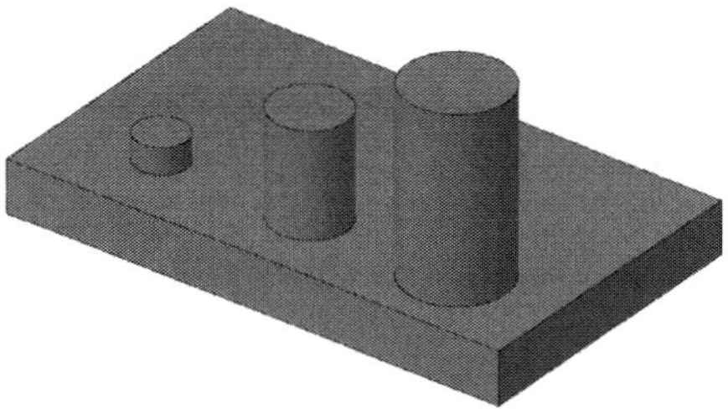

Figure 10 Simple pattern #4

Creating a Bi-directional Pattern

So far, we have not done anything with the second pattern direction. Select the pattern in the model tree and in the RMB pop-up select ***Delete Pattern***. This removes the pattern instances but leaves the leader. (Be careful not to pick ***Delete***, which gets rid of everything but asks for confirmation first.)

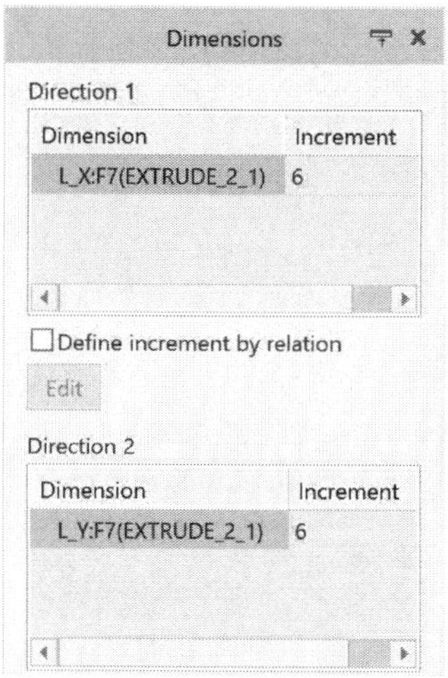

Figure 11 Pattern dimensions for pattern #5

Pattern #5 - Pick the cylindrical protrusion in the model tree, then in the pop-up menu, select ***Pattern***. Click on the horizontal dimension (4) and enter an increment of ***6***. In the dashboard enter the number of instances as ***3***. Now, still in the dashboard, click in the item collector for the

2nd Direction (lower box) that currently says, "Click here to add item." When you click on the collector, it turns green which means it is active. Now select the "**L_Y**" dimension (3) for the protrusion. Enter an increment of ***6*** and specify ***2*** instances (default).

The ***Dimensions*** slide-up panel (Figure 11) shows "***L_X***" for Direction 1 and "***L_Y***" for Direction 2, along with their increments. ***Accept*** the feature (Figure 12).

Let's modify the pattern one more time.

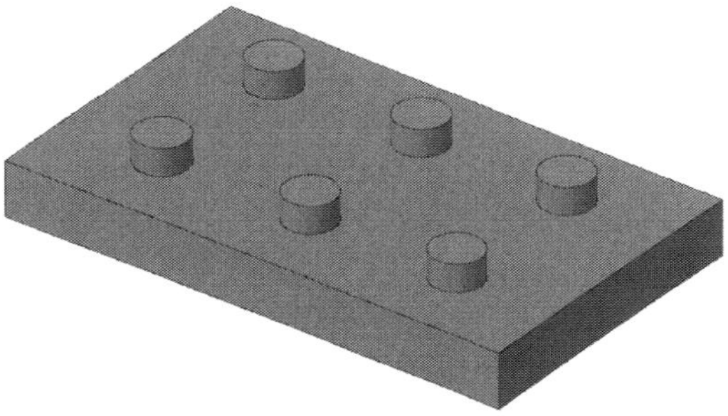

Figure 12 Simple pattern #5

Pattern #6 - Select the pattern in the model tree and ***Edit Definition***. Click on the **1st Direction** box that says, "1 Item". It becomes active. Hold down the CTRL key and add the diameter dimension (increment ***1***) and the height dimension (increment ***2***). Observe the colors of these dimensions. Now click on the **2nd Direction** item collector that says, "1 Item". Using the CTRL key, add the height dimension (increment ***4***). Open the **Dimensions** panel to see all the pattern dimensions. See Figure 13. Notice that a dimension, like "**H**" in this case, can be used in both directions of the pattern. ***Accept*** the pattern, which should look like Figure 14.

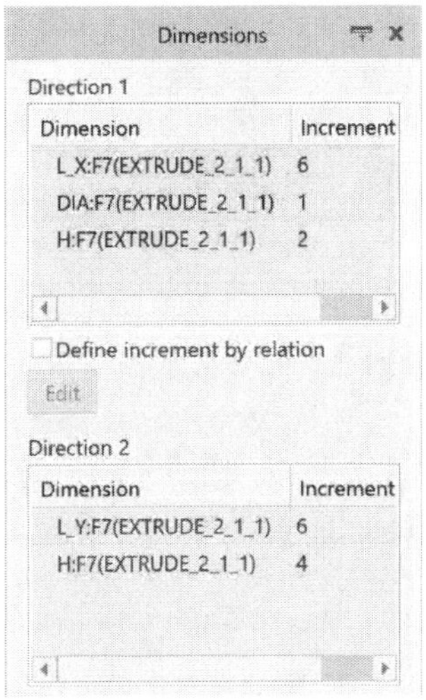

Figure 13 Pattern dimensions for pattern #6

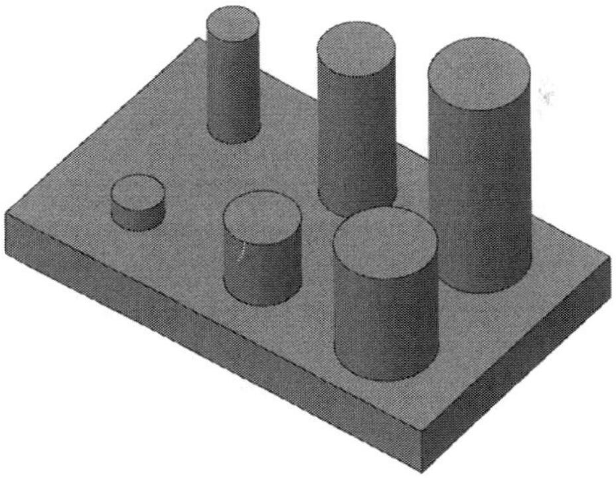

Figure 14 Pattern #6

Pattern Options

Pattern #7 - Select the pattern in the model tree and ***Edit Definition***. Open the ***Dimensions*** panel (Figure 13) and in the **Direction 1** collector, change the **DIA** increment to "**2**". In the **Direction 2** collector, change the **L_Y** increment to **4**. Accept the pattern. This results in two of the pattern members overlapping. This is allowed because of a setting in the ***Options*** panel in the pattern dashboard. Use ***Edit Definition*** to open that now. The default is ***General***, which allows pattern instances to change shape and overlap. Read the tool tip pop-ups and check out what happens to this pattern with the other two options - ***Identical*** and ***Varying***. We will explore this later but meanwhile you might check out the online help topic **"About Pattern Regeneration Options"**.

We'll leave this part now. ***Save*** it and perhaps come back to it later to explore more pattern options. After you have saved it, select ***File ' Manage Session ' Erase Current***. This removes the model from memory (takes it "out of session").

Creating a Simple Radial Pattern

Relations can be used to control patterns. We'll demonstrate this by creating a radial pattern of bolt holes. We'll explore the ***Hole*** dashboard to create a standard counterbored hole. Then we will set up a couple of relations to control the pattern geometry based on the specified number of holes.

Start a new part called ***flange*** using the "***inlbs_part_solid_abs***" template. (Note: Units ***must*** be in inches for this exercise!) Create the circular disk with central hole shown in Figure 15. We will need a central axis for the counterbored hole placement so you have a number of options: a) create a solid protrusion of two concentric circles, b) create a solid circular disk and add a coaxial hole, or c) revolve a rectangle around a central axis aligned with the datums. Each of these options will create the axis automatically.
Pick whichever of these options you are *least* familiar with - might as well get some practice! The outer diameter is ***16***, the hole diameter is ***8***, and the disk is ***3*** thick. Note that the disk is constructed on **TOP**.

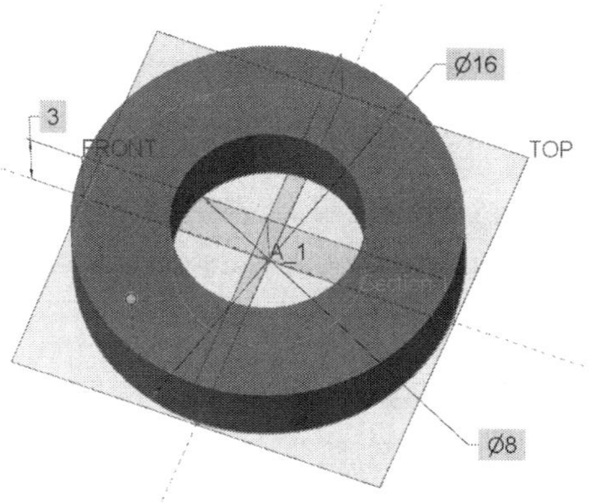

Figure 15 Base feature of flange

Now we'll create a single counterbored hole in the disk. This will be the pattern leader. In order to specify the pattern using an angular dimension, we choose a *Diameter* placement scheme (requiring an angle from a reference plane, and a diameter dimension for the bolt circle).

Select the ***Hole*** tool in the ribbon. Moving across the dashboard from left to right, do the following:

- under **Type**, pick the ***Standard*** button
- leave the ***Straight*** and ***Tapped*** buttons pressed
- select a **UNC** thread type (default)
- set size **1-8** (it's way down near the bottom of the hole list)
- set the depth to **Through All**

and finally

- press ***Counterbore***.

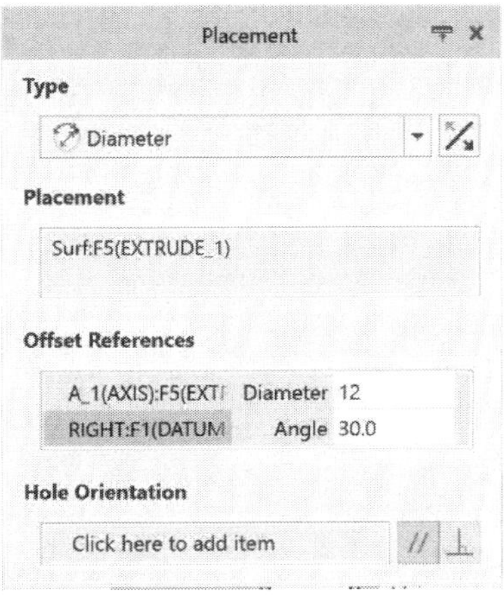

Figure 16 Placement panel for holes

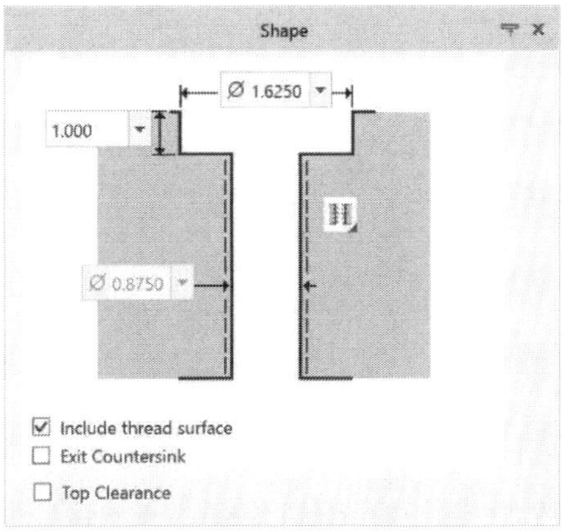

Figure 17 Specifying the counterbore shape and thread specification

Pick on the surface of the top of the protrusion at about the 5 o'clock position. A hole will be previewed. In the ***Placement*** panel change the placement type from *Linear* to ***Diameter***. See Figure 16. Click in the offset references collector to activate it. Now using CTRL, click on the axis of the protrusion (a diameter dimension appears) and on the RIGHT datum (an angle dimension appears). Change the diameter dimension to ***12*** and the angle dimension to ***30*** as shown in Figure 16.

Open the **Shape** panel. Note the default diameter and depth of the counterbore. Leave the checked box beside ***Include Thread Surface*** and change the thread depth to ***Through Thread***, as shown in Figure 17. ***Accept*** the feature.

Click on the note for Hole 2 or select Note under Hole 2 of the model tree. You can explore all the options available for editing the Note. This can eventually be displayed on the part drawing.

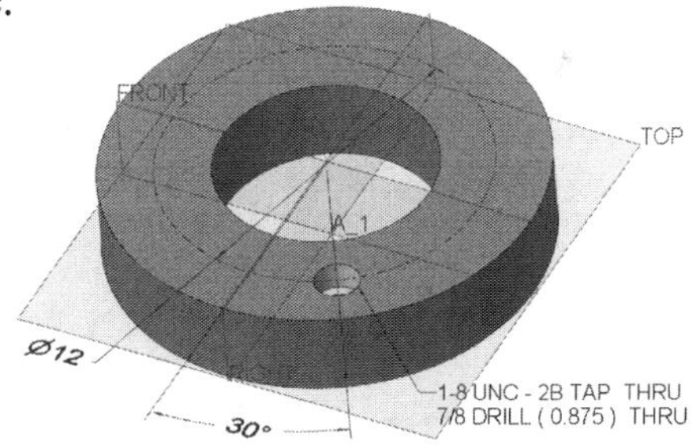

Figure 18 Completed hole with note

If you have been in shaded mode, go to ***Hidden Line***. Creo does not show the physical thread of a hole (that would take too much CPU resource!) but uses a visible indicator to show that a thread is defined. This internal cylinder is called a *cosmetic* thread, and indicates the thread major diameter. Also, if ***Annotation Display*** is turned on in the graphics toolbar you will see a note on a leader attached to the hole as in Figure 18. Notice that the counterbore dimensions are missing. Open the model tree and expand the entry for the hole. The note is listed there[3].

Incidentally, you can create 3D notes for other features as well. These notes can contain any text and are useful ways to attach documentation to the model. In the model tree,

[3] You may have to turn on the display of annotations in the model tree using ***Tree Filters*** and check the box beside **Annotations**.

select the base feature, and in the RMB pop-up menu, select ***Create a Note ➤ Feature***. This opens the **Note** window where you can enter some note text; maybe something like [***Flange base***]. Accept the text. Select the note in the model tree and in the pop-up pick ***Change Note Type*** to specify how and where to attach the note. Come back later and experiment with this. For now, just ***Delete*** the note and make sure ***Annotation Display*** in the graphics toolbar is turned off. You can also turn shading back on.

Now, back to our pattern of holes. The first hole becomes the pattern leader. We are going to make a pattern of 8 instances of the hole spaced equally around the flange. To create each instance we will increment the angular placement of each hole by 45°. This is another example of the importance of planning ahead: if you are going to use a dimension pattern, you must have a dimension to increment! For example, we could not create the bolt circle if we had used a linear placement for the pattern leader (or at least it would be very difficult) or especially not if we had aligned it to either FRONT or RIGHT[4]. As we create the pattern, follow the prompts in the message window. Select the hole and in the pop-up (or ribbon), select ***Pattern***.

This is actually a uni-directional pattern in disguise. There is only one dimension to increment - the angle to the pattern leader. Click on the 30 dimension, and enter an increment of **45**. Change the number of members to **8**. ***Accept*** the pattern. See Figure 19.

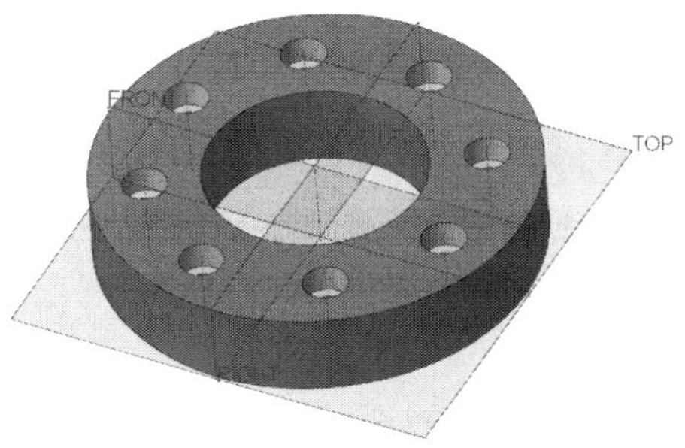

Figure 19 Pattern of holes

Open the model tree to see how the pattern is listed there. Turn on the 3D Annotations button. Does every hole inherit the thread note of the pattern leader? Examine the note carefully - it also contains the number of holes in the pattern.

Setting up Pattern Relations

Suppose we wanted to change the number of holes on the flange and still have them equally spaced - this would change the angular increment. Rather than figure that out and manually edit the increment, we can do this automatically with relations. In the **Model Intent** group of either the **Tools** or **Model** ribbon, select

Relations

Click on the 2nd hole in the bolt pattern (the one at about 3 o'clock). You should see all the dimensions that control the pattern as in Figure 20. Note that some hole dimensions have been removed from this figure for clarity.

---

[4] The ***Axis*** pattern type allows you to create a radial pattern around any axis or straight edge, even if the feature does not directly reference it, but does have some other restrictions. This pattern type is discussed in the *Advanced Tutorial*.

Take note of the symbols for the following dimensions (your symbols might be different): angular dimension between holes (*d17*), the angle of the first hole from RIGHT (*d12*), and the number of holes (*p18*).

With these symbols identified, enter a couple of relations as follows (using your own symbols, of course, and remember that you can easily pick off the dimensions on the graphics window just by clicking on them):

/* angular separation of holes
d17 = 360 / p18
/* location of first hole from RIGHT
d12 = d17 / 2

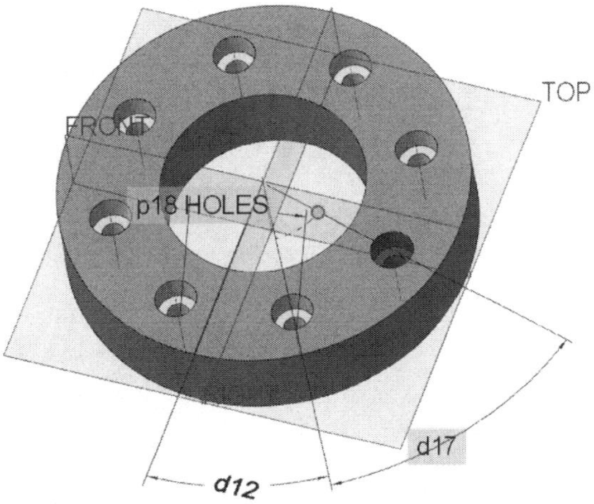

Figure 20 Critical pattern dimensions for setting up some relations

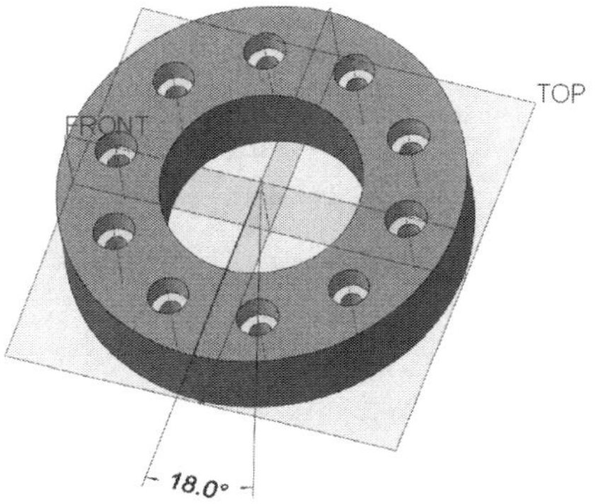

Figure 21 Hole pattern modified for 10 bolts

Note that the second relation uses a value computed by the first relation. In the database, all relations are evaluated top-down. Also, select the ***Verify*** button. Select ***OK***. Accept the Relations window with ***OK***, then ***Regenerate*** for the relations to take effect. The only dimension that will change this time is the angle to the pattern leader. This pulls the entire pattern around slightly. Hold down ***Alt*** and double click on any of the holes and change the number of holes to ***10***, then ***Regenerate*** the part. See Figure 21. If you select the pattern in the model tree, the ***Edit Dimensions*** command on the pop-up shows only the increment and number of instances. Check again for 6 and 5 holes. In each case, the separation and pattern leader placement are automatically determined.

Double click on any of the holes, and change the diameter of the counterbore to ***2***. ***Regenerate*** the model. Note that all pattern members change. With the diameter of 2, try to create a pattern of ***16*** holes, then ***20*** holes. What happens? Note that the pattern members are allowed to intersect each other. It appears that a ***General*** pattern is the default (and, it turns out, only!) hole pattern option. To turn off the display of the cosmetic thread, select any hole and ***Edit Definition***. Open the **Shape** panel and deselect ***Include Thread Surface***.

If you want to play with this part later, then ***Save*** it now (CTRL-s). Otherwise, select ***File* ➤ *Manage Session* ➤ *Erase Current*.**

A Pattern of Grouped Features

The patterns we have seen up to now have involved a single feature as pattern leader. Sometimes, a geometric shape that you want to pattern requires several features to create. In order to pattern these, they must first be grouped together.

We are going to create the part shown in Figure 22. The pattern leader is on the left in the front row. Each instance in the pattern actually consists of three features: a protrusion, a hole, and a round. We will use a pattern to set up two rows with the dimensions of the features incrementing along each row, and between rows. This is the same as the bi-directional pattern we did before, but this time involving three features simultaneously.

Figure 22 Pattern of grouped features

Open the part ***pattern1*** that we used before. Delete the previous pattern, keeping the pattern leader (the cylindrical protrusion). Create a ***Through All***, coaxial hole (diameter ***1.0***) on the axis of the protrusion, and a round (radius ***0.25***) on the edge around its base. The dimensions are shown in Figure 23.

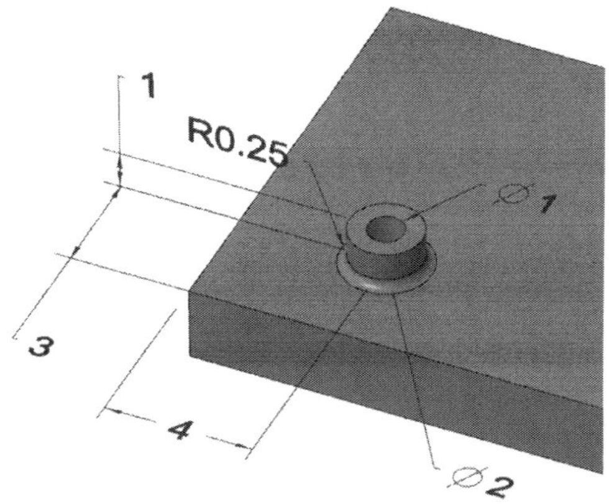

Figure 23 Pattern leader composed of circular protrusion, hole, and round

Before we can create the pattern, we have to group all the features (circular extrusion + hole + round) into a single entity - called (no surprise!) a *group*. Grouped features must be adjacent to each other in the model tree. Select the three features (with the CTRL key). Then select ***Group*** in the pop-up window

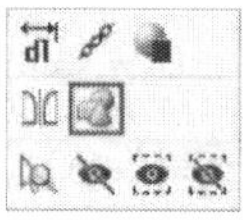

The three entries in the model tree will collapse under a single new entry called a LOCAL_GROUP. With this highlighted, in the RMB pop-up menu, select ***Rename*** and type in a new name like "holder". Expand the entry in the model tree to see the three original features. Note the icon that indicates a group.

With the group highlighted, select ***Pattern*** in the ribbon or pop-up menu. The pattern dashboard opens and you should see all the dimensions associated with the group as shown in Figure 23. We can now select the various dimensions we want to increment in the first and second directions. Select the following (remember to hold down the CTRL key while selecting dimensions to get them in the same collector!):

1st Direction

1. pick on the 4.0 dimension, and enter the increment ***6.0***. This will increment the location of the group along the plate.
2. pick on the diameter of the protrusion 2.0, and enter the increment ***1.0***
3. pick on the height of the protrusion 1.0, and enter the increment ***1.0***
4. pick on the diameter of the hole 1.0, and enter the increment ***1.0***
5. enter the number of instances ***3***

Now left click on the item collector for the second direction.

2nd Direction

1. pick on the 3.0 dimension, and enter the increment ***6.0***. This will increment the location of the group to the next row.
2. pick on the height of the protrusion 1.0, and enter the increment ***4.0***
3. pick on the protrusion diameter 2.0, and enter the increment ***1.0***
4. pick on the hole diameter 1.0, and enter the increment ***1.0***
5. enter the number of instances ***2***

For a summary of all this, open the **Dimensions** panel. ***Accept*** the pattern. It should appear as in Figure 22. Open up the model tree to see how a group pattern is represented.

What dimensions are available for modification (this may depend on which feature you pick)?

What happens here if you try to create a group off the end of the plate by extending the pattern? Change the direction 1 increment from 6 to ***10*** to find out. Fully expand the group pattern in the model tree (including the groups) to see how this condition is represented. How do you recover from this?

Before we save this part file, let's rename it. Select

File ➤ Manage File ➤ Rename

Be sure to select the ***Rename in Session*** option, otherwise the previously saved file on disk is also renamed. Call it something like *pattern2*. Now go ahead and save the renamed part in the usual way and erase it from the session.

Helpful Hint

When you use ***File ➤ Save As ➤ Save a Copy*** you are creating a copy with a new name in the designated directory. The active file in the Creo session still has the original name. This is not the usual Windows "Save As" behavior.

Radial Patterns of Shaped Features

A common modeling problem involves creating radial patterns of shaped (sketched) features. The radial hole pattern we did earlier was of an engineering (placed) feature and was pretty easy. For shaped features, we must be a bit more sophisticated in order to create an angular dimension which can be incremented to produce the pattern. The most common way of doing this is by using a make datum in the creation of the pattern leader[5]. We set up the feature so that the make datum is used as a reference for the sketch. The make datum is typically created using ***Through*** (the axis of the pattern) and ***Offset(rotation)*** (from a datum passing through the axis) so that we can increment the rotation angle to form the pattern. It is crucial when creating the sketch of the feature that there are **no** references, alignments, or dimensions to entities (datums, edges, surfaces) in the part other than the make datum, the axis of the radial pattern, or other axisymmetric features. References to other features will usually not survive a large angular increment of the patterned feature, causing the pattern to fail. This is the most common problem encountered when creating radial patterns.

There are two types of radial patterns, which are determined by how the make datum is used. In the first, the feature is sketched on a plane (the make datum) that goes through the radial pattern axis (or one parallel to this); the feature extrudes in a direction perpendicular to the axis. In this type, a make datum is used as the *sketching plane*. In the second type, the feature is sketched on a plane that is normal to the axis and extrudes parallel to it. The make datum is used as the *sketching reference plane*. We will see examples of both types here.

Radial Pattern using *Make Datum* as Sketching Plane

The first part we will make is (very approximately) the geometry of a turbocharger rotor (see Figure 29). We will take this opportunity to introduce a new feature variation - a *Thickened* extrusion, and a new curve type in Sketcher.

Start a new part called *turbo* using the ***mmns.part.solid.abs*** template and delete or ***Hide*** the default coordinate system.

---

[5] We use a make datum because we don't want the screen cluttered up by a bunch of duplicated datum planes - remember that a make datum is automatically hidden.

Our base feature is a revolved solid protrusion. Sketch this on **FRONT**. The sketch is shown in Figure 24. The ***Axis of Revolution*** is lined up with **RIGHT**. Sketch all the straight edges first.

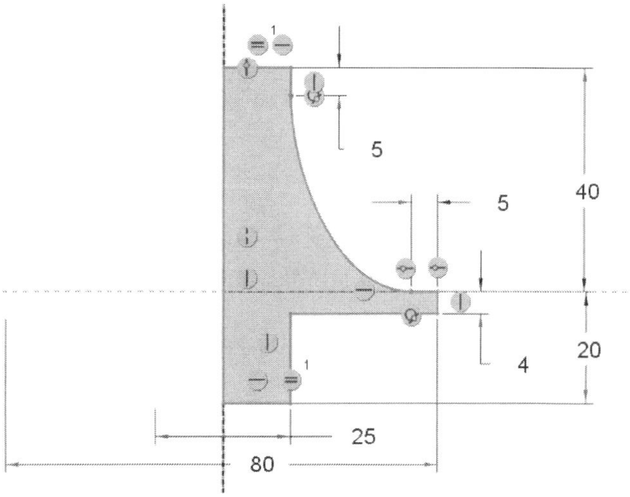

Figure 24 Base feature sketch for turbocharger

The curved edge in this sketch is a *spline.* Select the ***Spline*** Spline button in the Sketching group. Click at the top and bottom vertices of the spline. The curve will rubber band following the mouse. Middle click to indicate no more vertices, then again to leave the spline command. The spline will appear as a straight line between the vertices. You can force the spline to be tangent to the adjacent edges using a ***Tangent*** constraint - just select the two desired entities (spline + edge) and define the constraint. When the sketch is complete, revolve it through 360° to get the shape shown in Figure 25.

Come back later to play with the spline tool in Sketcher. You can control the shape of the spline parametrically by adding and applying dimensions to new vertices (sometimes called knot points) and/or by specifying constraints. The spline is a 3rd order curve, so it will have continuous derivatives up to the second, which means that the curvature changes smoothly. This is an important consideration in some product designs for both functional and/or esthetic reasons.

Figure 25 Base feature

Now we will make our first turbocharger blade to serve as pattern leader. We are going to create a sketching plane through the axis of the revolved base feature at an angle to **RIGHT**. We can then create the pattern instances by incrementing this angle. Since the sketching plane is through the axis, the extrude direction will be normal to the axis. We are also going to use a new type of feature that we haven't seen before - a *Thickened* feature (instead of *Solid*). All we have to sketch is a single line representing the cross section shape of the blade.

Select the ***Extrude*** tool, then select ***Placement* ➤ *Define***. We want to make a datum-on-the-fly here, so at the right end of the dashboard in the **Datum** drop-down panel pick the ***Plane*** tool. Pick on the axis of the revolve. This appears as a ***Through*** constraint.

Holding down CTRL, select the **RIGHT** datum. In combination with the previously selected axis, this gives us the correct ***Offset(Rotation)*** constraint. Change the ***Offset*** rotation value to ***60*** degrees. Accept this with ***OK***.

Back in the **Sketch** window, for the Sketch Orientation select the **TOP** datum with orientation **Top**. Then select ***Sketch***. We are now in Sketcher looking directly at the make datum. For our additional sketching references, select the revolve axis and the top surface of the base feature (we can use this surface because it will be constant for all instances in the pattern). Create the sketch shown in Figure 26. This consists of a single vertical line and a circular arc[6]. When the sketch is complete, select ***OK*** in Sketcher.

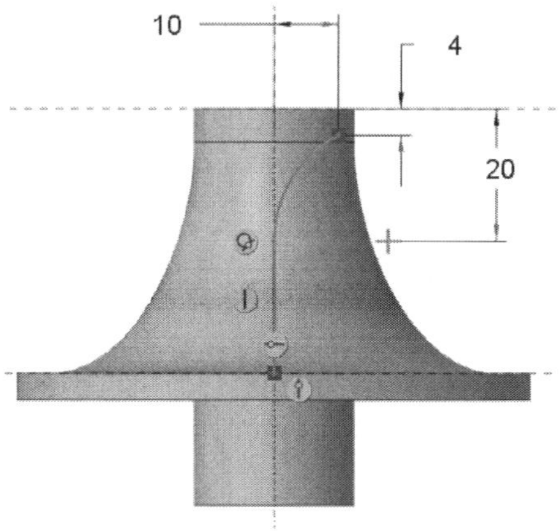

Figure 26 Sketch for ***Thin*** protrusion - turbocharger blade

Since this was an open sketch, you will have to pick the ***Un-attached*** preview option in the dashboard. The preview consists of a single surface. A direction arrow points away from this surface that indicates the side Creo Parametric will add material. This obviously won't work here to create a solid.

Pick the ***Thicken Sketch*** button Thicken on the dashboard[7], and specify a thickness of ***1.0***. Beside this data field is a button that lets you specify which side of the sketched line to add material (either side, or both). Watch the preview carefully as you cycle through these options. We want to thicken the sketch equally on both sides here. For the depth specification, select ***Variable*** with a value ***60***. Verify the blade shape (see Figure 27) and ***Accept*** the feature. In the model tree, select this extrude (note that it contains the hidden make datum) and rename it "**blade**".

Before we complicate the part with the pattern of blades, we will create a revolved cut. Launch the ***Revolve*** tool, then ***Placement*** > ***Define***. The feature is sketched on another make datum that is ***Through*** the revolve axis and ***Parallel*** to one of the vertical flat surfaces of the blade. Use the ***{Datum}:Plane*** command at the right end of the dashboard. For the sketching orientation reference, select ***TOP* ➤ *Top***. The orientation should be as shown in Figure 28. In Sketcher, select as references the top edge of the blade, the outside edge, and axis (for the axis of revolution). Create the sketch shown in Figure 28. You might wonder why it needs to extend past the end of the blade. Come back later to change the dimension 10 to 0 and regenerate - don't do this now! Revolve this cut through 360°. Don't forget the ***Remove Material*** button! Accept the feature.

---

[6] No claims are made here about the aerodynamic suitability of this blade shape, other than it's probably far from ideal!

[7] This is another recent change in terminology in Creo Parametric. Previously, this was known as a *Thin* solid.

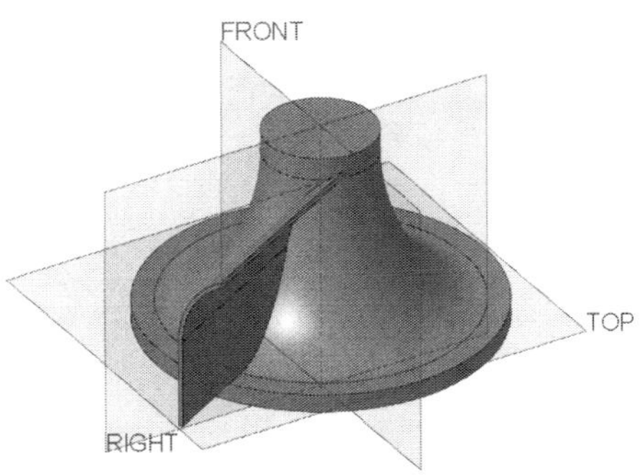

Figure 27 Completed blade

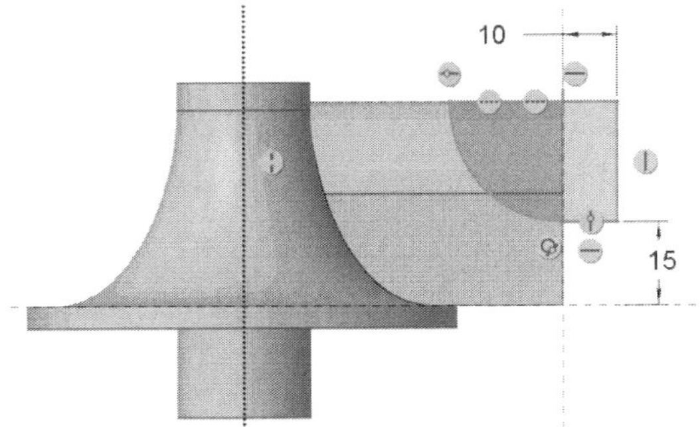

Figure 28 Revolved cut to trim blade

We are now ready to pattern the blade. In the model tree, select the blade and in the pop-up menu, select

Pattern

Select the angle (60) used to create the make datum sketching plane. Enter an increment of ***30*** for this dimension. In the dashboard, set the number of instances to **12**. That's all there is to this pattern. ***Accept*** the feature.

Open the model tree to observe where the pattern is. Recall that we created the pattern *after* we created the revolved cut, yet it appears immediately before the cut in the model tree. That is, patterns are created "in place" alongside the pattern leader. This was actually fortunate, since it allowed us to create the sketch for the revolved cut on a fairly simple view without a lot of additional clutter. Try opening the sketch of the cut now to see what it looks like with all 12 blades in view.

Figure 29 Completed pattern of blades

Why could we not use a vertical surface of the blade as the sketching plane for the revolved cut? Try that and see what problem arises. This will be more obvious if you make the blades thicker. Could we have used either the right or front datum plane? Save this part (CTRL-s) and then erase it from the session.

Let's look at another variation of radial patterns using a make datum.

Radial Pattern using *Make Datum* as Reference Plane

In this example (Figure 30), we will make the five cutouts in the web around the axis of the wheel. The feature extrusion direction for the pattern leader is parallel to the axis of the radial pattern. We must do something a bit different from the turbocharger. The main idea is the same - incorporate a make datum created using ***Through*** and ***Offset(rotation)*** into the sketch references. The angle parameter can then be incremented to produce the pattern. In the turbocharger, the make datum was the sketching plane; in this part, the make datum will be the sketching horizontal reference plane.

Figure 30 Wheel with radial pattern of cuts.

Start a new part called *wheel* using the ***inlbs_part_solid_abs*** template and start the ***Revolve*** tool. The sketch plane is **TOP** and the axis of the revolve goes through **RIGHT**. The sketch is shown in Figure 31. Hint: Try creating the top half of the sketch and then mirroring it across a construction centerline. Revolve the sketch through 360° and accept the feature.

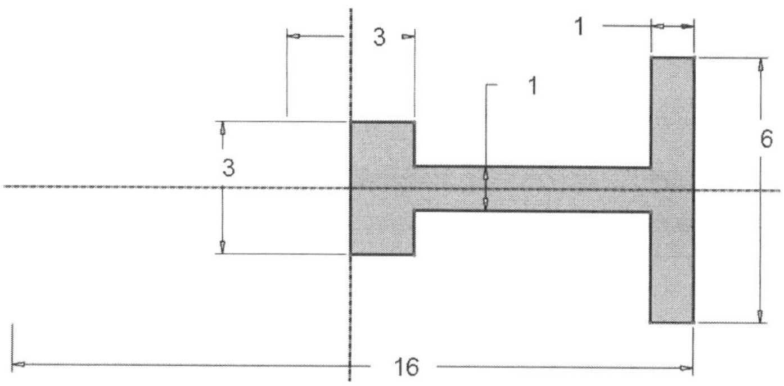

Figure 31 Base feature of **wheel** - revolved protrusion. (Constraint display turned OFF.)

The pattern leader will be a both-sides extruded cut, sketched on **FRONT** (the extrusion direction is therefore parallel to the axis of the wheel - compare this to the turbocharger blade where the extrusion direction was perpendicular to the axis). We want to pattern this feature around the axis of the base feature.

Select the ***Extrude*** tool and ***Placement ➤ Define***. For the sketching plane pick **FRONT**. For the sketch orientation reference plane, ***Remove*** the default RIGHT datum plane using the RMB pop-up in the *Reference* collector. Select the ***Datum Plane*** tool in the ribbon and create a make datum ***Through*** the axis of the wheel and ***Offset*** from the **TOP** datum. Enter an offset rotation angle of ***30*** degrees. Accept the make datum with ***OK***, make sure the reference is facing ***Top*** and proceed into Sketcher.

We are now looking at **FRONT** with the make datum **DTM1** facing the top of the screen. The edge views of **RIGHT** and **TOP** are rotated a bit. We must be very careful now about picking sketch references. **DTM1** is already selected. All we need to add is the axis of the wheel (when selected, you will see a small X there). You may have to spin the model a bit (or use the Search tool) to select the axis. Notice in the reference window that we are fully placed (press ***Update***).

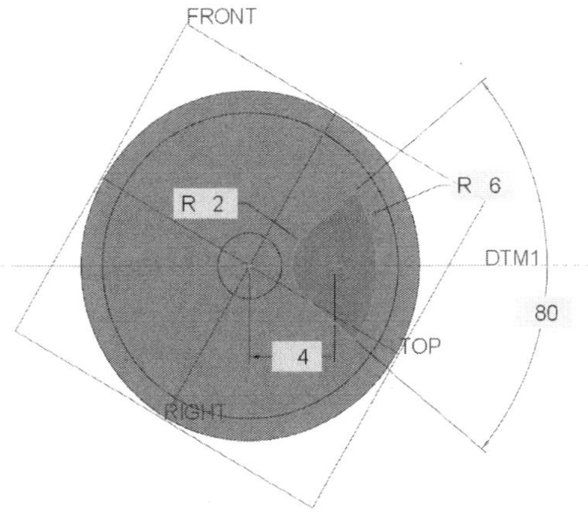

Figure 32 Sketch of pattern leader (note rotation of TOP and RIGHT). Constraint display is OFF.

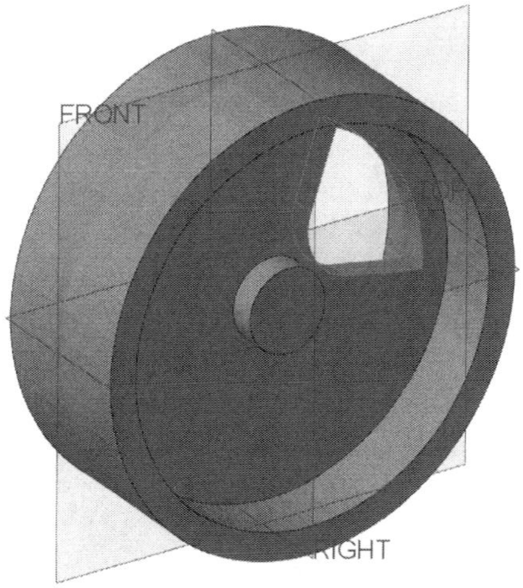

Figure 33 Pattern leader completed

The sketch for the cut is shown in Figure 32. Be sure to avoid any dimensions or constraints that involve the fixed datums **TOP** and **RIGHT**. If you turn them off, you don't have to worry about this since you can't select them by mistake.

When the sketch is complete, back in the dashboard **Options** panel set the Depth Spec ***Through All*** on both sides. Select the ***Remove Material*** button (this may happen automatically) and ***Accept*** the feature. See Figure 33.

We can now pattern the cut. Open up the model tree and select the extruded cut. This contains the make datum reference. In the pop-up menu, select ***Pattern***. In the dimensions on the screen, find the 30° dimension we used to locate the make datum reference plane from the TOP datum and click on it. Enter the increment *72°*, and up in the dashboard specify ***5*** members. There is nothing to do in the second direction, so the pattern definition is complete. Open the ***Options*** panel and select the ***Regeneration Option(Identical)***. This means that all instances in the pattern will have the same shape and use the same references. We are finished defining the pattern, so select ***Accept***. The pattern should now be created as in Figure 30.

What happens if you try to make a pattern of 3 cuts at 120° increments? How about 6 cuts at 60° degree increments? Would you say this is a *robust* model? Go back to the pattern dashboard using ***Edit Definition*** and in the ***Options*** panel, change the regeneration option to ***General***. At the expense of somewhat increased processing time, this allows pattern instances to change geometry (and intersect). Now try the pattern of 6 instances. Obviously, this wheel would not work, but we have successfully decoupled the pattern creation from the shape of the pattern instances, making a more robust model. We could now go back to the pattern leader and change its geometry without fear that the pattern will fail.

Save the part (CTRL-s) and erase it from the session.

This concludes our discussion of patterns. We have discussed **Dimension** patterns only here. There are several other types and many more things you can do with patterns, and some more advanced techniques. For example, instead of simply incrementing dimensions between instances, you can use *pattern relations* to develop formulas that will control the instance placement and geometry. Another powerful tool called a *pattern table* allows you to create instances at non-uniformly spaced locations driven by dimension values stored in a table like a spreadsheet. This is common for specifying hole locations and diameters in sheet metal parts. A *fill pattern* will duplicate features (like holes or slots) in a regular geometric pattern in order to cover a region bounded by a datum or sketched curve. Some examples are shown in the figures below.

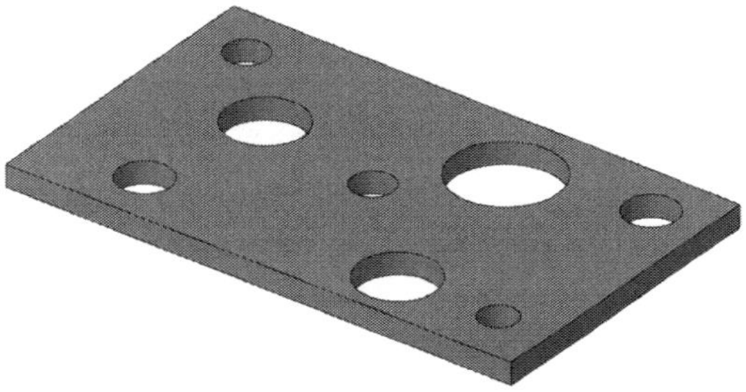

Figure 34 A **Table**-driven Pattern

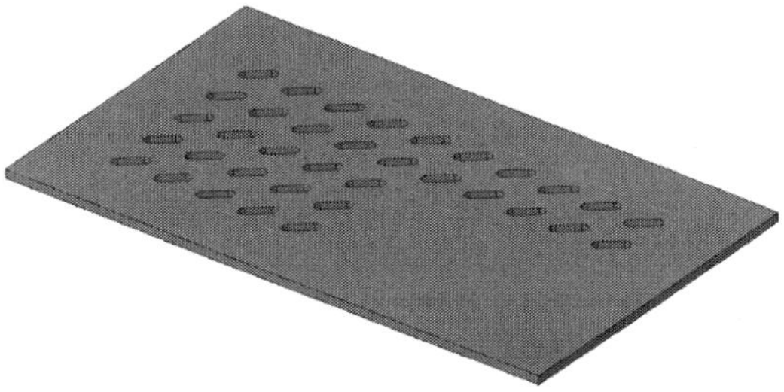

Figure 35 A **Fill** Pattern

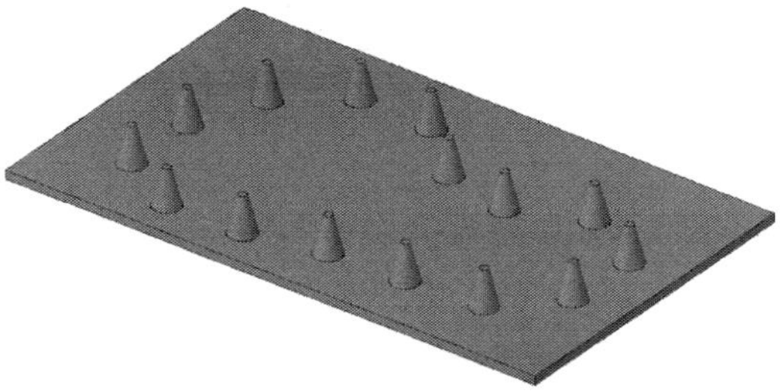

Figure 36 A **Curve** Pattern

Figure 34 shows a pattern of holes whose location and diameter are specified in a spreadsheet-like table. New holes are created simply by adding entries to the table. Existing holes are changed by editing entries. Figure 35 shows a pattern of instances bounded by a closed sketched curve. Fill layouts can be rectangular, hexagonal, or spiral. Parameters include minimum spacing from the boundary, orientation of the pattern, and spacing between instances. Figure 36 shows a specified number of instances arranged with equal spacing along a closed curve defined by a sketch. A curve pattern can also be set up with a specified increment. All these pattern variations can be used for individual features (extrusions or revolves), groups of features, bodies, and even (when we get to assemblies) patterns of parts. These advanced pattern functions are discussed in the *Advanced Tutorial* from SDC.

Copying Features with *Paste* and *Paste Special*

In the previous sections, we saw how to create a multiple-instance dimension-driven pattern of a single feature or a group of features. The pattern was created by incrementing one or more of the feature's existing dimensions. The **Copy** command creates only one copy at a time but allows more flexibility in terms of placement and geometric variation (you aren't restricted to the dimensions or references used to create a pattern leader, for example). Copying is not quite as simple as in a word processor. There are several options available with ***Copy*** and we will create several different simple parts to illustrate these. Note that the ***Copy*** command also allows copying features to a different part.

When only a single copy of a feature is desired, the first and main consideration is whether or not copying is actually more efficient than simply creating a new feature from scratch at the desired new location. If the feature is simple enough (like a simple hole), this may actually be quicker. However, if a copy is truly desired (especially a dependent copy), the following must also be considered:

- Should the copied feature be dependent or independent of the original feature?
- Will the copied feature use the same references as the original?
- Will the copied feature use the same dimension values or different ones?
- Which dimensions will be driven by the parent and which will be independent?
- Is the copied feature created by simple translation or rotation (or some combination) of the original?

As you can see, there are quite a few things to think about. We will explore a number of possibilities. The main option, after identifying the feature to be copied, is to choose one of the following operations:

Paste - creates an independent copy using the copied feature's dashboard interface to edit references. This is closest to how a paste operation works in a word processor. This is like creating a new feature but with the dashboard options already set.

Paste Special - allows choice of new references, variable dimensions, translation, rotation, and more. The copied feature can be independent, partially dependent, or fully dependent (default) on the original feature.

The fundamental difference between these is whether the copied feature will by default be independent (***Paste***) or dependent (***Paste Special***) with respect to the original.

Copying using *Paste*

Our first example of Copy/Paste will result in the part shown in Figure 37. The tab with hole on the left at the top is the original extrusion. The tabs on the top right and the right side are copies. This would be difficult to do with a pattern due to the change in reference surfaces and orientation of the two copied tabs. For this example, we will use the ***Paste*** command, which will result in the three tabs being independent of each other.

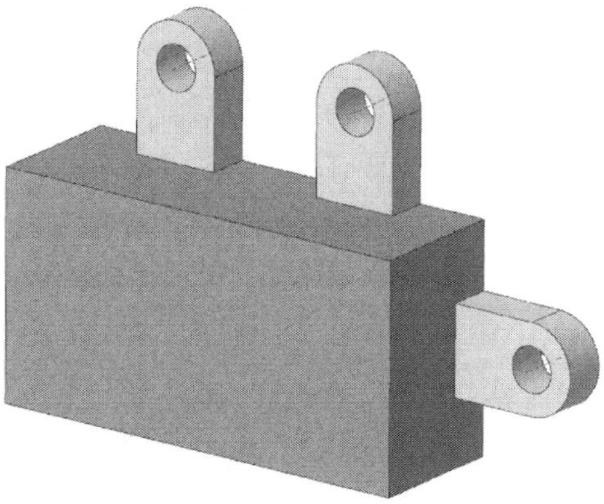

Figure 37 Part with original and two copied tabs

Start a new part [***copies1***] in inches. Create the extrusion shown in Figure 38. This is sketched on the FRONT datum and centered on the datum planes RIGHT and TOP.

Now create the first tab using an extrusion sketched on the FRONT datum. Use the top of the block as the top reference for Sketcher. The vertical reference is RIGHT. The sketched shape of the tab is shown in Figure 39. You will have to close the sketch across the bottom (you can't have a mix of open and closed curves in a single sketch). The feature has a variable depth of 3. The finished tab is shown in Figure 40.

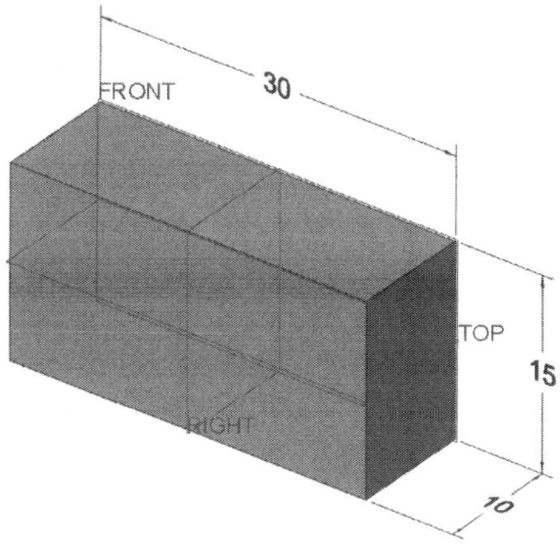

Figure 38 Base feature

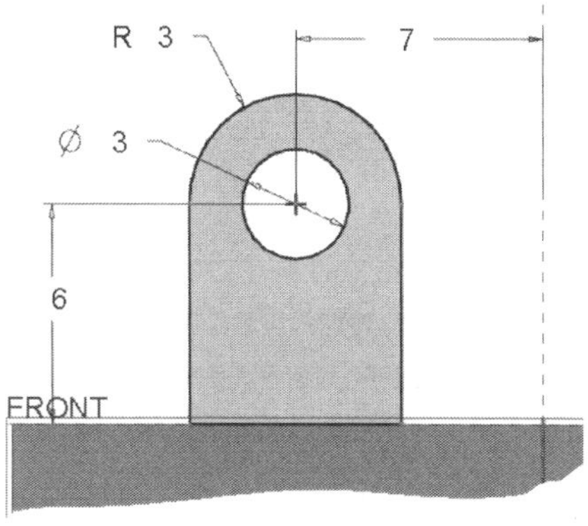

Figure 39 Sketch for first tab.
(Constraint display OFF)

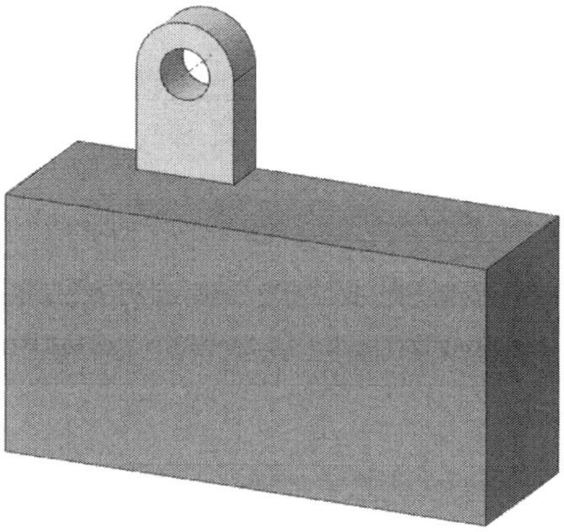

Figure 40 First tab completed

Select the tab extrusion (either in the model tree or on the screen). With the tab highlighted in green, select the ***Copy*** button in the **Operations** group in the ribbon or RMB pop-up (notice the standard Windows keyboard shortcut ***Ctrl+c***).

This places the feature in the clipboard. Now in the **Operations** group select the ***Paste*** button (or use the keyboard shortcut ***Ctrl+v***). Don't confuse this with ***Paste Special*** (***Ctrl+Shift+V***) which is right below it on the drop-down list or beside it in the pop-up. The difference between these buttons is sub-microscopic.

Because the feature we are pasting was created using the extrude dashboard, that now opens and we must specify some required references for the new pasted feature. The missing references are indicated by the red label on the **Placement** panel. Notice the red dot in the data field for the sketch. Select the ***Edit*** button. Missing references in the Sketch dialog window are indicated by the red dots. The first required reference is the sketch plane and the second is the sketching reference. These are both the same as before, so select ***Use Previous***.

We are now looking at the sketching plane. If you move your cursor, you will see a dark red outline of the sketch. Drag that to a position to the right of the first tab, and left click to drop it. We can now use the usual Sketcher tools to complete the placement. For example, constrain the bottom edge of the sketch to the top of the block and dimension the center of the hole to the RIGHT datum. See Figure 41. Accept the sketch and specify a blind depth of 3 as before (this is actually brought along from the original).

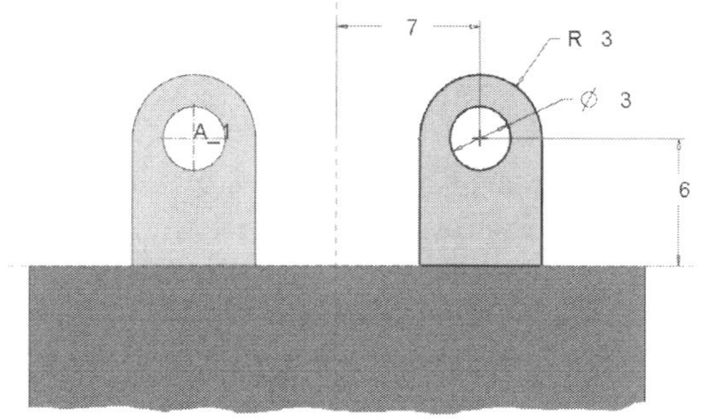

Figure 41 Pasting the copied feature. (Constraint display turned OFF)

Let's copy the original tab once more. It is still in the clipboard so all you have do is ***Paste*** (CTRL-v). Once again, the extrude dashboard opens with missing references indicated with red. Open the ***Placement*** slide-up panel (or use the RMB and select ***Edit Internal Sketch***). Select the FRONT datum as the sketching plane. Select the block's right vertical surface as the TOP sketching reference plane. This will maintain our sketch orientation. Now select ***Sketch***. The dark red outline of the sketch appears attached to the cursor. Place that so that the bottom edge aligns (more or less) with the surface of the block. Use the ***Constrain*** commands in Sketcher to fix the bottom edge onto the block surface, and align the center of the hole with the TOP datum. See Figure 42. Accept the sketch and the feature with the default blind depth of 3.

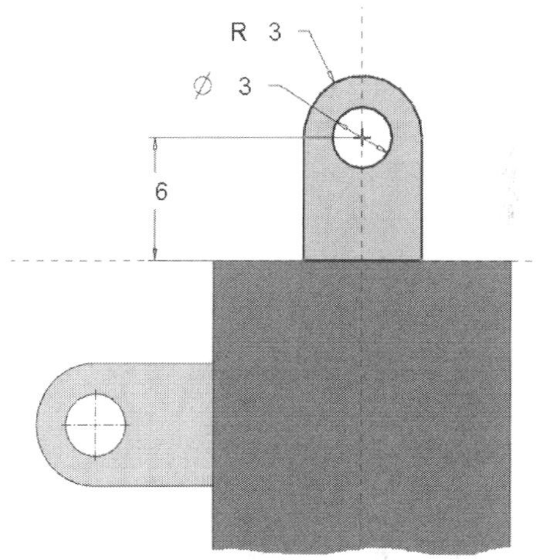

Figure 42 Pasting the second copied feature

You will have noticed that in both paste operations, the sketch stayed in the same orientation as originally created, and we had to manipulate the part under it to get the desired final orientation of the feature.

To show that ***Paste*** produces independent features, go to the first tab and use ***Edit*** to change the hole diameter from 3 to **1**. On the second tab, change the height dimension from 6 to **10**. On the third tab on the end, use ***Edit Definition*** to change **Variable** depth to **To Reference** and pick on the front surface of the block. After regeneration, the part should look like Figure 43. You can also examine the parent/child relations for each tab to check that they are independent of each other, or just suppress the first tab.

So, the ***Paste*** command uses the dashboard interface for the copied feature, and automatically creates independent copies. This may be useful in taking features from one model to another[8].

What if you do want all the tabs to be the same size and shape? It is possible to relate pasted features together either using relations or by using the geometry of the original feature as a reference in the pasted feature. For example, in our first copied tab, we could either specify a relation for the height dimension (setting it equal to the corresponding dimension in the original) or we could have selected the hole axis in the original as a reference for placing the sketch in the copied feature (thus eliminating the height dimension in the copy). This seems like a lot of trouble. So, if you want to make the copied feature dependent, it is probably advisable to use the ***Paste Special*** command instead.

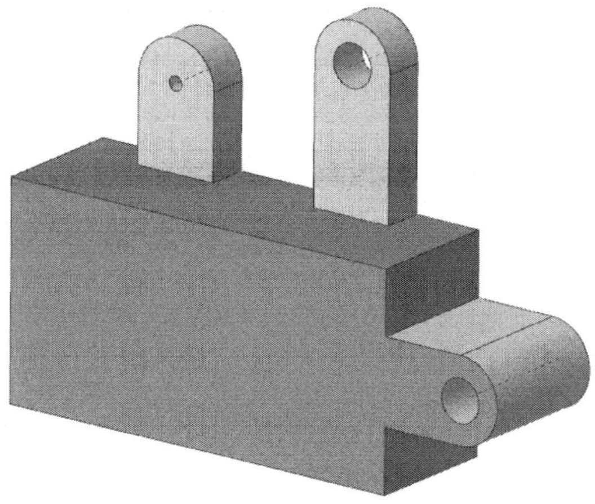

Figure 43 Exploring independence of copied features created using ***Paste***

Save the current part and remove it from your session.

Copying using *Paste Special*

Unlike ***Paste***, which is used to create independent copies, the default operation for ***Paste Special*** results in dependence of the copied feature on the original. This mostly involves the dimensions of the feature, although other elements (such as annotations) are also involved. There are several variations of ***Paste Special***, hence more options to consider, and we will have a look at some. The first copy we will make with ***Paste Special*** will use all defaults. An important point to note is that, once created, the dependent status can be changed to various forms of independent status. However, once declared to be independent, a copied feature cannot be redefined as dependent (unless additional steps are taken such as creating relations).

Create a new part [*copies2*] in inches. The base feature is an extruded protrusion, with the sketch created on TOP with the right reference being RIGHT. Once you are in Sketcher, select the ***Palette*** tool . This opens the window shown in Figure 44.

---

[8] If you are going to be using the same feature in many models, you may want to create a ***User Defined Feature*** (UDF) that can be stored in a feature library. UDFs operate much like the Paste portion of a Copy/Paste operation. UDFs can be set up to be independent, partially or fully dependent on a master. This is discussed in the *Advanced Tutorial*.

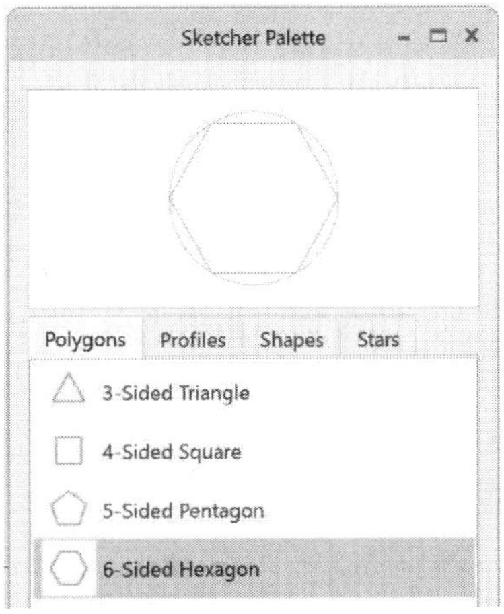

Figure 44 Selecting shapes in the Sketcher palette

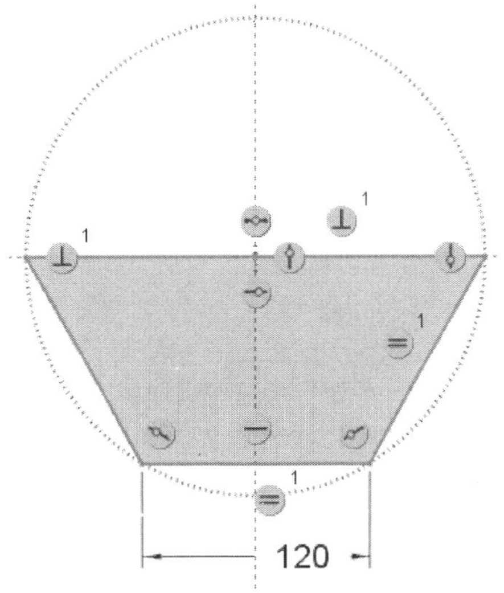

Figure 45 Modified sketch of hexagon

The **Sketcher Palette** tool allows you to select from a number of pre-defined shapes as indicated by the tabs. If you have previously saved sketch sections in your working directory, this will appear as an additional tab. In the **Polygons** tab, select the **6-Sided Hexagon** (that seems a bit redundant!) list entry with the LMB. This opens a preview of the shape at the top. Select the hexagon entry highlighted in Figure 44.

From the Sketcher Palette window, LMB click and hold down, the 6-Sided Hexagon shape and drag it into the graphics window, snapping it to the FRONT and RIGHT Datums. Check out the indicated drag handles for location, rotation, and scale of the sketch at the center, top right, and bottom right corners, respectively. You can also specify the scale and angle of the sketch in the **Import Section** dashboard. Enter a **Scale** value of **120** and middle click, and close the palette window. If you did not snap it to the center location, use the Sketcher ***{Constrain}: Coincident*** command to align the center of the hexagon with the sketching references. As shown in Figure 45, create a single line across the center of the hexagon, and delete the sketched lines on the top half of the hexagon. Accept the sketch, and (IMPORTANT) set the symmetric **Variable** depth of the extrusion to **120**. Accept the feature.

We will have need of a datum axis defined at the intersection of FRONT and RIGHT. Create that now.

The feature we are going to copy is an extruded cut created on the right surface of the base feature. Launch the ***Extrude*** command and select this surface for the sketching plane. Select the TOP datum as the top sketching reference. Once you are in Sketcher, select the datum axis for the second sketching reference. Open the ***Palette*** tool again. In the **Shapes** tab, select the **Cross** shape. Drag and drop the shape onto the sketching plane.

Enter a **Scale** value of **20**. Middle click, then ***Close*** the Palette window. Use the ***Coincident*** constraint to align the center of the cross sketch with the sketching references. The sketch should be driven by only two dimensions - see Figure 46 (constraint display has been turned off). Accept the sketch. In the Extrude dashboard, set the **Variable** depth to **20**. Pick on the magenta direction arrow to extrude into the part. The feature is automatically switched to a cut (check out the **Remove Material** button in the dashboard). The finished feature is shown in Figure 47. To make it easier to refer to the original feature, rename it to **Cross**. We will use this feature to explore a couple of variations of the ***Paste Special*** command, paying particular attention to the different levels of dependency between the original and copied features.

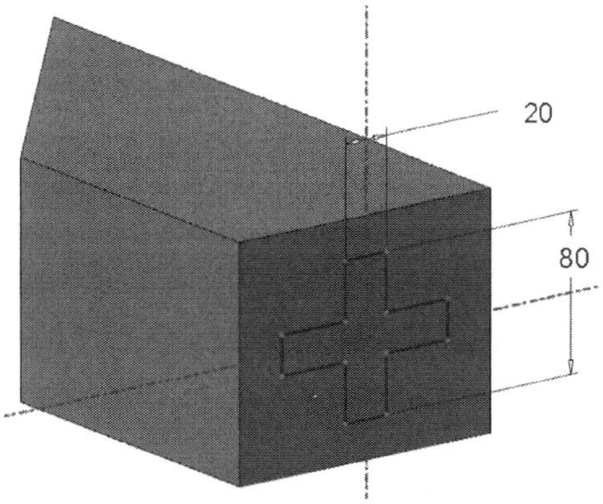

Figure 46 Sketch selected from the palette

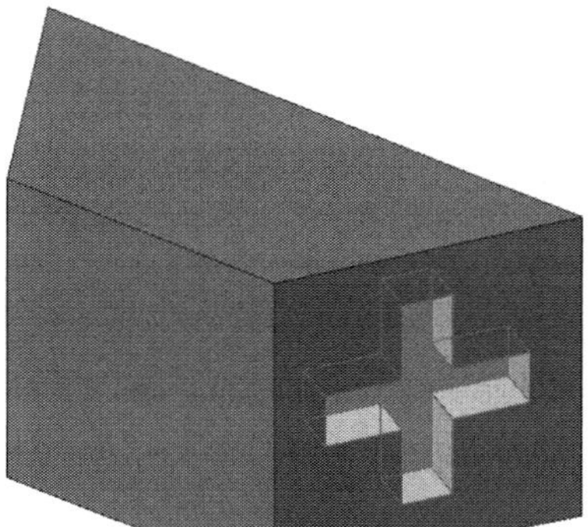
Figure 47 Original extruded cut

Select the cross feature (highlight in green), then pick the ***Copy*** command (or ***Ctrl+C***). Now select the ***Paste Special*** command in the RMB pop-up menu. It looks almost the same as the ***Paste*** command, so you will have to pay attention to the pop-up tool tips.

A new dialog window opens as shown in Figure 48. We will accept all the defaults in this window[9]. As you move your cursor over the window, a pop-up and the message window will give you a bit more information about each option. Do not change any settings yet, and accept the settings shown in Figure 48 with ***OK***. Come back later to explore some of the other options. These allow you to create copies that have some dependent and some independent references and dimensions.

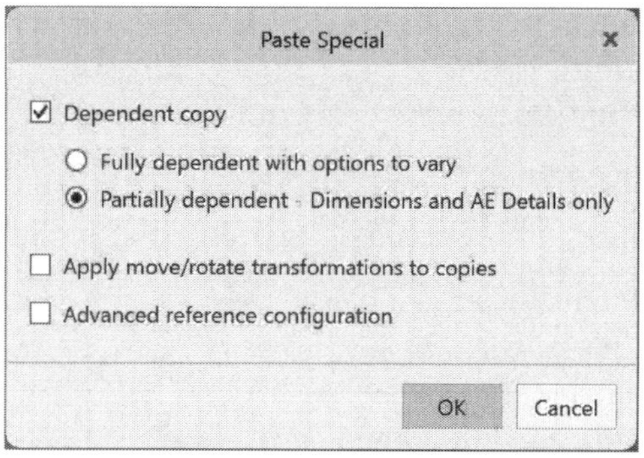

Figure 48 The **Paste Special** dialog window

[9] It is possible to change the defaults shown in Figure 48 using the system configuration file, *config.pro*. The ones shown are the unmodified default settings.

The feature dashboard opens and we must now specify the new references for the copied feature. Missing references are indicated using red. This behavior is the same as for ***Paste*** command. We will see at the end, however, that the copy will be dependent on the original (unlike ***Paste***).

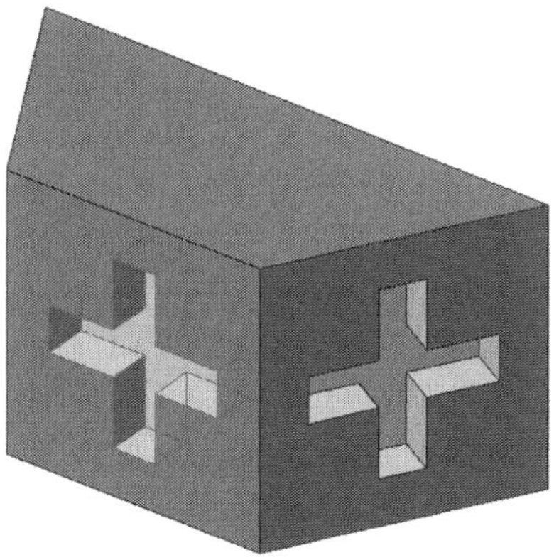

Figure 49 The copied extruded cut

In the **Placement** panel select the ***Edit*** button. Select ***Yes*** in the warning window. For the new sketching surface, select the front surface of the base. The sketching orientation reference is again the TOP datum, to face the top of the screen. Now select ***Sketch*** and you will now reorient to the sketch view. The cross sketch will show in dark red, and you can drop it on the sketching plane. Use the ***Constraints*** tool to align the cross centerlines with the sketch references. Accept the sketch. The original blind depth (20) has been carried over from the original. The final copied feature is shown in Figure 49. So far, this operates the same as ***Paste***.

Select the original cross, and edit the height dimension from **80** to **60**. Change the width from **20** to **10**. ***Regenerate*** the part. The new geometry is shown in Figure 50. Note that the copy is also changed, indicating its dependence on the original.

Figure 50 Dependency of the copied cut

Select the copied feature, and change the height dimension from 60 to **80**. Change the width from 10 to **40**. ***Regenerate*** the part. Both cuts change, indicating that the dependence is bidirectional. Curiously, this dependence is not reflected in the normal parent/child relations in the **Reference Viewer**. Check it out! What are the parents of the copied cut? Does the original CROSS have any children? Can you ***Suppress*** either of them without affecting the other?

Making Dimensions Independent

To break the dependency of selected dimensions in the copy, highlight the copy and select ***Edit Dimensions*** in the pop-up menu. Pick the blind depth dimension to highlight in green. Then in the RMB menu, select ***Make Dimension Independent***. In the Confirmation window pick ***Yes***, then double click the dimension again and enter a new value for the depth. ***Regenerate*** the part. The depths of the two cuts are now independent of each other, while the other dimensions remain dependent. Experiment with the dimensions of the two cuts to prove the dependent/independent status. We are finished with this part, so you can save it and remove it from your session.

Paste Special Using Translated and Rotated copies

To demonstrate a translated copy, we will use the part shown in Figure 51. The vertical plate with hole on the left is the original, and the one on the right will be a dependent copy. Start by creating a new part [*copies3*] with the default template. Create a rectangular solid protrusion on **TOP** that is **10 x 20 x 2 thick**. Line up the left face of the block with the **RIGHT** datum, and the back face with **FRONT**.

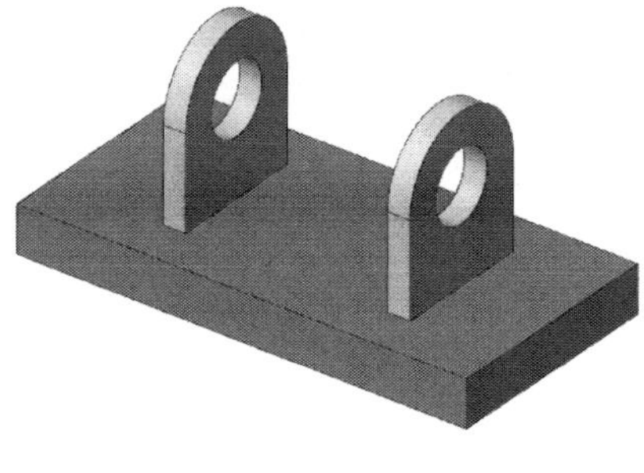

Figure 51 Part with copied feature

For the first vertical plate, the sketching plane for the extrude is a make datum that is ***Offset*** from **RIGHT** by ***5***. Use the selection filter if you have trouble picking this. Select the top of the rectangular base as the ***Top*** reference plane. Then sketch the protrusion as shown in Figure 52. Note the sketching references. Also, the sketch must be closed across the bottom since you can't have a mix of open and closed curves in the same sketch. The extrusion has a ***Variable*** depth of ***1***. The part should look like Figure 53.

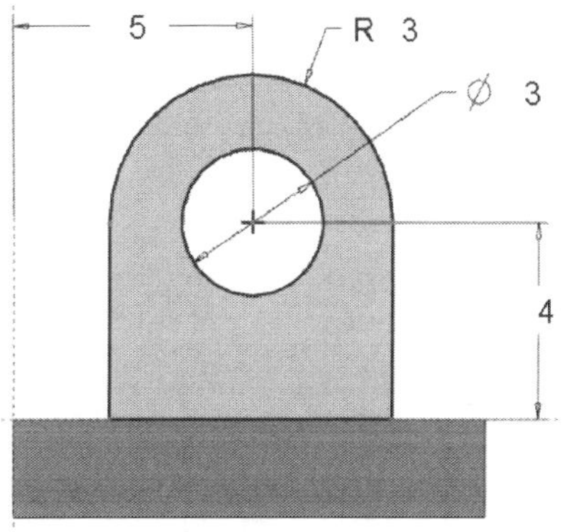

Figure 52 Sketch for vertical plate (Constraints OFF)

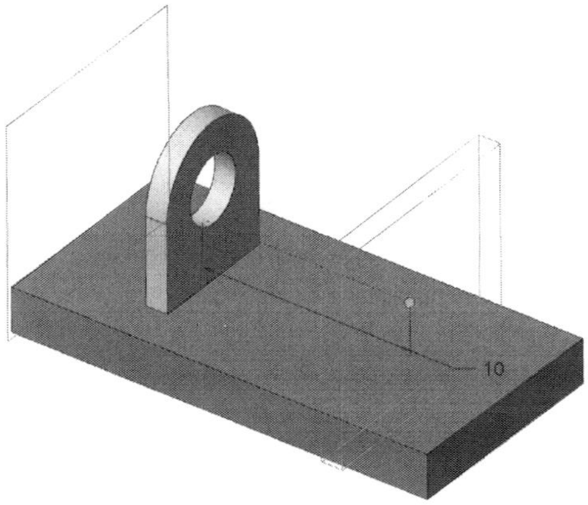

Figure 53 Making a translated copy

Now, we are ready to copy the feature. We want the copy to be translated 10 units to the right of the first. If the geometry of the first feature changes, we want the copy to change too. Highlight the vertical plate, then select ***Copy*** ➤ ***Paste Special***. Keep the **Dependent** ***Copy*** option selected, and check the box to ***Apply move/rotate transformations*** ➤ ***OK***. A new dashboard opens. On the dashboard, the two buttons at the left are for translating (default) and rotating the copied feature. The feature collector beside these is waiting for us to specify a reference for the translation. (Read the message line.) The translation will be normal to a selected surface, or along a selected edge or axis. Pick the RIGHT datum plane. A drag handle appears on the previewed feature - drag this out to 10 units as in Figure 53. Accept the feature with a middle click.

How does the copy appear in the model tree? What happened to the make datum in the copied feature? How can you change the translation distance? (HINT: Pick various elements in the model tree and use the ***Edit Dimensions*** command in the pop-up.) What happens if you try to change the hole diameter on the first protrusion? Or the height dimension on the copy? What happens if you suppress the original? The copy?

Select the **Moved Copy** item in the model tree and in the pop-up menu pick ***Edit Definition***. Open the **Transformations** panel. This lists our first translation. If you pick ***New Move***, you can now add a second translation by picking a new reference (plane or edge) and translation distance. Move the copy 2 units away from FRONT.

Can you make the copied plate independent of the original? (Hint: Select the copied extrude feature in the graphics window and in the RMB pop-up use ***Copied Feature > Make Section Independent***.) Can you suppress or delete the original?

You can now ***Save*** this part. Come back later to explore other options.

Now, we will use a rotated copy to create the part shown in Figure 54 - a large circular pipe with two pipes joining it off-axis. At the same time, we will see a situation where feature creation order can be used to advantage (or foul you up!). The original side pipe is on the left; the rotated copy is on the right. It can be obtained by a 180E rotation around the big pipe axis. This is not a mirrored feature.

Figure 54 Part with Rotated copy

Create a new solid part [***copies4***] using inches. Start by creating a circular solid ***both-sides*** (symmetric) protrusion from the sketching plane **TOP**. Use **RIGHT** and **FRONT** as sketching references. Sketch a circle with a diameter of ***20*** and set ***(Symmetric)Variable*** depth of ***20***. Do not add the inner surface of the pipe at this time - we will do that later. This may not be an obvious thing to do (for now!) but we have a situation where feature creation order is important as discussed below.

For the first side branch, create a one-sided solid protrusion. Use **FRONT** as the sketching plane (**Top** reference **TOP**) and sketch an ***8*** diameter circle ***aligned*** with TOP and with a center ***5*** from RIGHT (Figure 55). Check the feature creation direction arrow. Make the protrusion with a ***Variable*** depth of ***15.*** This will extend it outside the circumference of the main pipe.

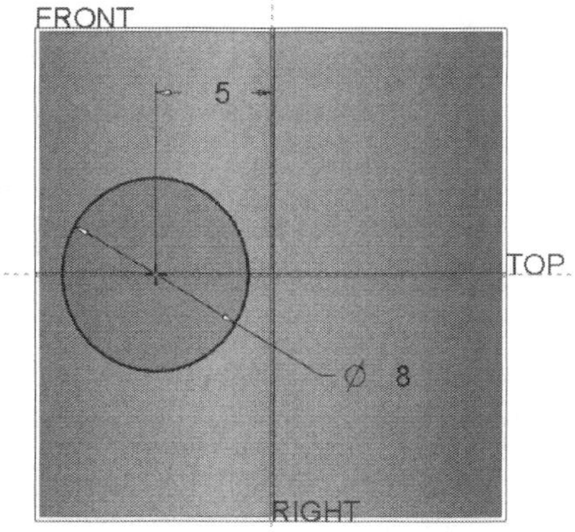

Figure 55 Sketch for side pipe

Since we are starting the extrude inside the existing solid, the ***Remove Material*** button may be selected automatically. If that happens, just switch it back.

Create a ***Simple, Coaxial hole*** on the axis of the side pipe. The hole diameter is *7.* IMPORTANT: Use the placement plane **FRONT**. We want ***Through All*** so that no matter how long the side pipe is, the hole will always exit the part. The part should look like Figure 56.

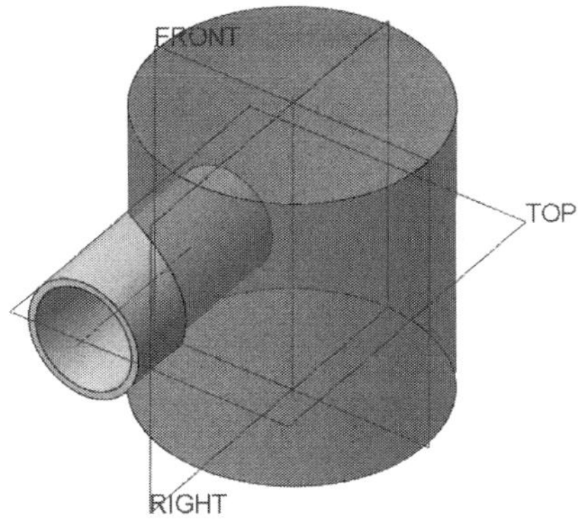

Figure 56 First pipe (Note the vertical cylinder is still solid)

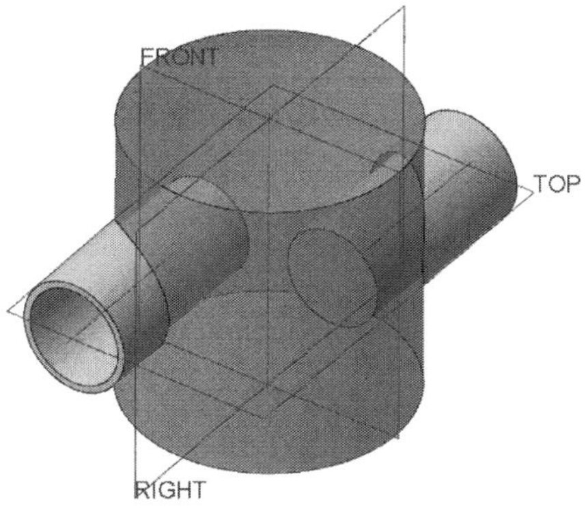

Figure 57 Rotated copy completed

Now we will create a rotated copy of both the sidepipe extrusion and the coaxial hole. Select these in the model tree (holding down the CTRL key), then select ***Copy* ➤ *Paste Special***. Turn off the ***Dependent copy*** option and check the ***Apply move/rotate transformations*** option, then ***OK***. In the dashboard that opens, select the second button on the left for a rotation. Now select the main axis of the large pipe as the rotation reference. Drag the rotation around to 180° and accept the feature. How do the two copied features appear in the model tree?

Since we have created an independent copy, check that you can change any dimensions of the copied features without affecting the other pipe. Notice that we did not have to group these features to copy them (as would be required for a pattern).

Now we can add the central ***Hole*** of the main pipe. Make it a ***Simple***, ***Coaxial*** hole from the placement plane **TOP**. Make it ***Through All*** in both directions, with a diameter of ***19.*** This completes the model, so ***Save*** it now. We are going to explore it a bit, which might mess it up.

Exploring the Model

Now, you may be wondering why we left the central hole until last. Let's experiment with the ***Edit Dimensions*** command, changing diameter dimensions of both the original and the copy. You can also modify the rotation angle. You should be able to modify both branch pipes with no problem. What happens if you modify the diameter of the main pipe to ***12*** and hole to ***11?*** The part will certainly regenerate but is clearly wrong. However, the error is relatively easy to fix. Consider what would happen if we had used the following "obvious" sequence (what is important here is the order that features are created - you might like to sketch each feature in the following sequence as it is added to the part [or actually make a new part] in order to visualize the problem that would arise):

1. **create main pipe** - same geometry as before.
2. **create central hole** - same geometry as before (but in a different order).
3. **create side branch** - We couldn't do this from **FRONT** since that would be inside the pipe (that now has the inner hole in it). We would have to create a **Make Datum** outside the pipe using an offset of 15 from **FRONT** and create the branch towards the main pipe using **To Next** or **To Selected** depth.
4. **create the side branch hole** - We could use the planar face of the branch as the placement plane for a coaxial hole with a depth specified as **To Next** (through the next part surface encountered, i.e. the inside surface of the big pipe). Note that a ***Through All*** would go out the other side.

These steps would create the same original geometry. However, we would have a big problem if we tried to reduce the diameter of the main pipe to anything less than 18, as we did above. Why? At step 3, the side branch solid protrusion would not totally intersect the surface of the main pipe as required by the **To Next/Selected** depth setting. The part would not regenerate at all, and we would have to spend some time fixing the broken features in the model. This is a more serious problem than we have with the current model, which is therefore more robust. Once again, we see the need to plan ahead!

Design Considerations

We have covered a lot of ground in this lesson and hopefully added a lot of ammunition to your modeling arsenal! We have also seen how the feature creation options can control the behavior of the model. So, now is a good time to say a few more words about part design.

One modeling decision you often have to make involves the balance or trade-off between using a few very complicated features or many simple ones. You must consider the following when trying to put a lot of geometry into a single feature:

- How easy will it be to modify the part/feature later? (Is it easier to suppress a feature or modify the shape of a sketch?)
- If the geometry is very complex, it is generally easier to create a number of simpler features that would combine to give the same resulting geometry.
- Using more, but simpler, features generally will give you more flexibility.
- Having a higher feature count increases the need for a carefully managed parent/child network.
- If you plan to do some engineering analysis of the part, for example a finite element analysis, then minor features such as rounds, chamfers, small holes, etc., will only complicate the model, perhaps unnecessarily. They will also lead to increased modeling effort downstream. These features are normally added last. We saw in Lesson #5 how they can be temporarily excluded from

the model (*suppressing* the feature), as long as they are not references (parents) of other features.

- If the entire part is contained in a single feature, some major changes to the part may not be feasible using that feature.
- What is the design intent of each feature? How should each feature be related to other features (via the parent/child relations)? Don't set up unnecessary interdependencies between features that will restrict your freedom of modification later.
- You must be very careful with references. Sometimes these are essential elements of the design intent; sometimes you will fall into the trap of using a reference as a convenience when setting up a sketch, where this is not in the design intent. If you try to modify the feature later, you may find that poorly chosen references will cause problems.

When creating the patterns and copies, we discovered the ways that duplicated features could be modified, either during feature creation or after the fact. We also saw some of the ramifications of feature order in the model.

These considerations should be kept in mind as you plan the creation of each new part. It is likely that there are many ways in which to set up the part, and each will have different advantages and disadvantages depending on your goals. The more you know about the Creo Parametric tools, and the more practice you get, the better you will be able to make good decisions about part design. Good planning will lead to an easier task of part creation and make it easier to modify the geometry of the part later. Like most design tasks, the model design is subject to some iteration. We discussed in Lesson #5 some of the tools that Creo Parametric provides (the three R's) to allow you to change the structure of your model if it becomes necessary or to recover from modeling errors.

Most importantly, since design is increasingly becoming a group activity, make sure your model will be easy for someone else on your design team to understand. They may have to make modifications while you are away on vacation and you want them to be happy with you when you get back!

In the next lesson we will see how to create an engineering drawing from a Creo Parametric part. This will include view layout, section and detail views, and dimensioning. We will also create a couple of parts that will be needed in our assembly in the following two lessons.

Questions for Review

1. When creating a revolved protrusion, does the sketch have to be open or closed or can it be either?
2. When creating a revolved cut, does the sketch have to be open or closed or either?
3. What essential element is common to all revolved features?
4. Suppose you are creating a revolved protrusion and you align a vertex of the sketch with an existing feature surface. What happens if you try to create a 360 degree revolve and the aligning surface doesn't exist for the full revolution?
5. What is the first feature in a pattern called?
6. What is meant by a "radial" placement of a hole?
7. In what order must you pick references for a coaxial hole? Can you create a hole from the inside of a solid going out?
8. How can you tell if a hole has a thread on it?
9. What dimensions are available for patterning a feature?
10. How could you create a spiral pattern of holes?
11. How could you create a radial pattern of rectangular cuts (extruded parallel to the radial pattern axis) so that the orientation of the cuts (a) changes with each instance to stay aligned with a radial line or (b) stays constant relative to the fixed datums?
12. What is the difference between independent and dependent copies?
13. What is the easiest way to create a pattern of several related features?
14. Is it possible to create a copy that is translated and rotated at the same time?
15. Comment on the rule of thumb in solid modeling: "Add material first, subtract material last." Do you think this would generally lead to good modeling practice?
16. What happens if you try to mirror one instance in a patterned feature?
17. What happens if you try to pattern a feature created using a make datum as a sketching plane?
18. What are the available variations for selecting references for specifying the translation direction and distance when making a ***Move*** copy?
19. Are there any dimensions of a pattern leader you cannot increment in the pattern?
20. What happens if you create a radial pattern with too many copies that they end up completely overlapping (for example 5 copies 90 degrees apart)?
21. What happens if you try to make a mirror copy that intersects the original feature? Is this even possible? Can you make a series of mirrored copies where the last one in the series intersects the first one?
22. Can you make a pattern of patterns? A group of patterns? A pattern of groups?
23. Are pattern options (***Identical***, ***Varying***, ***General***) available for hole patterns? Consider both ***Simple*** and ***Standard*** holes.
24. What is the main difference between ***Paste*** and ***Paste Special***?
25. What is the easiest way to create the following:
    a. An independent copy using the same references.
    b. A dependent copy with new references.
    c. A copy that uses some dimensions of the original but has some that differ.
    d. A copy that is always identical to the original but shifted sideways and vertically in the model.
26. How can you change a copied cut into a protrusion?

Exercises

Here are some parts to practice the features you have learned in this lesson - the top 4 use Patterns.

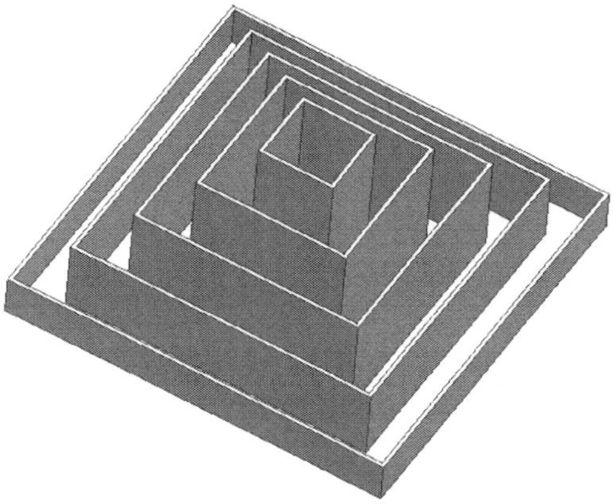

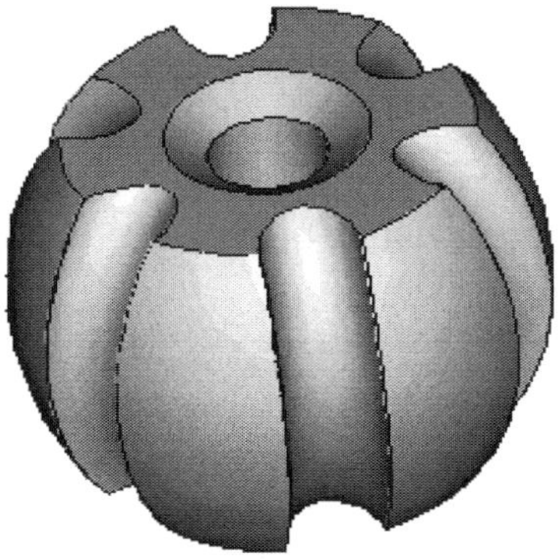

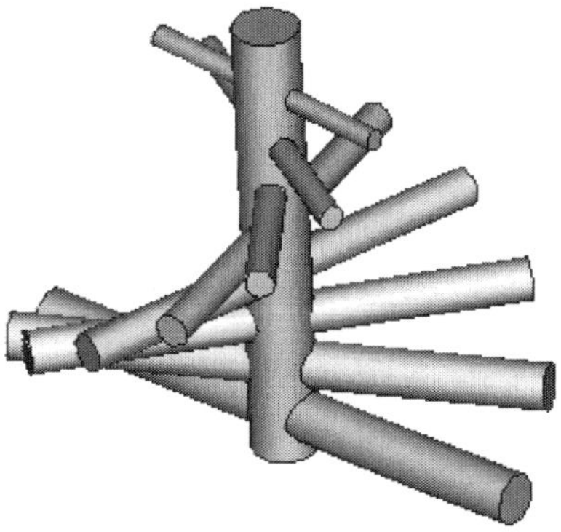

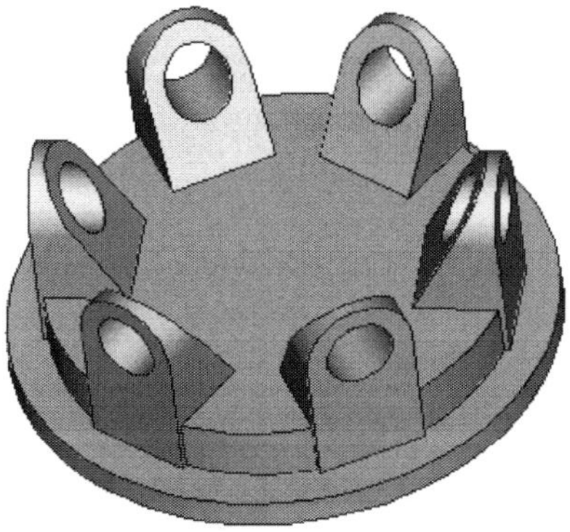

The figure at the right can be created using a copy of a copy.

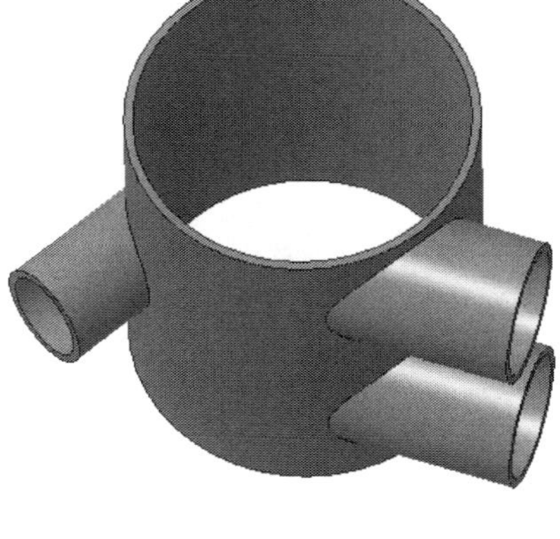

Project

This is the most complex project part. All dimensions are in millimeters. Some dimensions may be missing (because of implicit Sketcher rules or because the figures get too busy!). You can exercise some poetic license here and make a reasonable estimate for these. The important thing is that the assembly should fit together - you can easily edit dimension values later when you are putting the assembly together. Some additional figures are on the next page.

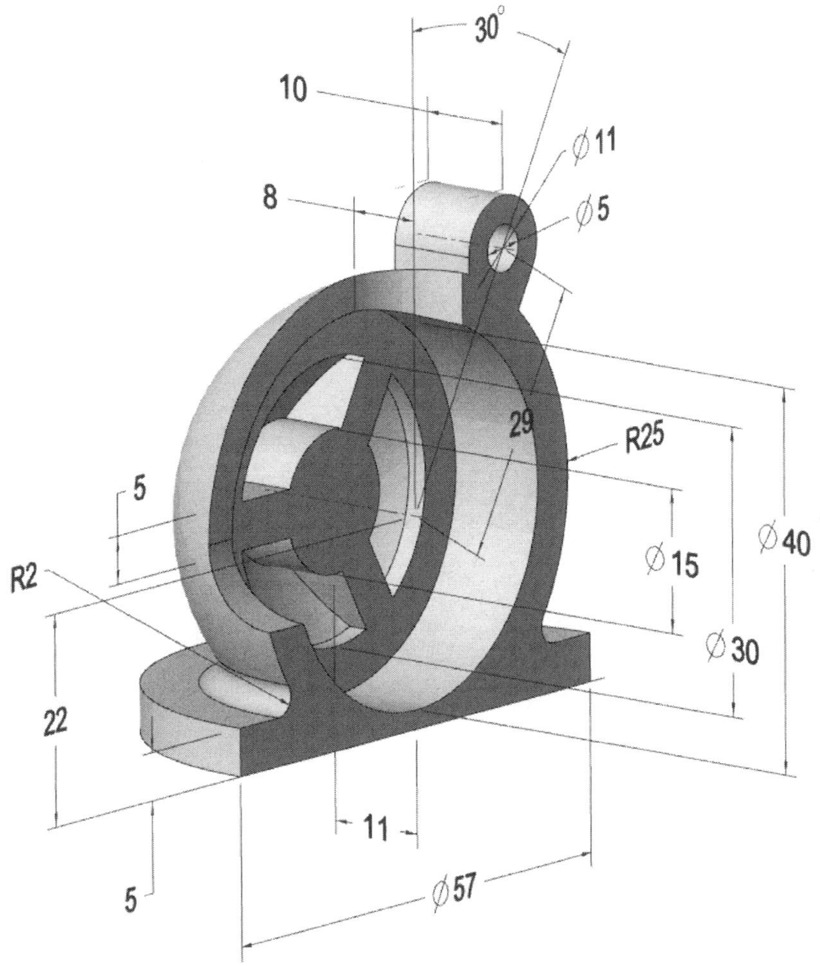

To help you visualize the part, these are close to full scale views of the Panavise part.

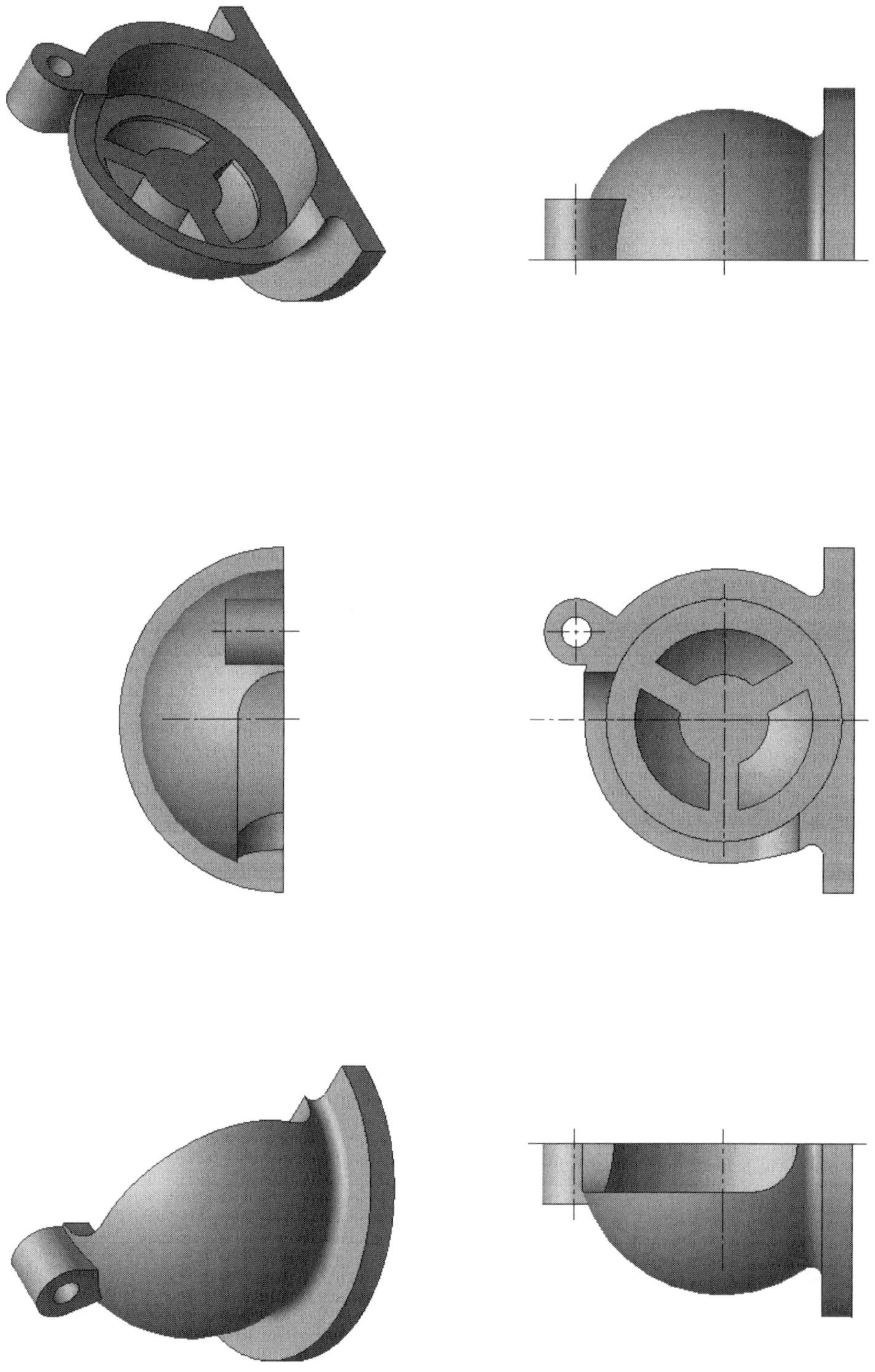

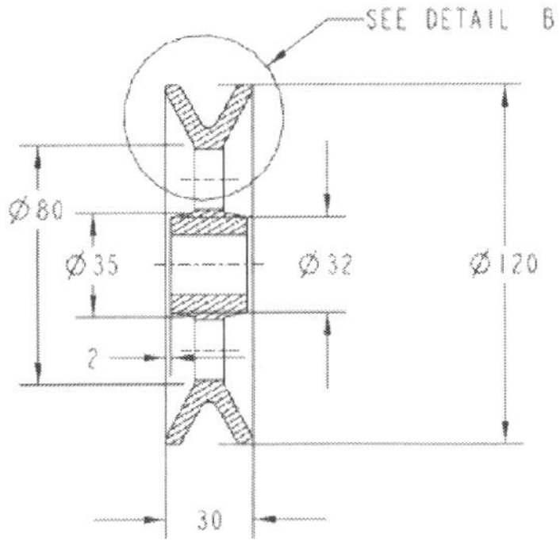

Lesson 8

Engineering Drawings

Synopsis

Overview of Drawing Mode environment. The Drawing Ribbon and Drawing Tree. Associativity. Shown vs created dimensions. View selection, orientation, and layout, section and detail views, dimensioning and detailing. Using a drawing template. Notes and parameters.

Overview of this Lesson

Although some changes in practice are becoming more commonplace, the primary form of design documentation is still the 2D engineering drawing[1]. The drawing must contain complete and unambiguous information about the part geometry and size, plus information on part material, surface finish, manufacturing notes, and so on. Over the years, the layout and practices used in engineering drawings have become standardized. This makes it easier for anyone to read the drawing, once they know what the standards are. Fortunately, Creo Parametric makes creating drawings relatively easy. You will find that all the standard practices are basically built-in - if you accept the default action for commands, most aspects of the drawing will be generally satisfactory. There are a number of commands we will see that are used to improve the "cosmetics" of the drawing.

Creating drawings with Creo Parametric lets you concentrate on *what* to show in the drawing instead of dealing with the drudgery involved in *how* to show it. For example, it is virtually impossible to create views of a part that is not physically realizable. Even if the shape is possible, we don't want any visible or hidden lines in the wrong position (or missing entirely) on the drawing. The solid model contains all necessary and sufficient

[1] Some product development processes can go directly from the computer model to the manufacturing plant floor. A standard (ASME Y14.41) was introduced in 2003 (updated 2019) for the annotation of 3D models, called Model Based Definition (MBD). This is introduced and discussed briefly in the *Advanced Tutorial*.

information in order to define the part geometry. While some aspects of creating drawings (like laying out views) are very easy, other aspects will require careful attention. For example, remember that when Creo Parametric interprets a sketch it fires a number of internal rules to solve the geometry. These rules are not indicated on the final drawing, and it may be necessary to augment the dimensions placed by Creo in order to complete the drawing to an acceptable industry standard[2]. So in addition to *showing* the dimensions already in the model, we may have to *create* additional dimensions in the drawing. It is also sometimes necessary to add graphical elements or notes to the drawing.

In this lesson, we are going to create drawings of two parts: an L-bracket support and a pulley. Both parts will be used in a subsequent lesson on creating assemblies, so don't forget to save the part files. We will also discover the power of bidirectional associativity, mentioned in the tutorial introduction. Here is what's on the agenda (these should be completed in order):

1. Exploring the Drawing Mode Environment
    - ▶ layout of the interface
    - ▶ exploring the ribbon tabs and drawing tree
    - ▶ dimension types (model *vs* draft)
    - ▶ exploring associativity
2. The L-Bracket
    - ▶ creating the part
        - changing part units
    - ▶ creating the drawing
        - layout (sheet setup)
        - layout (creating views)
        - annotation (showing dimensions)
        - annotation (cosmetic changes)
        - annotation (adding a note)
    - ▶ changing the part/drawing - exploring associativity
    - ▶ publishing the drawing
    - ▶ drawing templates
3. The Pulley
    - ▶ creating the part
    - ▶ creating the drawing
        - layout (selecting the sheet)
        - layout (creating a section view)
        - layout (creating a detailed view)
        - annotation (showing dimensions)
        - annotation (cosmetic changes)
        - annotation (using parameters in notes)
        - annotation (creating dimensions)

---

[2] This also applies to Model Based Definition (MBD) where the 3D model is dimensioned directly in the 3D display.

This will be a pretty basic lesson on part drawings, and we will only have time to cover the main topics. Even at that, this is a long lesson. The *Advanced Tutorial* from SDC contains more information on advanced drawing commands and functions, such as multisheet drawings, using multiple models, tables, repeat regions, formats, annotation elements, and creating drawing templates. As usual, there are some Questions for Review, Exercises, and some Project parts at the end.

Just as we did back in Lesson #1 for a part, we will first have a look at a finished drawing, the drawing user interface and tools for creating drawings. So, before we get started, make sure that the following files are available in your working directory:

creo_drwdemo.prt
creo_drwdemo.drw
creo_aformat.frm

These are available in the download package from the SDC web site.

The Drawing Environment

We will start by having a look at the various interface areas and drawing tools. In the working directory folder browser open the file *creo_drwdemo.drw*. (Note the "drw" file extension).[3] In the Graphics Window, turn off the datum display and set ***Shading With Edges*** on.

Drawing Interface

The drawing interface is shown in Figure 1. The main drawing area shows a simple drawing of a single part. Contents of this drawing are discussed below. The major changes to the screen are the *Drawing Ribbon*, and a *Drawing Tree* that appears in the Navigator. These are the major tools that you will use when in drawing mode. The ribbon contains ten tabs (***Layout***, ***Table, Annotate***, and so on). The operation of the ribbon tabs and the drawing tree are linked. We'll explore this some more in the next section.

At the bottom of the drawing area are some controls for drawing sheets and some information about the current drawing (drawing scale, type and name of the active model, sheet size). The Model Tree appears in the Navigator as usual. Both the model tree and the drawing tree can be detached from the Navigator pane and placed as floating windows anywhere in the graphics window, or even on a second monitor. The selection filter (lower right of the screen) refers to elements that would appear in the drawing.

---

[3] The part file must be available. We have placed this in the working directory along with the drawing. This is not strictly necessary, but having it elsewhere complicates things. The optional format file, on the other hand, is usually stored in a system formats directory along with some standard formats. Creating your own format file is a form of customization that is discussed in the *Advanced Tutorial*.

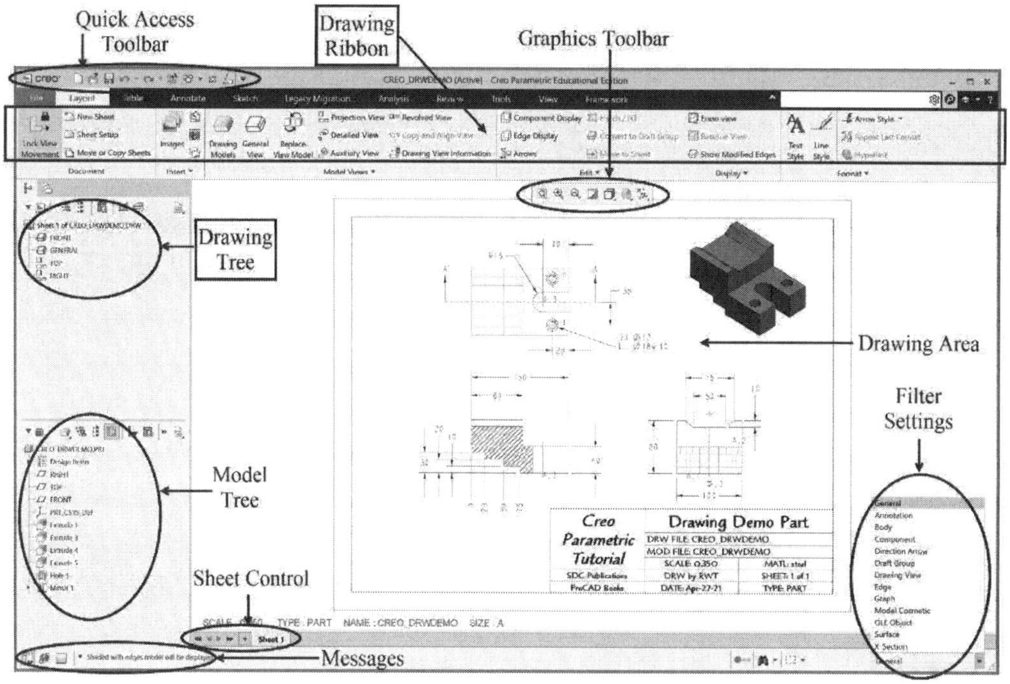

Figure 1 The Drawing Mode default interface

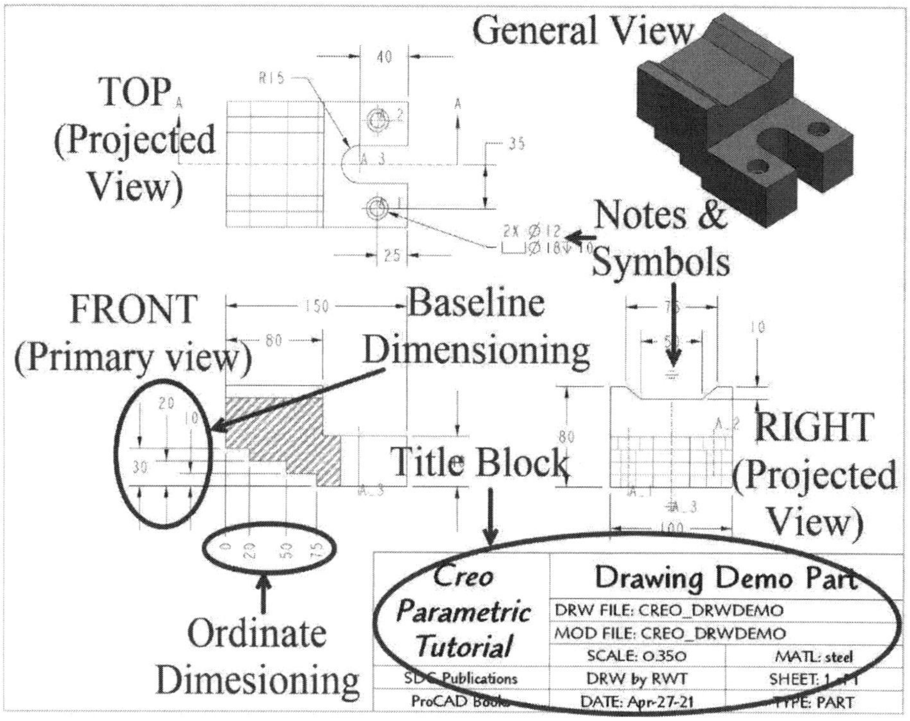

Figure 2 A simple drawing showing some typical drawing elements

A fairly simple engineering drawing appears in the drawing area (see Figure 2). The drawing sheet includes a title block that is defined in the drawing format file (creo_aformat.frm). Some of the information in the title block is filled in automatically using parameters contained in the model. The drawing itself shows the standard orthographic views (TOP-FRONT-RIGHT) as well as a shaded general view. The latter is becoming more common on engineering drawings in order to give the reader a quick look at the 3D shape of the object. The front view contains a full section, with a hatch pattern indicating that the part is made of steel (also indicated in the title block). The part has been annotated with dimensions and notes. There are several styles of dimensions shown here, including baseline and ordinate dimensions. Not immediately visible is the fact that some of the dimensions have come from the model, while others were created in the drawing. This very important point will be discussed later.

Try changing the model display (***Hidden***, ***No Hidden***, and ***Shading***) using the buttons in the Graphics toolbar or **View** tab (command ***Display Style***). You will find that only the general view on the drawing responds to these changes (you may have to do a ***Repaint*** after each, or just do a quick scroll wheel jog). This is because this view has been left to follow the environment (the default) while the three other views have been specifically set (to see how, double-click on a view, then select ***View Display***).

Mouse Controls in Drawing Mode

The middle mouse button function has changed a bit. Instead of controlling spin as in part mode, it is now used to pan. The scroll wheel controls zoom as usual. Try these out. If you are zoomed way in on the drawing, there is a ***Refit*** button in the Graphics toolbar.

Automatic pop-up toolbars will appear when most entities are selected. These contain commands relevant to the entity so are context sensitive, as is the RMB pop-up menu. Try picking some items on the drawing and see what is in those menus. If you accidentally make a change in the drawing, just use ***Undo*** or load a fresh copy from the working directory after removing the corrupt one from session.

Drawing Ribbon and Drawing Tree

We'll spend a few minutes exploring the functions in the various groups in the ribbon at the top. We will not be using all of them, but it may be helpful in the future to know what and where these commands are. As always, Creo allows us to launch commands in a number of ways (ribbon, pop-up or RMB menu on a selected element in the drawing or model tree). The commands available will change depending on the status of the ribbon (i.e. which tab is selected) and what is currently highlighted in the graphics window.

As mentioned above, the ribbon and drawing tree functions are linked. The information displayed in the drawing tree will depend on which of the ribbon tabs has been selected. Regardless of which tab is selected, the top-level organization in the drawing tree is the drawing sheet. It is common practice (and usually necessary) for drawings to contain multiple sheets; our demo drawing here has only one sheet. The next level of organization is the view. Each view will contain drawing elements that will change depending on which ribbon tab is currently selected.

There are ten tabs on the ribbon. You can select any tab at any time and are not restricted to a particular sequence. However, they are organized from left to right in generally the same order that your workflow will occur. Selecting each tab opens a new set of groups of commands. Within each ribbon group, some command buttons will display with an icon and text label. How these are displayed will depend on a number of factors (configuration settings, display size and resolution, customized settings, ...). When you mouse-over any of the buttons in the groups you will see a tool tip showing the command name and a brief explanation. Some groups also have an overflow containing infrequently used commands. As usual, buttons that are not relevant at any given moment are grayed out. The most-used tabs on the drawing ribbon are as follows:

Layout

When you first enter a drawing, you start by determining which model(s) you are going to display, details of drawing sheets (formats, sheet sizes), and setting up which views will appear on each sheet.

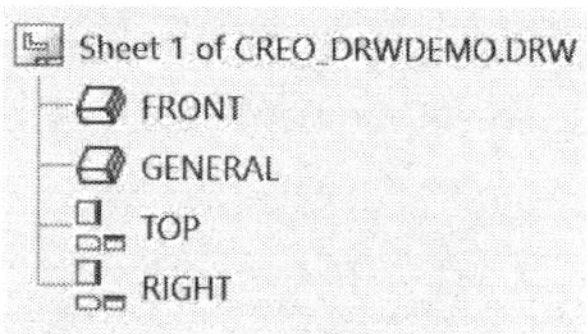

Figure 3 Drawing tree with **Layout** selected

When **Layout** is selected, the Drawing tree shows the views in the drawing, Figure 3. In the current drawing, the various views have been named to correspond to normal practice in order to make it easy to select the desired view. If a view in the drawing tree is selected, it will be surrounded by a dashed green highlight box in the drawing and a pop-up toolbar will appear.

Table

This tab is for creating things like the title block, but more often for drawing elements like Repeat Regions and Bill of Materials (BOM). When you select this tab, the drawing tree indicates the tables in the drawing (Figure 4). In the present case, the only table is the one forming the title block. We will not be doing any tables in this Tutorial, but this is covered in the *Advanced Tutorial*.

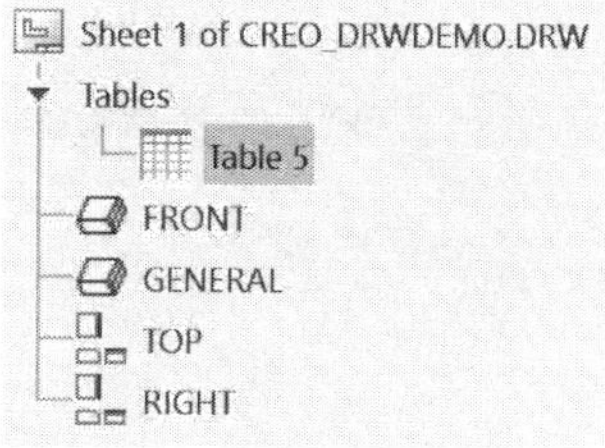

Figure 4 Drawing tree with **Table** selected

Annotate

You will be spending most of your time in this tab. It has the largest set of groups and individual commands. Selecting this tab changes the drawing tree to show the views on the drawing and, at the next level in the tree, the various annotations and datums associated with each view. This will likely be the most detailed set of data concerning the drawing you will see. For example, in the TOP view, open the annotations and datums groups. See Figure 5.

Selecting any of these entries in the drawing tree will highlight them on the drawing[4]. Note the symbolic names of the model dimensions given here. If you expand the annotations under the FRONT view, Figure 6, you will see that there are two types of annotations. **Model** annotations are based on dimensions and parameters that come from the 3D model itself. **Draft** annotations are created in the drawing. For example, the draft dimension **add9** in the Front view was necessary due to an explicit constraint being used in the sketch for the base feature of this part. (Can you figure out what that was?) Such constraints do not show on the drawing and we frequently must create some dimensions to make the drawing complete. The baseline and ordinate dimensioning schemes were also set up in the drawing itself (even though model dimensions already exist), which result in these becoming draft annotations. Note the symbolic names of these dimensions and the different icons in the tree. Other types of draft annotations are notes.

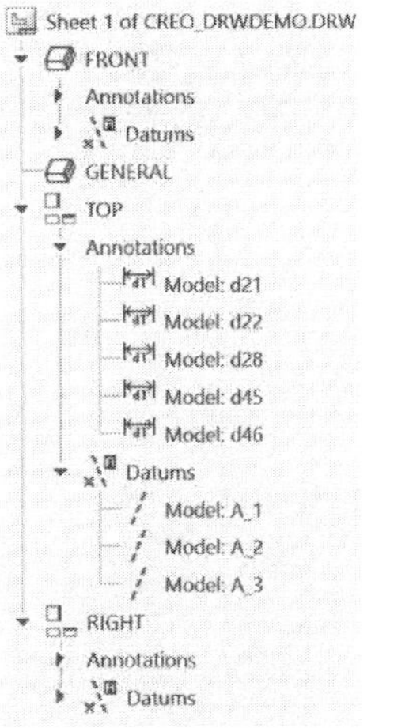

Figure 5 The drawing tree with **Annotate** selected

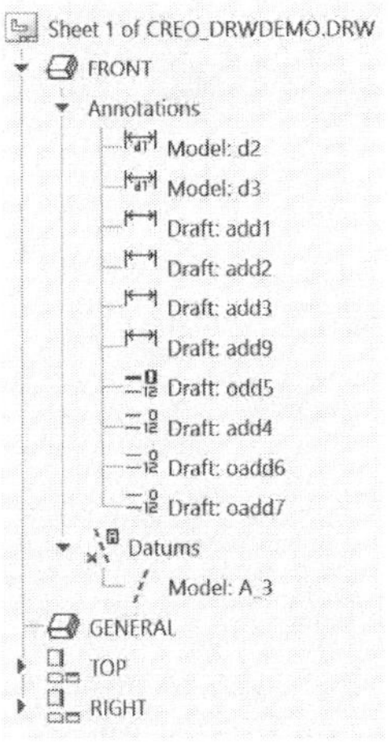

Figure 6 The drawing tree with **Annotate** selected

Sketch

These groups contain commands for creating and editing graphical drawing elements. The functionality of this is very similar to Sketcher. As you may expect, any entities you create here will be listed in the drawing tree. We will not be using these commands here, but you should come back some time to play with these functions.

Review

These commands let you check the status of the drawing prior to release. For example, you can make sure that information in drawing tables is updated, the model is regenerated, and so on. A handy command in the ***Query*** group is ***Highlight By Attributes*** which opens the window shown in Figure 7. Set the checkbox for ***Dimension*** in the **Item Type** area. When you select the ***Apply*** button at the bottom, the model and draft

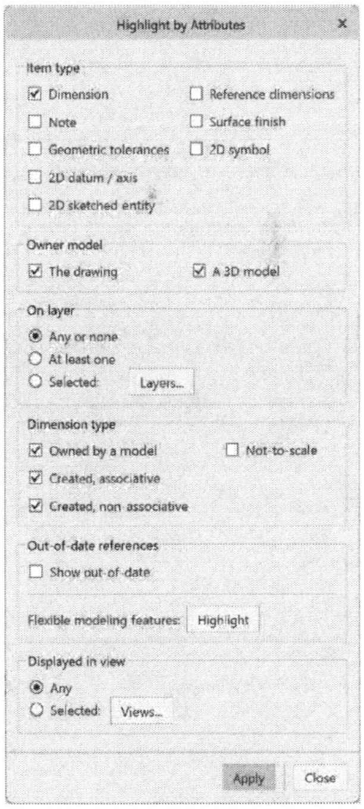

Figure 7 The **Highlight** menu

---

[4] If you turn on ***Preselection Highlighting*** using the ***Settings*** button in the Drawing Tree, you just have to mouse over the entries without selecting them.

dimensions will be highlighted in different colors. By changing the **Owner Model** settings in the dialog window, you can highlight only specific types of annotations. Close the **Highlight** dialog window.

This completes your introduction to some of the drawing ribbon tools. Before we start making our first drawing, there are a couple of important concepts we should go over: dimension types (model and draft) and associativity between the drawing and the model.

Shown vs *Created* Dimensions

The drawing in Figure 2 illustrates a number of ways for laying out the dimensions, including baseline and ordinate dimensioning, and the addition of notes and symbols. Although it is not obvious from the figure, most of these dimensions come from the part model itself where they are used in sketches or locating features like holes. When placing the annotations on the drawing, all we have to do is "show" these dimensions (with the opposite operation being to "hide" them). Such *model dimensions* cannot be created or deleted in the drawing, since they belong to the model. They are sometimes called *driving dimensions* since they control the model geometry. We will see this in the next section on the topic of associativity.

On the other hand, we may have to create some dimensions on the drawing that do not currently exist in the part. These exist only in the drawing and are called *draft dimensions* or, because their values are determined by the model, they are also known as *driven dimensions*. Draft dimensions could be necessary for a number of reasons, including the fact that we might prefer a different dimensioning scheme (ordinate, for example) than was used in the model or we must account for explicit constraints that have been used in feature sketches. This somewhat elaborate terminology is hopefully made a bit more clear in the following table.

Table 8.1 The Mysteries of Model vs Draft Dimensions

The type of dimension indicated in the drawing tree is L	Model	Draft
Where does the dimension value originate?	part or assembly	drawing
How do we get the dimension onto the drawing?	***Show***	***Create***
What is the relation of the dimension to the model?	Driving	Driven
What is the form of the symbolic name for the dimension on the drawing?	*dxx*	*addxx* or *oaddxx*

To see the symbolic names (last row in the table above), in the **Tools** tab select ***{Model Intent}:Switch Dimensions***. This causes all the dimensions (and parameters in the title block) to display on the drawing using their symbolic names. We can pick out which are model dimensions and which are draft dimensions. Select ***Switch Dimensions*** again to return the dimension display to numerical values. Another way to do this, as seen previously, is with the ***Highlight by Attributes*** button in the **Review** tab. Yet another way to determine the symbolic name is with a simple mouse-over - the driving dimensions will indicate which feature they come from while the driven dimensions do not (cannot since their references could be two different features!).

Dimension Properties

Consider the vertical dimension **80** that appears in the right view. As you mouse-over the dimension or extension lines, a pop-up should identify it as "d1:F5(EXTRUDE_1)", meaning symbolic dimension *d1* of feature #5 called "Extrude 1" - check the model tree. With this selected with a left click, the **Dimension** dashboard shown in Figure 8 appears. In the **Value** group, you can see the symbolic name and the nominal value. Check out the other options available (decimal places, tolerance, decimal/fractional format). Select the **Dimension Text** command. This lets you set non-default display properties. In particular, note the text string "@D" in the pane below. This string means, effectively, "print the nominal value of the driving dimension". The text string "@S" here would cause the symbolic value to be printed instead[5]. Right beside the **Dimension** dashboard tab is the **Format** dashboard tab (Figure 8). It contains settings for the text font, size, color, alignment, and so on. If you select another dimension, the contents of the dashboards will change to match. Both dashboards disappear when you click on the graphics window background to deselect the highlighted dimension.

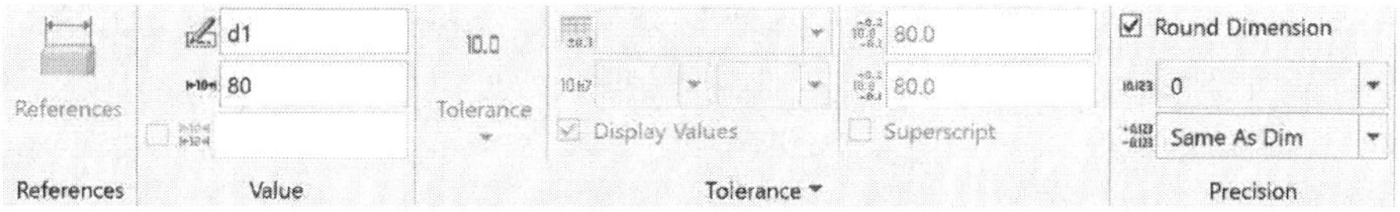

a) Dimension Dashboard (partial)

b) Dimension Dashboard (partial)

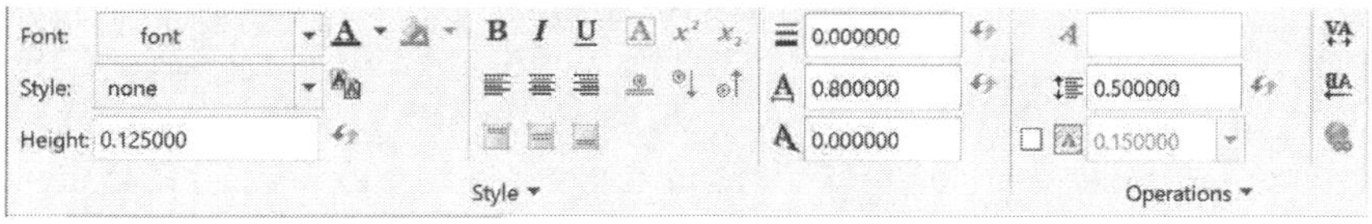

c) Format Dashboard (partial)

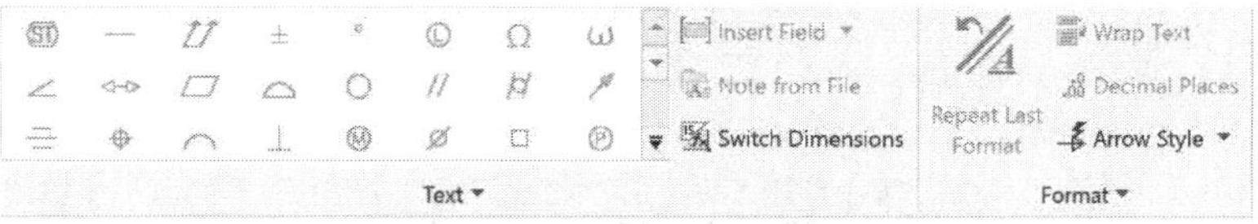

d) Format Dashboard (partial)

Figure 8 The **Dimension** and **Format** dashboards

[5] This is a handy way of identifying dimensions on a drawing for parts driven by **Family Tables**, as discussed in the *Advanced Tutorial*.

Move the mouse over the leader note on the hole in the TOP view. This indicates that the note is for dimension *d28* of the hole feature #9. This is actually the dimension for the diameter of the counterbore that the leader is touching (the outer circle). With the note selected, in the **Value** group, the symbolic name (d28) and nominal value (18) are indicated. Compare this to the note on the drawing. The value is buried inside the note text. How does this happen? Go to the **Dimension Text** window and check out the text shown. See Figure 9.

2X ⌀&d27
⌴@D↧&d26

Figure 9 Display text for the hole note

Observe that the "@D" notation occurs after the counterbore symbol on the second line. The other symbolic dimensions preceded with an ampersand, "&d27" and "&d26", are for the hole diameter and counterbore depth, respectively. The ampersand means "display the value" of the following model parameter. Note that only one "@D" can appear in the dimension text, since the dimension text is "owned" by only one model dimension (d28 in this case). This notation was created by entering the text directly in the pane, with help from the symbol palette at the bottom. You can put any text into this pane. Creo knows that this is a diameter dimension, so automatically puts the diameter symbol ⌀ in front of the value (although it is strange it doesn't do that for d27!). Close the window by clicking on the graphics window background.

In addition to the Dimension and Format dashboards that appear automatically, some other common commands are available on the RMB pop-up menu. Check this out. Since the menu is context sensitive, it will change depending on what you have selected.

Exploring Associativity

Our last exploration of the drawing environment involves the concept of associativity. This was mentioned in the Introduction, but we now get to see what it is all about.

Once again select (double click) the vertical dimension **80** in the right view (or highlight and select ***Modify Nominal Value*** in the RMB pop-up). Change this to **60**. If you clear the green highlight by clicking on the drawing background, the dimension should appear in blue. This is the indication that the nominal value has changed but the model has not been updated; see also the yellow warning light in the message window. Before you do that, have a look at the vertical dimension 40 in the front view. What dimension type is this? (Hint: use a mouse over). You can update the model in any of four ways: (1) in the **Review** tab, select ***{Update}: Regenerate Active Model***, (2) use the keyboard shortcut ***Ctrl+g***, (3) select ***Regenerate Active Model*** in the Quick Access toolbar, or (4) with nothing highlighted and the **Review** ribbon open, use an RMB on the background and select ***Regenerate Active Model***. Pick one!

When the model regenerates, the FRONT, RIGHT, and general view all change. This indicates that the dimension we chose to modify is driving the geometry. The drawing and the model are associated through the driving dimensions. Since the dimension is associated with both the model and drawing, changing its value in either place will result in a change in the geometry.

Meanwhile, the vertical dimension *add9* on the FRONT view (that was 40) is now 30. Try to select this dimension and change it back. You can select it but there is no way to modify its value. The nominal value data field is grayed out. It has been driven by the change in the model.

What other geometry changes can you make to this model in the drawing? This bidirectional associativity is one of the greatest strengths of parametric modeling (and one of the reasons why we go to all the trouble!). For it to work effectively, models must be planned and created carefully - remember our goals: *flexible* and *robust.*

This concludes our introduction to drawing mode. There is a lot to learn here. In the following we will create a couple of drawings from scratch. We will discover yet more options and drawing functions.

A final thing you should keep in mind is that although the drawing file is able to display and call up information about the model, the drawing file does not contain the model. This still resides in the part or assembly file. If you move the drawing file to another computer and don't take the part file with it, the drawing will not open correctly. Because the drawing and part are so tightly linked, most systems are set up so that if the drawing file makes changes in the model (using the driving dimensions) then the part file is automatically saved when the drawing file is saved. This way the drawing and model are always in sync. Eventually, you are liable to come across a situation when the drawing and model are out of sync (due to file mismanagement, for example) and you will receive warning and/or error messages.

Close the demo drawing window (and part window if you had that open), and remove everything from your session using ***File* ➤ *Manage Session* ➤ *Erase Not Displayed*.**

Now is a good time to take a break.

The L-Bracket

Creating the Part

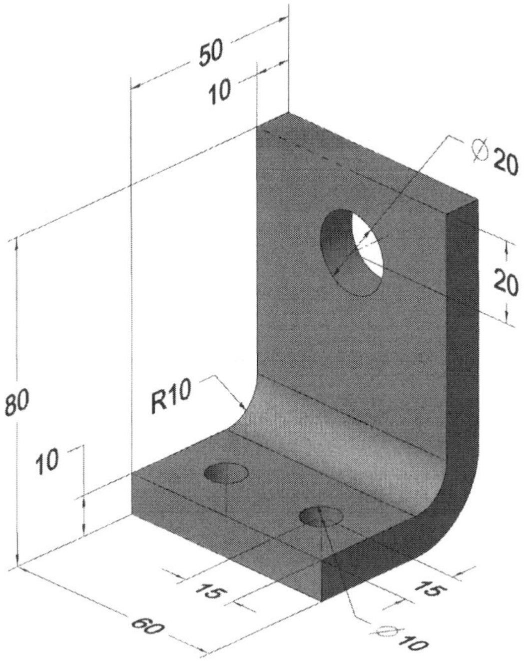

Figure 10 The L-bracket part (dimensions in mm)

First, we'll create the part shown in Figure 10[6]. Call this part **lbrack** and use the **inlbs** template for a solid part. (We're going to change the part units in a minute.) Study this figure carefully. When you create the part, make sure that the back surface of the vertical leg is aligned with **FRONT**, the lower surface of the horizontal leg is aligned with **TOP**, and the vertical plane of symmetry through the upper hole is **RIGHT**. An obvious choice for the base feature is a both-sides solid protrusion in the shape of a backwards "L" (as seen from the right side) sketched on **RIGHT**. This will allow us to mirror the second small hole and align the larger one with **RIGHT**. Notice the dimensioning scheme for the holes. Make the part now.

Changing Part Units

The model units are important so that the scale factor for showing the views on the drawing sheet will be correct. Creo will try to pick a reasonable scale based on the actual size of the object and the physical size of the drawing sheet.

Note that the units in Figure 10 are given in millimeters, whereas in a standard Creo Parametric installation, the default template uses inches. This is a common "oops" when creating a model, since the units are not topmost in our mind when we first start the part (or when you inherit a model from another source). Here's how to change the part units. Select (from the pull-down menu)

File ➤ Prepare ➤ Model Properties

This opens the **Model Properties** window that shows the current status of a number of attributes of the model. In the top group, in the row indicating the **Units**, pick ***change*** at the far right. The **Units Manager** window opens, as shown in Figure 11. This lists the common unit systems in Creo Parametric (and its companion Creo Simulate used for finite element analysis). The current units are indicated by the arrow pointer. Select the line containing the unit system

[6] If you are pressed for time, use the part *creo_lbrack.prt* in the download set.

millimeter Newton Second

and then ***Set***. When you change the units of a model, you have two options that will affect all linear dimensions:

Convert dimensions - This leaves the model the same real size as the original. For example, a 1 inch long bar will be converted to a 25.4mm long bar. The dimension number changes. This is normally what we mean when we "change units."

Interpret dimensions - This keeps the dimension numbers the same, but interprets them in the new units. In our example, the 1 inch long bar becomes a 1mm long bar. This is how we recover from the "oops" mentioned above.

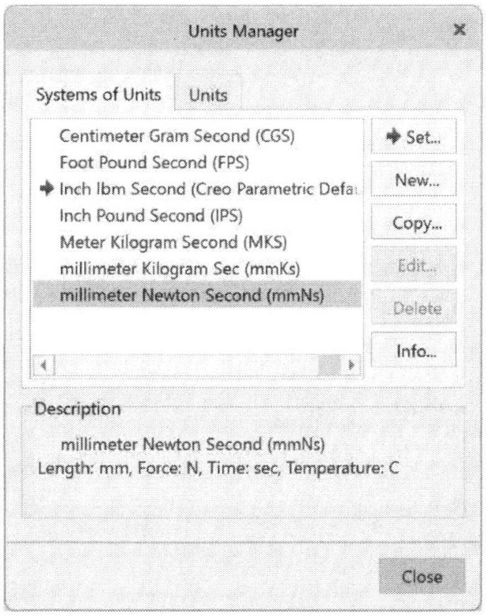

Figure 11 The **Units Manager** window

Managing units is especially important if you are going to produce an assembly of parts, as we will do in the next two lessons. It is also critical to be aware of units when you are working in a design group, since some people may be working in inches while others are in millimeters. Parts downloaded from the web also come in all varieties[7]. You may be aware of some classic engineering blunders that have occurred over the years due to mix-ups in the interpretation of units or people making assumptions about units.

If you have used the dimension values in the figure above (and were in inches), then you want to pick the second option here (***Interpret dimensions***) and select ***OK***. ***Close*** the Units Manager window, then ***Close*** the **Model Properties** window. When this is applied, double-click on the protrusion to verify that the dimension numbers haven't changed. Don't forget to save the part in the working directory! We are now ready to create the drawing.

Creating the Drawing of the L-Bracket

① Create the Drawing File

Leave the part in ***Hidden Line*** display mode, turn colors off, then select the following:

---

[7] You can actually assemble parts defined using different units and they will fit accordingly. That is perhaps an invitation to disaster or at least confusion. However, if you are working in *mm* and a supplier's parts are in *inches*, you are stuck with them! You can also define units for individual parameters, and Creo will automatically treat them correctly. For example, you could have a relation: d2 = d1 + offset, where d1 and d2 (part dimensions) are in inches, and offset (a parameter defined by you) is in millimeters.

File ➤ New ➤ Drawing

Turn off the option **Use default template**. We will deal with drawing templates a bit later. Leave the box checked to **Use the drawing model name** - we generally keep the drawing name the same as the part name. Select ***OK***.

The **New Drawing** menu will open up (see Figure 12). Note the currently active part is automatically selected as the drawing model. Keep the defaults for the template (**Empty**) and orientation (**Landscape**), but change the **Standard Size** option to **A** (an 8-1/2" by 11" sheet in landscape mode). When this window is complete, accept the entries with ***OK***.

A new window will open up with the title *LBRACK (Active)*. This will overlap or cover the part window, which is still open but pushed to the back. You can switch back and forth between the part and drawing windows using ***Windows*** (in the Quick Access toolbar, unfortunately right beside the irreversible ***Close Window*** command). If several windows are in view, only one of them will be active at a time (indicated by the word **Active** in the title).

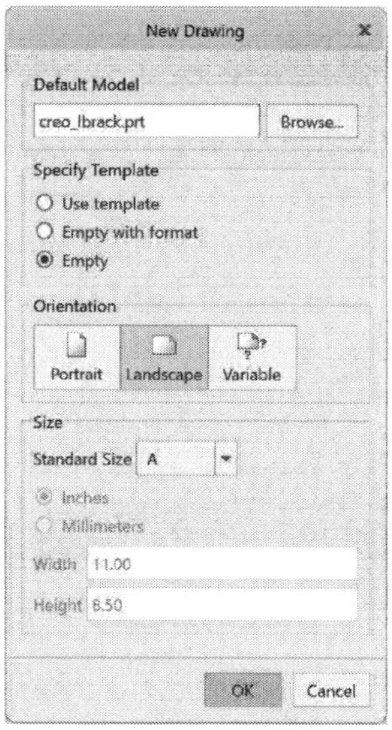

Figure 12 The **New Drawing** menu

In the drawing window, the drawing ribbon is now displayed, with the default **Layout** tab selected. The drawing tree shows that there is one sheet. The information area at the bottom of the screen shows the model type and name, and the sheet size. The scale value doesn't mean much yet since we haven't set up any views. Turn off the display of model tree columns, if necessary.

② Adding Views

Select ***{Model Views}:General View*** (or use the RMB pop-up menu, in either the drawing area or in the drawing tree, to select ***General View***). In the Combined State window, select ***No Combined State ➤ OK***. Read the prompt in the message window. The view we will place first will be our primary view. It will be the front view of the part, so select a center point a bit left and below the center of the sheet, as shown in Figure 13. The **Drawing View** dialog window also opens (Figure 14).

The **Drawing View** window contains all the options for creating (and changing properties of) views on the drawing sheet. This includes view type and orientation, section views, edge display, scale, and so on. Browse through the various items listed in the **Categories** area. As each is selected the window contents will change, showing the options available with that category. For example, note in the **Scale** category, a default sheet scale has been determined (in this case 1.0). We can come back later to change the sheet scale if desired. Don't change any of the default settings for now.

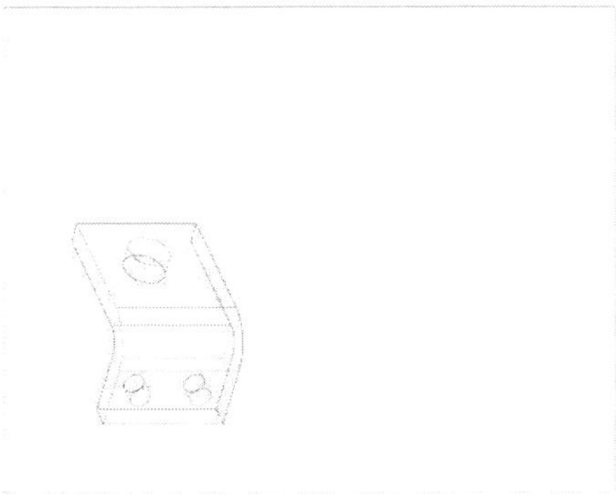

Figure 13 Placing the primary view

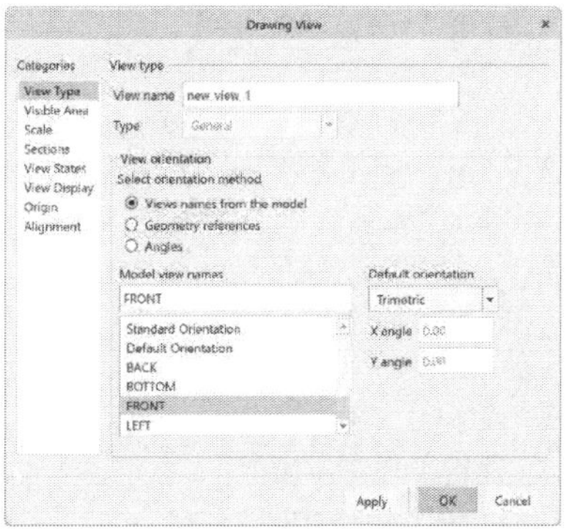

Figure 14 The **Drawing View** window

Select the **View Type** category. We want to set the desired orientation of the primary view. This will be the front view of the bracket. In the **Model view names** list, select **FRONT** from the list of views defined in the model (shown in Figure 14). Now select ***OK***.

Still in **Layout** tab, when you select the view, it has a dashed green box around it. This means that it is the currently selected view. Click anywhere else on the sheet to turn off the selection highlight. Turn off the datum display and ***Repaint***. Your drawing should look like Figure 15.

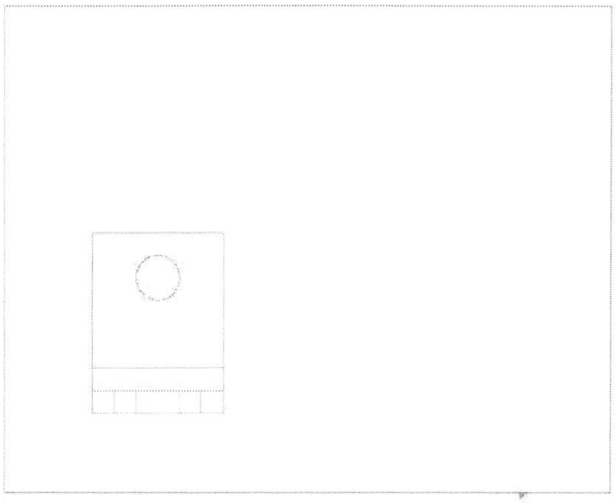

Figure 15 Primary view placed and oriented

Now we want to add the right and top views. These could be created using the ***General*** button and selecting the RIGHT and TOP named views in the **View Type** lists. However, this will not automatically align these new views with the primary view. We want to create projected views off the primary. Make sure you are still in the **Layout** tab.

Select the primary view so that it is highlighted with the green box. Now in the pop-up menu top row, select ***Projection View***. You will now see a sliding orange border that will always align with the parent view. Position the mouse somewhere to the right of the primary view and click. Voilà! The right side view appears. To create the top view, we once again select the primary view (this is what we want to project from) and use the pop-up command ***Projection View.*** Click above the front view to get the top view. Easy!

If you don't like the spacing of your views, you can easily move them. By default, views are locked in place. Select the right view. To unlock it either:

- deselect ***Lock View Movement*** in the **Layout** ribbon

or

- in the RMB pop-up, uncheck ***Lock View Movement***

It will be surrounded by a green border and have drag handles at the corners and center.

When you move views, Creo Parametric will ensure that your projected views stay aligned. Click anywhere on the view and drag to move the view. Try to move the view up, down, left, and right on the screen (you can't move up or down since the view must align with the front view). Left-click again to drop the view at the new location. Try moving the top view. Finally, try moving the front view. You should see the other views move to maintain the correct orthographic alignment.

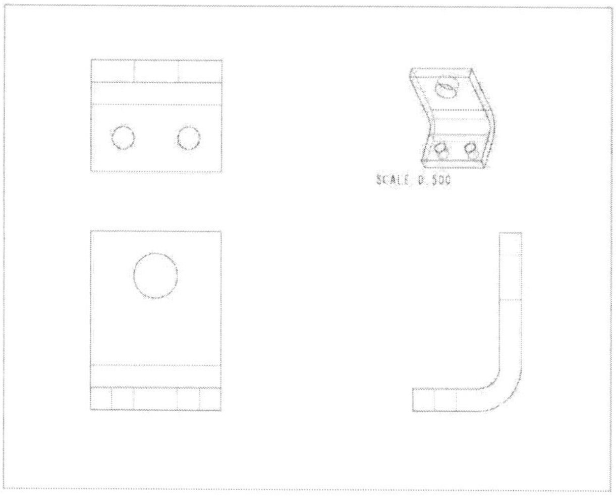

Figure 16 All views placed for L-brack

Click the **left**-mouse button on an open area of the screen (i.e. not on another view) when you are finished moving the views to turn off view selection.

Let's add a fourth view that shows the part in 3D. Note that this is not a projected view but a general one. We'll scale this one down to half size. In the RMB pop-up menu, select

General View

Once again accept the default "No combined state" with ***OK***, then click in the top right corner of the sheet. The view will appear. In the **Drawing View** window, under the **View Type** category, select the **Default Orientation**, then ***Apply***.

Select the category **Scale**, and change the **Custom Scale** to ***0.5*** and then ***OK***. See Figure 16.

③ Setting View Display Mode

The default view display (hidden line, wireframe, etc) is determined by the view display of the part in the model window (i.e. *Follow Environment*). Of course, we don't normally use shaded images in a drawing. To make sure hidden lines are treated correctly, select all four views (using the CTRL key), then in the pop-up menu, select **Properties**. (You can also do this one view at a time by double clicking on it.) In the **Drawing View** window (you are automatically in the ***View Disp*** category panel since you selected multiple views) set the following

Display Style (Hidden)
Tangent edges (None)

then press ***OK***. These view display settings are now fixed properties of the views and are not affected by the toolbar buttons or the display style in the part window. Click on the drawing background to deselect all the views.

④ Adding Dimensioning Detail

Select the ***Annotate*** ribbon tab. Then select ***Show Model Annotations*** (or just use the RMB pop-up in the drawing area to select ***Show Model Annotations***). Note in passing that this pop-up menu is different from the one we got with a different ribbon tab selected.

The **Show Model Annotations** window opens. The tabs at the top let you choose which form of annotation we want to process (dimensions, tolerances, notes, surface finish, and so on). The **Type** pull-down list (default to **All**) lets us select specific types of annotations of each form that might exist in the model. The message in the window directs us to select either views, components, or features, for which we want to show the model dimensions.

Let's start by selecting the entire part. Go to the Model Tree in the Navigator and select the part name. All of the part dimensions are now listed with check boxes in the window and appear in dark red on the drawing on various views. As you mouse-over the check boxes, the dimensions on the drawing will switch to orange. You can individually select check boxes for dimensions you want to appear. For now, pick the button at the lower left to select them all, as in Figure 17. The dimensions will turn purple. Then select the ***Apply*** button. The dimensions change to gray.

Now in the **Show Model Annotations** window, pick the tab at the far right. The window now lists all the datum and hole axes in the model. This includes axes in the general view at the top right. Once again, experiment with the various check boxes, turning on the ones you want to keep. You can also pick directly on the axes in the graphics window to select them. You likely won't need all of them since some may be aligned behind others in a particular view. Then select ***Apply***. The selected axes now appear in gray.

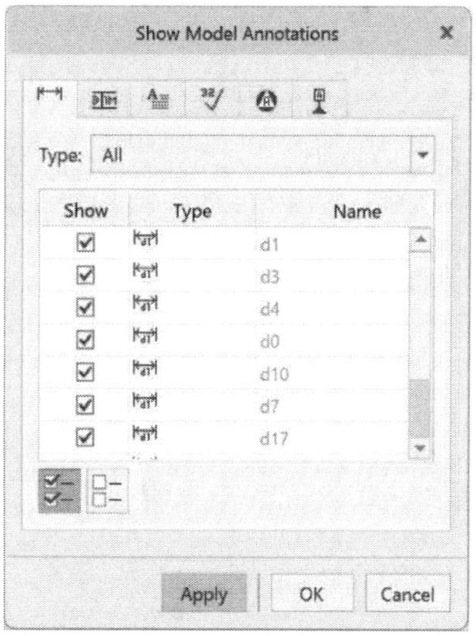

Figure 17 Window for **Show Model Annotations**

There is nothing left to show, so select ***Cancel***. Observe what has happened on the drawing and in the Drawing Tree. Each view now contains both annotation and datum elements.

Take a moment to think back to how you created the part. The dimensions put on the drawing using **Show** are exactly the ones you used in your features. These are (no surprise!) called *Shown* dimensions. Another type of dimension can be created in the drawing, naturally called *Created* dimensions, which we will discuss a bit later.

> ***Helpful Hint***
>
> We elected to show all the dimensions at once. This is a potentially hazardous thing to do - think of a part or assembly that might have hundreds of dimensions! We got away with that procedure this time because there aren't too many dimensions in this drawing. For more complicated parts, you might like to show the dimensions by individual feature, all dimensions in a given view, or a specific feature in a chosen view. Some experience with these options is necessary to make good choices here, otherwise you'll spend a lot of unproductive time cleaning up the drawing.
>
> In the next drawing, we will see a more systematic way of adding the dimensions rather than showing them all at once. This will almost always be the preferred way of defining annotations except for very simple parts.

The lesson here is to use (as much as possible) the dimensions in feature creation that you want to appear on the drawing[8]. So, you should know something about drawing standards and how you want the dimensions laid out in the drawing *before* you start to create the model - a point often missed by many. This is often a point of contention between designers (who make the models and are mostly concerned about function) and detailers (who create the drawings and are sticklers for drawing standards). Another layer of complexity is added with the use of 3D Annotations in the model, part of Model Based Definition (MBD). These can contain both driving and driven dimensions and, with the right configuration options, be shown automatically on the drawing. We will not have time (or space) to explore that topic here. Consult the online Help for further information.

⑤ Dimension Cosmetics

Although all the dimensions are now on the drawing (except maybe the hole on the left side created using Mirror), and Creo does the best job it can to determine where to place the dimensions (which view, and so on), there is a lot we may need to do to improve their placement and appearance.

For example, some of the dimensions may be a bit crowded. To fix this, select the ***Cleanup Dimensions*** button Cleanup Dimensions in the **Edit** group or use the same command in the RMB pop-up menu. This opens the window shown in Figure 18. We have to identify which dimensions we want cleaned. Draw a selection box around the entire drawing with

---

[8] This is a good topic for a philosophical discussion. Some users differentiate between "design" intent (dimension scheme used to create the computer model) and "manufacturing" intent (dimensions appearing on the drawing), maintaining that sometimes these are different. If you want to start a lively discussion among hard-core Creo Parametric users, ask them if it is better to use "Shown" dimensions or "Created" dimensions in a drawing.

the left mouse button, then select middle click. The number of dimensions affected will appear at the top of the **Clean Dimensions** window and these will appear in dark red on the drawing. Notice that dimensions using leaders are not affected (check the hole diameters). The default spacings (an offset of 0.5 is the spacing in real inches from the edge of the part to the first dimension line, the increment of 0.375 is between parallel dimension lines) are drawing standards. Pick on the ***Apply*** button. All the affected dimensions should spread out and appear in dark red. The faint dashed gray lines are called the *snap lines*. As you proceed to modify the drawing layout, the dimensions will snap to these locations to help you maintain the spacings set in ***Clean Dimensions***. These snap lines are a convenience only and will not be printed with the drawing.

Depending on your view placement and dimensioning scheme, Creo Parametric might have some trouble with dimension placement (for example, too little room between views). Look in the message window for error messages and warnings. Check what is on the **Cosmetic** tab, then ***Close*** the **Clean Dimensions** window.

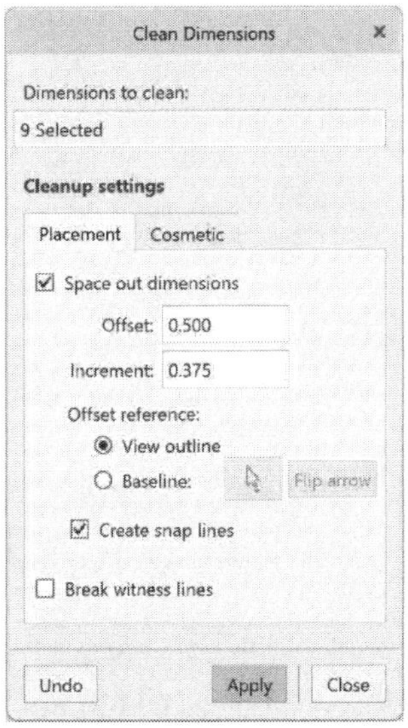

Figure 18 Cleaning up the dimension layout

You can come back and launch the ***Clean Dimensions*** command whenever you want - it is not a one-time deal.

The drawing should look something like Figure 19 (your dimensioning scheme may be slightly different from this, depending on how you created your model).

There is a lot more we can do to modify the display esthetics of the dimensioning detail. To start with, some of the dimension placement locations chosen by Creo may need to be touched up a little. It is probably necessary to switch some of the dimensions to a different view, and you may want to modify spacing and location of dimensions on views, direction of dimension arrows, and so on. For example, the location dimensions for all the holes should be on the view that shows the circular shape of the hole. For the two small holes, this is the top view. For the large hole, this is the front view. Most of these cosmetic modifications can be made using the mouse buttons as follows.

If you want to move a dimension to another view, after you have picked out the dimension, select ***Move to View*** from the pop-up menu, then left-click on the desired new view.

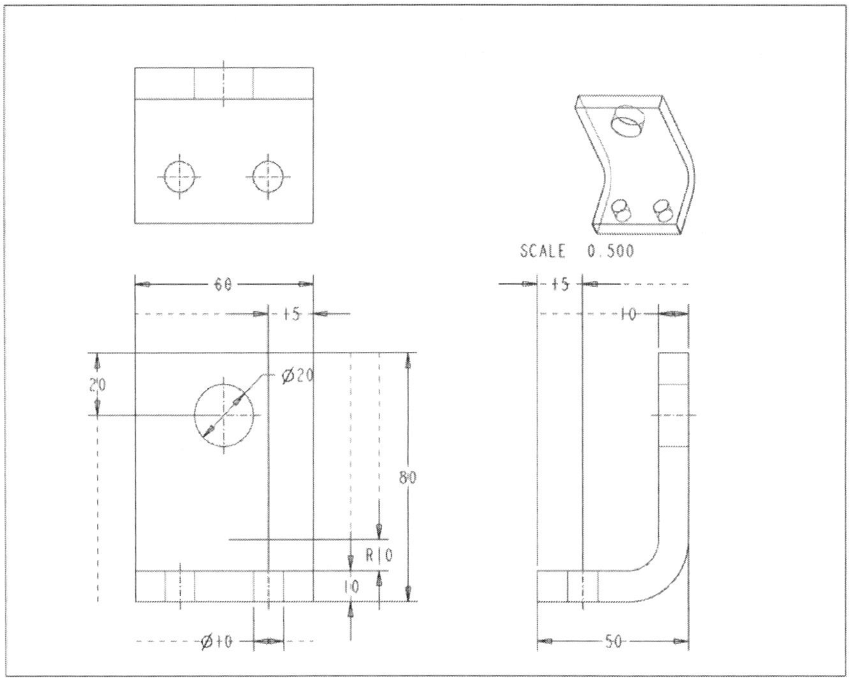

Figure 19 Dimensions placed and ***Clean*ed**

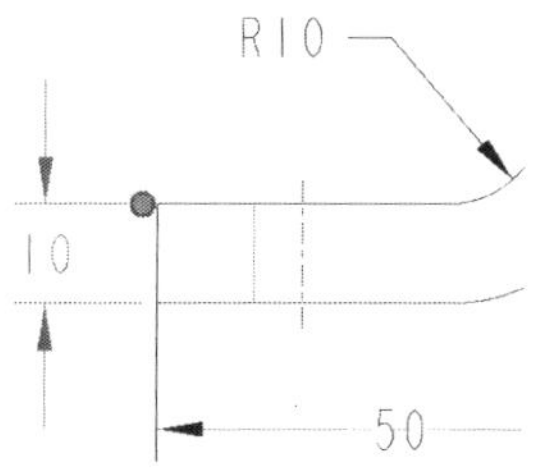

Figure 20 Handle for modifying witness line

To change the appearance of a dimension, select it, for example, the dimension giving the thickness of the plate as shown in the right view. See Figure 20. If you select the dimension value, you can drag it practically anywhere. The extension lines and arrows will automatically follow. If you select a witness line, you can pick a drag handle (blue circle in Figure 20) and move it off the part. Notice the effect of the snap lines.

While dragging the dimension text around, if you want to flip the dimension arrows (i.e. put them inside/outside the extension lines), just ***Right-click*** (while you are holding down the left button). When you select a dimension with a leader (like a radial or diameter dimension), clicking the RMB will cycle through a number of arrangements for the leader line and dimension arrowhead placement. What happens (this may depend on what is selected) if you momentarily hold down the SHFT key while dragging with the left mouse button?

When the dimension is where you want it, release the LMB to drop it. You can continue to left-click on the handles to move the dimension, extension lines, dimension line, and arrows until you get exactly the appearance you want. ***To accept the final placement and format, click the left mouse button somewhere else on the screen or select another detail item.***

The final configuration might look something like Figure 21.

To modify more dimensions, continue the sequence:

- left-click on a dimension,
- pop-up toolbar to move to another view,
- drag on the text or handles as desired using left-mouse,
- right-click to flip arrows if desired,
- left-click on the background to accept

until you are satisfied with the layout.

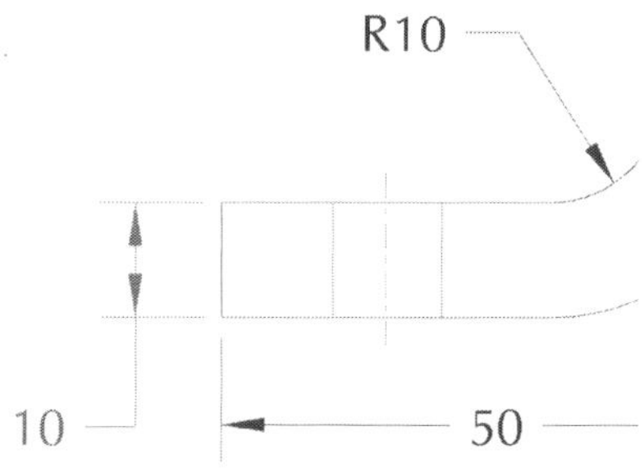

Figure 21 Modified dimension cosmetics

Helpful Hint

You can ***Erase*** or ***Delete*** a shown dimension in the drawing. The safest is ***Erase***. If you change your mind later and want the dimension back, you will find it in the **Annotations** in the Drawing Tree. It will appear as a hidden item, which you just ***Unerase*** by RMB. If you ***Delete*** the dimension, you have to show it again, which is more steps. However, note that a shown dimension can appear in only one place in the drawing tree, even if it is erased. For example, it can't be erased in one view and shown on another view or sheet.

Try to lay out all the dimensions so that your drawing looks similar to Figure 22. The dashed offset (snap) lines created when we cleaned the dimensions can be removed by selecting ***Delete*** from the RMB menu. Selecting snap lines depends on which ribbon you are in and the setting of the **Selection Filter** (should be set to ***General***). You don't really need to do this for hard copy, since Creo Parametric will not print snap lines.

Do not be concerned at this time if the witness lines are touching or crossing the model. As you probably know, this is a "no-no" in engineering drawings. Although you can clear that up on the screen using the witness line drag handle, Creo will clean up the witness lines automatically when a hard copy is generated. Creo will also look after all the line weights and line styles (for visible and hidden lines, center lines, dimension and extension lines, and so on) according to standard engineering drawing practice.

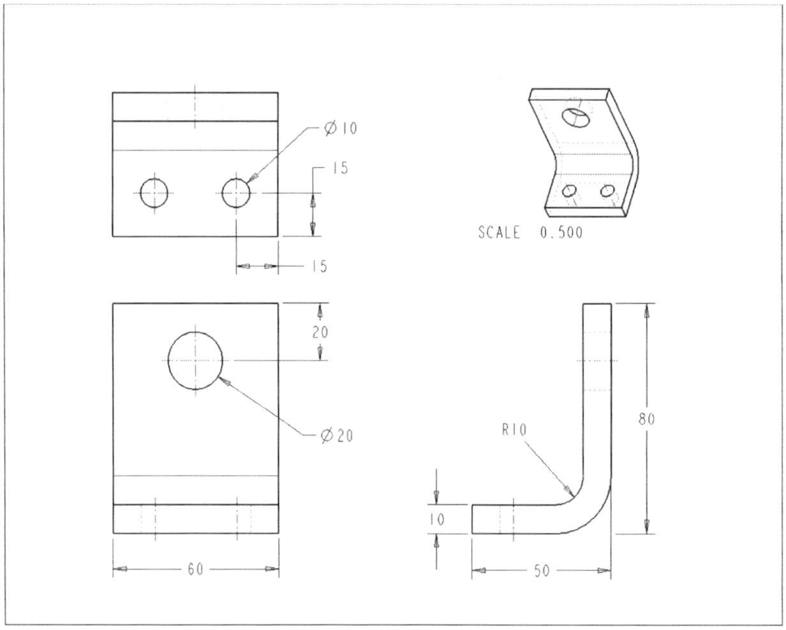

Figure 22 L-bracket drawing in progress

⑥ Creating a Note

Let's add a short note on the drawing (we will talk about title blocks in the next section). You may have to move the other views up a bit to fit this in (you can do that after the note is created, if necessary). In the **Annotate** ribbon, select the ***Note*** button Note (this is the default option ***Unattached Note*** in a drop-down list). A smaller window opens that gives controls for determining the placement of the note. The default is an unattached (aka free) point, that is, not tied to a specific drawing entity.

Read the message prompt and observe the setting (*Free Point*) in the **Select Point** window. Then select a location a little below the right side view. An input box opens where you can type in the note text. Also, the **Format** dashboard appears which has controls for font, size, color, alignment, and so on, and includes a palette of special symbols (Figure 23). Type in something like the following (don't worry if this extends outside the drawing border - we can move it later):

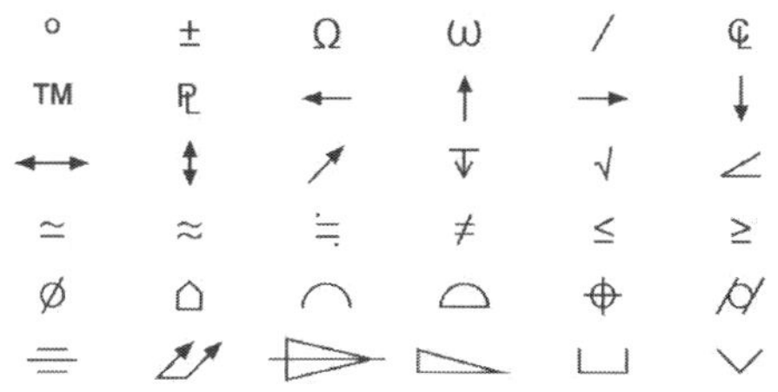

Figure 23 The Symbol palette

> ***ALL DIMENSIONS IN mm***
> ***Drawn by Art O'Graphic***
> ***01 February 2026***

Double click on the drawing background to finish. If you select the note, the **Format** dashboard tab reappears and you can move the note box around by dragging. Use the green dot drag handle to rotate the note. There is lots available using the RMB to quickly change font and alignment of selected text. Also, remember that ***Undo*** works here too!

Save the drawing using the default filename; Creo will automatically append a *drw* extension to the file name.

> ***File ➤ Save***

Exploring Associativity

One of the most powerful features of Creo is its ability to connect the part model and the drawing in a bidirectional link. This is called *associativity*. Here is a scenario where this is very useful.

It's late Friday afternoon and your boss has just reviewed the design and drawings of the L-bracket and has decided that a few changes are needed as follows (before you go home!):

- the height must be increased to 100 mm
- the diameter of the large hole must be changed to 30 mm
- the top of the bracket must be rounded in an arc concentric with the large hole
- the manufacturing group wants the drawing to show the height of the large hole as 70mm from the bottom of the part

Hmmmm... You could do this by going back to the part and modifying/changing. BUT..there is an easier way! To really see the power of what you are about to do, resize the drawing and part windows so that both are visible. Close both Navigator panes. Make sure the **DRAWING** window is active. If not, either use the ***Windows*** drop-down list in the Quick Access toolbar, or click on the drawing window.

Double click the diameter dimension of the large hole (or select it and in the RMB pop-up menu, select ***Modify Nominal Value***). Enter a new value of ***30***. Click somewhere off the dimension and it will show in blue. Similarly, select the height dimension and change it to ***100***. Now, in the Quick Access toolbar select (or use the shortcut key CTRL-g)

> ***Regenerate Active Model***

The drawing should change to show the new geometry. Even better, as soon as you activate it, the part window also shows the new geometry. In the part window double click on the protrusion and change the width of the bracket from 60 to ***80***, then ***Regenerate***. Change back to the drawing window and activate it - it shows the new shape too. These actions show that there is a *bidirectional* link between the drawing and the part. If changes are made to an item in either view of the model, the other is automatically updated. The same holds true when you deal with assemblies of parts, and drawings of those assemblies.

* VERY IMPORTANT POINT *

New users often find that changes they have made to a drawing get lost somehow. The following will help you avoid the "What happened to my drawing that I spent the last hour editing?" syndrome.

Consider what Creo needs to maintain associativity between part and drawing. In order to update the part when a dimension is changed on the drawing, then the part itself must be "in session" (but not necessarily in its own window; for example, it could be in an assembly window) whenever the drawing is[9]. Therefore, to bring up the drawing (read it from disk), the part file must also be available (on disk or already in session). The part will be brought into session, if necessary, even if it is not displayed in its own window. Think of the drawing file as a set of instructions for how the part model should be represented in the drawing. **The drawing file does not itself contain the part.** Furthermore, when the drawing file is read from disk, the part file must be in the same location relative to the drawing file (and have the same name) as it was when the drawing was created (usually the same directory) - otherwise Creo doesn't know where it is and cannot load it. You cannot load the drawing as an independent object[10]. The same holds true for assemblies: when you want to work on an assembly, all its constituent part files must be available to be read in as required. You can see that for complicated projects, file management might become an issue.

Consider also that if you change a dimension on the drawing, this makes a change in the part file loaded in session. It does not immediately change the part file on disk. To make the change "stick" you have to save the part file.[11] The same holds for some other drawing entities like section definitions.

Before we forget, change the width of the bracket back to ***60***. In the drawing, select the dimension to highlight it, right click to bring up the pop-up menu, select ***Modify Nominal Value***, enter in the new value ***60***, select ***Regenerate Active Model*** as before.

One thing we can't do in the **DRAWING** window is change the basic features of the part (like creating new solid features, or changing feature references). For that you have to go back to the **PART** window. Do that now, so that we can add the cut to round off the top of the bracket and change the dimension reference for the hole.

---

[9] Check out ***File ➤ Manage Session ➤ Object List***

[10] You can create model-less drawings that consist of "dumb" 2D entities like lines, circles, notes, and so on. This is what CAD was like many years ago.

[11] A configuration option *save_modified_draw_models_only* (default = Yes) lets you automatically save the part file whenever you save the drawing file.

First, if necessary, ***Reroute*** the large hole (select it and in the pop-up menu select ***Edit References***) so that the horizontal dimension reference is the **TOP** datum instead of the upper surface of the bracket. The distance above this reference should be ***70***. If the hole disappears off the bottom of the part (the axis is still visible) you can still select it in the model tree. Modify its dimension value to -70. If you scroll back a few lines in the message window[12], you will see a warning that was produced when the hole was regenerated (something like "Hole is entirely outside the model").

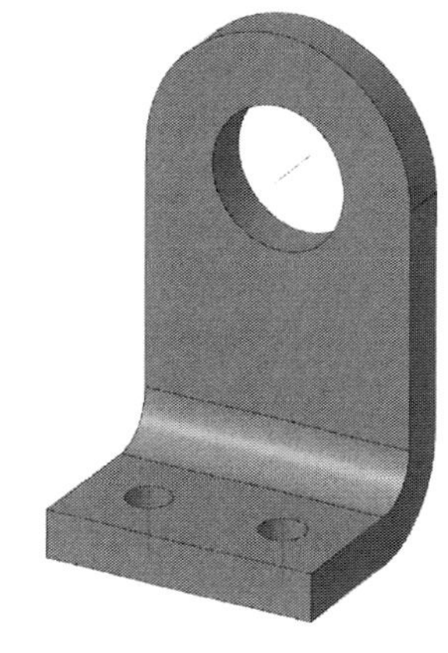

Figure 24 The modified L-bracket

Now create a circular arc cut, concentric with the large hole and aligned with the left and right sides of the bracket. The part should look like Figure 24 when you are finished. Don't forget to save the new part.

Now we have to touch up the drawing a little. Change back over to the drawing window and activate it. The new circular arc should be shown there.

First, the drawing scale is a little too big for the sheet. Double-click on the ***Scale*** value shown on the bottom line of the graphics window and enter ***0.8***. You might like to reposition the views. Notice that this change in scale affects the orthographic views (that are using the sheet scale) and not the 3D general view (that has its own scale).

Next, you may note that the large arc isn't dimensioned. Actually, a dimension isn't needed for the arc since we know the block width. And anyway, because of the way the feature was created, it has no dimensions in the model, so there is nothing to ***Show***! Since the dimension is not strictly necessary, we will create a *reference dimension* in the drawing. In the **Annotate** ribbon select ***Reference Dimension*** (in the **Annotations** group overflow) and then left click on the arc in the front view. Use the MMB to place the dimension - it will appear with the symbol REF to indicate it is a reference dimension[13]. Select ***Cancel*** in the **Select Reference** menu. If you mouse-over the new dimension you will see its symbol is "rdx" where the "r" is for reference. You might like to clean up the dimension cosmetics a bit. In Creo language, the dimension we just added is called a *created* dimension (as opposed to a *shown* dimension). These do not have to be reference dimensions (which is why there are three versions of the ***Dimension*** command).

There is possibly a redundant dimension on the drawing (depending on how you constrained the arc). This is usually not allowed! Can you find it? You have a choice of three. Select it and use ***Delete*** in the RMB pop-up.

---

[12] In the **Tools** ribbon, check out ***{Investigate}: Message Log***

[13] Your system may be set up to show reference dimensions in parentheses.

You should also change the text in the note using your system editor. Left click on the note, then in the pop-up toolbar select ***Properties***. The text will appear in the **Note Properties** window. Change the first line to something like

SCALE 0.8, DIMENSIONS IN mm

Check out the available settings in the **Text Style** tab. Select ***OK*** when you are finished. The drawing note should update. Now is a good time to save the drawing.

Printing and Exporting the Drawing

In addition to saving the drawing on your hard disk, you will likely need to create images of the drawing for distribution or printing. Obtaining hard copy may depend on the details of your local installation (for example, to obtain large format plots). See your system administrator for information on this. However, there are two possible ways that might work. To start, open the **Print Preview** ribbon with

File ➤ Print ➤ Print

Select the ***Settings*** button. If you are running under Windows, in the **Destination** tab of the new window, select ***Printer(MS Printer Manager) ➤ OK***, then in the **Model** group make sure that ***Plot (Full Plot)*** is set.

Now select ***Preview***. This will confirm exactly what will be printed. Use this along with the **Display** group options (**X-Offset** and **Y-Offset**) to set up the print margins. Now pick the ***Print*** button. This should bring up your normal Windows print control dialog. Use it as you usually would to select the printer and printer properties (destination, quality, speed, color, page size, etc.). Some experimentation may be required here to get margins, orientation, and so on set just right. Select ***Close Print Setup***.

If you do not have a plotter attached directly, wish to archive the drawing image, or want to send the drawing to someone who cannot read the native Creo Parametric file you can print directly to a pdf file (in the working directory with the same name as the part file) by using

File ➤ Save As ➤ Quick Export (*.PDF)

This will automatically print the entire sheet. Check out the options available under

File ➤ Save As ➤ Export

The ***Configure*** group is where you specify what you want - a DXF or IGES format drawing file, a PDF format document, a TIFF image, and so on - and create the settings for the output. The content of the ***Settings*** window will depend on the type of output requested so select the output type first, then the ***Settings*** button. These include the print resolution (dpi), color choice, treatment of hidden lines, security features (like passwords), and so on. For PDF output, you should probably go for the highest resolution possible (disk space is cheap!). The default output file will be *lbrack.pdf* in the current working directory (unless this has been over-ridden by your system administrator). Once you have the settings you want, then print as usual using the ***Export*** button.

Using Drawing Templates

For our first drawing, we did a number of operations manually. Many of these are common to all part drawings. Fortunately, there is a way to do much of this tedious drawing creation automatically.

First, make sure the current drawing has been saved, then remove it with ***File ➤ Manage Session ➤ Erase Current***. Note that this does not delete the drawing from your hard disk but just removes it from the current session (takes it out of memory). You should be back in the part window (and it should be active).

Create a new drawing with ***File ➤ New ➤ Drawing***. Uncheck both boxes at the bottom to **Use default template** (we'll pick one) and **Use the drawing model name**. Then replace the default name with [***lbrack2***] and select ***OK***. In the **New Drawing** dialog window, check the button beside ***Use template*** and select an **A** sized drawing by picking ***a_drawing*** in the **Template** area. This does the following:

- creates the drawing sheet (A size)
- orients the model
- places the standard views (top, front, right) for a multiview drawing
- scales the views to give you room for detailing

When you select ***OK*** and enter the drawing window, everything should be set up for you as shown in Figure 25.

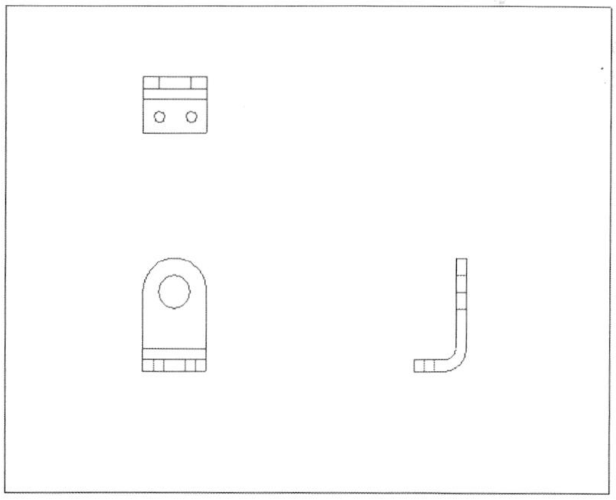

Figure 25 Drawing created using the default template

How does Creo Parametric know what standard views you wanted? The views that are shown on the drawing are determined by the layout of the template[14]. The drawing template refers to standard views defined in the part (that were likely created with the part template). The drawing views are based on the default datum planes TOP, FRONT, and RIGHT and the associated Saved Views. The orientation of the part in the drawing is therefore determined by how we orient the geometry of the part relative to the datums. If your part is upside down in the model, then the drawing views will be upside down too. The template can also specify a default arrangement for section views, dimension

[14] Creating your own templates is discussed in the *Advanced Tutorial*.

preferences, and much more. All of this information must be created in the model with the knowledge that the template will be using it. Another good reason to plan ahead!

Now, there may be a good reason to have the orientation of the part different in the drawing than in the model. If you still want to use the part and drawing templates, here is how to reorient the drawing views created automatically. Double-click on the front (primary) view. In the **Drawing View** window, select ***Category(View Type)***. In the view orientation area, select the view **LEFT**, then the ***Apply*** button. A confirmation window will ask if you want to modify the orientation of the children views (these are projections of the primary view). Select ***Yes***. All three views will change, maintaining the specified orthographic projection[15] relation between views. Change the primary view back to our original **FRONT** view.

With the views created, go ahead and finish detailing the drawing for practice. Try to do this on your own, but refer back to our previous procedures if necessary.

On to the second part - a simple pulley. Here we will concentrate on creating a section view, a detail view, and a quick look at drawing title blocks and customization tools.

Now is a good time to take a break!

The Pulley

We're going to use this part (Figure 26) in the next lesson (on assembly). We will make it now to see how to create a drawing with section and detail views, title block and border. Most importantly, we will look at a more orderly way of getting the dimensions onto the drawing instead of showing them all at once as we did for the bracket. As mentioned previously, when parts get even moderately complicated that strategy will take more work than necessary.

Figure 26 The pulley

First, let's get on with creating the pulley model[16].

Creating the Pulley

The main interest in this part is the cross sectional shape. The key dimensions of this shape are illustrated in Figure 27. We could create the base feature as a revolved

---

[15] Incidentally, the default is *3rd Angle Projection*. European users may be more used to *1st Angle Projection*. There is a setting to control that - see the drawing detail option **projection_type**. Drawing option settings are discussed briefly on page 8-37.

[16] Once again, if you are pressed for time, use *creo_pulley.prt* in the download set.

protrusion using the shape in Figure 27. However, this single feature would require a fairly complicated sketch. Instead, we'll create this geometry using three features then add the holes and rounds.

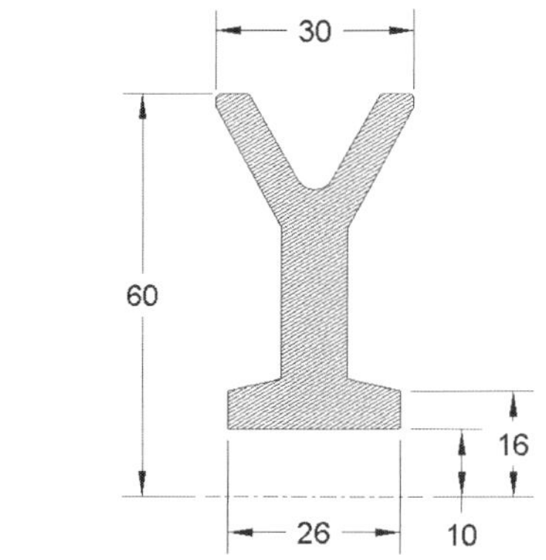

Figure 27 Pulley cross section (dimensions in millimeters)

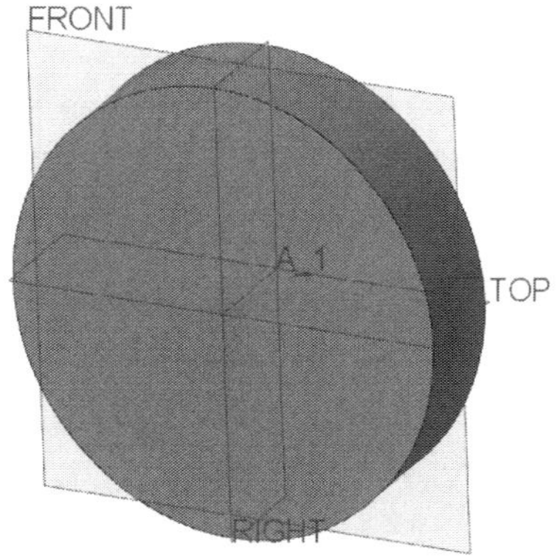

Figure 28 Base feature of pulley

Start by creating a new part called ***pulley*** using the ***mmns_part_solid_abs*** template. In the appropriate parameter fields, enter something like ***[Tutorial pulley]*** for the *Description* and your initials for *Modeled_by* . You can also enter (or modify) these values using ***Parameters*** in the Tools ribbon.

Create a circular disk (both sided protrusion off **FRONT**) aligned with the origin. Look ahead to Figure 35 to see why we want this orientation. The disk has a diameter of ***120*** and a thickness of ***30***. The disk should look like Figure 28.

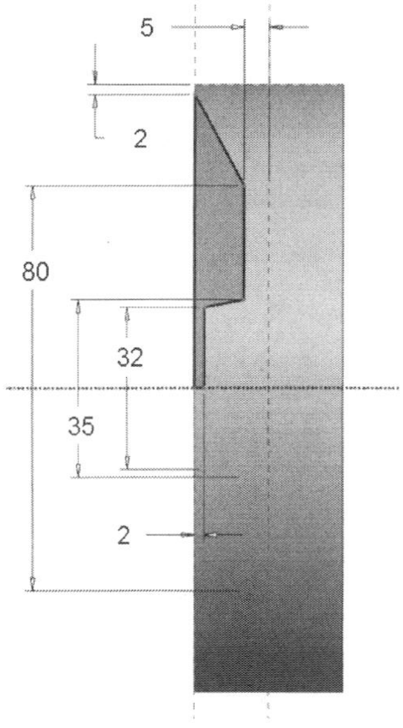

Figure 29 Sketch for revolved cut (constraints not displayed)

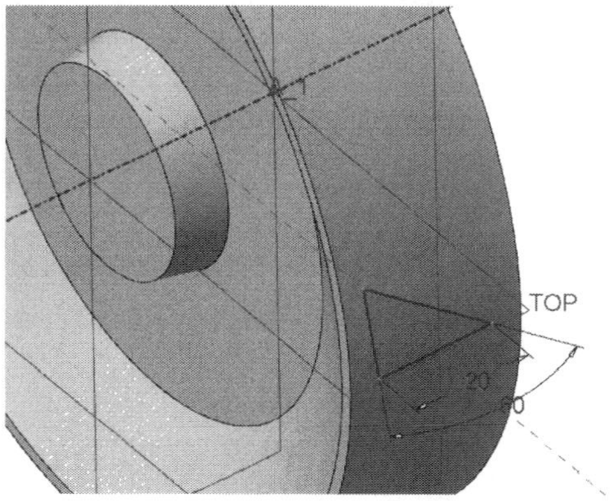

Figure 30 Revolved cut to make pulley groove (constraints not displayed)

Now, create a 360 degree revolved cut on the front side of the disk. The sketching plane is **RIGHT**. The dimensions are shown in Figure 29. The revolved cut can be mirrored through **FRONT** by (with the feature highlighted) selecting the ***Mirror*** command, selecting the FRONT datum plane, then middle click.

Create the pulley groove around the outer circumference as another revolved cut. Just make a symmetrical 60E V-shaped groove as shown in Figure 30. The vertex at the bottom of the V aligns with **FRONT**.

Add a round at the bottom of the pulley groove with a radius of ***3***.

Add the central hole for the pulley axle. This can be created as a ***double-sided, coaxial*** hole off **FRONT** with a diameter of ***20.*** The depth is ***Through All*** in both directions. See Figure 31.

Now we'll create the pattern of holes arranged around the pulley. We start by creating the pattern leader. Again, use **FRONT** as the placement plane and go ***Through All*** in both directions. Create the hole using the ***radial*** option (***28.5*** from pulley axis). Measure the angle ***30*** from **TOP**. This is the angle that we will increment to make the pattern. See Figure 32.

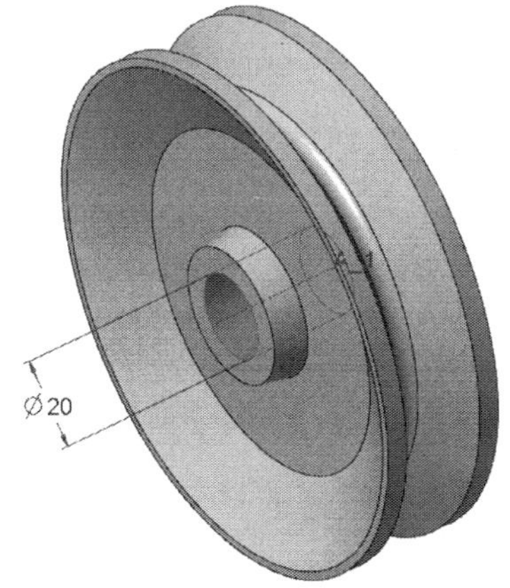

Figure 31 Central hole added to pulley

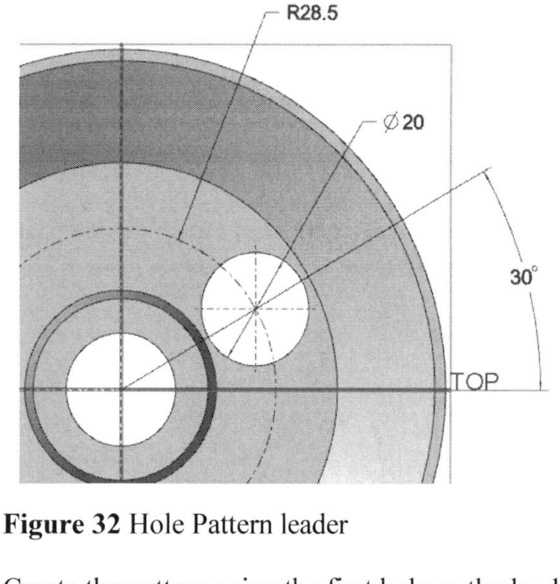

Figure 32 Hole Pattern leader

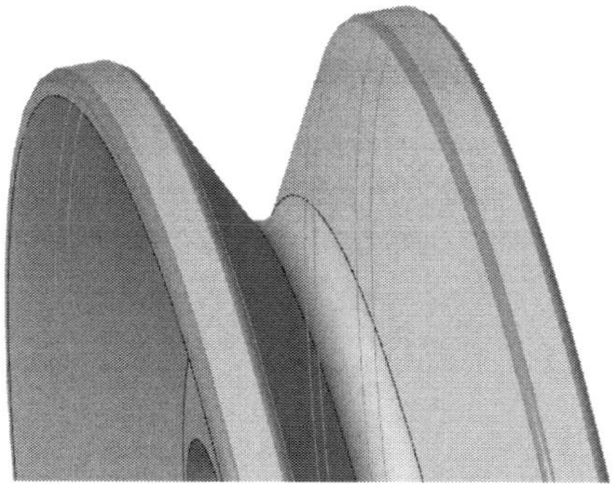

Figure 33 Rounds added to outer edges

Create the pattern using the first hole as the leader. Increment the angular dimension by ***60*** and make a total of ***6*** holes.

As a final touch, add some ***rounds*** (radius ***1***) to the outer edges as shown in Figure 33. All four edges can be included in the same feature (using CTRL, not separate selection) - don't create four separate round features or sets!

That completes the creation of the pulley. Before we go on to the drawing, don't forget to save the part.

Creating the Drawing

① Selecting a Formatted Sheet

For this drawing, we will use a pre-formatted sheet with a title block. Start a new drawing with

File ➤ New ➤ Drawing

You can leave the two boxes at the bottom checked. In the **New Drawing** window that opens up, select ***Empty with format***. In the **Format** area, select ***Browse.*** This takes you to the directory on your system that contains drawing formats. We are looking for a file called **a.frm**. The default location is (for Windows systems with a "generic" Creo Parametric installation) **<proeCreo_loadpoint>/formats/a.frm**. If you can't find it, either consult your system administrator, or carry on without the format by canceling the command. In the **New Drawing** window, select ***OK***.

Assuming you were able to load the format, the drawing window will open with an ANSI standard title block and border already drawn on the A-sized sheet as shown in Figure 34. The drawing tree opens showing that we have one drawing sheet.

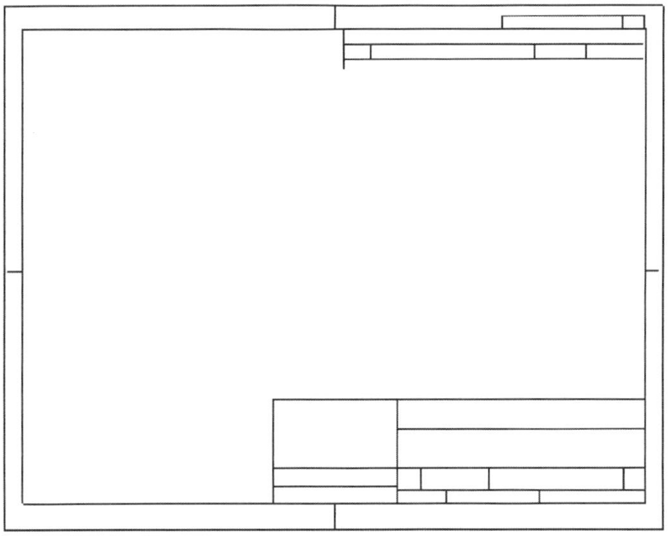

Figure 34 Formatted drawing sheet (A size)

② Creating the Primary View

Since the **Layout** group is open by default when we enter a new drawing, use the RMB pop-up menu and select the command

General View ➤ *No Combined State | OK*

Click to the left of center of the sheet. In the **Drawing View** dialog, select **View Type** and set the orientation to **FRONT**, then ***Apply***. You may have to change the sheet scale (indicated at the bottom of the graphics window) to ***0.5***. Close the **Drawing View** dialog window. Your screen should now look like Figure 35. If the view is shaded and/or with datums displayed, don't worry - we will take care of that later.

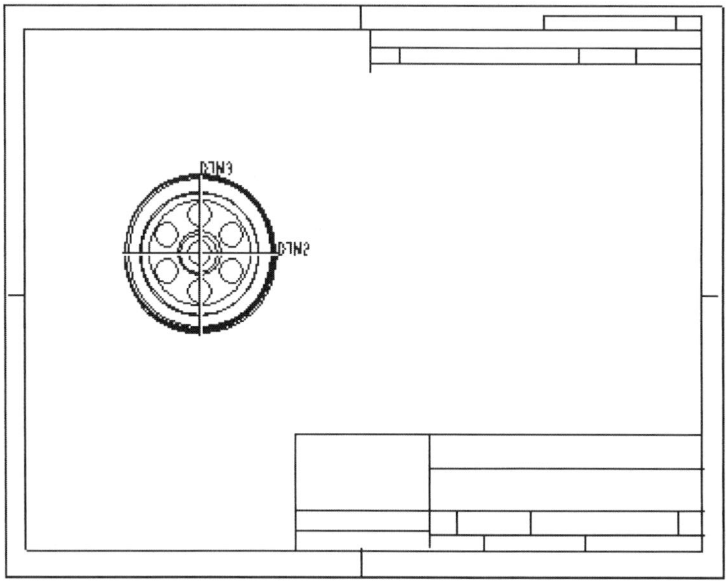

Figure 35 Primary view placed and oriented

③ Add a Full Section View

We will create a full section to the right of the primary view. The section can be created either in the model (using the ***View Manager***) or in the drawing. Either way, the section definition will be stored with the model (another reason to keep the model and drawing files together!). The preferred way is probably to create the section definition in the model, which is what we will do in Lesson #10 using a tool for creating sections. For now, since the drawing is open, we will do it here.

To create a section in the drawing, we have to specify the type of view, the location of the view, where the section is to be taken, and on what view to indicate the section cutting plane and view arrows. Turn datum plane display on since we will use a datum to define the section plane. Select the primary view (note the green border) and in the pop-up toolbar, select

Projection View

Drag the mouse to the right (orange border shows where the view will be) and place the view.

Now we have to modify some properties of the view and specify where we want the section taken. With the new view selected (green border), select ***Properties*** in the pop-up toolbar. In the **Drawing View** window (Figure 36), under **Categories** select ***Sections***. Select the button beside **2D cross section** then click the ***Add*** ("+" sign) button. One of the old-style cascading menus will appear (XSEC CREATE), possibly partially hidden behind the Drawing View window. The default to create a new section is a ***Planar*** section using a single datum plane. In the XSEC CREATE menu, select ***Done***.

Read the message prompt (it makes sense that if the environment is set to Shaded, the section view will have to override this). In the text entry window, enter the single character "A" so that our section will be identified on the drawing as *Section A-A*. The **SETUP PLANE** menu opens so that we can specify the cutting plane for the section. We want to use a vertical plane through the pulley. If a datum plane doesn't exist for this, you could create a make datum. In our case, **RIGHT** will do just fine. Select it in the front view or model tree, then back in the **Drawing View** window select ***Apply***.

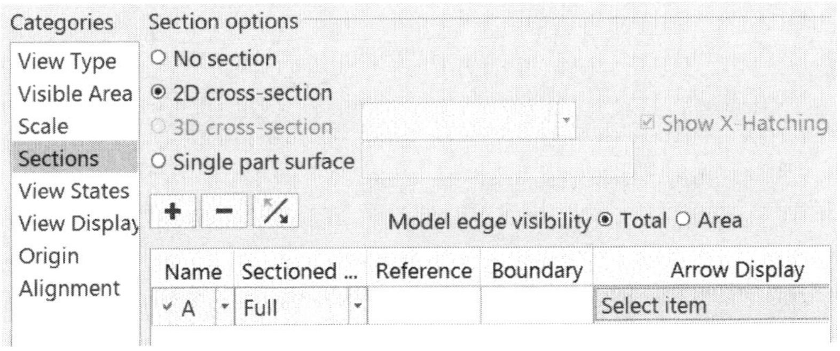

Figure 36 Drawing View dialog for creating section views

In the **Drawing View** window, slide the horizontal tab all the way to the right to expose the **Arrow Display** collector (see Figure 36). Click in the blank space and read the message window. We want the section arrows to appear on the front view, so click somewhere on the front view, then ***Apply***. The arrows appear showing the direction of view onto the section. Close the **Drawing View** window. We are finished with the datum planes, so you can turn them off now. Your drawing should look like Figure 37.

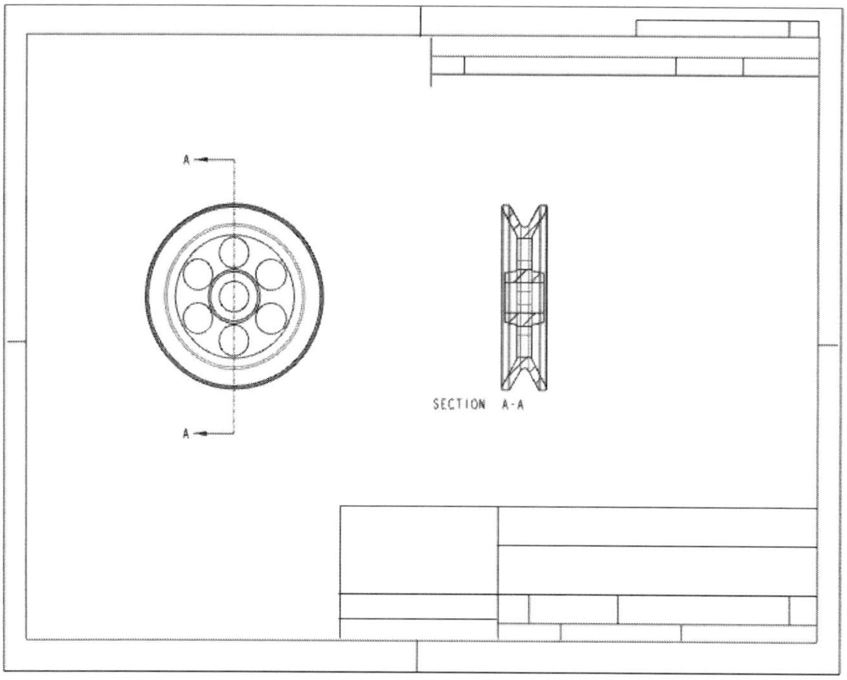

Figure 37 Section view placed

④ Modify the Section View Display

Section views generally do not show any hidden edges. Let's turn them off. Double-click on the section view at the right (this is the same as selecting ***Properties***). In the **Drawing View** window, select ***View Display*** in the **Categories** list. Then set

Display style (No Hidden)
Tangent edges (None)

then select ***OK***. For the FRONT view, set the view display to

Display style (Hidden)
Tangent edges (None)

⑤ Adding a Detail View

We'll add a broken out detail view of the pulley groove. This will be useful for dimensioning and showing the rounds. We'll also draw this at twice the scale of the drawing. With neither current view selected, in the **Layout** ribbon select

Detailed View

(or use ***Detailed View*** in the RMB pop-up). Read the message window prompts carefully as you do this. Pick a point near the bottom of the pulley groove in the section view (you can zoom in to help locate this). A large green X appears. We now want to indicate the area around the pick point to be included within the detailed view. As you click with the left mouse button, a spline curve will be drawn. See Figure 38. Make sure this encloses the groove (four or five points should be enough). When you have fully enclosed the area to be drawn, click with the middle mouse button.

A note appears ("See Detail A") on the drawing and the region is surrounded by a circle roughly around the area you identified. The message prompt now asks you to select a center point for the new view on the drawing. Pick a point where there will be enough space for the view (we can always move the view later if this point doesn't work out).

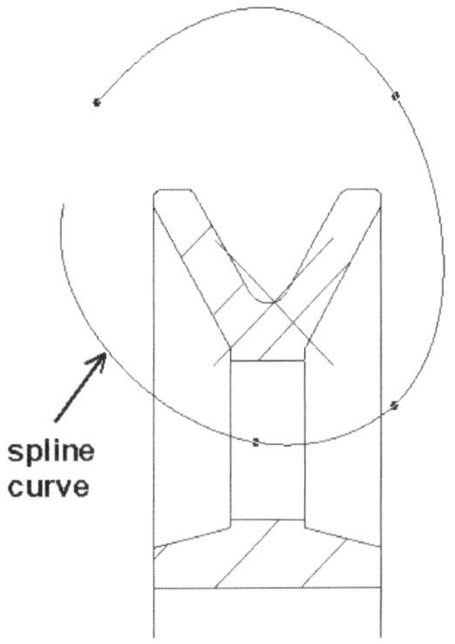

Figure 38 Defining boundary of detailed view

With the new detailed view highlighted, open the pop-up menu and select ***Properties***. In the **View Type** category, change the view name to "***B***". Check out the various boundary types. Don't forget you must ***Apply*** the settings before going to a different category. Now select **Scale** category and enter ***1.2***. You should now have a scaled-up detailed view something like Figure 39. You can move the views around by unlocking them and dragging them wherever you want. The notes can be moved by simply dragging them to the desired location. Try center justifying the note "DETAIL B SCALE 1.200".

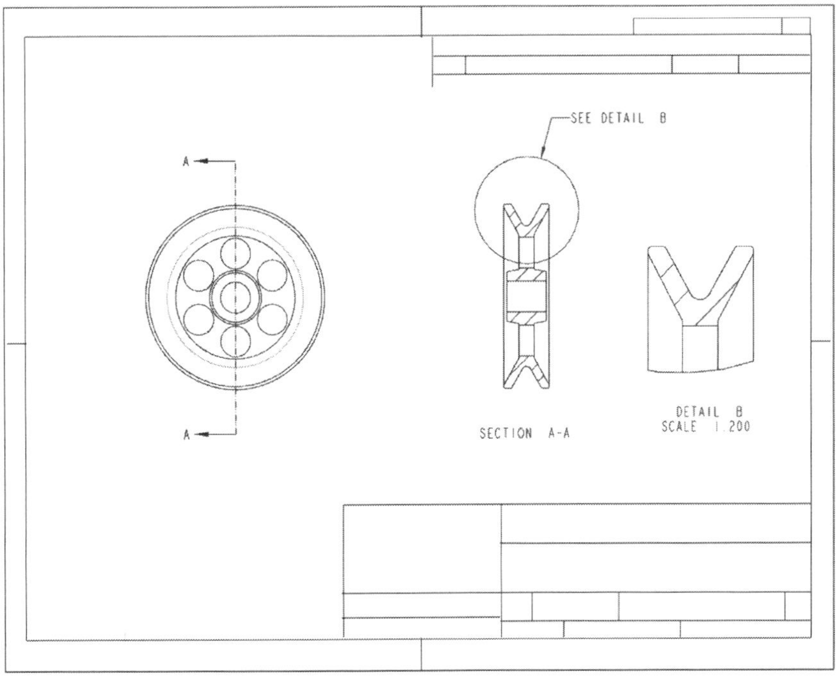

Figure 39 Detail view of section added

⑥ Adding Dimension Details

Instead of getting Creo to show us all the dimensions at once as we did for the previous drawing, we will be a little more selective since this part has quite a few dimensions. This will give us more control about initial placement of dimensions, which means fewer changes later (hopefully!). We are going to deal with individual features in the model tree. This was another reason to use several simple features - less to deal with at once when we show the dimensions.

In the model tree, select the base feature (the extruded disk) and in the RMB pop-up menu, select ***Show Model Annotations***. The dimensions for only the selected feature appear in dark red and are listed in the **Show Model Annotations** dialog window (for this feature, look for the diameter and thickness). Check the boxes for both of these and then select ***OK***. You can easily manage the cosmetic changes for these. Look ahead to Figure 40 for some drawing layout ideas.

Right click on the next feature listed in the model tree (the revolved cut) and then ***Show Dimensions by View*** in the RMB pop-up. These will all go into the RIGHT (section) view. You will want to show all of these except the 360° for the revolve. Clean these up using the methods we saw previously (including ***Cleanup Dimensions*** in the RMB pop-up menu if things get really messy!). Continue moving down the model tree, picking one feature at a time and showing the dimensions. This is a foolproof way of making sure that all the model dimensions are shown on the drawing. For the hole pattern, dimension the pattern leader and change (using ***Properties*** in the RMB pop-up) the diameter dimension text as shown in Figure 40. Use the **Dimension** and **Format** dashboards as required, for example to change the number of displayed decimals.

You may find that picking features out of the model tree (if you know exactly what they are in the drawing) is the easiest way to manage the creation of the drawing. In a complex model, it really helps if the features are all named *and* you have thought about how you want to lay out the dimensions before you start! Following standard drawing practice, Creo Parametric will place each shown dimension only once (this will be important in a minute or two!), so if you want a dimension in a particular view you must either first show it there, or use ***Move Item to View*** (in the pop-up toolbar) later.

Helpful Hint

When you think you have captured all the model dimensions and shown them on the drawing, use the RMB pop-up to select ***Show Model Annotations***, pick the dimension tab, and then pick on the part name in the model tree. You may get a surprise about something you accidentally skipped over. Just select the (hopefully few!) dimensions you have missed.

At this point, you should probably save the drawing. Make sure it goes in the same directory as the part file.

⑦ Improving the Esthetics

If you haven't already, use the drag handles and the right mouse button to modify/move the dimension details as required to get a better layout.

Let's change the crosshatch pattern in the section and detailed views. Since the section is closely related to the view definition, select the **Layout** tab (if you are in the **Annotate** tab, you cannot pick the section hatch). Then select only the hatch in the section view so that it is highlighted in green. This opens the **Edit Hatching** dashboard. Have a quick look over the various groups. The ***Hatch Gallery*** allows you to select from standard ANSI and ISO hatch patterns (the default is ANSI31) and some custom hatch patterns shown at the bottom. You can even create your own custom hatch patterns. In the **Editing** group, you can change the hatch orientation and scale, translate the hatch horizontally or vertically, and/or change the hatch color. For our hatch, change the **Rotate** angle to ***30*** (default 45) and the **Scale** to ***50***. Note that the hatching changes in the detail view as well. Lots to experiment with here. We'll see more use for this dashboard when we do an assembly drawing in a later lesson.

Next, we'll add all the centerlines for circular features. These may already be visible (with labels); however, if you turn off datum axis display they will disappear, indicating that they are not yet placed on the drawing. Select the **Annotate** ribbon. In the model tree, select the base feature and in the RMB pop-up select the ***Show Model Annotations***. Because all the dimensions for this feature are already being shown, the dialog window indicates there are none available. Pick the tab on the furthest right to show datums. Two axes are indicated (actually the same axis A_1 in two views) - check them both, then ***OK***. Obviously, we could have done this previously when we were showing dimensions. There are no additional axes required for the revolved cuts or center hole - the drawing is already showing those (i.e. they are the same as the revolved protrusion).

When you get to the pattern of holes, use CTRL to select all the hole features in the model tree, then in the RMB pop-up, select ***Show Model Annotations***. Pick the tab at the far right. A large number of axes will show in the collector window. You will have to be careful about which ones you want to appear on the drawing. For example, there are a number of axes for holes that are either in front of or behind the section plane in the right view. These should probably not be shown. If you accidentally create one of these, no problem - it can be selected and deleted individually later. It is easiest to select these directly in the graphics window. When all the boxes you want are checked, select ***OK*** to accept.

⑧ Changing Drawing Options

It is likely that the centerlines on the hole pattern in the front view look slightly different from Figure 40. The display of these centerlines is controlled by a drawing option. Other options include text height, arrowhead size and style, tolerance display, and many more.

To see the options, in the pull-down menus at the top-left select

File ➤ Prepare ➤ Drawing Properties ➤ Detail Options | Change

This brings up a long list of options. They are sorted **By Category** (see the setting at the top right of the window). Default values/settings are indicated with an asterisk "*". This information was read from a default file on your system when you created the drawing. Any values you change here will affect only the current drawing (unless you can change and over-write the system file) and are stored with it.

Browse down this list to get a feel for what you can do in terms of drawing customization. Go ahead and change things like *text_height* (0.125), *arrow_style* ("filled"), *draw_arrow_length* (0.15), *draw_arrow_width* (0.05), and so on.

About 2/3 of the way down the list (or sort the list alphabetically), look for the option

radial_pattern_axis_circle

and in the **Value** list at the bottom set it to ***Yes***. Then select ***Add/Change ➤ Apply ➤ Close ➤ Close***. The centerline layout should now be circular and radial as in Figure 40.

There is some further discussion of the drawing option file in the *Advanced Tutorial*.

⑨ Adding Notes with Parameters

Finally, add some text to the title block. You can, of course, use notes to create simple text within the title box. You may want some notes to change if the model changes. You can do this with parameters. Do you remember entering a value for the parameter *DESCRIPTION* when creating the part using the template? The text was something like "Tutorial pulley". In the **Annotate** ribbon select ***Note*** (actually ***Unattached Note***) then pick a point in the appropriate cell in the title block (see Figure 40). Type in the following text in the prompt area (without the square brackets):

[&description]

Click (twice) on the background. The value of the part parameter will appear at the insertion point - this is what the "&" symbol does when used with parameters. You can move the note later to center it in the box. Put a note for the *MODELED_BY* parameter in another box in the title block using the text string

[Modeled by &modeled_by]

Mouse over the diameter dimension for the axle hole to find its symbolic name, something like *dxx*. Enter a note in the title block with the following text (using the dimension symbol on your drawing for the diameter):

[Pulley shaft &dxx]

The diameter symbol will appear automatically since Creo knows this is a diameter dimension. When you accept this, what happens to the dimension on the drawing? Why does this happen? Look for the dimension in the Drawing Tree under Annotations for the primary view. Where is it now listed in the Drawing Tree?

Can you change the value of the dimension in the note we just created? (You may have to Regenerate to see the effect.) In any case, it is not good practice to put part dimensions in the title block, so highlight this note and delete it. What happens to the hole dimension in the drawing tree and on the drawing?

Can you enter the note to display the drawing scale? Take a guess for the name of the built-in parameter, and don't forget the "&" symbol.

Add some other notes to fill in the title block. Your company or school will likely have some custom formats that will have title blocks that will automatically be filled with part and drawing parameter information. These formats will typically be set up to work with your installation's part templates.

Your final drawing should look something like Figure 40. Here is a test of your drawing-reading abilities: what is the missing dimension in Figure 40?

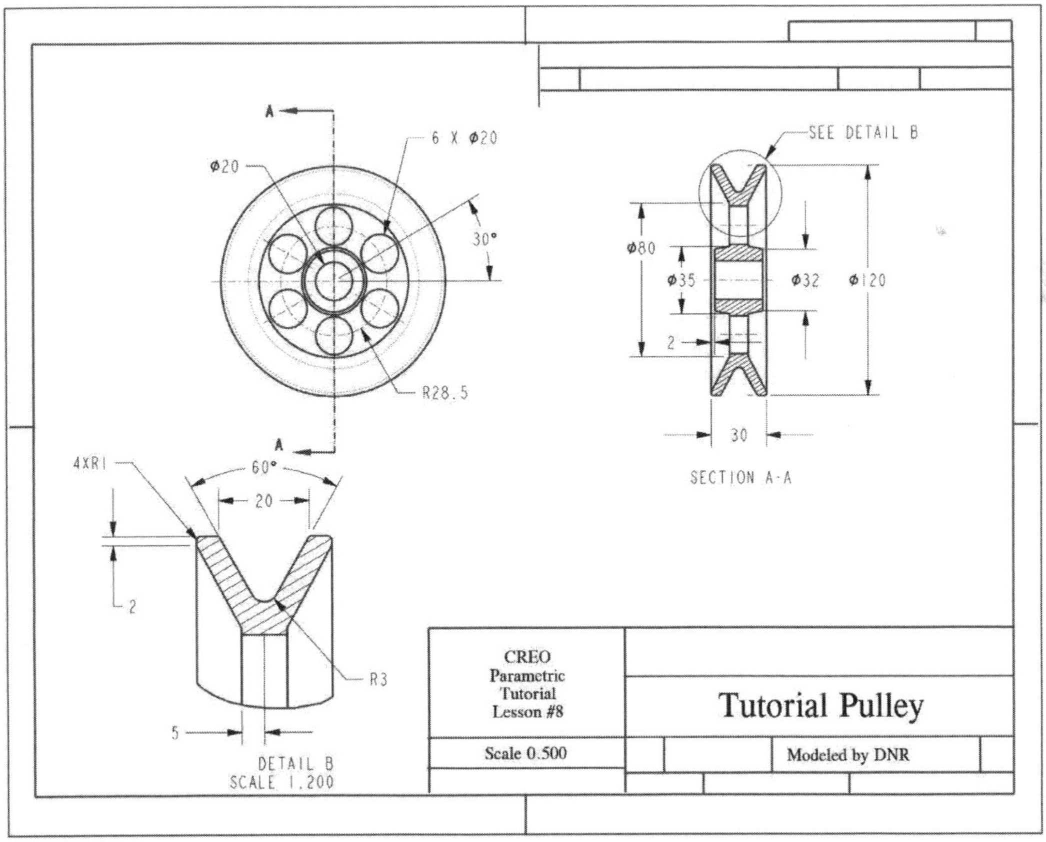

Figure 40 Finished drawing of pulley. Can you find the missing dimension?

⑩ Creating Dimensions

As mentioned earlier, the dimensions placed by Creo Parametric are the ones used explicitly to construct the model and all you have to do is show them. From time to time, you may have to create some dimensions manually. You do this using the dimensioning tools available in the **Annotate** ribbon. The main dimensioning tools here are

Dimension

Ordinate Dimension

Reference Dimension

There are some others in the **Annotations** group drop-down menu (the Reference Dimension is located there). These tools are fairly self-explanatory (follow the message prompts) and will be easy to pick up by anyone who has done 2D CAD. The main thing to remember, however, is that dimensions you create cannot be used to drive the geometry - they are strictly lines on the drawing. As discussed earlier these are called "*driven*" dimensions whose value cannot be modified in the drawing (but will change if the geometry of the part changes).

The **Annotate** groups also contain a number of other tools for completing the drawing: surface finish, symbols, geometric tolerances, jogs and breaks, changing the arrow style, managing text styles and format, and so on. You will need to do a lot of exploring to cover all this!

If you want, make a hard copy of the pulley drawing. If you have zoomed in or out on the drawing, make sure that the plot setup is set to ***Full Plot*** before creating the plot file.

Don't forget to save your drawing.

Conclusion

As you can see, although Creo Parametric handles most of the work in creating the geometry of the drawing, there is a fair amount to be done manually regarding the esthetics of the drawing. It is for this reason that you need to be quite familiar with drawing practices and standards. Creo gives you a lot of tools for manipulating the drawing - we have only scratched the surface here. There is actually an entire volume of Creo Parametric documentation (several hundred pages) devoted expressly to creating drawings! All this information is available in the Help Center on-line. Some additional drawing tools and techniques are discussed in the *Creo Parametric Advanced Tutorial* from SDC, including multisheet and multimodel drawings, tables, BOM (Bill of Materials), drawing tools, custom templates and more.

The most important lesson here is that the engineering drawing is a by-product of the 3D solid model. We don't so much "make a drawing" as much as just "show the existing model". We observed how bidirectional associativity works. It is this capability that gives

Creo Parametric and all its related modules so much power. If several people are working on a design, any changes done by, for example, the person doing the part modeling, are automatically reflected in the drawings managed by the drafting office. As you can imagine, this means that in a large company, model and file management (and control) becomes a big issue. Creo Parametric contains a number of other utilities to make that management easier, but we will not go into them here.

A second lesson is that the dimensions that will automatically show up in the drawing are those used (for example, in Sketcher) to create the features of the model. Therefore, when creating features, you must think ahead to what information you want to show in the final drawing (and how). This involves your identification and understanding of the design intent of the features in the part. A part kludged together from disorganized features will be very difficult to present in an acceptable drawing, probably using a lot of non-functional created dimensions.

We will return to creation of drawings for assemblies in Lesson 10. There we will see some more tools and techniques to expand on the ones covered here.

In the next lesson, we will see how to create an assembly using the L-bracket and pulley you created in this lesson. We will also have to create a few small parts (washers, shaft, base plate).

Questions for Review

1. When creating a new drawing, how do you specify which part is going to be drawn?
2. Is it possible to create a 2D drawing without a part? What advantages/disadvantages might this have?
3. The first view added to the drawing is called the _____.
4. How do you set the orientation of a view? Consider both the first view and subsequent views.
5. How do the mouse buttons function for dynamic view control in a drawing?
6. What is the meaning of the following toolbar icons?

a)  b) 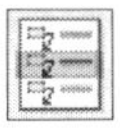 c)  d)  e) 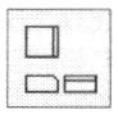

7. What is the easiest way to move a view on the drawing sheet?
8. Is it possible to delete a view once it has been created? What about the primary view?
9. How do you show the axis of a revolved feature? Can you find an alternate way?
10. How can you edit the text contained in a note?
11. How do you change a section hatch pattern to a standard material?
12. When you select ***Show Model Annotation***, in what color do dimension details first appear?
13. Explain the functioning of the three mouse buttons as used to modify dimension cosmetics.

14. How do you select a drawing template? What does it create for you automatically? Can you do this after the drawing is already created?
15. Describe two methods to move a dimension from one view to another. When might you want to do this?
16. How do you create a text note (for example, to put in a title block)?
17. Can you move or delete views that were created with a drawing template?
18. What happens if you change the value of a dimension in the drawing?
19. What is the difference between shown and created dimensions?
20. Is it possible to add new features, or redefine existing features, when you are in drawing mode?
21. How can you produce hard copy of a drawing?
22. When you want to create a section view, at what point in the command sequence for adding the view do you designate it to be a section view?
23. What four items of information are required in order for Creo Parametric to generate a section view?
24. How can you turn off hidden lines in a section view?
25. What boundary options are available for creating a detail view?
26. What information appears in the drawing tree? What does that depend on?
27. How can you change the spacing and angle of a hatch pattern?
28. How is the design intent reflected in a drawing, and how does this relate back to the part?
29. Do you think it would be possible to have a completely automatic system for creating a fully dimensioned drawing?
30. What symbol is used in a note to tell Creo Parametric to display the value of a parameter?
31. What other parameters (other than the three we used) are built into the system?
32. How are drawing options set?
33. What is the difference between a *template* and a *format*?
34. What are the advantages and disadvantages, as far as drawings are concerned, of constructing a model using many simple features instead of a few complicated ones?
35. Can a section view be "un-sectioned", that is converted back to a normal view?

Exercises

Here are some parts to practice producing detailed engineering drawings. You may have created the models for the first two at the end of a previous lesson. Do these again, keeping in mind what you want to show on the drawing - you may make the model differently this time!

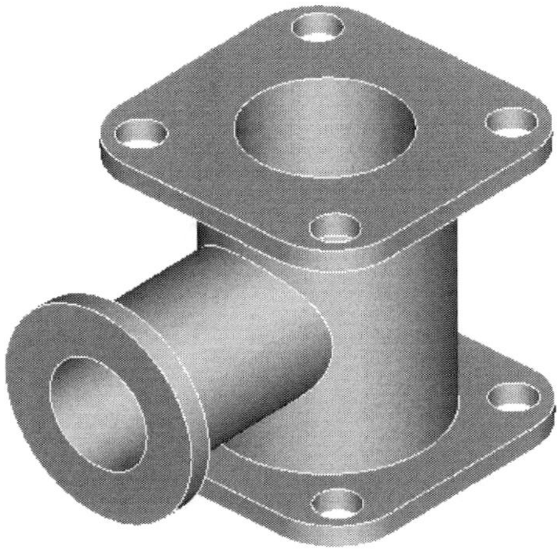

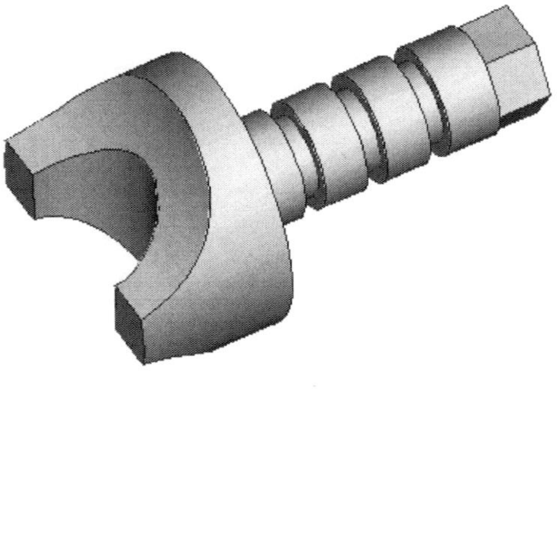

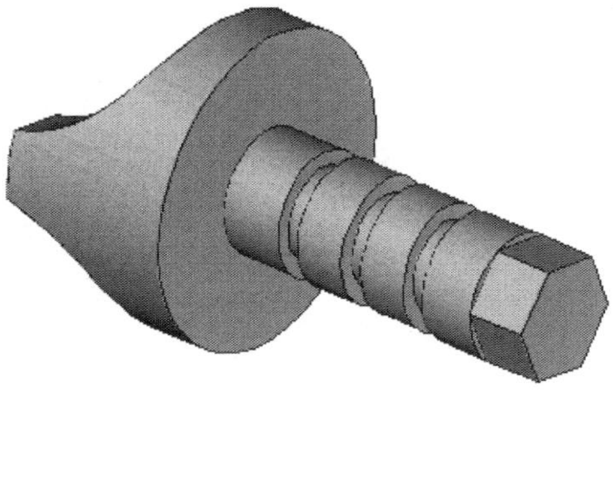

Project

Here is the final major part of the project. Some dimensions will have to be inferred - use a reasonable estimate. It is not necessary that you reproduce this geometry exactly, but use it to explore the feature creation tools. Some geometry modification may be necessary later when the entire project is assembled. Some careful planning for this part will pay off in reduced modeling time. There are a couple more figures showing this part on the next page.

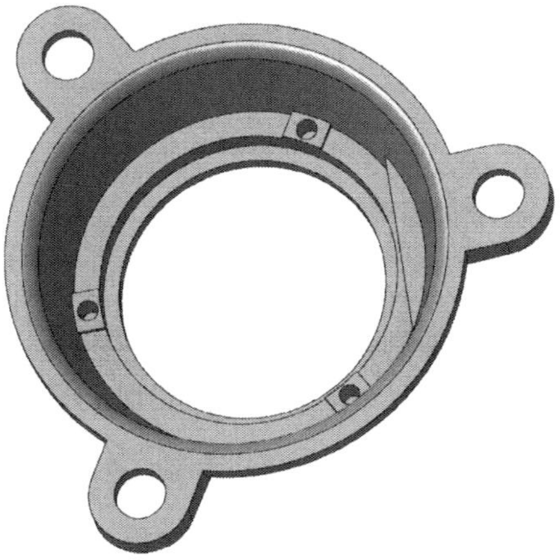

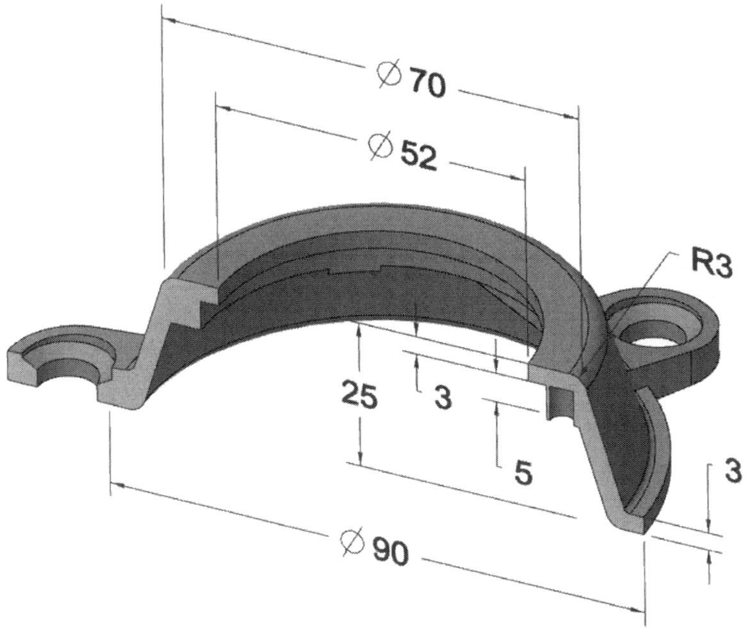

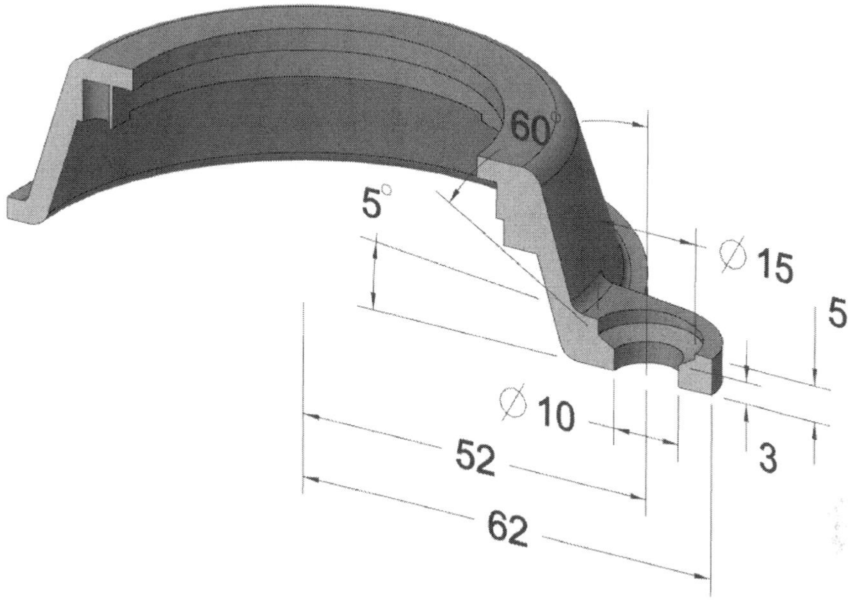
60°
5°
∅15
5
∅10
3
52
62

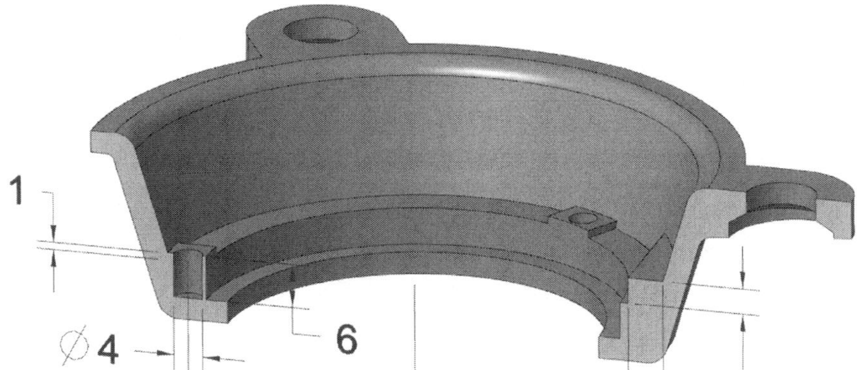
1
∅4
6
32
5
4

This page left blank.

Lesson 9

Assembly Fundamentals

Synopsis

Introduction to assembly mode; assembly constraints; subassemblies; screen layout and assembly options; copying components in assemblies; assigning appearances.

Overview of this Lesson

In this and the next lesson, we are going to look at how you can use Creo to create and modify an assembly of parts. You should have already created the parts involved in previous lessons. The finished assembly is shown in Figure 1. An exploded view of the components is shown in Figure 2 (we'll do the explode in the next lesson).

Figure 1 Final assembly containing 13 parts (some repeated)

Figure 2 Exploded view of final assembly

The lesson is organized as follows:

1. Collecting the Components
2. Discussion of Assembly Constraints
3. Assembly Design Issues
4. Assembling the Components
   - ▶ Creating a Sub-Assembly
   - ▶ Creating the Main Assembly
5. Assigning Appearances

As usual, there are Questions for Review and an exercise at the end of the lesson.

Collecting the Assembly Components

We will have need of the following parts, all of which have been made in previous lessons as indicated:

bplate.prt	page 3-33	bushing.prt	page 2-36
washer.prt	page 2-36	axle.prt	page 4-26
bolt.prt	page 4-27	pulley.prt	page 8-28
lbrack.prt	page 8-12		

Before you start assembling these components, make sure they are all in your working directory and make sure all your parts have units set to millimeters.

As mentioned above, you should have created the pulley in Lesson #8. One thing we forgot to do then was add a keyway to the central hub of the pulley. Do that now: the keyway is ***5mm*** wide and about ***3mm*** deep. Create the keyway as a ***both sides cut*** off **FRONT**. Put the keyway at the 3:00 o'clock position when viewed from the front (symmetric about **TOP**). The keyway should look like Figure 3.

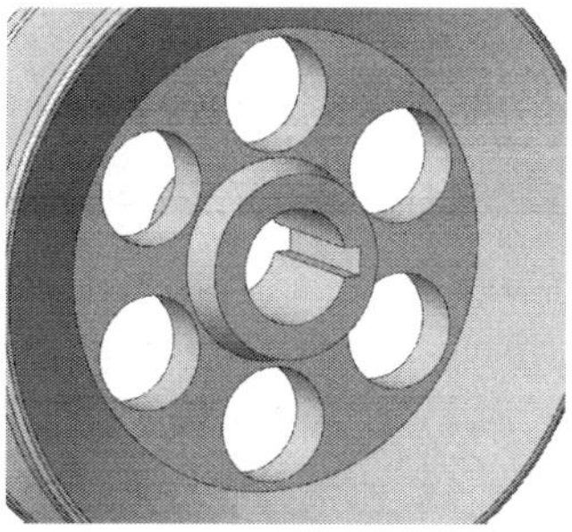

Figure 3 Pulley with keyway added

You can leave the pulley in session when you move on to the assembly. As a safety precaution, you should save it now.

Assembly Constraints

Creating an assembly is actually a lot of fun and not too difficult. Your main challenge will be display management as the screen gets more cluttered with objects. Creating an assembly involves telling Creo how the various components fit together. To do this, we specify *assembly constraints*. A component that is fully constrained in the assembly is called *placed* or *assembled*. It is possible to leave a component not fully constrained, in which case it is called *packaged*. Creo will be able to tell you whether a new component is packaged or assembled [1].

The geometric relation between any two parts has six degrees of freedom (DOF): 3 translational and 3 rotational. These are represented in a tool that will appear on the screen when you are assembling a part or component - the 3D Dragger shown in Figure 4. This shows three directions for translation (the arrows), and three directions for rotation (the arcs).

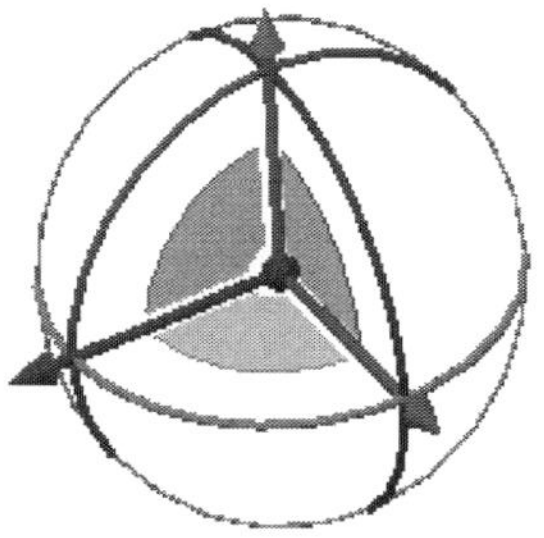

Figure 4 The 3D dragger

In order to completely define the position of one part relative to another, we must constrain (provide conditions for) all these DOF. As a part is constrained, the dragger will indicate the constraints by changing color for the constrained DOF.

Once we give Creo enough information it will be able to tell us when the part is fully constrained and we can assemble the part. We proceed through the assembly process by adding another part, and so on. Setting up the assembly constraints leads to a hierarchical structure of the assembly (essentially an "assembly tree" but still called the model tree). Within this structure, different components are related by parent/child relations formed by specifying the constraints.

Packaged components (that is, components that are not completely constrained) will be easy to identify with a special symbol in the model tree. We normally avoid leaving components in this condition since their behavior is sometimes unpredictable when the assembly is regenerated. Components that have at least one constraint to a packaged component are automatically considered as packaged only (even if they are fully constrained to the packaged component).

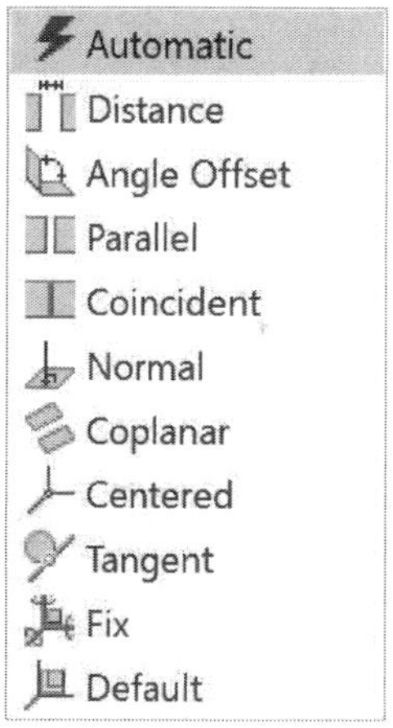

Figure 5 Assembly constraints

There are a number of constraint types that we can specify - see Figure 5. Each of these has numerous variations depending on the type of references chosen (surfaces, axes, datum planes, and datum points of the components) in the assembly. Within each component, these references can come from different bodies, that is, the component does not have to be a single merged solid. Each constraint will produce some combination of constrained DOF, or in a couple of cases, will totally constrain the component. In this lesson, we will use a few of these constraints. The rest should be pretty easy to figure out on your own.

An example of a simple assembly containing two parts is shown in Figure 6. The

---

[1] To further complicate things, components in a moveable assembly (a *mechanism*) are joined together by *connections* which are special types of constraint that allow motion along specific degrees of freedom. We will not be discussing mechanisms here. For further information go to the Help Center and look in **Simulation ➤ Mechanism Design and Mechanism Dynamics**. Or see the **Creo Mechanism Tutorial** from SDC.

constraints (type and shaded references) used to constrain part B to part A are shown in Figure 7.

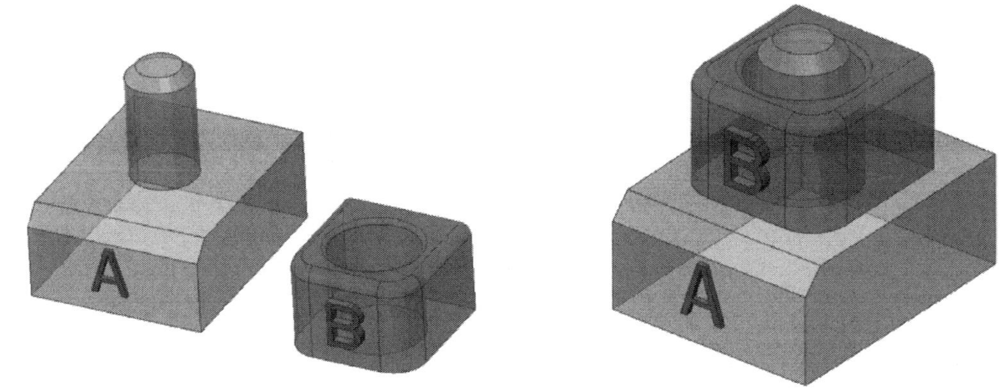

Two components   Components assembled

Figure 6 Two components forming an assembly

For this example, three constraints were required as follows:

- **Coincident** - axis of cylindrical shaft on A is coincident with axis of hole on B
- **Parallel** - front surface of B is parallel to front surface of A
- **Coincident** - bottom surface of B is coincident with top surface of A

The assembly was created by specifying the surfaces involved in each constraint. Creo automatically determined the type of constraint to use. The common constraints are discussed below.

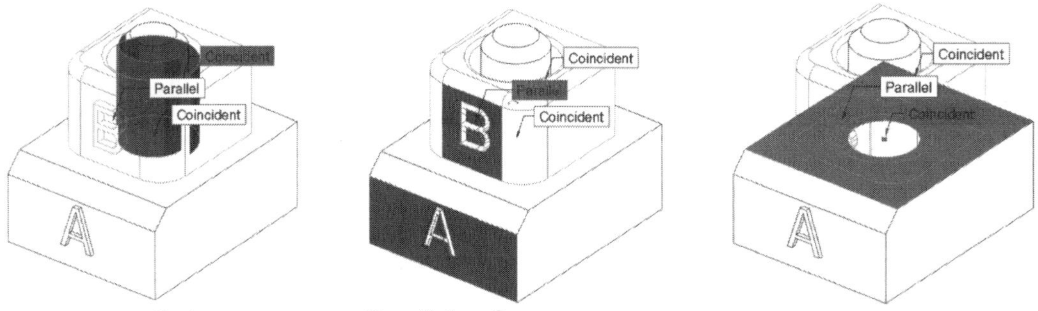

Coincident cylinders   *Parallel* surfaces   *Coincident* surfaces

Figure 7 Constraints used in assembly of Figure 6

Default Constraint

The component is placed so that the default position in the component matches the default position in the assembly. Think of this as aligning the (TOP-FRONT-RIGHT) datum planes of the component and assembly. This is most useful for the first component brought into the assembly and reinforces the requirement for advance planning when creating the component so that it is constrained in the orientation that you want.

Coincident Constraint

This may be the most common type of constraint since it has so many possible variations. It can be used with solid surfaces, datum planes and axes, cylindrical surfaces, edges, vertices, and more. Basically, the constraint requires that two references in the component and assembly must align. Some examples are shown in Figure 8. You can frequently use a ***Flip*** command to reverse the sense of a reference (for example, which side of the datum should be aligned or which direction an axis should point). For cylindrical surfaces, the coincident command basically forces the axes of the cylinders to align - the surfaces do not have to be the same diameter. Coincident edges become colinear and not necessarily overlap. The number of degrees of freedom that are constrained depends on the references chosen. For example, coincident planes or vertices constrain 3 degrees of freedom, while coincident cylinders or edges constrain 4. Can you figure out what they are?

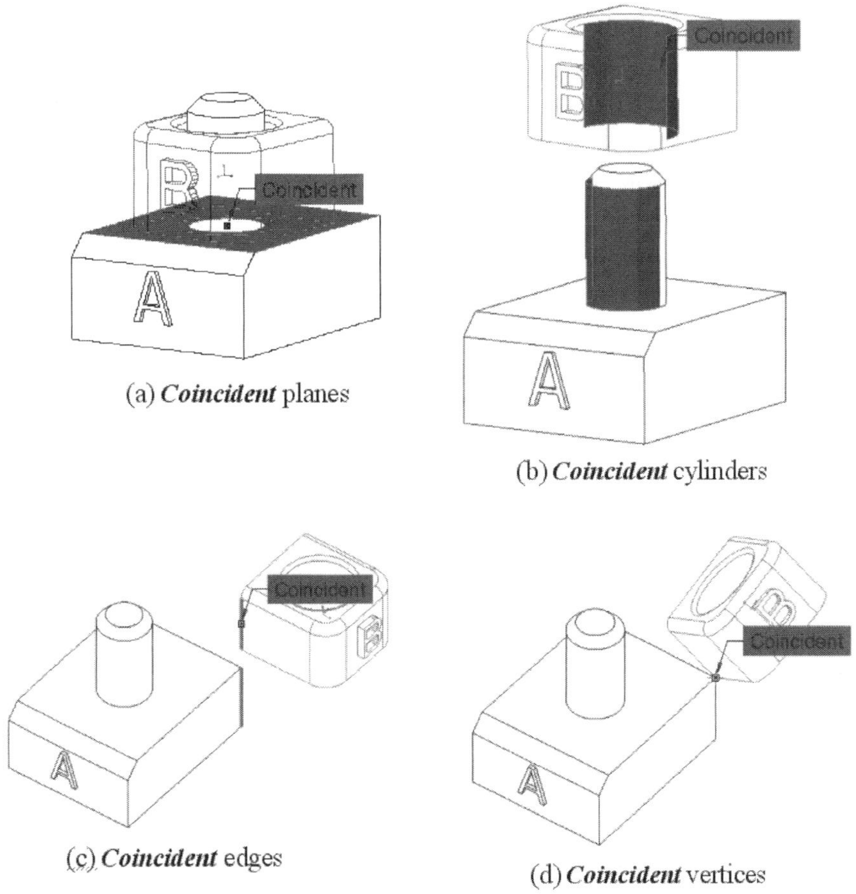

(a) ***Coincident*** planes

(b) ***Coincident*** cylinders

(c) ***Coincident*** edges

(d) ***Coincident*** vertices

Figure 8 Variations of the ***Coincident*** constraint

Normal Constraint

Two planar surfaces or datum planes, or an edge and a surface or datum, or two edges or axes, become perpendicular. See Figure 9. The constrained degrees of freedom depend on the references chosen. For example, two normal planes constrain only one DOF, while an edge-plane pair constrains two, and an edge-edge pair constrains two.

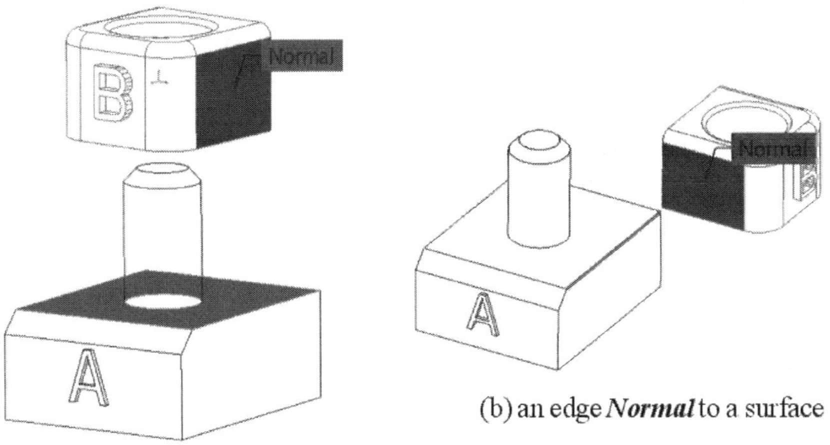

(a) *Normal* surfaces

(b) an edge *Normal* to a surface

Figure 9 Variations of ***Normal*** constraints

Distance Constraint

Two planar surfaces or datums are made parallel, with a specified offset distance, and face in opposite directions. See Figure 10. If two axes or edges are picked as reference, they must be parallel. For planes, the offset dimension can be initially set negative to measure the distance in the opposite direction; once set, the distance becomes positive. A ***Flip*** command is available to reverse the normal orientation of surfaces.

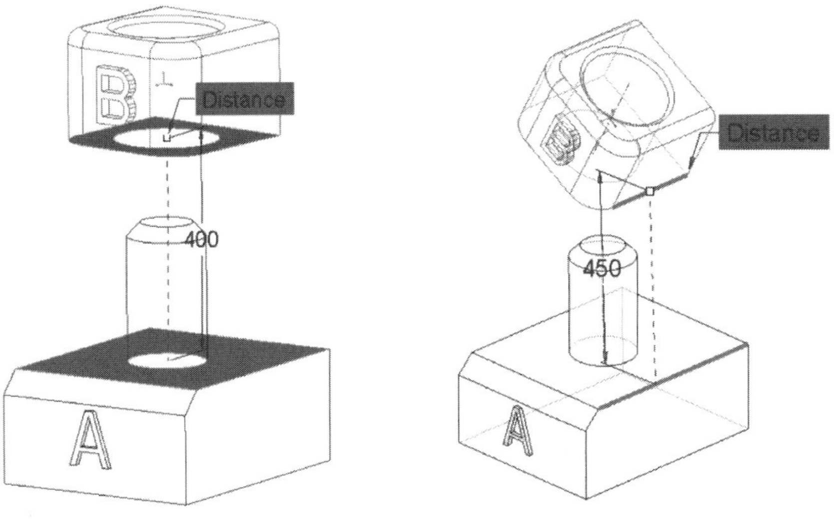

Figure 10 Variations of ***Distance*** constraint

Parallel Constraint

This can be applied to planar surfaces (Figure 11) and datums, and to edges and axes, or any combination. For two planes, this constrains three DOF.

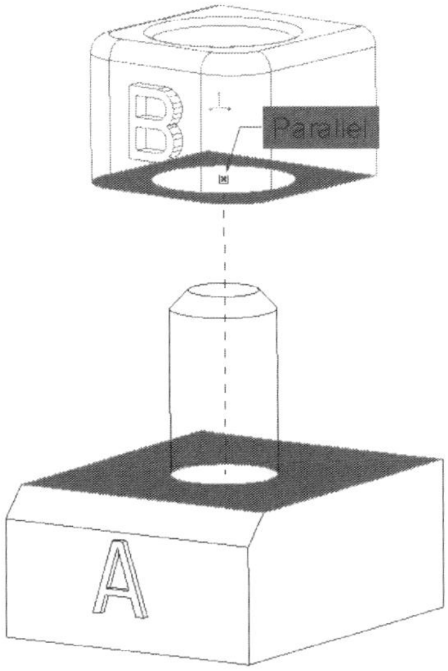

Figure 11 ***Parallel*** constraint

Angle Offset Constraint

This can be applied to planar surfaces (Figure 12), datums, edges, and axes. For two surfaces or datum planes, the angle is measured from one plane to the next, around an axis formed by the intersection of the two planes.

Tangent Constraint

Two chosen surfaces become tangent (Figure 13).

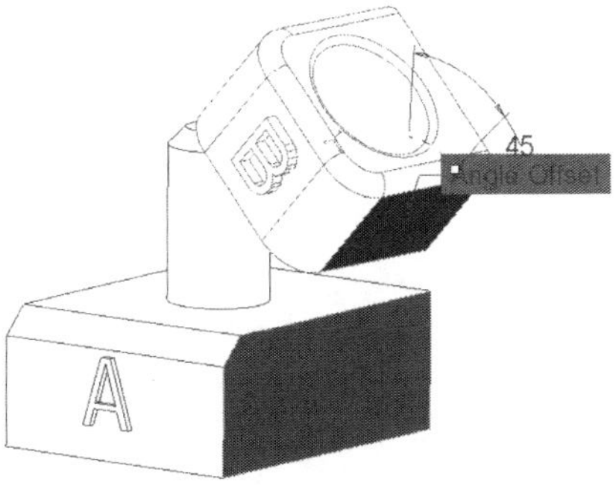

Figure 12 ***Angle Offset*** constraint

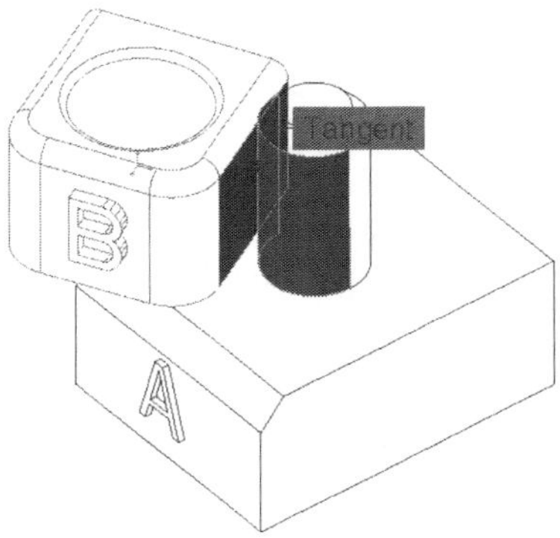

Figure 13 ***Tangent*** constraint

Fix Constraint

This is not really a proper constraint, since it does not require any references in the assembly so the component is not constrained *to* anything. Rather, it is a temporary "Just leave the component here but consider it fully constrained, while I attend to something else" kind of placement. It freezes all six degrees of freedom of the component in the current position/orientation and is non-parametric. If mixed with parametric constraints, this will over constrain the component and can cause unpredictable behavior in the assembly when it is regenerated.

Assembly Design Issues

Before beginning an assembly (or even before you create the parts), you should think about how you will be using these constraints to construct the assembly. As we saw in Lesson #1, the logical structure of the assembly is reflected in the model tree. Like designing the features of a part, the chosen assembly constraints and references should reflect the design intent. It is possible to create an assembly that fits together, but if the chosen constraints and references do not match the design intent, changes that may be required later could become very difficult. Creo does provide tools for dealing with this (the 3 R's also work in assemblies on the assembly references of components), but you should really try to think it through and do it right the first time. Obviously, it takes a lot of practice to do this. The "I-should-have-done-it-*that*-way!" realization is common among new (and even experienced) users.

This is a good time to mention that when you are placing a component into an assembly, it does not matter what order you use to define the placement constraints for the component, since they are applied simultaneously. Creo will tell you when you have constrained the component sufficiently for it to be assembled. The order you use can be chosen strictly for convenience. You can also enable and disable constraints at any time.

Do not confuse constraint selection order with the order that components are brought into the assembly. Like the base feature in a part, the choice of the first component in the assembly is important. It would not make sense, for example, to start the assembly of a car with a bolt that holds on the roof rack!

It is possible to create assembly features (like datum planes and axes) that will exist only in the assembly. This would allow you, for example, to use an assembly parameter like an angle or linear dimension between datums to control the assembly geometry. In this way, if you used the assembly feature as a constraint reference for a number of component parts, you could change the position of all parts simultaneously in the assembly by modifying that parameter. We saw a couple of examples of this in Lesson #1.

Assemblies do NOT contain bodies as such - those only exist within individual part files. In a sense, components are to assemblies what bodies are to parts. For example, two components can be involved in the same types of Boolean operations (***Merge***, ***Intersect***, and so on) that can be used on bodies in a part, with the same effects.

Assemblies can also contain solid features (primarily holes and cuts) and it is possible to set the affect of these features on individual components. For example, a cut can be set up to remove material only from selected components. Furthermore, holes through components in an assembly can be set so that they are visible (or not) when the part is loaded by itself or used to create a drawing.

In the context of this lesson, an assembly consists of a number of components that are rigidly constrained to each other - no moving parts! There are additional functions and software modules in Creo that allow you to create assemblies of moving parts, that is, mechanisms. This extension lets you create different types of joints and connections (pin, slider, ball, cam, and so on) between components, and drivers to control the degrees of freedom. You can then analyze the motion of components and even create animations of the moving system.

Finally, you should note that Creo will happily let you assemble two components that interfere with each other (have partially overlapping solid volumes). Although the parts are solids, Creo does not prevent you from assembling them with interference. Sometimes, in fact, this is the desired result (as in a shrink fit, or designing built-in snaps and catches in plastic parts). More usually, interference happens when dimensional values are not correct (or finalized) in a design. Creo has a simple tool that will tell you if any components are interfering (and by how much). It is usually a simple matter to correct the model to remove interference once the assembly is put together.

Assembling the Components

We will start by assembling the L-bracket, a bushing, and a washer into a subassembly. This will save us some time, since two copies of this subassembly must be inserted into the final assembly. Once created, a subassembly is treated exactly the same way (in regards to subsequent placement constraints) as a single part.

In the following, it is assumed that you have color display turned on using

File ➤ Options ➤ Entity Display ➤ Show Colors

Creating a Subassembly

Select *New* from the **Home** ribbon, **File** menu, or Quick Access toolbar. Then

Assembly | Design ➤ [support]

Turn off the **Use default template** option and select ***OK***. In the **New File Options** window, observe the default assembly template for your system. Since all our parts are in millimeters, we might as well be consistent so choose the ***mmns_asm_design_abs*** template. Enter values for the parameters DESCRIPTION and MODELED_BY and select ***OK***. The template has default datums (notice the names) and the usual predefined views. Unless you have a really good reason not to, you should always use a template or at least create the default datums. The advantage of the template is the saved views plus it gives us more freedom in defining the orientation of the first component we bring into the assembly[2].

---

[2] Without assembly datums to constrain to, the first component brought into the assembly can only be placed in its default orientation. It is not a good idea to limit your options this way.

The assembly ribbon tabs and contents are somewhat different from the part ribbons. Spend a minute or two to look around. Many of the commands are the same, but there are some here that are unique to assemblies. In the **Model** ribbon, select

Assemble 

Select the part **lbrack.prt** and then ***Open***.

The part appears in the graphics window, either in magenta shaded form or magenta hidden line depending on your display setting. The 3D Dragger also appears. At the top of the screen is the *Component Placement* dashboard, Figure 14. This contains virtually all the options required for constraining components brought into an assembly. We will be discussing these options throughout this lesson.

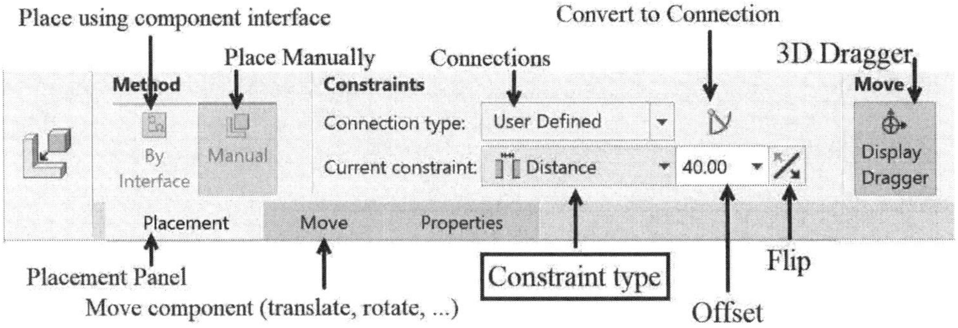

Figure 14 The *Component Placement* dashboard

On the left side of the dashboard, the first two icons on the left are grayed out - they have no use in the current context or with the current part. The top pull-down list will contain options for creating connections in mechanisms. Leave this at the default, ***User Defined***. The lower list contains the constraints discussed earlier in this lesson, plus some additional ones. The default setting is ***Automatic***. This is where you will usually leave this option. With this setting, all you have to do is supply (pick on) the desired references in the component and assembly and Creo will figure out what kind of constraint you (probably) want. For example, if you pick two planar solid surfaces, it will assume you want a *Distance*, or perhaps *Coincident*, constraint; two axes might be assumed to be *Coincident*, *Distance*, or *Normal* depending on their current spacing and orientation. The constraint chosen automatically will depend on the current assembly and component orientation and the relative positions of the chosen references. These assumptions are also based on any existing constraints, that is, Creo will not assume anything that is inconsistent with existing constraints. If an offset distance is specified, its value will appear in the text field beside the selector. The ***Flip*** command can be used to reverse the sense of a coincident or aligned reference. Finally, there is a toggle to turn the display of the 3D Dragger on and off.

For this support subassembly, we can afford to place the bracket in the same default position in the assembly as it is in the part. Therefore, in the **Constraint type** list (or RMB pop-up), select ***Default*** (bottom of list). The bracket will move over so that its datum planes align with the assembly datums. Its color will change from magenta to gold, and the status is now listed in the dashboard as **Fully Constrained**. Accept the component placement with the green check on the right end of the dashboard, or simply middle click.

Open the model tree. The tree may show just the name of the assembly (**support.asm**) and the component (**lbrack.prt**). If so, select the ***Tree Filters*** button. Set the Display options for ***Features***, and ***Placement Folder***. Apply the new settings and expand the tree for the L-bracket component. The model tree will appear as in Figure 15. This shows the assembly datums and individual components. The placement settings are listed along with all the features of the bracket part.

You can ***Hide*** the datum planes and coordinate systems display if you like. Now we'll add the bushing. Select the ***Assemble*** button again, find the component **bushing.prt**, and ***Open.***

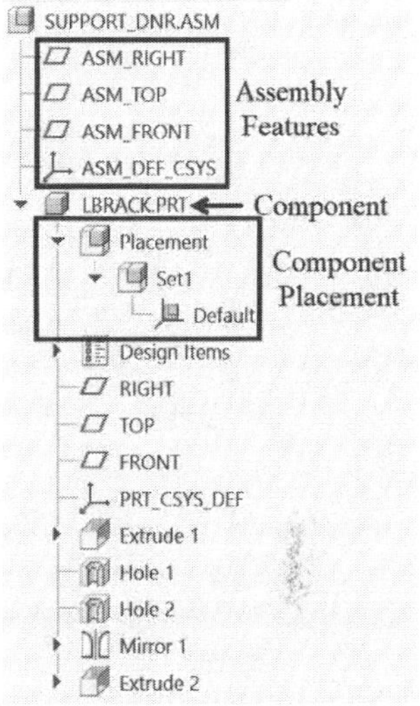

Figure 15 The assembly model tree

The bushing will appear in magenta beside the bracket along with the 3D Dragger. The **Component Placement** dashboard (Figure 14) will open up. We'll have a quick look at some new functions available on the dashboard.

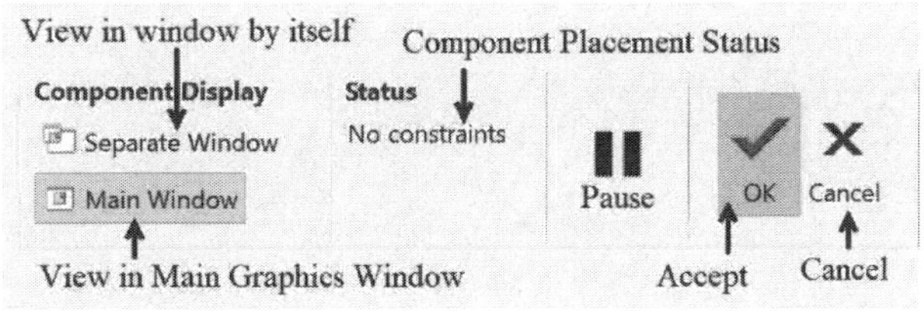

Figure 16 The *Component Placement* dashboard (right side)

There are two main display modes when you are doing assembly. These modes are set by the two buttons to the right of the 3D dragger toggle; see Figure 16. The default is ***Main Window***. To see the difference, make sure that only the ***Separate Window*** option is selected. This puts the current assembly in one graphics window, and the component being added in another. Having both windows open makes it easy to locate references and gives us independent viewing control over the zoom/spin/pan in the two windows. This is useful, for example, when dealing with a small component in a very large assembly. To close the separate window you must turn on the ***Main Window*** button and deselect the ***Separate Window*** button in the dashboard. We will see that there are a number of useful controls available to manipulate the new component, so you will probably not need the separate window very often. We will do all our assemblies tasks in the same window. The ***Pause*** button on the dashboard allows you to temporarily pause the assembly operation so that you can do other things with the model (for example, create a necessary datum). Once you are ready to restart doing the assembly, look for the ***Resume*** button at the same location on the dashboard.

The 3D Dragger

Let's take a few minutes to check out the 3D Dragger shown in Figure 17. Every visual element of this device has a function as follows:

- arrows - Selecting one of the red/green/blue arrows lets you drag the component in the indicated direction.
- rings - Selecting one of the red/green/blue rings lets you rotate the part around the arrow of the same color.
- planes - Clicking on one of the planar segments between the arrows lets you translate in the plane defined by the two arrows.
- Clicking on the ball at the center of the dragger lets you translate the object in an arbitrary direction in the view plane. Be careful of doing this, since if you rotate the assembly a large amount, the position of the incoming component is unpredictable and may be out of view.

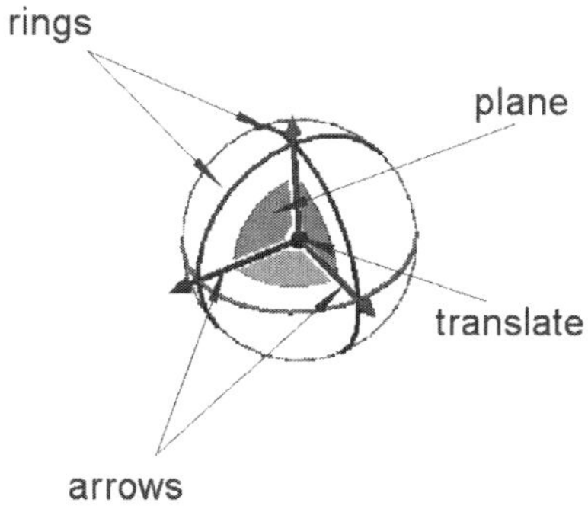

Figure 17 Elements of the 3D Dragger

Right click on the ball at the center of the dragger and select ***Drag Along Geometry***. Pick on one of the edges of the bracket (use the filter or RMB query to select an edge). The dragger will jump to that edge and one of the dragger direction arrows will align with the edge. The dragger movement is now relative to that edge. The allowed movement will be consistent with current constraints. To return to normal dragger mode, right click on the ball and turn off the ***Drag Along Geometry*** option.

As an alternative to using the 3D dragger to move the component, check out what happens if you drag with the left, middle, and right mouse buttons while holding down **Ctrl+Alt**. This is sometimes easier/quicker to do than using the 3D dragger or if it is turned off. This movement also obeys all currently defined constraints.

Let's continue with assembling the bushing. **Read the following few paragraphs before proceeding since a lot happens here automatically:**

Constraining the component involves three steps:

1. Select the constraint type.
   In the **Constraints** list of the **Component Placement** dashboard, you can select the desired constraint. Selecting ***Automatic*** (the default) allows Creo to determine the type of constraint you probably want based on the type of

entities you choose for references. The one Creo uses will depend on the relative position and orientation of the references. You can, of course, override this constraint, but you can also guarantee the constraint you want by dragging the component to the approximate position in the assembly prior to picking references.

2. Select the appropriate constraint reference in the component.

3. Select the matching reference for the constraint in the assembly. This can be assembly datums or any other assembly geometry.

4. If desired or necessary, you can override the automatically chosen constraint type, as long as the one you specify is consistent with existing constraints.

As you add constraints, the position of the component in the assembly will adjust according to the specified constraint. Keep your eye on the **Status** area in the dashboard. You will be shown when you have provided enough constraints for the new component to be placed in the assembly: the component color switches to gold and the status is indicated as **Fully Constrained**. The dashboard will let you exit without fully constraining the component; as mentioned earlier this is called "packaging" the component. Unless you really want to do this, make sure the **Fully Constrained** status appears before leaving the dashboard[3]. Also, remember that the order of creating the constraints does not matter, nor does the order of picking references on the component or assembly (that is, steps 2 and 3 can be done in either order).

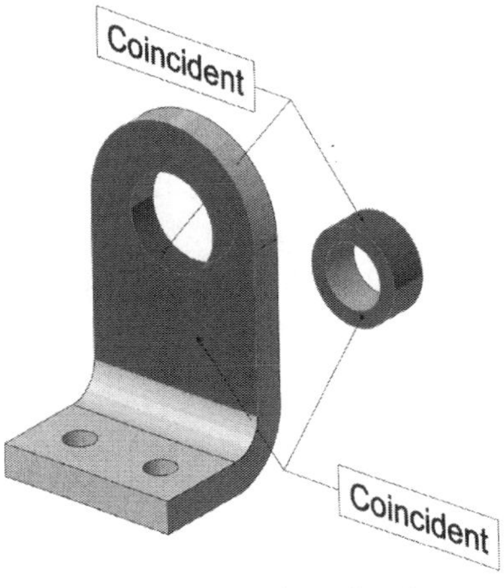

Figure 18 Constraints for the bushing

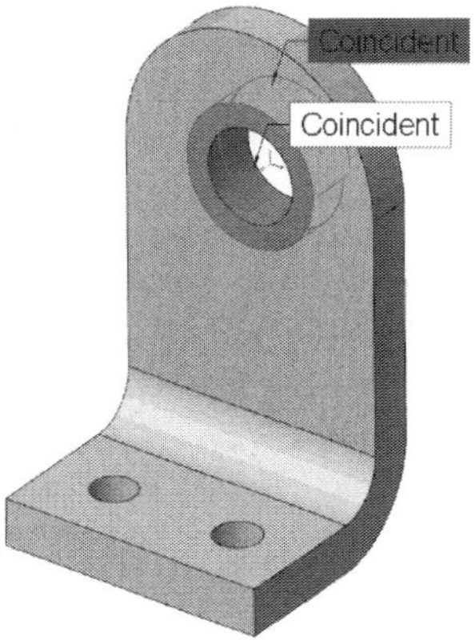

Figure 19 Bushing assembled to the L-bracket

For the bushing, we want to set the constraints shown in Figure 18. One ***Coincident*** constraint makes the cylindrical outer surface of the bushing line up with the surface of

[3] In older releases of Creo, it was not possible to overconstrain a component, that is, continue to add constraints after the status became fully constrained. That is now possible (as long as the constraints are consistent).

the hole[4]; the other ***Coincident*** constraint keeps the face of the bushing even with the surface of the bracket. The final position of the bushing is shown in Figure 19. Before proceeding with applying these constraints, resize/reorient/move the bracket and bushing displays so that you will be able to easily pick on the appropriate entities.
Now we'll proceed with the assembly. Make sure the following is selected in the dashboard:

Constraint Type (Automatic)

and pick on the outer surface of the bushing. It highlights in green. A small flag "**Automatic**" appears with a leader pointing at the bushing surface and a dashed green line follows the cursor. Now pick on the inner surface of the large hole in the bracket. The bushing moves over to line up with the hole, the flag changes to "**Coincident**", and the component status is "Partially constrained". The bushing can still slide along, and rotate around, its axis. Notice that some of the DOF on the 3D dragger are turned off (grayed out). Slide the bushing approximately to its final position.

Now pick on the flat face of the bushing. Again the "**Automatic**" label appears. Then pick on the flat surface of the bracket. The bushing turns gold and the status message will inform you that the component is fully constrained. The constraint type has been selected by Creo - we want a ***Coincident*** constraint. If Creo has chosen ***Distance*** with some offset dimension, then change the constraint type shown on the dashboard (see Figure 14) to ***Coincident***. The bushing will move so that the two reference surfaces line up.

Using *Allow Assumptions*

Although the status indicates fully constrained, is that actually true? The answer is no (!), since with just these two constraints the bushing is still technically free to rotate around its axis. Creo has determined, with an assumption, that this degree of freedom doesn't matter for this part. This is a common occurrence for axisymmetric parts such as bushings, washers, bearings, bolts, nuts, and so on, which are very common in mechanical assemblies.

Check out the **Placement** panel on the dashboard. See Figure 20. In the pane on the left, our existing Coincident constraints are listed as part of a

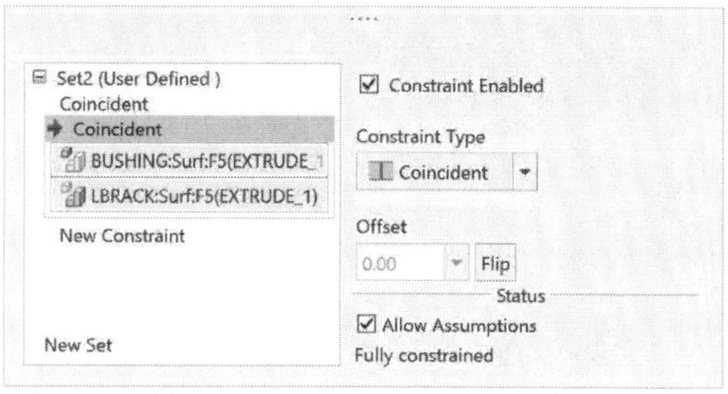

Figure 20 The **Placement** panel in the **Component Placement** dashboard

[4] As mentioned before, the two surfaces we select for ***Coincident*** do not have to be physically in contact with each other.

constraint set[5]. As you select each of these, the associated references are listed and highlighted on the model, and the constraint details are given in the panel area on the right. Each individual constraint can be toggled using the **Constraint Enabled** checkbox at the top right. At the bottom right, in the Status area, the option **Allow Assumptions** is checked by default. This will often be the case for axisymmetric parts constrained the same way as the bushing. Deselect that now. Creo tells you that, indeed, the bushing is not fully constrained, the color changes back to purple, and the rotation DOF on the 3D dragger is now active. What would be required to complete the constraints? (Hint: turn the datums back on.) Don't bother adding this constraint now, since the assumption isn't going to hurt us. Turn the **Allow Assumptions** box back on.

The bushing should be even with one side of the bracket and protrude slightly from the other (since it is a different thickness than the bracket).

If you make a mistake in specifying the type or references of a placement constraint, you can select it in the **Placement** panel. For example, click on the second ***Coincident*** entry. The associated references on the component and assembly are shown in green. These are easiest to see in wireframe or hidden line display. Then, either ***Delete*** the constraint (use the RMB pop-up), select a new type, or pick new component and/or assembly references. Try doing that now, by picking different vertical (parallel) surfaces on the bracket. What happens if you pick on one of the side surfaces of the bracket?

If you are experimenting with different constraints, remember that Creo will not allow you to use inconsistent constraints. Instead of deleting ones you aren't sure you want, it is possible to disable individual constraints by unchecking the box at the top of the panel.

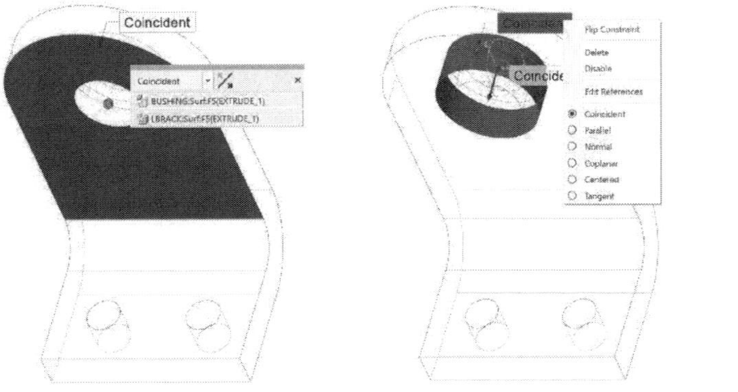

a) constraint reference options b) more constraint options (RMB)

Figure 21 Selecting options in the graphics window in assembly mode

---

[5] The existence of constraint sets implies that more than one set is possible for a given component. One set might involve all assembly constraints, while another might have mechanism connections. The same assembly could then be used in either a fixed state or as a movable mechanism by enabling one or the other of the constraint sets.

You can also modify the constraint references by double clicking on the constraint flags in the graphics window (Figure 21a). Selecting a flag and using the RMB pop-up in the graphics window always brings up something interesting and useful (Figure 21b).

Sometimes when you are experimenting with or modifying the existing constraints, you will break out of the normal sequence for specifying constraints using the Automatic setting. If that happens and you want to create a new constraint, open the **Placement** panel and select the ***New Constraint*** entry, specify the type you want (or ***Automatic***), and pick on the desired component and assembly references. This button is automatically selected unless you have interrupted the normal flow of assembly steps.

If you are happy with the bushing placement, accept it with the middle mouse button.

Next we will place a washer on the outside of the bushing. Select the ***Assemble*** button and pick the **washer.prt**, then ***Open***.

We will create the placement constraints shown in Figure 22. These constraints are between the washer and bushing axes, and washer and bushing faces. Both of these are ***Coincident*** by default. Leave the constraint type in the dashboard set to **Automatic** and observe the constraint labels on the model as we go through this.

Figure 22 Placement constraints for the washer

First - the axes. Pick the washer axis (easier to see if you turn off the display of the 3D Dragger). For the assembly reference, how can you make sure you are picking the axis of the bushing and not the hole in the bracket since these coincide in the assembly window? Right click on the bushing/hole axes - one of them will highlight. Now hold down the RMB and select ***Pick From List***. This opens a window that lists both axes at the pick point. Select the axis for the bushing and then ***OK***. If this comes up automatically as a **Distance** constraint, select the constraint flag in the graphics window and in the RMB pop-up select ***Coincident***. If the washer gets buried inside the bracket, just use the 3D dragger to pull it out along its axis.

For the next constraint, just pick on the flat surface of the washer (use the RMB and ***Pick From List*** again, if necessary), then on the end surface of the bushing. Set the offset type to ***Coincident***. If you select the wrong surface of the washer, you may have to ***Flip*** the constraint (see the Placement panel).

The washer should now be shown in gold and its status indicated as **Fully Constrained** (we will leave **Allow Assumptions** turned on). Accept the placement with the middle mouse button when you are satisfied. We are finished creating this subassembly, so select (and make sure this goes into your working directory)

File ➤ Save ➤ OK

Now we will move on to creating the main assembly, with the bracket/bushing/washer sub-assembly treated as a single component.

Creating the Main Assembly

Leave the subassembly window open, and create a new assembly called **less9** using an assembly template as follows

File ➤ New ➤ Assembly | Design ➤ [less9]

Deselect the **Use default template** option and select ***OK***. Pick the assembly template ***mmns_asm_design_abs***. Enter data for the parameters and then ***OK***.

Bring in the first component, the base plate, using the ***Assemble*** button in the ribbon toolbar and selecting the part ***bplate.prt***. We need to constrain this component to the assembly datums. We will experiment with that a bit, showing three ways of doing it. For this assembly, each will result in the same position/orientation of the part in the assembly. For other assemblies, two of these variations will allow more freedom in determining how the component is placed.

The first (and easiest) constraint, as we did for the bracket in the subassembly, can be obtained by changing the constraint type to ***Default***. Do not do this now so that we can check out some other methods for constraining the first component. If you have created the default constraint, open the **Placement** panel, select the constraint, and delete it using the RMB pop-up menu.

Use the 3D Dragger to move the part away from the assembly datum planes.

To see a second method of constraining the plate, select ***New Constraint*** and pick on the coordinate systems in the component and the assembly (unless they have been deleted). You may have to use a selection filter, or select in the model tree. This creates a ***Coincident*** constraint between the two coordinate systems. Pretty easy! The coordinate systems in either the component or assembly do not have to be at the origin of the default datum planes and can be oriented in any way in the model. Delete this constraint so that we can see a third option - hold the cursor over the **Coincident** flag in the graphics window, and select ***Delete*** in the RMB pop-up menu. Once again, move the part away from the assembly datums using the dragger (or CTRL-ALT-RMB drag).

Confirm that the constraint type has automatically reset to **Automatic**. Pick any one of the datums in the component then pick on the corresponding datum in the assembly (component **RIGHT** and assembly **ASM_RIGHT**, for example). Creo automatically sets up a ***Distance*** (or possibly ***Coincident***) constraint. Do this for each pair of datums (**FRONT** and **ASM_FRONT**, **TOP** and **ASM_TOP**) creating a ***Distance*** constraint for each pair. No other mouse clicks are required. After selecting the three pairs of datums, the component should be fully constrained. We could, of course, have created any correspondence with the three datums, as long as they were consistent, to reorient the base plate however we like.

You can accept the component placement by clicking the middle mouse button. The offset distances can be changed by selecting the component in the model tree and then selecting ***Edit Dimensions*** in the RMB pop-up. Set all three to **0** for now, then ***Regenerate*** the assembly. To change the ***Distance*** constraints to ***Coincident***, use ***Edit Definition***, then select the constraint flag in the graphics window and use the RMB pop-up menu. We don't need the datum planes or coordinate systems any more, so you can turn off their display. Leave the axes turned on.

Adding a Subassembly

Now bring in the bracket subassembly. We add this just as if it was a single component. We will set up the placement constraints for the subassembly shown in Figure 23. Other constraint systems would also assemble the subassembly here, but the ones shown in Figure 23 are closer to the physically meaningful design intent (two bolts and mating surfaces).

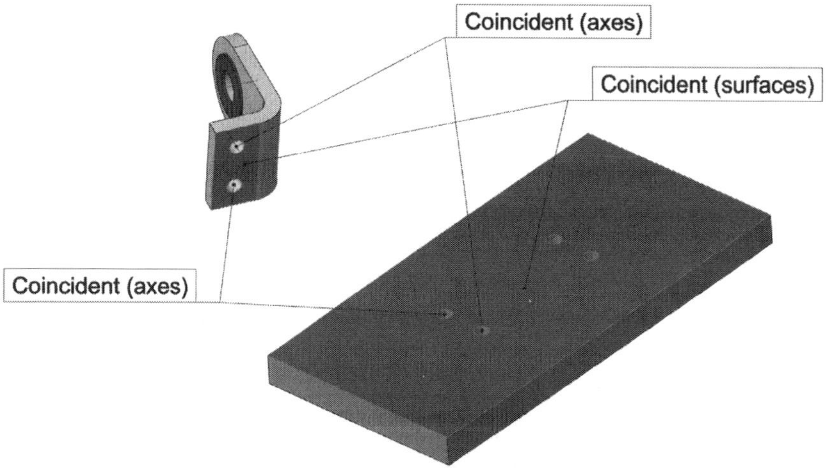

Figure 23 Placement constraints for the subassembly

Select the ***Assemble*** button and pick the ***support.asm*** sub-assembly that is currently in session. Use the 3D Dragger to move it to the location approximately as in Figure 24 - having it closer to the final position means that reasonable assumptions will be made about the type of constraint we want.

First, pick the lower surface of the bracket and the upper surface of the base plate. The default constraint for two solid surfaces is ***Coincident***. That is fine here. Then select the axis of one of the bolt holes in the bracket with the axis of the appropriate hole in the base plate. The default constraint here is ***Coincident***. If **Allow Assumptions** is turned on, you will get the message that the component is fully constrained.

However, if you reorient the model to have a close look at the other set of holes, you may see that they do not exactly line up. To fix this, open the RMB pop-up menu and deselect the **Assumptions** option. You should now be able to rotate the subassembly around the axis of the first hole (Figure 25).

Figure 24 Reposition of subassembly

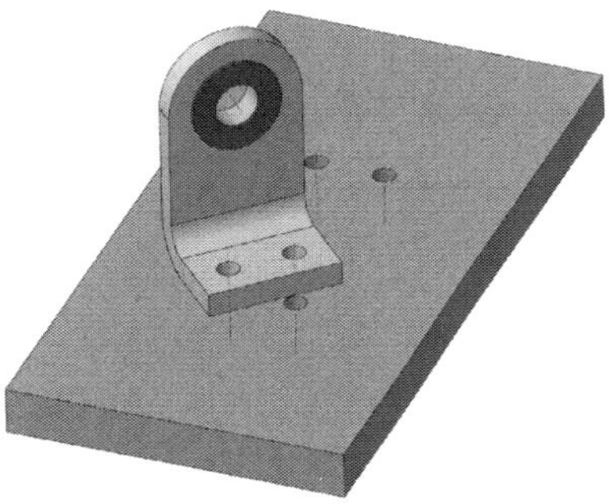

Figure 25 First Coincident (axes) and Coincident (surfaces) constraints applied to subassembly

Hmmm... not exactly what we want. With **Allow Assumptions** turned off, the bracket is not actually fully constrained yet, since it can still rotate around the hole. Select ***New Constraint*** in the **Placement** panel and pick the other bolt hole axis in the bracket and the appropriate axis on the base. This may produce an ***Oriented*** constraint. This is not coincident to allow for the possibility that the distances between the two axes on the bracket and base plate are not exactly the same. If that was the case, the position would be set so that the misaligned axes would be in the same plane as the first hole. Since we know that is not an issue, you can change the constraint to ***Coincident***. The sub-assembly is now fully constrained without any assumptions. Accept the placement with a middle click.

We'll now bring in another copy of the support subassembly and attach it to the base plate so that it faces the first one as shown in Figure 29. This would be easy to do with the ***Copy*** command (which we will use later), but we will use the normal procedure so that we will use a slightly different screen display and options. Select the ***Assemble*** button and once again bring in the component ***support.asm***[6] (Figure 26). For the first constraint, select the axes of the holes that will end up at the front of the plate. This will produce a ***Distance*** constraint. Double click on the constraint flag and change it to ***Coincident***.

---

[6] A good question at this point is: Why not just mirror the existing support? When you mirror a component in an assembly, you create a new component. In this case, it would make a new sub-assembly. It would also make mirror copies of all components in the sub-assembly. This is actually a valuable tool to know if that is really what you want. In this case we don't - we want the same component used twice. This will keep the necessary number of components and files to a minimum and will result in the appropriate entries in the assembly Bill of Materials.

Figure 26 Subassembly brought into session

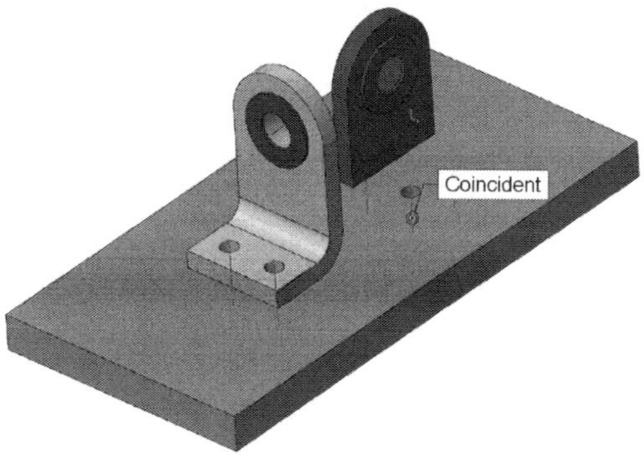

Figure 27 First hole axes aligned - support buried in base plate!

Notice that in Figure 27 there is overlap (interference) of the bracket and base plate. After you get to this stage, observe the effect of the different view options (wireframe, hidden line, no hidden, shaded) as you spin the assembly. The displays may not be what you expect and will even change with your view orientation if you spin the model. You should be able to recognize these view effects as symptoms that you have interfering components[7].

When you reach the configuration shown in Figure 27, hold down the CTRL and ALT keys simultaneously while you drag with the mouse (middle button produces rotate, right button produces translate). You can, of course, also do this with the 3D Dragger (unless it is turned off). This lets you reposition the support manually while maintaining the existing constraints. Drag it above the base plate. Select ***New Constraint*** in the RMB menu then pick the bracket lower surface. Then pick on the top surface of the base plate. The ***Coincident*** constraint will occur automatically. See Figure 28.

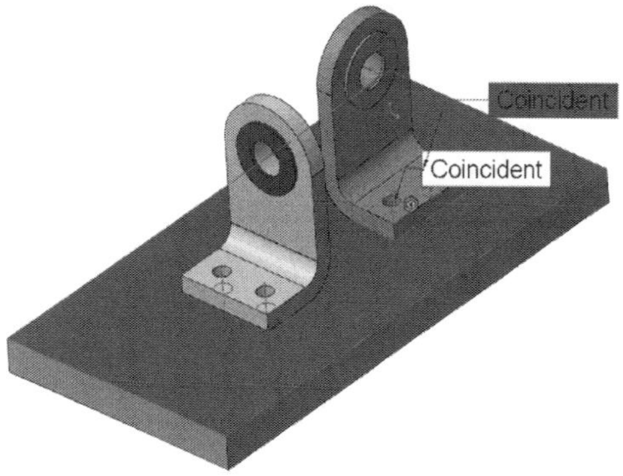

Figure 28 Surfaces mated. Fully constrained with assumptions. No overlap.

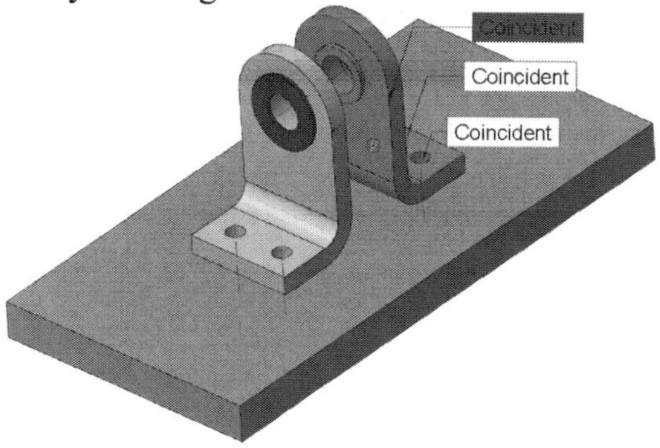

Figure 29 Second bolt axis aligned with other hole on base plate

You can now rotate the bracket around the first hole axis to position it as shown in Figure 29. The existing constraints are honored. Again, the status may be showing as fully constrained if assumptions are being allowed. If you turn the assumptions option off, it is

[7] Creo has other much more sophisticated tools for detecting interference.

possible to leave the assembly dashboard with the bracket looking like it is in the correct position. However, recall that since it is not fully constrained, it is called "packaged" only. This could have undesirable effects later, so we should make sure the bracket is aligned exactly and fully constrained.

Finally, select ***New Constraint*** (unless **Automatic** is already showing) and pick on the second hole axes on the bracket and the base plate. This should get you to the final configuration shown in Figure 29 with the bracket fully constrained. Accept the placement and then ***Save*** the assembly.

Helpful Hint

When you bring a component into the ***Assembly Window***, holding down the **Ctrl** and **Alt** keys at the same time lets you move it around the assembly using the **RMB**. Using **Ctrl+Alt+MMB** lets you spin the component being assembled independently of the assembly itself. The 3D Dragger offers similar controls, although rotation and translation directions are predefined by the Dragger directions.

Now we'll assemble the axle using the constraints shown in Figure 30. Before we do the assembly, let's review our constraints. One ***Coincident*** constraint is between the bottom of the axle head and the outer face of the washer. The second ***Coincident*** constraint could be with any of the inner surfaces of the bushings or washers on either support. The design intent will be best served if you pick a surface of a bushing. In either case, this constraint will allow the component to be placed, but it will still be able to rotate around its own axis. We'll add another constraint to prevent this by making the lower surface in the keyway parallel to the upper surface of the base plate.

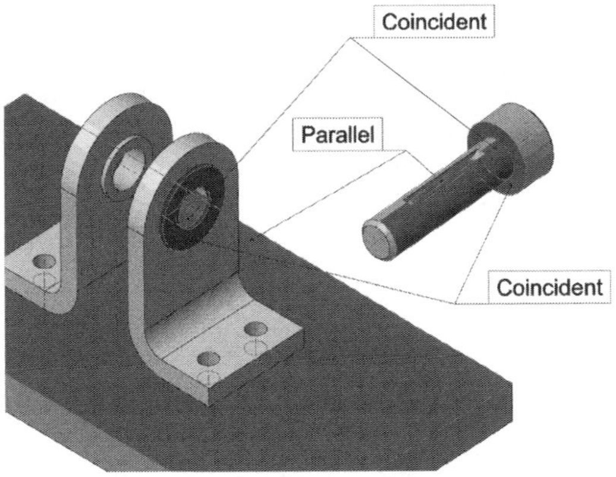

Figure 30 Placement constraints for axle

Figure 31 Axle moved using 3D Dragger

Bring in the axle using ***Assemble***. Use the 3D Dragger to move it to the position shown in Figure 31. The closer you get to the desired position and orientation, the better the assumptions will be concerning the automatic constraints. At the same time as you are dragging the component, you can control your view (spin, zoom, and pan) using the dynamic view controls as usual. This gives you considerable control over what you see on the screen. You will need to experiment with these controls for a while before you will be comfortable with them and find which ones work best for you.

Helpful Hint

At some point, you will probably accidentally leave the **Component Placement** dashboard with an inadvertent middle mouse click which will leave the component in a "packaged" state. A special notation will appear in the model tree for such a component. When that happens, just select the component in the graphics window or model tree and select ***Edit Definition*** in the pop-up menu. This brings you back to the assembly dashboard.

When your display shows you a convenient view of the axle and the assembly together, proceed to set up the assembly constraints indicated above. We are going to apply the constraints in a particular order although as mentioned earlier, the order of creating these constraints doesn't matter to the final placement. You will find some sequences easier to implement than others. For example, try to avoid the "buried" phenomenon we encountered earlier (for the support subassembly) that makes it hard to select references. The ***Coincident*** constraint between the cylinder of the axle and the inside surface of the bushing should not be any trouble. Set that up first: pick on the shaft of the axle and the inside of one of the bushings.

We will apply the second constraint (orientation of the keyway) while we can still see it clearly. Pick the surface at the bottom of the keyway on the axle and the top surface of the plate. The status might show "Constraints Invalid" depending on what constraint was assumed (possibly ***Coincident***), or a ***Distance*** constraint might be used. Regardless, change the constraint type to ***Parallel***. The axle will rotate to the desired position.

Finally, in the RMB pop-up, select ***New Constraint***, then pick on the two surfaces for the second ***Coincident*** constraint (face of axle, face of washer). We have saved this one for last so that the surfaces are clearly visible and easy to pick out. Change the constraint type to ***Coincident*** if required.

The final fully constrained position of the axle should be as shown in Figure 32. Notice the position of the keyway.

If everything is satisfactory, accept the placement. Otherwise, double click on a constraint flag, select either the constraint type, component reference, or assembly reference, and make the appropriate corrections.

Now is a good time to save the assembly.

Figure 32 Final placement of axle

We can now bring in the pulley and attach it using the constraints shown in Figure 33. Use the ***Assemble*** button and select the **pulley.prt**. You might like to experiment with ***Separate*** and ***Assembly*** window displays, and possibly use shaded or wireframe views to help identify surfaces. This is useful when the assembly starts to get crowded with visible and hidden edges, datum planes and axes, and so on.

Helpful Hint

When your assembly starts to contain a lot of components that you don't immediately need, you can either ***Suppress*** them or ***Hide*** them using the pop-up menu. The least dangerous of these is ***Hide***, since suppressing a component may accidentally suppress an assembly reference and then child components will also disappear (usually the ones you need!).

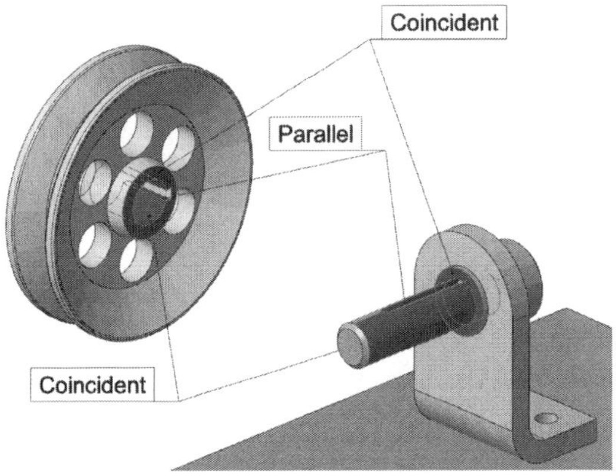

Figure 33 Placement constraints for pulley

Figure 34 Final position of pulley

Once again, bring the pulley close to the placement position using the 3D Dragger. The pulley could be placed with just the two ***Coincident*** constraints and allowing assumptions. But, we want to make sure the keyway lines up with the axle. The ***Parallel*** constraint can be used with a side surface of the keyway on the pulley and axle. It is probably best to set this constraint up first, otherwise it will be difficult to see the appropriate surfaces hidden within the model. Conversely, the previous constraints can be disabled in the **Placement** panel, and the pulley moved out in the open (with Ctrl+Alt+RMB) to expose the surfaces. When the align constraint is set up, the previous constraints can be re-enabled. When the pulley is assembled, it should look like Figure 34.

Sometimes, when you pick an axis or surface alignment, Creo decides to place the component 180° from where you want it. In that case, select the constraint and try the ***Flip*** button. This is sometimes available on the RMB pop-up. You might have to disable other constraints temporarily while you do this, then enable them again.

Finally, bring in the four bolts to attach the brackets to the base plate. We'll bring in the first one and then use the ***Copy*** command to place the rest. There are a number of advanced assembly commands that would allow you to create a pattern of bolts that would match a pattern of bolt holes. This would allow the assembly to automatically adjust, for example, if the pattern of bolt holes in the base plate was changed (including changing the number of bolts in the assembly). To place a single bolt, use the constraints shown in Figure 35. Do that now, using ***Allow Assumptions***.

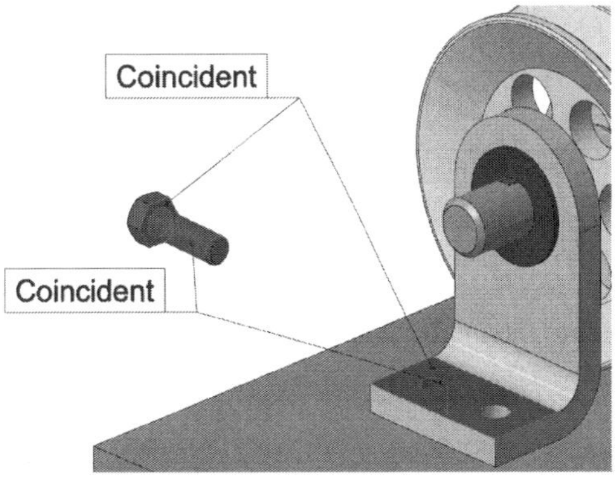

Figure 35 Placement constraints for bolts

Figure 36 Completed assembly

Using *Copy* with Components

Select the assembled bolt in the model tree, then the ***Copy*** command in the **Operations** group in the ribbon (or RMB pop-up or CTRL-C). We have ***Paste*** and ***Paste Special*** commands at our disposal that work in a similar manner as for features. Try the ***Paste*** (CTRL-V) command first. This opens the **Component Placement** dashboard and the **Placement** panel - this is not strictly necessary, but lets you keep track of proceedings. All we need to specify are the two assembly references (since the references on the bolt are already known[8]). Notice the labels in the graphics window. The first ***Coincident*** constraint is highlighted (easier to see in wireframe). All you need to do is pick the corresponding hole surface in the bracket or base plate for the copied bolt. Now the second ***Coincident*** constraint is highlighted (and the corresponding surface on the bolt is highlighted). Pick on the corresponding surface of the bracket or baseplate. Presto! The bolt is copied to the new location. Middle click to accept the placement.

Try this again. Since the bolt is still on the clipboard, just select ***Paste*** again to put the copy at the third location. You should be able to place a bolt at a new location with just

[8] The set of assembly references on a component is called its "interface." The interface can be stored with the part. When the part is used again, the component references are already known and all that is required is to specify the matching references in the assembly.

four mouse clicks (***Paste*** command, pick cylindrical surface on bracket or base plate, pick top surface on bracket, middle click).

Add the fourth bolt, and check out how the ***Paste Special*** command works. The final assembly is shown in Figure 36.

We are finished with our first look at assembly functions in Creo. We will continue in the next lesson. You should continue to experiment with constraint selection, view controls, and display modes. Most users prefer the **Assembly Window** mode of operation for specifying constraints; however, the **Separate Window** is very useful if you are having trouble picking references in a part, or if you are assembling a very small part to a very large assembly. Some advanced techniques for creating and managing assemblies (repeating and replacing components, component interfaces, drag-and-drop placement, and more) are discussed in the *Advanced Tutorial* from SDC.

Save the assembly. Open the model tree and explore the information presented there.

Assigning Appearances to Components

When a component or assembly is displayed in shaded mode, or when a rendered image is created, its appearance is determined by several settings. The most important of the appearance settings is color. Other aspects of an appearance definition are texture, transparency, reflection, shadows, and so on. Appearances can be applied to entire objects or individual surfaces. Defined appearances can be stored in files. A default appearance definition file (*appearance.dmt*) is loaded at program start-up. You can create your own palette of appearances and store it in your startup directory. Once an appearance has been assigned to a component, its definition is contained with the component file. This means if the component is sent to another system the component's appearance will be taken with it.

In the following, we will deal only with color.

IMPORTANT: Make sure that colors are turned on using

File ➤ Options ➤ Entity Display ➤ Show Colors

We will assign appearances in two steps: first we define (or choose) the colors we are going to use, then we apply the colors to the desired components or surfaces. The extent to which you can do this will depend on the specifics of your Creo installation and your hardware.

Select the **Appearance Gallery** pull-down button in the Model ribbon. The appearance gallery opens as shown in Figure 37. This contains three areas. The one at the top (**My Appearances**) contains appearances loaded when you started up Creo for this session[9]. This is often called the *color palette*, but as was mentioned above each appearance contains much more than just color. By default, these appearances are stored in a file called *appearance.dmt* in the directory *<loadpoint>/graphic-library/appearances*. If that file is missing, only one color - gray - is defined. Each appearance will have a name (use the small button at the top right to control display settings in this menu - including showing the appearance names). The name of the default appearance in Figure 37 is **ref_color1**. If you have loaded another assembly that has defined appearances, they will be displayed here.

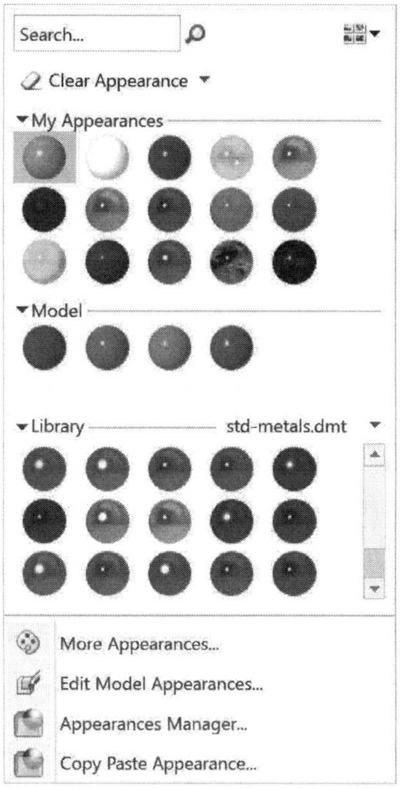

Figure 37 The **Appearance Gallery** with *Tut_AppearanceFile.dmt* opened

The middle pane on the menu shows appearances assigned in the current model. The lower pane shows appearances stored in the system library. A pull-down list in the Library area lets you choose from among the many system libraries for metals, plastics, and appearances for advanced rendering for producing high-quality rendered images of your models.

To assign an appearance to a component, just select it from any of the areas on the menu. The **Appearance Gallery** will disappear and you can then select a component to assign the chosen appearance. When you are finished, select ***OK*** or middle click. To assign it to additional components, just click the button in the dashboard, then select the component and middle click. If you get carried away, the Gallery has a ***Clear Appearance*** button that allows you to unassign some or all of the appearances (you must open the pull-down list to select what you want to clear).

Let's define some more colors. At the bottom of the **Appearance Gallery**, select ***More Appearances***. This opens the **Appearance Editor** window (Figure 38). In the **Properties** tab, appearances are first organized by **Class** (**Generic**, **Metal**, **Plastic**, **Paint**, and so on). For each class there are a number of sliders that control such effects as the lighting on the model (ambient light, shine, and highlight intensity), edge color, transparency, reflectivity, and more. Leave the **Class** set to **Generic** and click the color patch button just above the top slider at the right. This opens a new window called the **Color Editor**.

---

[9] The file *Tut_AppearanceFile.dmt* is discussed in a couple of pages.

The **Color Editor** contains three different ways to select a color. The default method is to select values of red-green-blue (RGB) (ranging from 0 to 255) in the new color. You move the sliders until you get the right mix of RGB for the new color, or enter integer values in the range 0 - 255 in the boxes on the right. As you use the sliders to change the value of any color component, the new color is shown at the top of the **Color Editor**, and on the shaded ball back in the **Appearance Editor**. Another method of setting color is to use the **Color Wheel**. This shows a disk containing all the available colors. The current color setting is shown by a small faint (almost invisible!) cross. You can click within the color wheel to pick the new color - this automatically sets RGB values for you. Note that there is also an HSV (hue - saturation - value) method of selecting color. If you select this method with the color wheel open, you can see the effect of the three HSV sliders. The **Hue** slider moves the cursor in circles on the color wheel. The **Saturation** slider moves the cursor radially on the color wheel. The **Value** slider sets the range between black and the color on the color wheel.

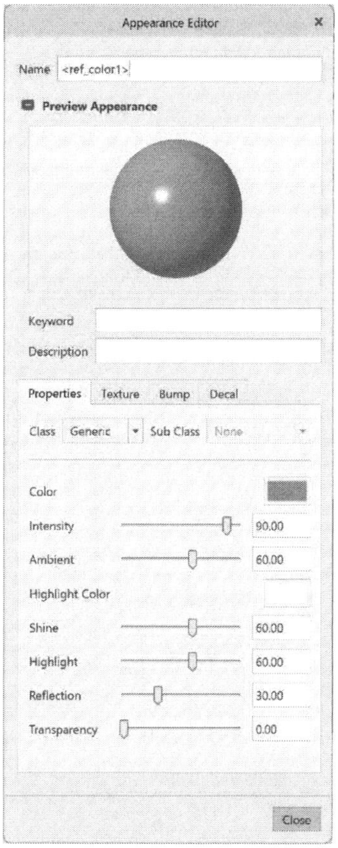

Figure 38 The **Appearance Editor**

Once you have selected a color in the **Color Editor**, select ***OK***. The new color will show up in the palette at the top of the **Appearance Editor** window where you can type in a name for the new appearance. When you close the Editor, you can then assign the new color/appearance to the model.

To experiment with this some more, open the Appearance Gallery and select

Appearances Manager

This opens a combined window showing the **Gallery** and **Appearance Editor** at the same time. With the mouse in the **My Appearances** area, hold down the RMB and select ***New*** several times to create a number of copies of the currently selected color, named "<ref_color1>". Select one of these and then launch the **Color Editor** by selecting the color swatch. Set up a new color, rename it (make sure you hit the ***Enter*** key after renaming), and repeat for each of the new copies. Define the appearance colors in the table shown at the right (these are the "primary" colors) or make up your own color mix.

Color	RGB Composition		
	Red	Green	Blue
Red	255	0	0
Green	0	255	0
Blue	0	0	255
Yellow	255	255	0
Cyan	0	255	255
Magenta	255	0	255

Color selection is a matter of personal preference. However, in general, lighter colors work better than darker ones, and saturated colors (as given) are also a bit "harsh." You might modify each of the colors in the table by increasing the other color components from 0 to, say, 128. Appearance names might correspond, for example, with different materials (steel, aluminum, plastic, ...) You might experiment with the lighting and transparency options as well, as well as the HSV system for specifying colors.

Additional options in the **Appearances Manager** allow you to create advanced settings using bump maps, texture files, and decals. These are used with the advanced photo-rendering functions to produce life-like images.

When you have created the colors, ***Close*** the Color Editor. Back in the **Appearances Manager** window, you can use

File ➤ Save As

to save your newly created appearances. Remember that if you want this loaded automatically, the file must be called *appearance.dmt* and be located in the default startup directory or in *<loadpoint>/graphic-library/appearances*[10]. You can use ***File ➤ Open*** in the **Appearances Manager** window to read in previously saved appearance files and either overwrite or append colors in a stored file into the current session. There may be a system restriction on the total number of colors you can define - see your system administrator for details.

A sample dmt file (*Tut_AppearancesFile.dmt*) with 27 colors is included in the file download set from SDC. See Figure 37. A search on the internet will yield dozens of custom created color palette files, some containing hundreds of colors.

To apply color to the axle, select a color in the palette, pick on the axle then middle click. Pretty simple.

Choose different colors and assign them to the pulley, the base plate, and the four bolts. The final assembly might look something like the first couple of figures in this lesson.

When we are in assembly mode, we can individually color the components in the subassembly **support.asm**. For example, the two identical washers could be different colors. However, colors defined at the assembly level are not known at the part level - if you opened the washer part, it would not be either color. Therefore, you should note that appearances are preferably defined and assigned at the individual part level. These are carried with the part into the assembly where they will stay unless overridden. For multiple occurrences of a part (like the bolts), it is easier to assign colors at part level, where you only have to do it once! Appearances assigned at the highest level in the assembly tree take precedence. If a part has an appearance assigned, it will be overridden by an assembly appearance unless the assembly appearance is ***Cleared***.

---

[10] There is a configuration option which allows you to set a search path for multiple *dmt* files. See *pro_colormap_path* in the on-line help.

See how the display changes for wireframe, hidden line, and shaded displays. In wireframe display the edges of each part are shown in the assigned color. This might be awkward if you want to do any editing of the part, since line color is so important in representing information like highlighted edges, constraint surfaces, parent/child relations, and the like. You may find yourself toggling the color display state often enough that you might like to add a button to the Quick Access toolbar to do that. See the Appendix on customizing the toolbars.

Completely separate from the Appearance definition, we can also modify the model Transparency using controls we saw previously. For example, expand one of the supports in the model tree. For the bracket, expand the **Design Items** and **Bodies** areas. Select **Body 1** and in the pop-up menu, pick the ***Make Transparent*** command. This keeps the same color. The level of transparency is determined by the ***Transparency Control*** in the graphics toolbar. Note that this control simultaneously affects all components declared to be transparent. This is different from individual appearance definitions that can have unique transparency levels.

You will probably want to come back later to play with the options for rendering of the model. The entry point for this exploration is the **Render Studio** command in the **Applications** ribbon. If your system has a license for this, you can have fun playing with the ***Scene Editor***, the rendering ***Room*** (walls, ceiling, and floor that can have superimposed images), ***Lights*** (and shadows) of various kinds, and special effects like fog and perspective. In conjunction with the appearance definitions, it is possible to create photo-realistic images in very high resolution.

We are finished with the first lesson on assemblies. Don't forget to save your assembly - we'll need it in the next lesson. An important thing to note is that when you save the assembly, any component that has been changed in the session will also be saved automatically.

You will also note that the keyway in the axle extends beyond one of the support bushings. Also, the base plate is quite large. In the next lesson we will see how to modify an assembly and its component parts. This will involve creating assembly features (i.e. specific to the assembly), as well as making changes to the parts themselves. It is also possible to create new parts while you are in assembly mode (we'll make the key this way, to make sure it fits in the assembly). We'll also find out how to get an exploded view of the assembly, and set up an assembly drawing.

Questions for Review

1. If several identical parts are required in an assembly, do you need a separate part file for each one?
2. What are the main assembly constraints?
3. What is a "packaged" component? How does this restrict what you can do?
4. What degrees of freedom are constrained by each of the main assembly constraints? Draw a sketch and illustrate the constrained and unconstrained degrees of freedom.
5. What entities can be used when specifying assembly constraints?
6. What is the difference between applying assembly constraints to individual components versus a subassembly?
7. What is the difference between ***Separate Window*** and ***Assembly Window***? Where are these options located?
8. How do you select the constraint types?
9. If you are in the process of constraining a component and you make a mistake, how can you a) delete, or b) edit a constraint?
10. What does the 3D Dragger do, and how is it different from using CTRL+ALT and the mouse buttons? How is it related to applying assembly constraints?
11. Does it matter what order you create assembly constraints? When you are picking references does it matter if you select component references first?
12. Find out how many colors you can define on your local system.
13. How can you specify the colors of individual components in a subassembly?
14. How do you turn off color display in wireframe mode?
15. What aspects of the display are controlled by "appearances"?
16. What (and where) is the default appearance file? What does it do? How can you modify it?
17. How do you change the color of an object?
18. Which takes precedence - an appearance assigned at the part level or at the assembly level?
22. What are the special mouse button functions when assembling a component?
23. What is meant by a *connection* in an assembly?
24. What are the advantages and disadvantages of creating an assembly with and without using an assembly template?
25. What is contained in your system's default assembly template?
26. What functions are available with the RMB pop-up? For what mode of operation do these appear?
27. How does the 3D Dragger indicate the available DOF of a component?
28. How do you temporarily attach the 3D Dragger to the assembly?
29. When copying a component, what is the difference between ***Paste*** and ***Paste Special***?
30. How can you get the component placement information to appear in the Model Tree?
31. What does ***Allow Assumptions*** do? How do you turn this off?
32. What is a constraint set? How do you set them up and use them?
33. What do you think will happen if you drag a bolt onto the support subassembly in the model tree?
34. Can you assemble a part to another copy of itself in an assembly? What does this do

to the parent-child relationship shown in the **Reference Viewer**?

35. How do you assign an appearance to a single surface of a part or assembly?
36. What happens if you (1) assign an appearance to a part, (2) assign a different appearance to the part surfaces, and then (3) cut away some of the part with an extrude?
37. What is the effect on having several bodies in the components of an assembly? What happens to the assembly constraint references if, after the assembly is finished, the bodies are ***Merged*** in the parts?

Project

Start assembling the vise with this subassembly. This is placed as a unit within the upper level assembly, so it makes sense to constrain these parts to each other. Think about an assembly strategy and which constraints you are going to use before you start actually doing anything. What orientation do you want the components to be in this subassembly? Look ahead to the finished product to see how you will be able to constrain this subassembly into the vise.

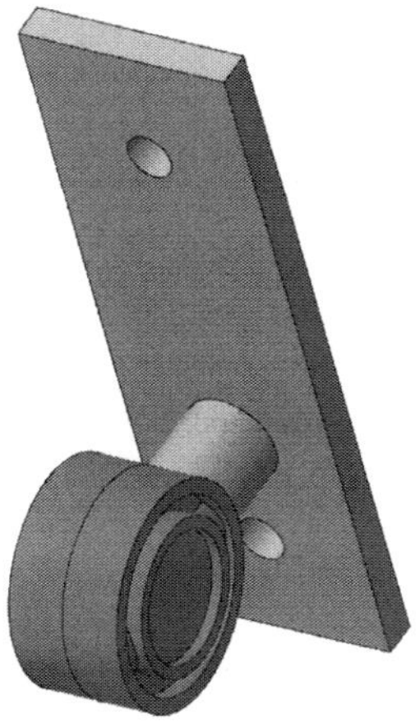

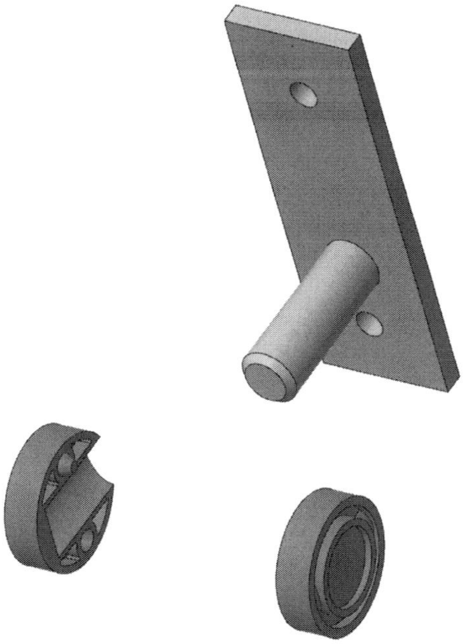

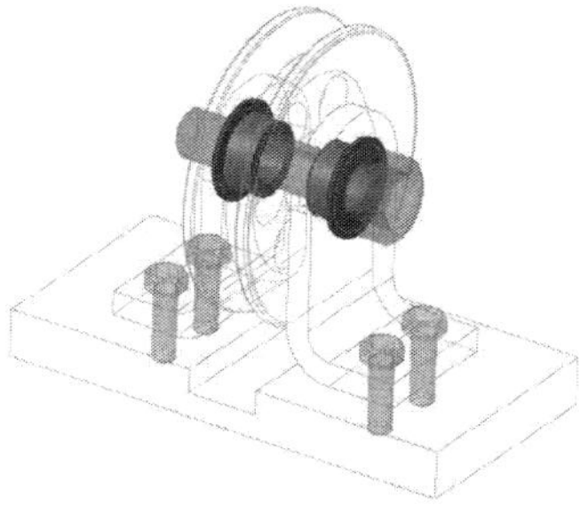

Lesson 10

Assembly Operations

Synopsis

Examining the assembly database. Declaring active components. Modifying parts in an assembly. Creating parts in assembly mode. Assembly features. Exploded views and display styles. Creating sections. Assembly drawings.

Overview of this Lesson

In this lesson, we will continue to work with the pulley assembly we created in Lesson #9. We start by using some Creo utilities to get information about the assembly (model tree, assembly references, assembly sequence). We will find out how to add features to the assembly, modify the parts by changing dimensions and adding features, and create a new part to fit with existing parts in the assembly. We will see how to get an exploded view, and modify it, and how to set up section views and a drawing of the assembly. This seems like a lot, but there's actually not much involved with each topic. Here are the sections of this lesson:

1. Assembly Information
2. Assembly Features
3. Assembly and Part Modifications
4. Part Creation in Assembly Mode
5. Exploding the Assembly
6. Modifying the Component Display
7. Creating sections
8. Assembly Drawings

To get started, make sure all the part and assembly files you created in Lesson #9 are in your working directory. Then start Creo and load the assembly:

File ➤ Open ➤ less9.asm

or open your working directory in the Browser and double click on **less9.asm**. Shut off all the datums, colors, and set no hidden lines. Close the Navigator panel (i.e. model tree) with the button at the extreme lower left of the screen.

Assembly Information

In this section we will look at some commands to dig out information about the assembly. We saw some of this way back in Lesson #1, so this will be a bit of a review. Go to the **Tools** ribbon and in the **Investigate** group overflow select

Feature List* ➤ *Top Level* | *Apply

The Browser opens with a list very similar to the feature list of a single part. Resize the Browser so that you can see the model. For an assembly, the list identifies all assembly features (like the assembly datums) and components, their numbers and IDs, the name, type (feature or component), and regeneration status of everything in the assembly. Clicking on a listed component will highlight it in the graphics window (use **Repaint** to clear the highlight). In the small **Feature List** window at the top right, select ***Subassembly*** and pick on the L-bracket in the graphics window, then ***Apply***. This lists the components in the subassembly containing the bracket. Finally, select ***Part*** and click on the same bracket, then ***Apply***. This lists individual part features. You can see that we can easily dig down quite deeply into the model structure. ***Close*** the Browser window and the **Feature List** window and ***Repaint*** to clear any color highlights.

To see how the assembly was put together (the regeneration sequence), in the **Tools** ribbon select

Model Player 

Select the rewind button to go to the beginning of the model. Proceed through the regeneration sequence with the step forward button. As you step forward, find out what information is available using the ***Show Dims*** and ***Feat Info*** buttons. The former may not work as you expect. ***Show Dims*** does not show feature dimensions within parts. Rather, it shows assembly dimensions, such as would be used in an offset constraint or a component pattern increment. Continue in the model player until you have reached the last component. Then select ***Finish***.

If you want to find out more information about how the assembly was put together, in particular the placement constraints, select (in the **Investigate** group):

Component 

Pick on the axle. The **Component Constraints** window will open as shown at the right. This gives you a list of the placement constraints used to position the axle in the assembly. Place the cursor over one of the lines in the table - a pop-up will

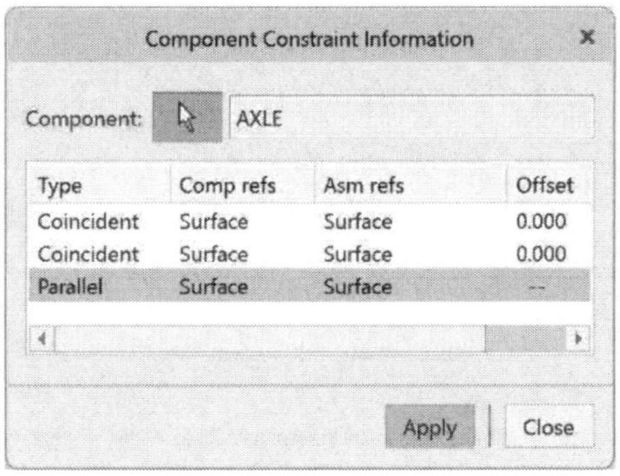

Figure 1 Displaying component constraints

describe the constraint (and remind you of the importance of selecting useful part names!)[1]. If you click on the line, the reference surfaces will highlight in purple and blue on the model (this will be easier to see if you turn colors and shading off). ***Repaint*** the screen, select the arrow button beside **Component**, and pick on another component, like one of the bolts to see similar information. ***Close*** the window.

Figure 2 The assembly model tree

Another way of looking at the logical structure of the assembly is, of course, with the model tree. Open that now. Figure 2 shows the tree with the filter settings **Placement Folder** and **Features** both turned ***OFF***. See ***Tree Filters***. Using ***Tree Columns***, add the following columns to the model tree: ***Feat #*** and ***Feat Type***. Click on the small ▶ sign in front of the *support.asm* entries. Note the individual components organized in levels.

Now select ***Tree Filters*** and check the box beside **Placement Folder**, then ***OK***. Expand the entries for each component as shown in Figure 3. These are the assembly constraints used for each component. Clicking on a listed constraint in the model tree will cause the references involved to highlight on the model. This is a pretty simple and useful way to explore the assembly database and model structure.

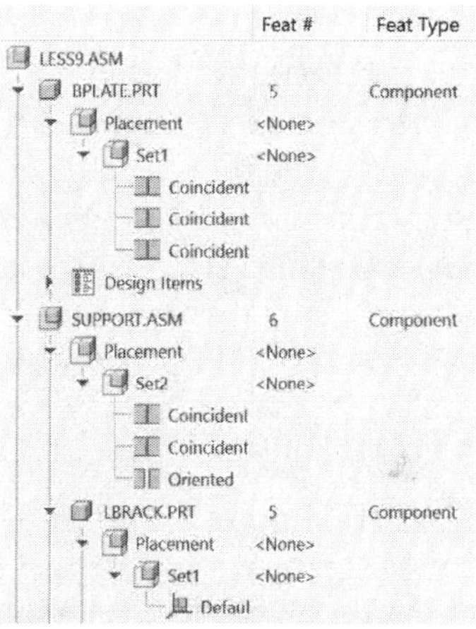

Figure 3 Model tree with **Placement Folders** open to show constraints

Select ***Tree Filters*** again and check the box beside **Features**, then ***OK***. Now select one of the ▶ signs in front of a part. The model tree shows all the features in the part. Click on any of these features and it will be highlighted on the assembly model. If you select any feature, the pop-up menu will appear with a number of the utility commands we have seen before: ***Edit Actions(Edit Dimensions, Edit Definition, Edit References)***, ***Suppress***, and so on.

For the top level assembly (*less9.asm*), right click and select

Information ➤ Model Information

This data has automatically been written to a file (*less9.inf*) - see the message window. The Browser controls can be used to save or print this model information. This is useful for model documentation (in particular, noting the path to each component file). Close the Navigator and Browser windows.

---

[1] As of this writing, the pop-up contained an error in naming the components involved in the constraint. Presumably that will get fixed!

Assembly Features

Creating Assembly Features

An assembly feature is one that is created and resides in the assembly. You can only create them when you are in assembly mode, and they will generally *not* be available to individual parts when you are in part mode unless you explicitly allow it. Like features in part mode, assembly features will involve parent/child relations (either with other assembly features or with part features) and can be edited, patterned, suppressed and resumed. Our assembly contains several assembly features - the default datums.

Assembly features can also be solid feature types (extrudes, sweeps, holes, ...) but can only remove material. Go to the **Model** ribbon and open the pull-down ***Cut & Surface*** overflow to see the possibilities. We will create a couple of assembly features in this lesson. The first is composed of a longitudinal cut through the entire assembly in order to show the interior detail[2].

In the **Model** ribbon, select the ***Extrude*** tool. This opens the extrude dashboard that we have seen many times before, with one addition - the **Intersection** panel - that we will get to in a minute or two. Select ***Placement ➤ Define***. For the sketching plane, pick the right face of the base plate (assuming you are in the default orientation). For the **Bottom** sketching reference, pick the bottom face of the base plate, then ***Sketch***. Select your references as the lower surface of the plate (or **ASM_TOP**), the outer edge of the pulley (zoom in to make sure you get the outer edge, and not an edge of the round), and the vertical datum **ASM_FRONT**. Sketch a single vertical line from the top of the pulley to the bottom of the base plate. This should be aligned with **ASM_FRONT**. Your sketch should look like Figure 4.

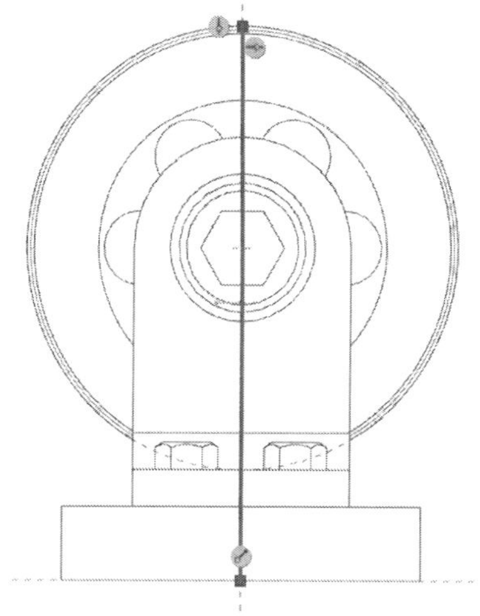

Figure 4 Sketch of first assembly feature (a vertical edge to create a one-sided cut)

Leave Sketcher and make sure the material removal side is on the left of this sketched line (on the front side of the model), and select the depth as ***Through All***. Open the **Intersection** panel. This shows which components in the assembly will be intersected by the cut. Notice that ***Automatic Update*** is checked (on) by default. We can remove components from the list by deselecting this option, then using the RMB pop-up menu to ***Remove*** selected components so they won't be cut by the feature. Don't do that now - leave ***Automatic Update*** checked.

---

[2] A much more elegant and flexible way of displaying a section will be explored later in this lesson.

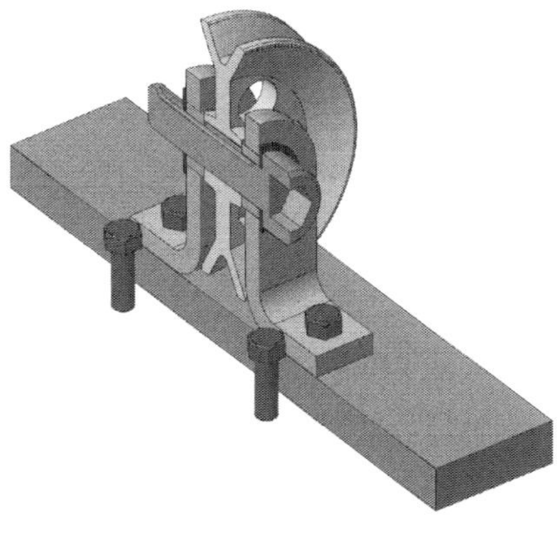

Figure 5 Cut complete - shaded view

Open the other slide-up panels to see their contents - they are the same as the part mode extrude feature.

The ***Preview*** button on the dashboard is not available, so we are finished with the dashboard for now. ***Accept*** the cut feature. Turn the part colors back on, and shade the display. The assembly should look like Figure 5. (Why are two bolts left hanging out in space?) Note that the keyway in the axle is too long - extending into the bushings in both directions. We will fix this a little later.

We are going to come back to the creation of features in a little while, after a slight diversion into display management functions. Meanwhile, select the final extrude feature in the model tree (the assembly cut) and use the RMB pop-up to suppress it.

Assembly Display Management

Creo can handle assemblies containing hundreds (even thousands) of components. When you are working on such an assembly, you may want to temporarily remove some components from the model for two reasons. First, it will remove some of the screen clutter, allowing you a better view of the things of current importance. If you are creating an assembly of a car, for example, and you are working on the rear suspension, you might want to get rid of all body panels that are blocking your view. Second, removing some (or many) components will improve the performance of your system (faster graphics performance, faster model regeneration, less memory required). In the car example, while working on the rear suspension, you could get rid of the engine and steering components. Creo has special tools to handle very large assemblies which we won't have time to look at now (Hint: look up *Simplified Reps* in the on-line help). We will look at some simpler ways, that you have already seen, to find out how they can be used to solve this problem.

Since we want to do some modifications on the keyway in the axle, we would like to remove some of the other components from the screen that are blocking our view of this feature. We have two options to do this: ***Suppress*** and ***Hide***. (The ***Make Transparent*** command applies only to bodies.) The differences between these are as follows:

Suppress - takes the component (and all the references it provides) out of the assembly sequence. This removes it from the model (temporarily) and can speed up regeneration. Any children it has will be suppressed by default or will require some special handling. The component can be brought back into the assembly by ***Resume***.

Hide - keeps the component in the assembly sequence but makes it invisible. It can continue to provide assembly references (so we don't have to worry about its children) and is regenerated with the rest of the model when required. The component is made visible again by ***Show***.

Let's suppress all components in the assembly except the axle and the pulley. This is not strictly necessary to do the modification to the keyway, but it will remove the visual clutter from the screen. Since line color will be important here, turn the colors off.

In the model tree, select the first SUPPORT sub-assembly (the one on the left in the model). Using the pop-up menu, select ***Suppress***. The sub-assembly will highlight in green and the two bolts (which reference a surface of the L-bracket) will highlight in blue. Creo wants to know if it should suppress the bolts too. Select ***OK***. Observe the entries in the model tree, in particular the small black rectangles beside the component name.

Now select the second sub-assembly - the one on the right in the model. Use the pop-up menu and select ***Suppress*** as before. This component has four children: two bolts, the axle, and the pulley. Since we want to keep the axle and pulley to work on, we will have to handle them in a special way. In the **Suppress** window, select ***Edit Details***. This opens the **Children Handling** window that lists the four children (Figure 6). ***Repaint*** the screen. As you select components in this window, they will highlight in blue on the model. Leave the two bolts as **Suppress**, but change the axle and pulley components to **Fix** (click on the cell in the **Status** column and open the pull-down list). When the correct status has been set, select ***OK***. You should now see the axle and pulley all by themselves floating above the base plate. Check the model tree display - there is a special symbol for components that are children of suppressed components. The **Notification Center** window at the lower right opens to indicate a warning. If this doesn't dissolve automatically (the default), just close it. Notice the yellow warning flag on the message line; click it to open the Notification Center.

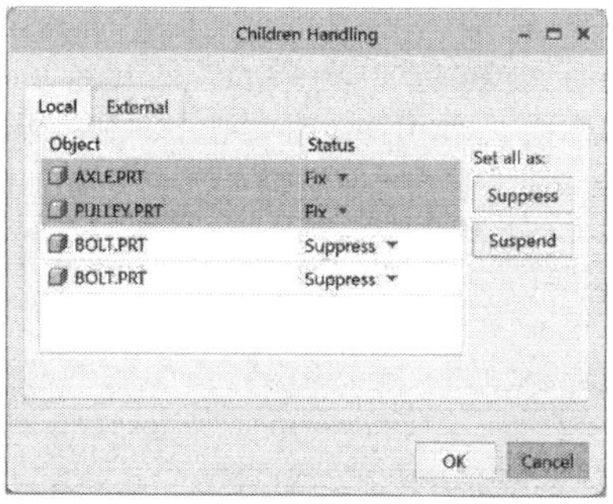

Figure 6 Handling the children of a suppressed component

The **Fix** option we just used will lock the component in its current location - with its parents suppressed some of its references will be missing. **Fix** adds a new constraint set to the placement of the component. Open the model tree **Axle(Placement)** entry to see the new constraint set. The previous constraint set is still there, but the constraints are shown as suppressed (small black squares). If you use ***Edit Definition*** on the component and open the **Placement** panel, you will see both constraint sets, with the initial one disabled because references are missing (note the red dots). Use ***Cancel*** to get out of the **Edit Definition** ribbon.

Notice the keyway extending past the edge of the pulley hub on both sides. In the next section, we will modify the dimension of the keyway and add some other assembly features.

Before we move on, let's see how ***Hide*** works. To get all the components back into the assembly, in the **Model** ribbon, **Operations** group overflow select

Resume* ➤ *Resume All

Everything should now regenerate (including the assembly cut - suppress it again). However, the new **Fix** constraint is still being used for the axle and pulley. You can remove this constraint by selecting the component and then in the pop-up menu selecting ***Unfix Location***. Notice the small white squares beside the components that indicate they are *packaged* (not fully constrained) instead of *placed* (fully constrained). It would be nice if you could just Resume the previous constraints (say with an RMB pop-up), but that is not available. You will have to use ***Edit Definition*** for the axle and pulley, and open the **Placement** panel to check the **Set Enabled** box for the previous constraint set. This seems like a long way around and is why the next method using ***Hide*** is probably easier in most cases than ***Suppress***.

Our second option for removing screen clutter is hiding components. Select one of the L-brackets either in the graphics window or in the model tree (expand the subassembly), and then pick ***Hide*** from the pop-up menu. It's that easy! Notice that the Hide status applies to the L-brackets in both sub-assemblies. Also, since the component is still regenerated, we don't have to worry about losing references for its children (bushing, washer, bolts). Bring both L-brackets into view with ***Show***. What happens if you select a complete subassembly in the model tree, then ***Hide***? Does this make sense? Hide the other one, and carry on!

Figure 7 Assembly with hidden components

Hiding will work well as long as the assembly is not too complex (since we are still paying for the overhead of regeneration). Select the axle and pulley (using CTRL) and in the pop-up toolbar, select ***Show Only***. Observe how hidden components are indicated in the model tree.

Assembly and Part Modifications

Active Components and Visibility

Creo gives you considerable flexibility in making changes in the assembly that involve components or features. The most common modification is to change one or more dimensions of feature(s) in a part, for example, to remove an interference. However, modifications can also include creating new features or components. Some changes will affect only the assembly, some will affect only sub-assemblies, while others will be felt all the way down to part level. Anything done at the part level (like modifying dimensions) is automatically visible in any higher level subassemblies or the top level assembly. We will see that the converse is not necessarily true. So, when you are working in the assembly, you have to be careful about what exactly you are modifying or creating:

- part features and dimensions
- subassembly features and dimensions
- assembly features and dimensions

There are two principal factors that determine the effect of what is done in the assembly. These factors involve consideration of the *Active Component* and the *Visibility Level.*

The active component is the one which is the current focus of operations. If an individual part is the active component, the next created feature will be added to the part (at the part's insert location). If a subassembly is active, the next component that is brought in (or feature that is created) will be added to the subassembly (at its insert location). By default, the top level assembly is the active component when the assembly is first created or loaded. Thus, in what we have done so far, each new component or feature was added at the bottom of the model tree as the assembly was put together. We will see in a few minutes how we can change the active component.

The concept of visibility level refers to features created in a multi-level assembly. By default, the feature (for example the long cut in our assembly) is visible only at or above the level where it was created. So, if you open the pulley by itself, it does not show the cut - it is visible only in the top level assembly. We will see later how to change the visibility of features like the cut so that they are visible in the part itself. This is handy for creating features like holes at the top level in the assembly and have them visible in individual parts when they are opened alone, or used in a drawing for detailing.

Before we do anything drastic, let's take care of the keyway dimensions in the axle. We could, of course, load the axle in a separate window and make the dimension changes there. We don't have to do that because, even in assembly mode, changes to dimensions defined in the part will be made **in the part**. Thus, these will show up if you bring up the part in **Part** or **Drawing** modes. Let's see how that works.

Changing Part Dimensions

We need to shorten the keyway on the axle, and we want to make this a permanent change in the part (i.e. reflected in the part file). First, we need to pick the cut feature that created the keyway so that its dimensions are visible. You can use Preselection Highlighting (it may help if the **Selection Filter** is set to ***Features***) to locate this. Double-clicking on the feature should make its dimensions visible. Alternatively, you can find the feature in the model tree, and use the pop-up menu to pick the ***Edit Dimensions*** command. The dimensions will show up something like Figure 8. If you need to move them to make them clearer, select an individual dimension and open the RMB menu. Then select ***Move*** or ***Move Text***. Click on the screen where you want the dimension placed. You can also change the ***Nominal Value*** of the dimension while you are doing this.

Change the following dimensions (double click on the old value and enter the new value):

- length of keyway (between the centers of the curved ends) was 50, new value = ***15***
- distance from shoulder of the axle was 12.5, new value = ***20***
- radius of rounded end was R10, new value = ***R5*** (both ends)

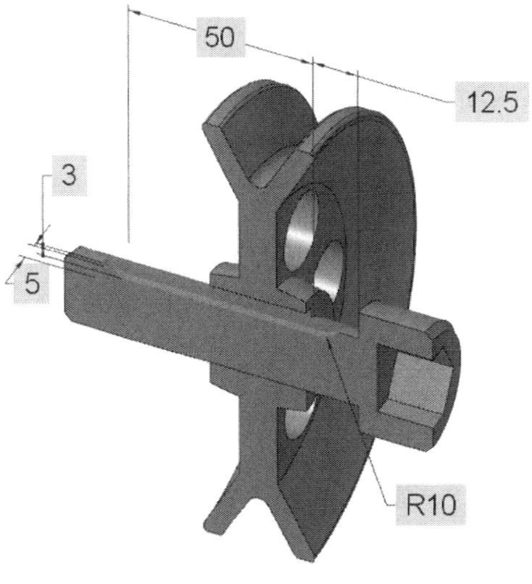

Figure 8 Original dimensions of keyway

The dimensions will stay blue until you ***Regenerate*** (use ***Ctrl+G***). Spin the axle/pulley to verify that the keyway does not extend beyond the end of the pulley hub. To get another view of the new keyway, go to the **View** ribbon, **Visibility** group and select

Unhide All

This will bring all the components back into view. Now, resume the assembly cut. If you shade the display, it should look like Figure 9.

Figure 9 Axle keyway with new dimensions

To see what has happened to the *axle* part file, we will bring it in by itself. Click on the axle in the model tree to select it, then select ***Open*** in the pop-up menu. The complete axle should show up in a new window. And, voilà, the keyway has changed. If we also had a drawing of this part and brought it up in **Drawing** mode, we would find that it has also been updated. Close the part window by using ***Window Close*** in the Quick Access toolbar at the top (shortcut ***Ctrl+W***).

While we are dealing with part modifications, change the dimensions of the base plate. Use the selection filter to pick the base plate protrusion. Then use the RMB and pick ***Edit***. Change the following dimensions:

- overall length was 300, new value = ***200***
- half-length was 150, new value = ***100*** (if necessary)

The new base plate dimensions are shown in Figure 10.

Select ***Regenerate***. Since these changes were made to the part, they will also be reflected in the original part file. Before we continue, suppress the assembly cut. You can select it in the graphics window by picking on any of the cut surfaces.

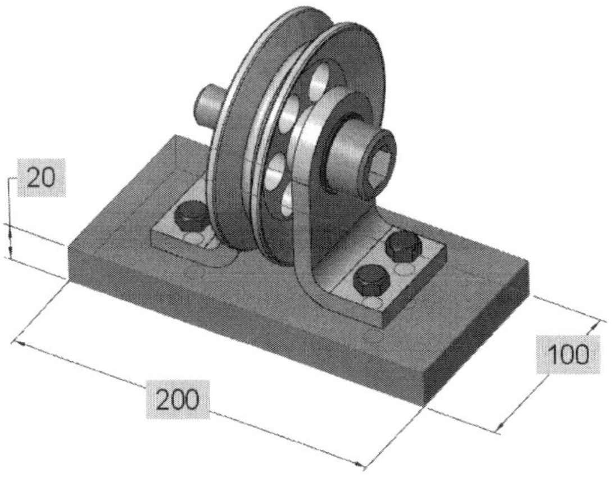

Figure 10 New base plate dimensions

Adding another Assembly Feature

We are going to add a U-shaped cut to the base plate in between the L-brackets. We want to do this in the assembly so that the width of this new cut will be determined by the placement of the two L-brackets. We can't open the base plate part and create this cut there since it requires references that are defined in the assembly. So, we must create the cut in assembly mode. We still have several options to deal with; each will result in a different model structure. We will look at three variations.

For our first variation, we'll create the new cut the same way we created the other assembly cut. Select the ***Extrude*** tool and ***Placement* ▾ *Define*** (or use the RMB and ***Define Internal Sketch***). For the sketch plane, select the long front face of the base plate. For the ***Top*** sketching reference plane, select the upper surface of the base plate. To control the width of the cut, add sketcher references by selecting the inside vertical surfaces of the L-brackets. Make a closed sketch as shown in Figure 11. Accept the sketch.

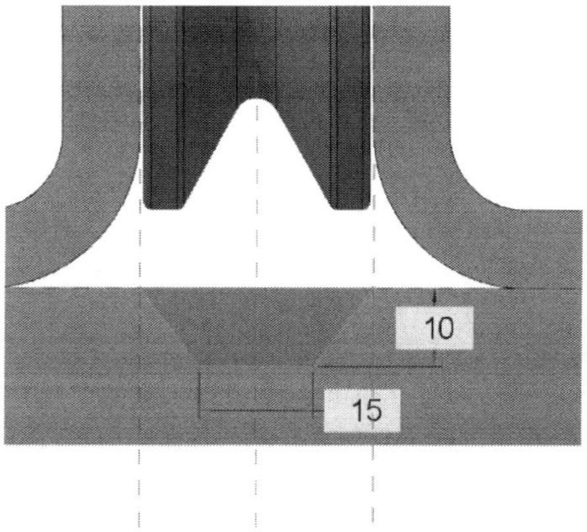

Figure 11 Sketch of second assembly feature (constraints off)

Back in the extrude dashboard, for the depth, select ***Through All***. The ***Remove Material*** button is selected automatically. Make sure the removal direction is inside the U. Open the **Intersection** slide-up panel and turn off **Automatic Update**. Select (using CTRL) the two brackets, then use the RMB and pick ***Remove***. Notice the **Display** column for the intersected models indicates our top-level assembly. Also, the pull-down list for the **Set Display Level** is set to ***Top Level***. Accept the feature. The assembly should now show the cut in the base plate as shown in Figure 12.

Go to the model tree and make sure that

Tree Filters

has checks beside all display objects. You should see that a second extrude feature has been added to the assembly at the bottom of the tree. It was placed here because, by default, the top-level assembly was the active component when we made the cut. The previous feature listed is the big cut we made before and is currently suppressed as indicated by the small black square. Expand the base plate listing in the model tree. You will see that both assembly cuts are also listed there (one is suppressed).

Now ***Open*** the base plate in its own window. You should find that the dimensions have changed (since that happens at the part level), but neither cut appears in the part or model tree. Recall that the second factor which determines the extent of a modification is the visibility of the new feature. By default, our U-shaped cut was visible only in the top-level assembly. Let's see how to change that. Select the ***Windows*** command in the Quick Access toolbar and switch back to the assembly window.

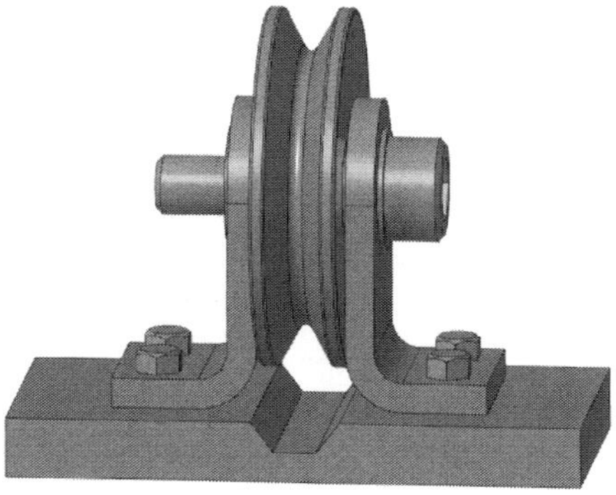

Figure 12 Completed cut in base plate

Changing Feature Visibility

In the assembly, we need to redefine the new cut so that it is visible at the part level. Select the cut in the top-level assembly model tree and in the pop-up menu, pick ***Edit Definition***. In the extrude dashboard, open the **Intersection** slide-up panel. In the pull-down list near the bottom of the panel, change the **Set Display Level** setting from ***Top Level*** to ***Part Level*** and check that this is set in the Display column. ***Accept*** the feature in the assembly. It should look the same as before. However, open the window containing just the base plate. There is the cut! It is now visible in the part, and in any drawings of the part. Check out the model tree of the part and the feature information for the cut - this tells you where it was made ("*Feature was created in assembly LESS9.*"). When you are in the part window, can you edit the dimension of the cut? Where is that dimension value actually stored?

Creating a feature in the assembly and controlling its visibility can be very useful. For example, the assembly feature might represent a custom design modification of a standard part. You may not want this modification to be reflected in the part model or its drawings (which could be used in other assemblies as well), so you would set the visibility to the top-level assembly. Conversely, if you want the feature to show up in the part model, set the visibility to the part. Creating the feature in the assembly also means that you can selectively affect many components in the assembly. For example, if you want to create a bolt hole through several parts, you can do that in the assembly and the holes will always line up perfectly. To make sure the hole shows up in each part, set its visibility to part level and pick all the parts where you want the hole to appear[3].

---

[3] How would you dimension the location of these holes in each part drawing?

Select the cut in the base plate's model tree and ***Delete*** it. If you do this in the base plate, it will disappear from the model tree there but still exist in the assembly model tree. Return to the assembly, select the cut, then ***Edit Definition***. If you check the ***Intersection*** panel, you will see that there are no intersected parts listed. This shows you can remove a part from the visibility list of an assembly feature by deleting it within the part. Now ***Delete*** the cut in the assembly. There is a third option we should consider for creating the U-shaped cut.

Changing the Active Component

Suppose we want to create the U-shaped cut in the base plate model (that is, not as an assembly feature and without intersecting any other components in the assembly). We want to add the cut to the base plate itself and not have it defined in the assembly at all, although it will certainly be visible there. However, we can't create it in the part window because we need the references provided by the L-brackets in the assembly. We have to make it in the assembly, but we want it added only to the part. Since the top-level assembly is the current active component, any cut we create will appear in the top-level assembly model tree. The solution to this dilemma is to change the active component to the base plate.

Select the base plate in the model tree (or in the graphics window), and use the pop-up to pick ***Activate*** . Several changes will occur on the screen. First, in the graphics window, the non-active components have become translucent and a label **Active Part : BPLATE** will appear. In the model tree a small green diamond appears on the lower right corner of the active component's icon and all the other components are grayed out. The components aren't automatically hidden because it is likely that we want to use references from them. Finally, you will note that some of the toolbar icons have changed - it doesn't make sense to have assembly-related icons in view when a part is active. This final screen change does not occur if you have set a subassembly as active.

Now, select the ***Extrude*** tool and create the U-shaped cut exactly as we did before. Use the L-bracket surfaces for the width references. Make a ***Through All*** cut. Notice that the **Intersection** slide-up panel does not exist, since we are creating this feature in the part. Also, unlike creating the extrusion in the assembly where the default was a cut, at part level the default is a protrusion. Make sure this is a cut and accept the feature.

Open the base plate part window. Our cut is visible. Select the cut and use RMB to open the ***Information*** ➤ ***Feature Information*** window in the Browser. This will tell you that the feature was created in the assembly. What is the implication of this? To see that, you will need to save the current assembly (this will automatically save the base plate) and then erase the assembly (and all components) from the session. Notice that when you returned to the assembly window from the base plate window, the active component had reverted to the top-level assembly. Now, with nothing else in session, open the base plate by itself and call up the ***Reference Viewer*** window for the cut. Try to use ***Edit Definition*** and ***Placement*** ▾ ***Edit***. Because the rest of the assembly is not in session, there are a number of missing references ("external references"). These can be a major cause of problems in complicated assemblies if file management and model organization are not well thought out. If you get into any trouble here, back out of the feature without making any changes (select ***No*** and ***Cancel*** as required).

Bring the complete assembly back into the session by opening it, and the problems with editing the U-shaped cut disappear.

What would happen to this cut if we suppressed either L-bracket? What if we used ***Hide***?

To see more implications of what we have done, open the base plate in a window by itself and change the location of the holes. Referring back to Figure 6 in Lesson 9, change the dimension 50.5 to ***65***. This should move the brackets apart and widen the cut. Assuming that you created the other holes by mirroring, that's all you should need to do. Now ***Regenerate*** the part. The cut does not change (yet!). Go to the assembly window and select ***Regenerate*** there. The cut widens as the two brackets follow the holes in the base plate. Still in the assembly, change the dimension back to ***50.5*** and ***Regenerate***. Can you explain the sequence of events that occurs resulting in the new display (follow the sequence of parent/child relations from the position of the first hole to the width of the cut)? This is one case where a feature (the cut) or component (the base plate) can have a parent (the brackets) that appears below it in the model tree, which may require regeneration to be done twice[4].

Part Creation in Assembly Mode

When you are working with an assembly, you may want to create a new part that must exactly match up with other parts in the assembly. You could, of course, do this by creating parts separately (as we have done up to now) and very carefully keeping track of all your individual part dimensions and making sure they all agree. You might even use relations to drive part dimensions by referencing dimensions in other part files. Here, we will find out how to create a new part using the assembly geometry as a guide and a constraint, much like the assembly features (the cuts) we created earlier.

We are going to create the key for the axle/pulley. To simplify the environment, hide all the other components and assembly features except the axle and pulley (Hint: use ***Show Only***). Turn on the datum planes and hidden lines. To create the new part, select the ***{Component}:Create*** button  in the **Model** ribbon. In the window that opens, select the options ***Type(Part)*** and ***Subtype(Solid)***. Observe the other subtype options[5]. In the **Name** area, enter a new part name "key" then ***OK***.

In the **Creation Options** window, check out the options for the creation method. A tool-

---

[4] The need to ***Regenerate*** a model twice in order for modifications to propagate through the model does occur from time to time.

[5] The ***Mirror*** option that used to be here is now a standalone command in the **Component** group.

tip will provide more information. These options basically give you different starting conditions for the new component you are creating such as default features and constrained location within the assembly. Select the **Create Features** option, then ***OK***. Our key will consist of a single protrusion based entirely on assembly references so this will give us the least cluttered part model.

Now the graphics window indicates that the active part is KEY. If you are in shaded mode, the other components will become translucent. In the model tree, a new part *key.prt* has been added to the assembly and is the active component (notice the green diamond). (NOTE: in the top toolbar and ***File*** pull-down menu, the ***Save*** and ***Save as*** buttons are grayed out when the top-level assembly is not active.) The ribbon contains the regular part-level feature creation tools. Select the ***Extrude*** tool, then ***Placement* ➤ *Define***. For the sketch plane, select **ASM_RIGHT**. For the ***Top*** reference plane, select **ASM_TOP**. It is probably better to use assembly datums for these references, since they are less likely to change or be suppressed. Select ***Sketch***, then in Sketcher, zoom in on the central hub of the pulley and turn on hidden lines.

We will use the existing edges of the keyway in the axle and pulley to create the sketch for the key. Close the **References** window. If you get a message about missing references, you can ignore it. Selecting existing edges will create references automatically. Select the

Project 

button in the Sketching group. In the graphics window, Ctrl+click on the edges of the keyway in the axle and pulley (these are hidden), then middle click. Be sure to select them all (six picks) to create a rectangular, closed section as shown in Figure 13.

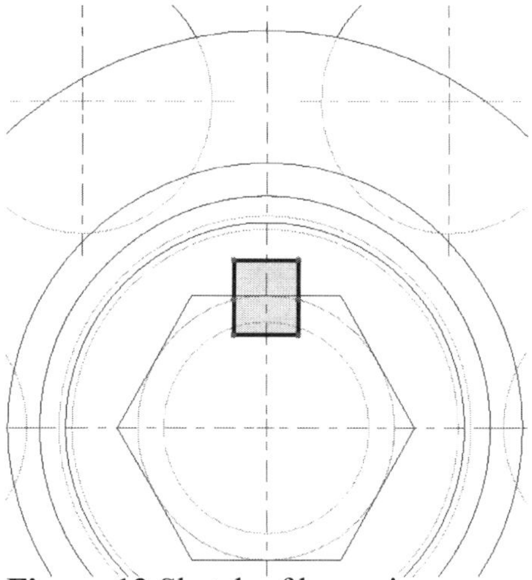

Figure 13 Sketch of key using ***Project*** (constraints not shown)

We didn't have to provide any dimensioning information for sketcher - it automatically reads the dimensions from the previous parts. This means that if we change the keyway size in the pulley, the key will automatically change shape. Note that we have not explicitly connected the width of the keyway in the pulley to the width in the axle. You might think about how you could do this. If we didn't do this, what would happen if you increased the keyway width in the axle but not the pulley? How could you avoid this potential problem? (Hint: Recall our discussion of assembly features.)

Accept the sketch. For the depth of the key protrusion, select ***Symmetric***, and enter a value of ***15***. Accept the feature.

Check out the model tree - you should see that a feature has been added to the *key.prt*. Make the top-level assembly active (select it and use the pop-up menu). Resume the longitudinal cut to see inside the assembly (select the cut in the model tree, use the pop-up and pick ***Resume***). ***Show*** all the components and shade the display. You can now see our rectangular key (Figure 14). Why hasn't the key been cut along with all the other parts?

Figure 14 New *key.prt* in assembly

Now is a good time to save everything.
You will find that the new part file **key.prt** is automatically saved for you - when you select ***Save*** in assembly mode, every object that has changed since the last save is also saved. Open the key into its own window; the only dimension shown on the part (with ***Edit***) is the length. All the other dimensions are determined by the edges used in the assembly, and therefore can't be modified within part mode[6]. What do you think would happen the next time you start Creo if you try to open the key part by itself?

Helpful Hint

If you are going to create parts in assembly mode, try to arrange as many *size and shape* dimensions as possible to be contained within the part. Use other assembly features only for *locational* references (like alignments, or dimensions to locate the new part). The fewer external references you have, (generally) the better.

If you load a part containing assembly references, these can sometimes be hard to track down. Let's see what we can dig out for the key. Select the extrude in the key part model tree, then in the RMB pop-up pick

Information* ➤ *Reference Viewer

The **Reference Viewer** opens. Notice the parents are identified by feature name and location. These include both assembly features (like ASM_RIGHT) and features in other components. Close the **Reference Viewer** window.

With the first feature in the key part created, you can continue to add new features to the key either in its own part window, or in the assembly window. Remember that by making the key the active component, the next feature(s) you create would automatically be added to the key. Return to the assembly window and activate the top-level assembly.

---

[6] Incidentally, while the key part is open, find out what its units are. Recall that when we created it, we did not designate a part template.

Exploding the Assembly

A useful way of illustrating assemblies is with exploded views. Creating these is very easy. First, suppress the longitudinal cut and unhide or resume all the components. In the **Model Display** group select

Exploded View 

All the parts will be translated by some default distance. You should see something like Figure 15. The assembly has been exploded in directions, and by distances, determined by Creo. For a better view arrangement, we can change the explosion positions as follows. In the **Model Display** group select

Figure 15 Default exploded view

Edit Position 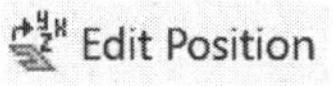

This opens the **Explode Tool** dashboard (Figure 16). We modify the exploded position of each component by specifying a motion type (***Translate***, ***Rotate***, or in the ***View Plane***) and a motion direction reference (solid or datum plane, edge, axis, or coordinate system). Once the motion reference is defined, we select one or more components and drag them in the chosen direction.

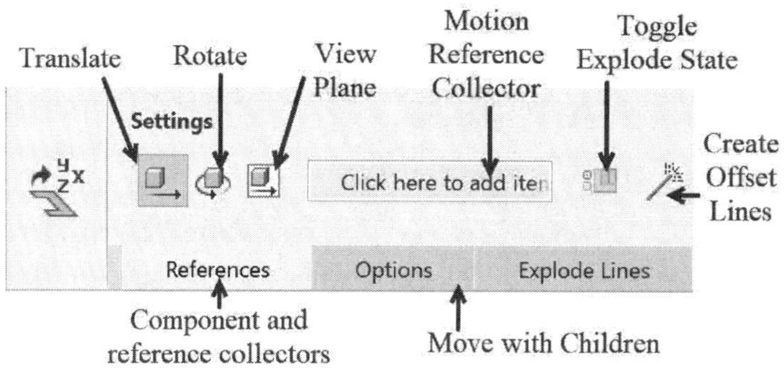

Figure 16 The **Explode Tool** dashboard

Let's try this out. Make sure the **Translate** direction button is selected in the dashboard. Then click in the motion reference collector to activate it, then select the top surface of the base plate. It will highlight in cyan. By default, this will result in explode directions normal and parallel to this plane. Open the RMB pop-up and select **Components to Move**. Now click on the pulley. A temporary direction triad will appear at your pick point, with the X-axis normal to the base plate surface. If you picked an edge for the direction reference instead, the X-axis would be parallel to that edge. If you mouse-over one of the three orthogonal directions (X, Y, or Z), that direction indicator will show as a bold orange arrow. This will indicate the drag direction. Go ahead and drag the pulley with the LMB. Look ahead to Figure 17 for its final resting place. The dynamic view controls on the MMB are still available, so you can reposition the model orientation however you like[7]. If you click on another component, the XYZ direction triad will switch to it, and you can then drag it around.

Open the **References** panel. You should see the currently movable component listed there. Select it and in the RMB pop-up select ***Remove***. Now hold down the CTRL key and click on all four bolts. The move triad will appear on only one of them, and all four bolts will be dragged the same direction and distance.

Remove everything from the **Components to Move** area with ***Remove All*** in the RMB menu. Open the **Options** panel. Check the box **Move with Children**. Now select one of the L-brackets and drag it sideways. You should also see the bushing and washer move with the bracket. Similarly, if you select and drag the axle, the pulley should move with it. If you select the *support.asm* entry in the model tree, the entire subassembly and its children (bolts, too!) will move.

Sometimes you don't want to explode all the components. Clear the box for **Components to Move**. Select (with CTRL) one L-bracket, bushing, and washer on one side. Now select the ***Explode State*** toggle in the dashboard. The three components move to their unexploded position[8]. Selecting the toggle again will explode them.

You can probably pretty accurately guess how the ***Rotate*** and ***View Plane*** motion direction options work. Give them a try.

See if you can position all the components as shown in Figure 17. When you are finished, middle click to accept the final explode positions. These will become the new default explode positions for this session. However, we will have to do something further to ensure that our explode positions are stored permanently in the assembly. The indicator that the current explode state has not been saved is the "+" sign, as in "DEFAULT EXPLODE(+)" that appears in the graphics window.

To save the current explode positions as the default, select the ***View Manager*** button  in the graphics toolbar.

---

[7] One thing you cannot do is select a view in the ***Named View List***. These saved views contain explode status and position data. So if you selected the FRONT view, for example, you would get the unexploded FRONT view. This is actually a handy way of defining and saving different exploded view configurations and view directions.

[8] You can add a column in the model tree that indicates the explode state of all the components in the assembly. Check it out!

Figure 17 Modified explode distances

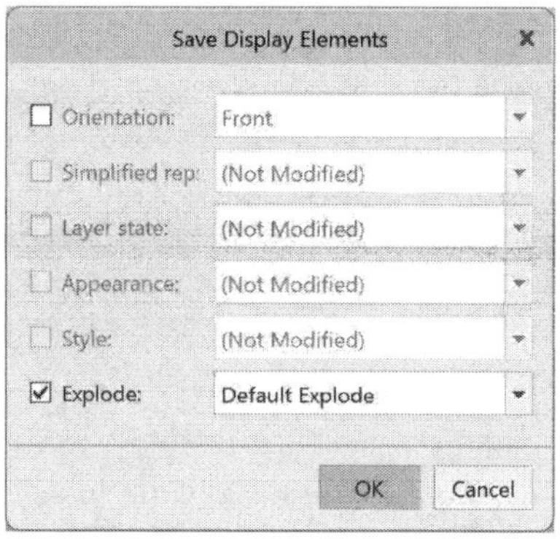

Figure 18 Saving the explode state

In the view manager window (see Figure 20), pick the **Explode** tab. The default explode state is listed. Select that and in the RMB pop-up select ***Save***. This opens the **Save Display Elements** window (Figure 18). Make sure the Explode box is checked and select ***OK***. You are asked to confirm that you want to update the default explode state. This will reset the default explode positions to the ones we set up in the previous section - exactly what we want. Select ***Update Default*** then ***Close*** the View Manager window.

Before we continue to the next section, unexplode the assembly. In the View ribbon, select the ***Exploded View*** button to toggle it off. If you do a lot of work in assembly mode, you will probably find it useful to add the ***Explode*** command button to the Quick Access toolbar. It will allow you to quickly toggle between exploded and unexploded views. Leave the assembly un-exploded before you move on.

Helpful Hint

Because we have saved the explode positions with the current assembly file, there is no need to create another assembly file (for example using *Save A Copy*) containing the exploded assembly. A single assembly file can contain multiple sets of explode positions, for example to show an assembly sequence.

Component Display Style

Display styles provide another tool for dealing with the display of very complex assemblies with many components. So far, we have seen the ***Suppress*** and ***Hide*** commands to deal with the problem of screen clutter. In addition, the graphics toolbar buttons control all related entities on the screen in the same way: datums are either on or off, components are either shaded, wireframe, hidden line, and so on. Here is an easy and very flexible way of setting the display of individual components in a complicated assembly. Before we start, expand the *support.asm* components in the model tree. We are going to set the display appearance of each component individually. Turn ***Shading*** on and colors off.

Modifying Component Display Styles

Select the two L-brackets (using CTRL). Then in the **View** ribbon, **Model Display** group overflow select

Component Display Style ➤ Hidden Line

The two components now appear in hidden line instead of shaded. This allows us to see "through" the brackets without taking them right out of the model (or making them transparent, which requires an appearance definition and the use of color or turning on transparency for the bodies involved). The graphics window also contains the notation for the **Style State : Master Style(+)**. We'll find out what this means in a moment.

Now select the base plate and pulley and pick

Component Display Style
No Hidden

Finally, pick the axle and the four bolts (with CTRL) and select

Component Display Style
Transparent

The screen should look like Figure 19. Now, use the graphics toolbar buttons (***Shaded***, ***Hidden Line***, etc.) to change the display state. These buttons now affect only the parts we have not included in the style definition (bushings, washers, and key).

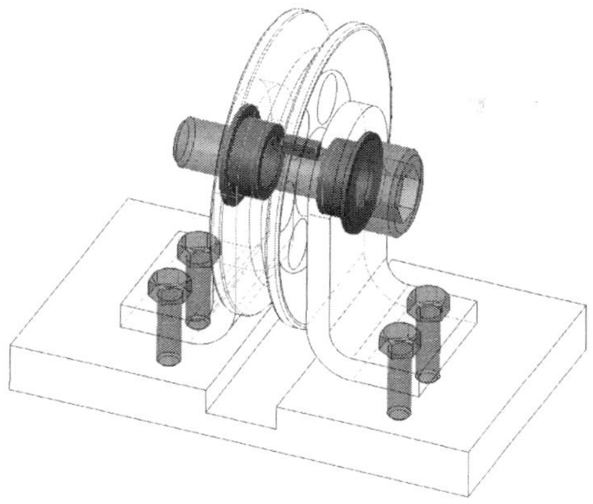

Figure 19 Display style

We would like to set up the model so that we can get to this display style easily whenever we want. This is accomplished with the ***View Manager***.

Select ***View Manager*** button and pick the ***Style*** tab. You will see the style **Master Style(+)**. The "+" sign indicates that the master style has been modified. Select this and in the RMB pop-up, select ***Save***. This opens the **Save Display Elements** window in which you can typeover the default name (*Style0001*) of the new style with something like "***Mystyle***". Select ***OK*** in the **Save Display Element** window. You should now be back in the **View Manager** window as shown in Figure 20.

If you want to change back to the original Master Style display state, just double click it in the **View Manager** window (or select it and choose ***Activate*** in the RMB pop-up). You can easily move back and forth between these (or any) defined display styles.

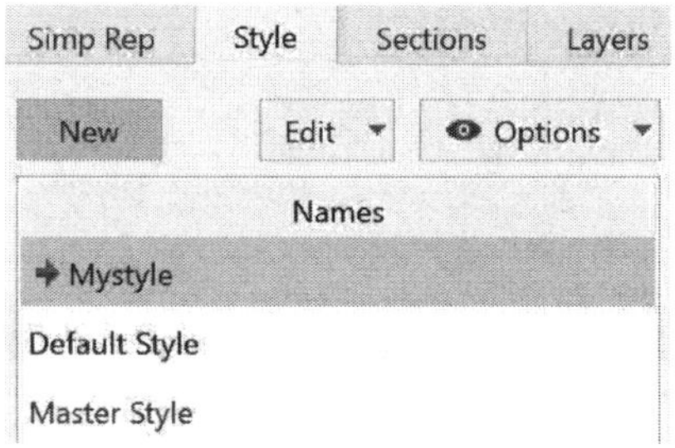

Figure 20 The **View Manager** window

If you want to change any settings for a display style, try this: highlight our newly created ***Mystyle*** in the **View Manager** pane, and select ***Edit Definition*** in the View Manager ***Edit*** pull-down menu. This brings up the EDIT window (Figure 21) which allows you to edit the display. If you select the ***Show*** tab, you can select a desired display method, then pick on components in the model tree or the graphics window. This also temporarily changes the contents of the model tree to show the display state of all components as defined in the selected style. In addition to the four states we have seen so far (wireframe, no hidden, hidden, and shaded) we can also ***Blank*** a component. This is essentially the same as hiding it. With the **Blank** tab selected, pick the key in the model tree, then ***OK***. Press ***Preview*** to see the result.

The buttons at the top of the EDIT window are ***Undo***, ***Reset*** (to remove a style setting from a component), ***Show Selected*** (Navigator will show only those components that have a defined style), ***Show Info*** (about selected components), and ***Rules*** (for automatically determining a display state).

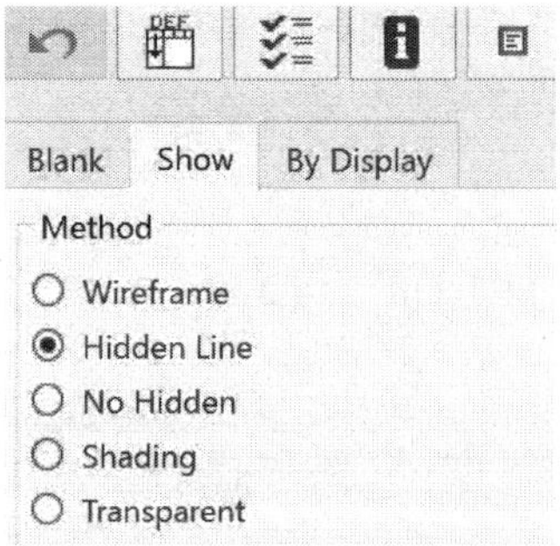

Figure 21 Editing display styles

Accept the current style definitions using ***OK*** in the EDIT window. This returns you to the **View Manager**. At the bottom of this window, select ***Properties***. This brings up a display (Figure 22) of all the components in the assembly that have a style defined. There are buttons across the top that correspond to each possible style. Select a component and pick on the desired style to make a change. Which button corresponds to ***Blanking*** the component?

Try out the various ***Properties*** options to set up another display style as shown in Figure 23.

To return to the listing of display states, select the ***List*** button in the **View Manager.** The small "+" sign that appears indicates the style definition has changed. If you want to keep these changes, in the RMB pop-up select ***Save***. In the **Save Display Elements** dialog window, change the style name to something new, like ***Mystyle2***. Once the style is saved, the "+" sign disappears.

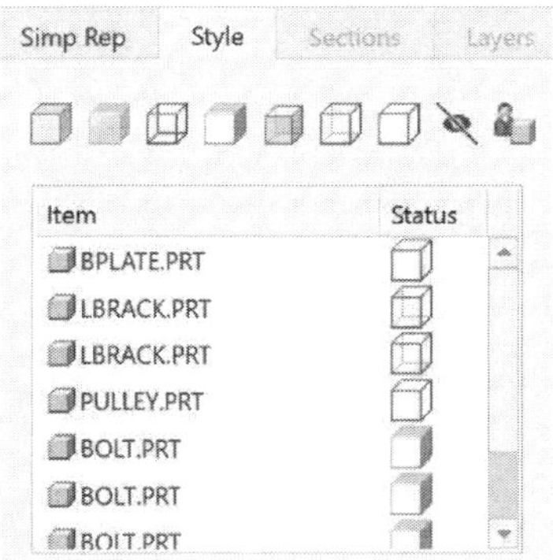

Figure 22 The View Manager Properties pane

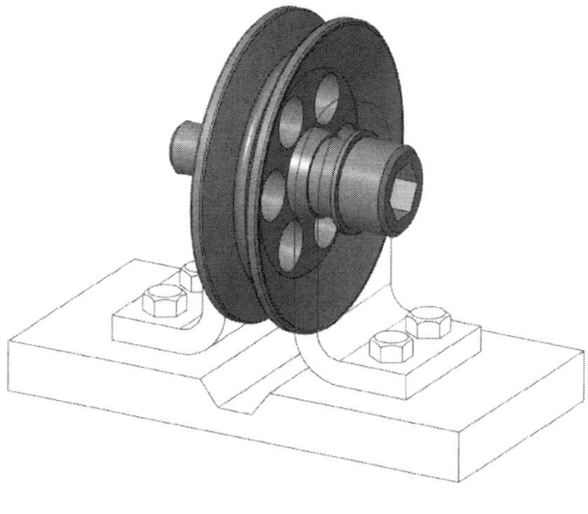

Figure 23 Component display state #2

Modifying the Explode State

Pick the **Explode** tab in the View Manager window and select the listed style "Default Explode". In the RMB pop-up, select ***Explode,*** then (again in the RMB pop-up) ***Copy***. This creates a new explode state - call it something like "***Myexplode***". In the **View Manager** window, double click on it to make it active (with the small green arrow). Now pick

Edit ➤ Toggle Explode Status

A small menu appears at the right and another column (Explode Status) is shown in the model tree. Expand the model tree entries for the two support sub-assemblies. In each sub- assembly, pick on the three components (LBRACK, BUSHING, WASHER) either in the model tree or the graphics window. Toggle each of them until all of them have their status as *Unexploded.* In the small **Explode Status** window at the right, select ***Done***. The model now appears as in Figure 24; note the location of the bushing and washer components in the bracket.

Figure 24 Explode state with modified explode status of some components

With the Explode tab still selected, at the bottom of the **View Manager** window, select ***Properties***. In this window (see Figure 25), you can toggle the explode status of components in the assembly, edit their explode position, and add or modify the explode offset lines using the buttons just below the tab at the top. Return to the previous **View Manager** window using ***List***.

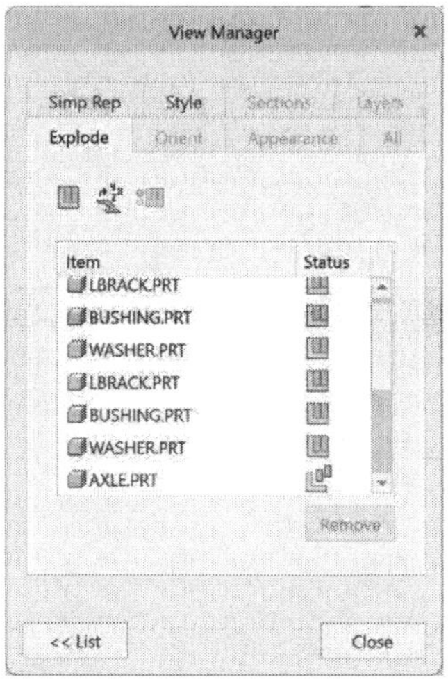

Figure 25 Setting explode properties in the View Manager window

We will not explore it here, but the ***All*** tab in the view manager window allows you to set up combinations of view states, including component display style and explode status, among other options. You might like to experiment with these functions. An advantage of these view state definitions is that they will be available easily to other users, and in the creation of drawing views of the assembly (especially useful for putting exploded views on the drawing). Return the model to the unexploded master style (not *mystyle*) using the **View Manager**. Close the **View Manager** window, save the assembly, and set a shaded view display.

Sections

Cross-sections, or just sections, are very common in assembly drawings (which we will get to next). Although sections can be created in drawing mode (as we did with the pulley in Lesson #8), it is often handy to have them available when you are working in assembly mode. This can help you check for fit and/or interference between components, measure clearances, and just generally help understand how an assembly fits together. Sections are very easy to create. You can define multiple sections in the same model, of different types and at different locations, and display as many of them at the same time as you like.

Remind yourself of the orientation of the default assembly coordinate system - that will be relevant for what follows. Set your display as follows: shading on, colors off, datums off.

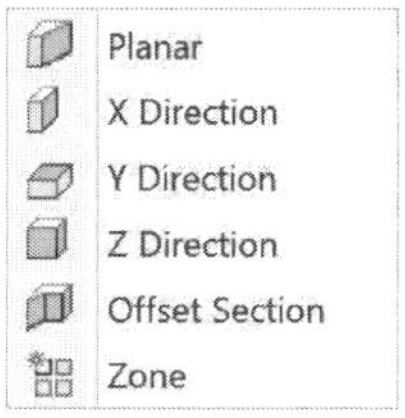

Figure 26 The ***Section*** pull-down menu

We are going to recreate the view of the assembly obtained with the assembly cut earlier in this lesson. When using an extruded cut to do this, our only options are to suppress or resume the cut and to determine which components were intersected by the cut. The ***Section*** command allows us to do a lot more, and much more easily. The section functionality in Creo works with parts as well as with assemblies.

Go to the **Model** ribbon. In the **Model Display** group, click on the **Section** pull-down menu. This is shown in Figure 26. The ***Planar***, ***X***, ***Y*** and ***Z Direction*** commands will create planar sections, with the first being the most general. The ***Offset*** option allows you to create a section using a sketched curve to define the 2D shape of the cutting plane.

Select the ***Z Direction*** command. The model appears with a cutting (or clipping) plane whose normal is the Z axis direction of the assembly default system. The reference coordinate system is shown in the **References** panel on the dashboard. This is the same view we created earlier using the cutting plane. An offset distance value is given (default 0) that can be changed by dragging on the arrow normal to the section surface or double clicking on the numeric value. Try that now, setting the offset to **20**.

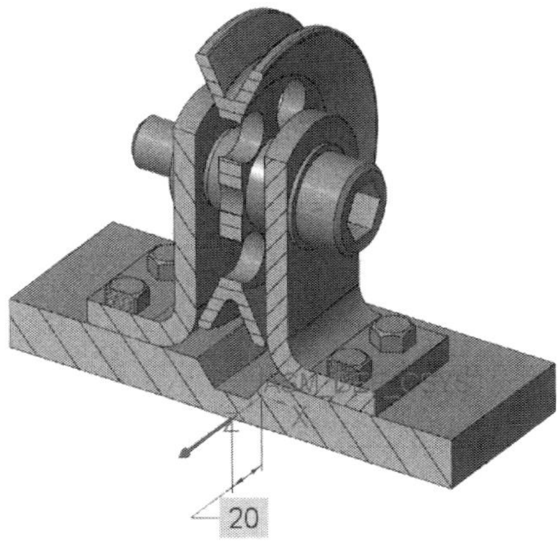

Figure 27 Section defined (Z direction, offset 20) with hatching

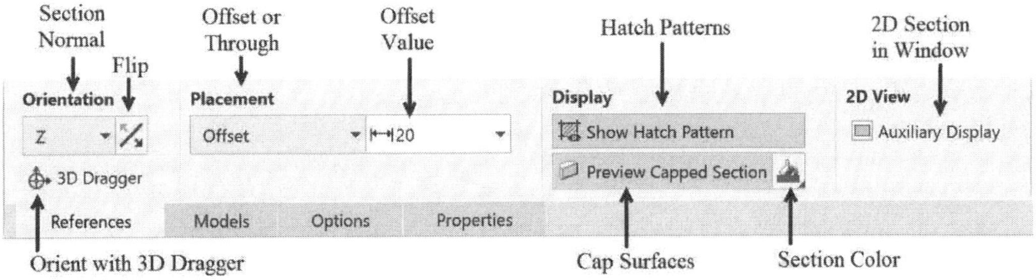

Figure 28 The ***Section*** dashboard

Let's have a closer look at the dashboard (Figure 28) - feel free to try these commands out as we go. Starting at the left, the orientation of the section normal is given in a drop-down list. The ***Flip*** button allows you to select which side of the section plane is removed. Just below that is the ***3D Dragger*** for free positioning of the section plane. Select that now. Dragging on the arrows or rings allows us to change the position and orientation of the clipping plane. See Figure 29. If you turn off the dragger, the clipping plane returns to the originally selected (Z, in this case) direction. The **Placement** option ***Through*** will let you choose any planar surface or datum plane to specify the section. The ***Capping Surface*** button (on by default) toggles the surface created on the section plane. If the surfaces are capped, the next button lets you set their color (default, model color, or other). The ***Hatch*** button does just that - toggles the presence of a hatch pattern on each component (Figure 27). Notice that the hatch on each component is different. We will see later how these hatch patterns can be changed. Finally, the ***2D View*** button opens a new window to show the section plane in true shape (looking normal to the section). Select the ***2D View*** button in the dashboard again to close this window.

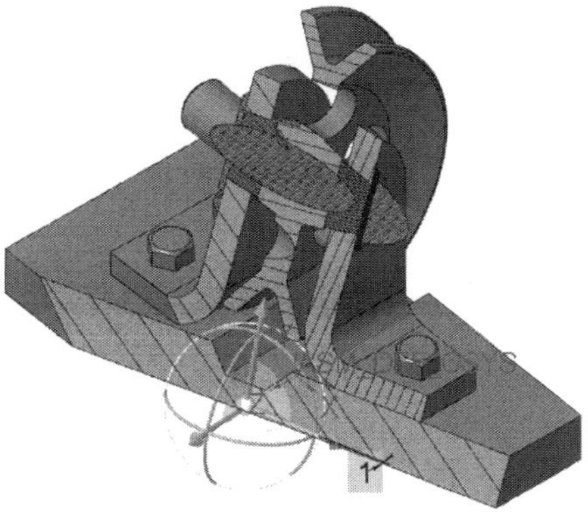

Figure 29 An oblique clipping plane created using the 3D dragger

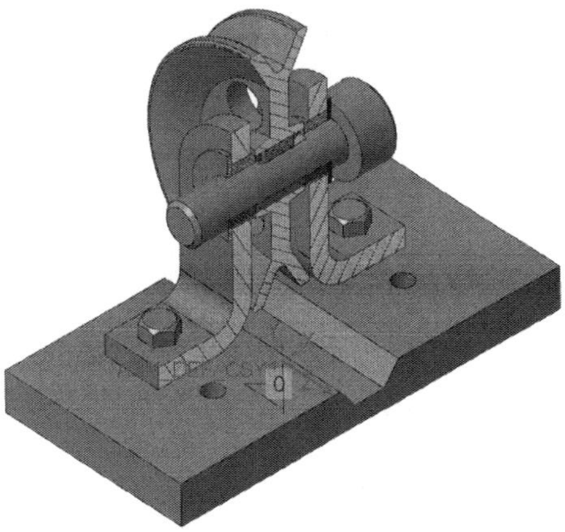

Figure 30 Base plate and axle components **Excluded** from section

Let's explore some of the section options in the dashboard tabs. The **References** tab lets you set the coordinate system or surface used to determine the section normal. The **Models** tab allows you to select which components to include or exclude from the section. Select the **Exclude** option and pick (with CTRL) the base plate and the axle. These components are now not clipped (Figure 30). Come back later to reset this to **Include all items**. The next tab, ***Options***, lets you toggle the visibility of interfering components. You can set the color of the interfering volume. Finally, the ***Properties*** tab lets you specify the name for the section, as it would appear on a drawing. Type in "***A***" for the name of this section.

In the **Placement** pull-down, select ***Through*** and then pick on the ASM_FRONT datum. Notice that the **Orientation** box is now grayed out (no options there with a selected surface). After setting all the section options, a middle click will accept the section definition. It is now listed at the bottom of the model tree and is active (small green diamond). Clicking on the section entry will show the outline of the sectioned parts. If you use the pop-up menu, you can select ***Deactivate*** to make the section invisible. The ***Show/Hide Section*** toggle option in the pop-up menu controls the visibility of the hatch patterns. With the section selected and activated, the RMB pop-up menu in the graphics window allows you to execute several of the commands in the section dashboard.

Select the bushing component in one of the subassemblies, then open it using the pop-up menu. Change the outer diameter to 35 and ***Regenerate***. Return to the assembly window, ***Regenerate***[9], and ***Activate*** the section. Use the RMB and select ***Show Interference***. You should see a red highlight of the interference between the bushing and the L-bracket as shown in Figure 31.

Note that in the normal and RMB pop-up menus for sections you have options to pattern the section, flip the clip direction, edit the hatch patterns and offset distance. You can also ***Rename***, ***Suppress***, or ***Delete*** the section.

---

[9] Can you figure out why the assembly must also be regenerated? (Hint: section)

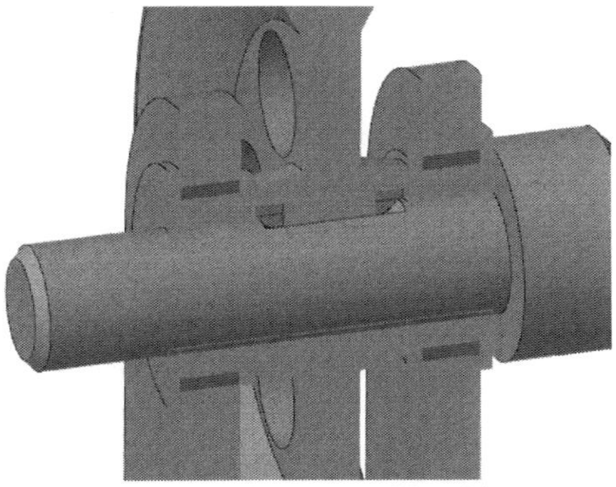

Figure 31 Interference of bushing outer diameter and hole in L-bracket

Another way to deal with sections is with the ***View Manager***. Open this window and select the ***Sections*** tab. You will see the section "A" we created previously. You can use the ***Edit*** and ***Option*** pull-down menus (and RMB pop-up) to access a number of commands for activating, editing hatch patterns, and so on. You cannot change the position or orientation of the section. Using ***View Manager***, you can use defined sections as part of the display style. Close the View Manager.

Use the pop-up and RMB menus to ***Deactivate*** the section and turn off ***Show Section*** and ***Show Interference***. While still in the assembly, edit the bushing diameter back to the original value (**30**) and ***Regenerate***. Don't forget to ***Save*** the assembly.

Assembly Drawings

Our last task is to create a drawing of the entire assembly. We will not do any dimensioning here, just lay out the views and provide some leader notes. Turn off the display of all datums (planes, axes, points, Csys). Then select

File* ➤ *New* ➤ *Drawing* ➤ *[less9asm]

Deselect the **Use default template** and **Use drawing model name** options, and use an empty A-sized drawing sheet. The drawing ribbon appears at the top of the screen[10]. Select (in the RMB pop-up)

General View (or select the ***{Model Views}General View*** button)

In the **Select Combined State** window, select ***No Combined State* ➤ *OK***.

Pick a view center point on the left side of the sheet (look ahead to Figure 32 for the placement, although you can always move the view later). For the view orientation, in the **View Type** category select RIGHT, then ***OK***. The default display style (shaded, hidden, etc.) of the view is controlled by the buttons in the graphics toolbar (use ***Repaint*** after changing settings). Modify the scale of the drawing by double-clicking on the scale legend at the bottom of the drawing and entering the new value (**0.5**).

Now we'll add a section view that uses the section "A" that we defined previously. Select

---

[10] If you haven't done Lesson #8 on creation of drawings, you should probably go back and do that now. A lot of the drawing ribbon functions will not be discussed here, only some of the ones relevant to assemblies.

the primary view we just created, then in the pop-up mini-toolbar select

Projection View

Make the center point of the view to the right of the main view. Now we have to set up the section. With the new view highlighted (green border), in the pop-up toolbar, select ***Properties***. In the **Drawing View** window, select the **Sections** category, then ***2D cross-section***. Press the ***Add*** button ("+"). Our previously created section "A" appears in the list. Move the list slider to the far right and click in the collector under **Arrow Display**. Now pick on the primary view. Select ***Apply***. The section view appears with the sectioning arrows on the first view.

Before you finish with the section view, change the **View Display** panel to

Display style (No Hidden)
Tangent edges (None)

Notice that by explicitly assigning a display style to the view, it becomes unaffected by the graphics toolbar buttons. Select ***Apply*** and ***OK***.

Let's add one more view - the exploded assembly. You may have to unlock and move the two existing views down a bit to fit this one in. Then select (in the ribbon)

{Model Views}:General View

or just use the RMB pop-up and select ***General View***. This time, for the presentation select ***DEFAULT ALL*** ▾ ***OK***. Place the view near the top of the sheet. In the **Drawing View** window for this new view set (or check - some were determined by the presentation option) the following:

View Type - orientation set to Default orientation
Scale - set a Custom scale of 0.25 (don't forget to ***Apply*** it)
View State - select ***Explode components in view*** and in the pull-down list pick the named state ("Myexplode") we created previously, then ***Apply***
View Display - set to ***No Hidden*** and turn off the tangent edges

We're almost finished. You might like to modify the hatching in the section view (pick the hatch on one of the components, then in the RMB menu select ***Properties*** and play with the spacing, angle, and hatch pattern). When you are finished setting the hatch for the component highlighted (blue border, green hatch), select ***Next*** or ***Pick*** to move on to the next one. You can set the hatch pattern for each component separately. Any hatch properties set in the assembly will be carried over to the drawing. This includes (in the **Hatch Gallery**) standard patterns for common materials (iron, steel, aluminum, and so on), hatch spacing and angle, color. As you would expect, the hatch styles of the components are independent of each other. Finally, select the **Annotate** tab in the drawing ribbon and add some leader notes. Your final drawing might look something like Figure 32. This is a keeper!

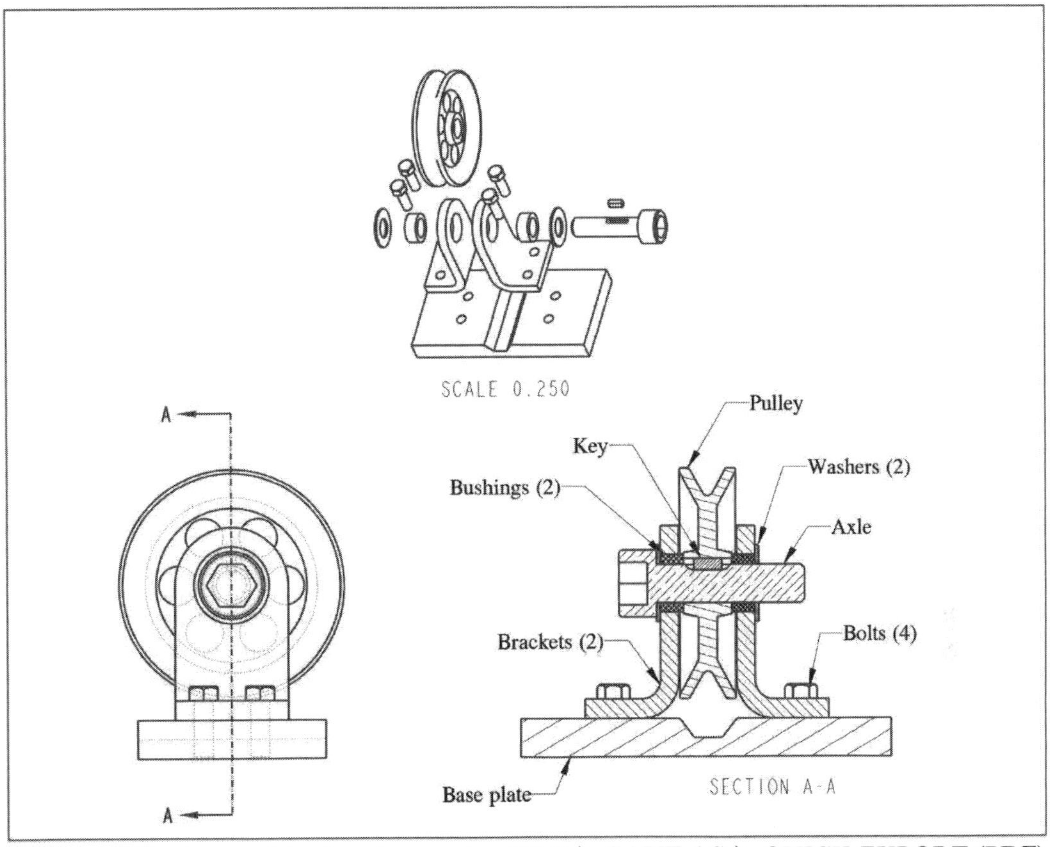

Figure 32 Final assembly drawing—using FILE ➤ SAVE AS ➤ QUICK EXPORT (PDF)

There is lots more you can do with assembly drawings. For example, you will typically use multi-sheet drawings for an assembly. The first couple of sheets would show the entire assembly, possibly with exploded views and sections, and a Bill of Materials (BOM). Subsequent sheets would show individual components. All these sheets can be contained within a single drawing file. And, of course, all drawings are associative to the original solid models. If the assembly represents a mechanism, you can create multiple positions of the system and store them as "snapshots". Each snapshot can be placed on the drawing to show the sequential positions of the mechanism. These topics are discussed in the *Advanced Tutorial.*

This concludes our discussion of assemblies. We have seen a number of the tools available for dealing with assemblies, assembly features, and some utilities to help you work with assemblies. What we have not had time for here is a discussion of some very important ideas concerning how to use assemblies most effectively in a design environment. You have probably heard of terms like "top-down design, bottom up implementation". Because of associativity, Creo is very powerful when used in this way. As you become more experienced with Creo, you will discover ideas about "skeleton models", "layouts", and others. The *Advanced Tutorial* presents more information on some of these topics.

Before we leave, here are some final points you should know about:

Helpful Hints

➊Using ***Save A Copy*** in Assembly mode is somewhat tricky. You must remember that the assembly file does not contain the parts themselves, only references to other files. When making a copy of the assembly file, you are given options about what to do with the individual part files (reuse them, copy them, rename them, etc.). Consider what Creo must do with the parent/child relations between components in an assembly if some of them get renamed and/or moved. **Do not use *Save A Copy* in Assembly mode unless you really know what you are doing!** That is, practice this some time with a simple assembly (like the bracket subassembly we made here) when you have enough time to experiment and with parts/assemblies that you can afford to lose or damage. Don't try this the night before the big project is due.

➋ If you want to change the name of an assembly (or part) file, you must use the ***File ➤ Rename*** function in Creo - do not do that using Explorer, for example. For assemblies, you must do it while the assembly is in session so that the new name can be registered in the assembly. If a part is used in an assembly, renaming the part file will automatically update the assembly if it is also in session (even if it isn't being displayed). **If you copy, move, or rename files outside Creo, you can expect trouble!** One of the most common error messages you will get when retrieving assemblies is "Component not found" because the component's name or location has been changed without the assembly knowing about it.

In the next, and final, lesson in this book, we will return to the creation of features, with an introduction to sweeps and blends. These are among the more complicated features available in Creo, even in their simplest forms. However, they can produce a very wide variety of geometric shapes, and should be in your basic modeling "toolkit."

Questions for Review

1. How can you find out the order that components are brought into an assembly?
2. How can you determine the assembly constraints used for a particular component?
3. What happens if you left click on a component in the model tree? Right click?
4. In the assembly model tree, is it possible to find out what individual features were used to create an individual component?
5. What is an *assembly* feature?
6. Can assembly features refer to individual part features, or only to other assembly features?
7. What kind of features can be created as assembly features?
8. What are the differences between using the default and the empty assembly template?
9. Can you modify individual dimensions of a part while in assembly mode? Is this a

permanent modification (that is, is the part geometry changed if you load it alone)?
10. In assembly mode, how can you add a feature to a part so that it becomes a permanent feature in the part? What advantages would this have? What is the alternative?
11. If you change a feature dimension on an assembly drawing, what happens to the part containing that feature a) by itself, and b) in an assembly?
12. Describe two ways that you get a cut-away view of an assembly. What are the advantages of each method?
13. How do you explode an assembly?
14. What does ***Automatic Update*** do when creating a cut through an assembly?
15. How can you explode some components and not others?
16. Draw a graphical representation of the model tree for the pulley assembly, and trace all the parent/child relations in the assembly.
17. What is contained in an assembly template? Does it have units? Why?
18. There are a couple of ways you can set up the display so that some of the assembly components are transparent. What are they and what are their advantages and disadvantages?
19. If you want a section view in an assembly drawing, does the assembly model require a cut feature along the sectioning plane?
20. What does the command ***Activate*** do for components and sections? How do you reverse this?
21. How do you set up a display style? What options are available? Where is the style stored?
22. Can you combine display styles and explode states in random combinations?
23. Can a model have more than one section at a time? Can you display more than one at a time?
24. How is the active component indicated in the model tree?
25. Assembly sections can only be defined using ________.
26. Under what circumstances would you set the visibility of an assembly feature to Top Level? What about setting it to Part level?
27. What is the difference between ***Suppress***, ***Hide***, and ***Blank***?
28. What happens if a sub-assembly is the active component when you select the ***Create Component*** tool?
29. Can you drag and drop the insertion point in the assembly model tree?
30. Can you reorder components in the assembly tree using drag and drop?
31. How can you change the assembly references of a component?
32. Can you select a Hidden component for a feature operation (like ***Edit Dimension*** or ***Suppress***)?
33. Is a section created using a clipping state in View Manager visible at part level?
34. Suppose an assembly feature is visible in a part. What happens if you delete it (a) in the part, and (b) in the assembly?
35. What is the difference between intersection and visibility in regard to assembly features?
36. How do you change the intersection settings and visibility settings for an assembly feature?
37. Can you drag and drop components in the assembly model tree? What happens if you try to drag the bolts from the top-level assembly into the bracket sub-assembly?

38. How does the model tree indicate components that were **Fix**ed during Child Handling? What does this mean?
39. When a feature like the base plate cut is made at the top-level assembly level but set to be visible at the part level, what happens if you suppress the feature (a) at part level, and (b) at assembly level?
40. Just for fun, try the ***3D Box Select*** command just to the right of the message area. Follow the prompts in the message area. What is the effect of component status (hidden or suppressed) on this selection method?
41. What symbol is used in the model tree to show a child of a packaged component?

Project

Complete the vise assembly using the parts you made at the end of each lesson. There is one more part to make - the ring shown below. It fits underneath as shown in the second figure. Create this part in assembly mode, using the existing features for dimensional references. A cutaway and exploded view are shown on the next pages.

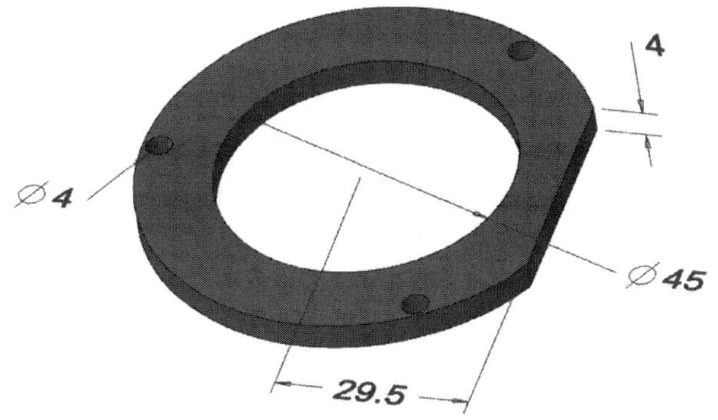

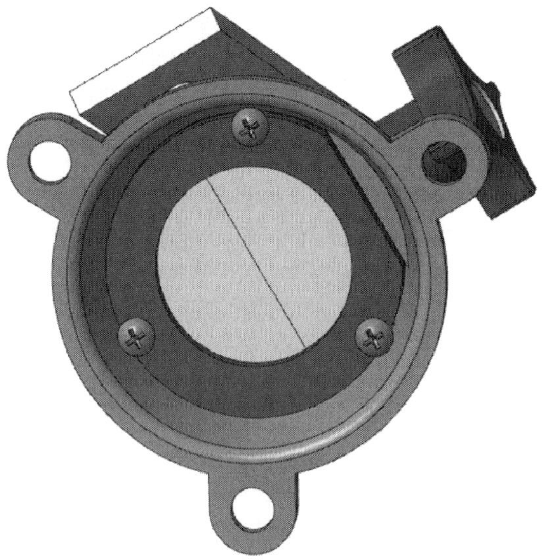

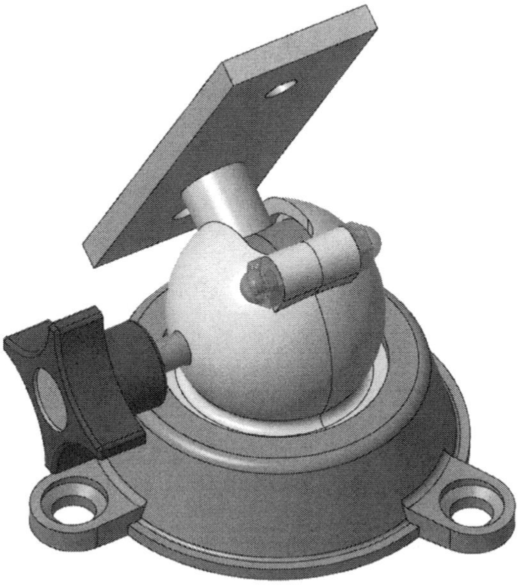

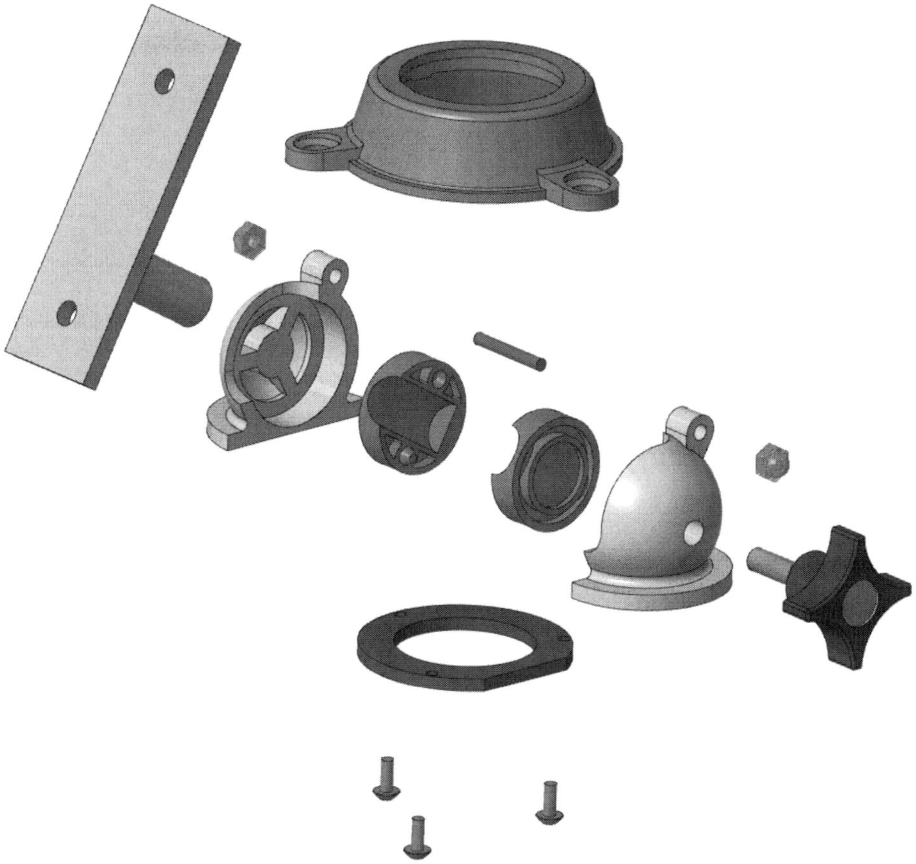

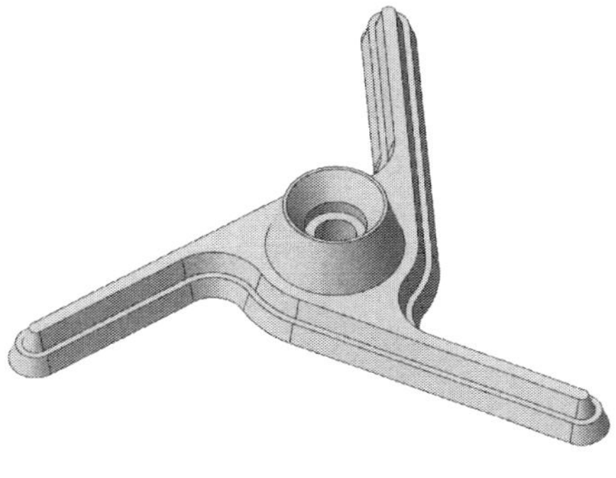

Lesson 11

Sweeps and Blends

Synopsis

Simple sweeps and blends; axis pattern; sketched holes; the ***Shell*** command.

Overview of this Lesson

So far in these lessons, we have, surprisingly, only used two types of sketched features (extrudes and revolves) to create most of our geometry. There are actually several more types of sketched features and this lesson will introduce you to the simpler versions of two of them. These can both be used to either add or remove material (protrusions and cuts) and can pack a lot of geometry into a single feature. Even these simple versions are quite complicated, which is why they have been left until last! The two features are:

Sweeps - a feature that sweeps a section along a specified trajectory (like an extrude that follows a curved path). If the section is closed, the sweep can create either a solid, a thin solid, or a surface. The sweep is controlled by one or more trajectories which can be datum curves, sketched curves, or solid edges; trajectories can be either planar or 3D. The section can either have a constant shape, or vary along the sweep.

Blends - a feature that uses two or more specified cross sections (separated in space) and creates a smooth transition to join the sequence of cross sections (like an extrude or revolve with a varying cross section). This is called, in other systems, a *loft*.

Sweeps and blends are advanced modeling features with many options. We will have a look at simple versions of these to create four different parts. These are totally independent of each other, so you can jump ahead to any one of these:

1. Sweeps
   - a protrusion using a sweep along a datum curve
   - a cut using a sweep along a part edge chain
2. Blends
   - Parallel Blend

- Rotational Blend

Advanced sweeps created using the ***Variable Section Sweep*** and ***Helical Sweep*** tools are discussed in the *Advanced Tutorial* from SDC.

Along the way we will also create a new style of pattern (an *axis* pattern), what is called a *sketched hole*, and discover the ***Shell*** command. As usual, there are some Questions for Review and Exercises at the end of the lesson.

Sweeps

There are a number of different sweep geometries and many options available in Creo, all launched with the same ***Sweep*** command in the **Model** ribbon. The same sweep command can be used to create solids, thin-walled solids, or surfaces (aka quilts). We will create two solid sweeps (a protrusion and a cut) using a couple of options for defining the basic shape. Even for simple solids, sweeps can produce a wide range of geometries, as illustrated in the figures below, including three dimensional sweeps. The swept section can change size, shape, and orientation as it moves along the sweep, controlled by section parameters or additional trajectories. If the section changes, the feature is called a *Variable Section Sweep* - we will not deal with those here.

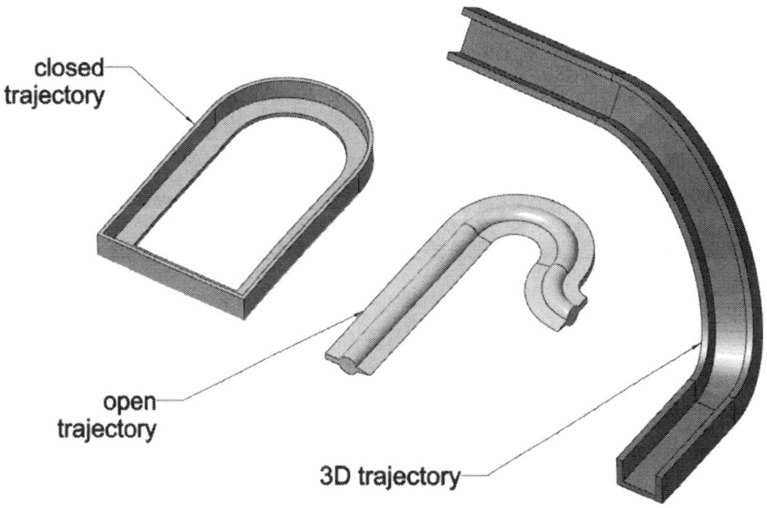

Figure 1 Simple examples of constant section sweeps

Sweep #1 - The S-Bracket

The first part we are going to create is shown in Figure 2. The part consists of two features: the solid protrusion block at the left, and the S-shaped sweep coming off to the right.

Start a new part called **s_brack** using the mmNs template. First create a symmetric both-sides protrusion using **FRONT** as the sketching plane and **RIGHT** as the **Right** reference. The sketch dimensions are shown in Figure 3. Use a **symmetric blind** depth of **120**. The right edge of the sketch **aligns** with **RIGHT** and the lower edge is on **TOP**. Accept the feature. Select the two vertical edges on the right corners and place an **R30** round.

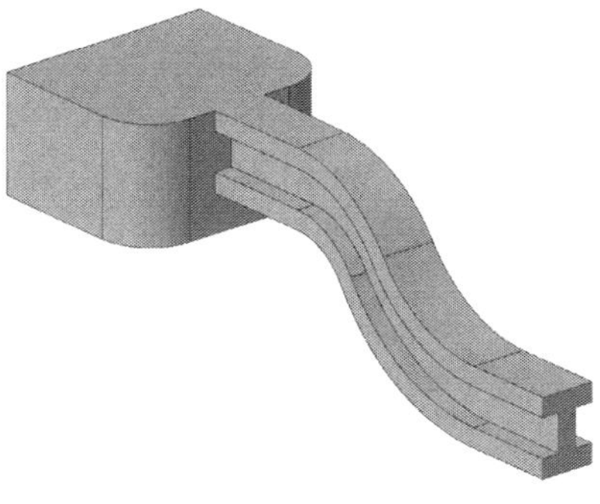

Figure 2 The S-bracket

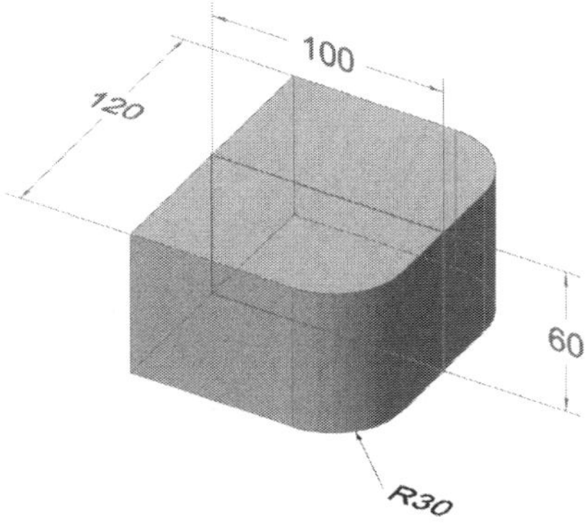

Figure 3 Base feature and rounds

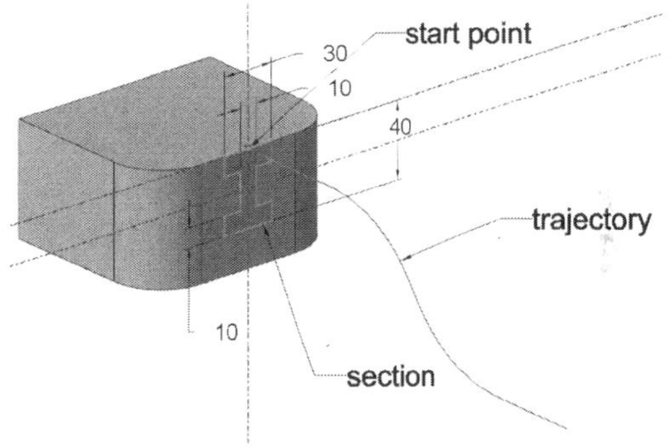

Figure 4 Elements of a simple sweep

Now we will create the sweep. The elements of a simple sweep feature are shown in Figure 4. The major elements are the *origin trajectory* and the *section*. The origin trajectory is the path followed by the section as it is swept - it can be either an open or closed curve. In this case, the trajectory is *open*. For default sweeps, the section stays perpendicular to the trajectory ("*Normal to Trajectory*"). The trajectory can be either an existing edge or it can be sketched as we will do here. The cross section of this sweep is like an I-beam. It is created on a sketching plane located at the *start point* of the trajectory. The section sketch does not have to lie on the trajectory at the start point, that is, it can be offset from the origin trajectory within the plane of the sketch.

Defining the sweep is done in two steps: creating or selecting the sweep trajectory, then creating the cross section. The geometry is shown in Figures 5 and 7. We have several options for the procedure we follow to define the trajectory. If we plan to use a sketched curve for the origin trajectory, we can either create the curve first and then launch the sweep command, or launch the sweep command first and create the curve. For our first sweep, we'll create the curve first[1].

---

[1] For advanced sweeps that use multiple trajectories in addition to the origin trajectory, these are almost always created first, before the sweep command is launched. Then they are merely selected for the appropriate function within the sweep. See the *Advanced Tutorial* for examples of sweeps that use multiple trajectories.

Select the ***Sketch*** tool. Pick the **FRONT** datum as the sketching plane. The default **Right** reference is the **RIGHT** datum. When you get into sketcher, select the top of the block as a reference (and you can remove the TOP datum plane reference). Create the sketch shown in Figure 5.

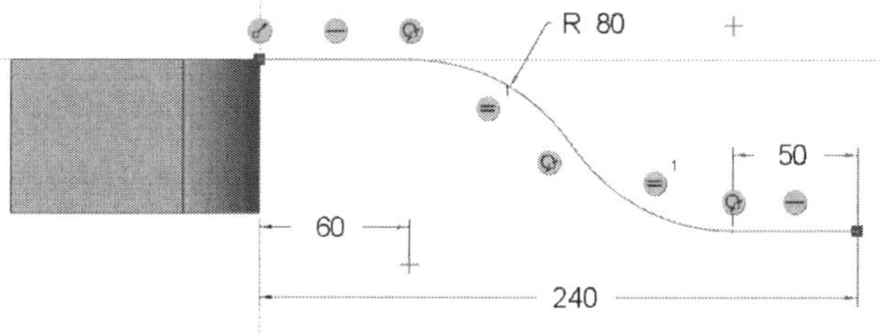

Figure 5 Sketched curve (origin trajectory)

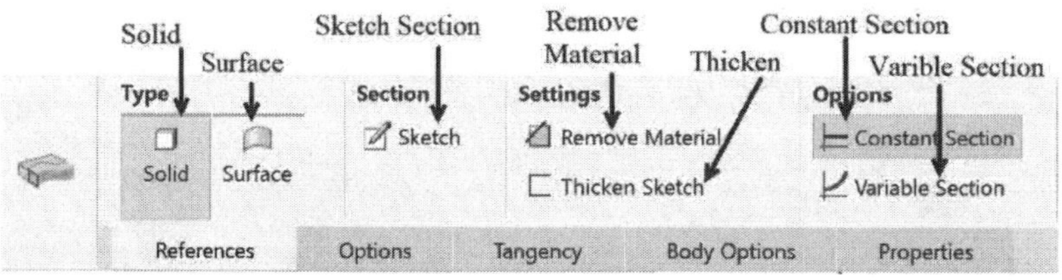

Figure 6 The ***Sweep*** dashboard

With the datum curve highlighted in green (as the last feature created), select the ***Sweep*** button in the ribbon. The sweep dashboard opens (Figure 6) and the sketched curve becomes highlighted with a heavy green line (with a label *Origin*) (Figure 7). At each end of the trajectory are two round yellow handles and parameters (currently set to 0) which we will discuss later. At one end of this trajectory, there is a magenta arrow pointing along the trajectory. This defines the *start point* of the trajectory. Position the cursor on each of these elements and hold down the RMB to see the options in the pop-up menus. We want to have the start point at the left end as shown in Figure 7. If it is at the other end, just left click on it - it will jump to the other end. Or, you can hold down the RMB and select ***Flip Chain Direction***.

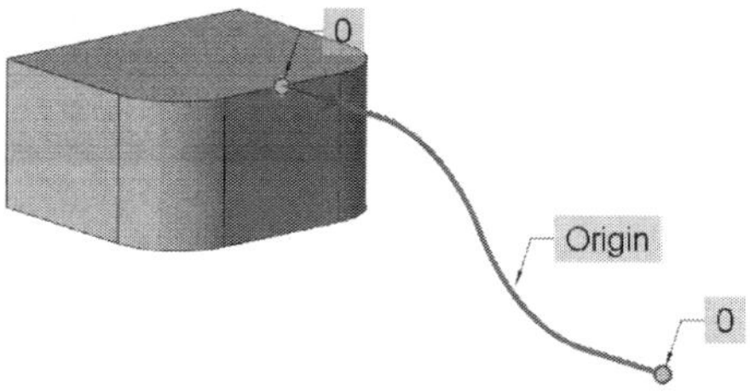

Figure 7 The sweep trajectory

In the dashboard, notice that the default is to create a swept solid. We have seen the ***Remove Material*** and ***Thicken*** options before. Two new ones are ***Constant Section*** and ***Variable Section***. As their names imply, these determine whether the swept section will change size/shape as it moves along the sweep trajectory. The default is to use a constant section. In the ***References*** panel the origin trajectory (our sketched curve) is listed in the top pane. Notice the default **Section plane control** option *Normal to Trajectory*. We will leave all the options in this panel alone for now but will come back later to explore the ***Merge Ends*** and ***Placement Point*** options in the ***Options*** panel.

Now, we move on to the second step - creating the cross section. Select the ***Sketch*** button in the dashboard or pop-up menu. In passing, note the other options in the RMB menu. The view reorients so that you are facing onto the right face of the block. The screen should show you a dashed cross hair that automatically defines your sketch references. This is centered on the *start point* of the trajectory with the sweep coming toward you. You might like to rotate the view a bit to see the orientation of the sketch that is determined automatically by Creo. Use the Sketcher tools (or select the shape in the Sketcher palette. Hint: use a scale of 10) to create the cross section shown in Figure 8. The constraint display has been turned off in the figure - can you figure out what constraints are active?

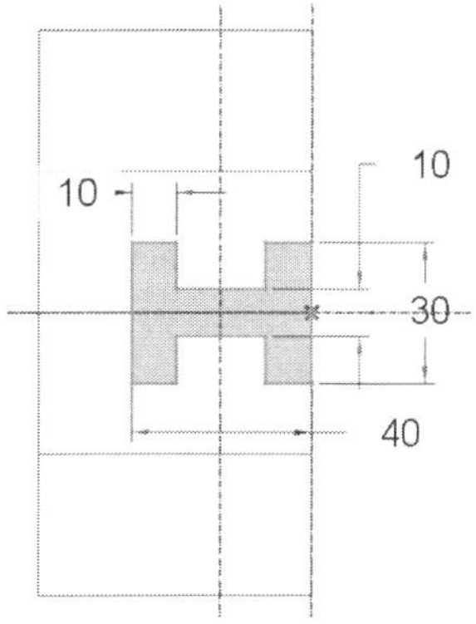

Figure 8 Sketch of sweep cross section

When the section sketch is complete, you can select ***OK*** in the Sketcher menu. The sweep will preview. If everything is satisfactory, you can ***Accept*** the sweep.

If you double-click on the sweep, you will see all the dimensions for the swept section. To change the dimensions of the trajectory, select the sketched curve in the model tree and select ***Edit Dimensions*** in the pop-up menu. See Figure 9. Not all combinations of dimensions are guaranteed to work, however. For example, if you increase the height of the section from 40 to 60, then 80, then 100, the feature will eventually not regenerate. Try to figure out why. (A hint is given at the end of this section!)

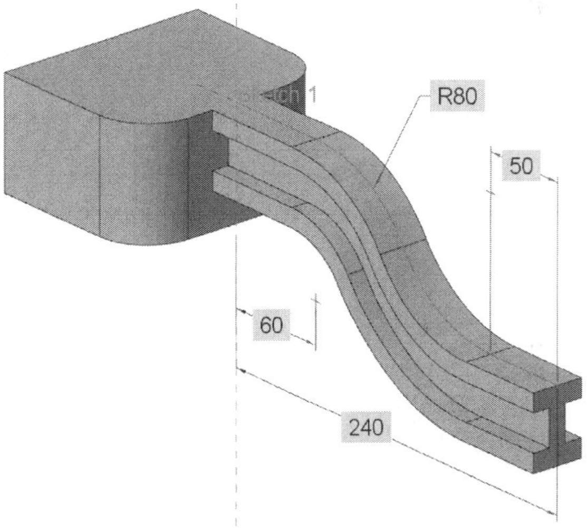

Figure 9 Completed sweep

Finally, notice the model structure in the model tree. The sketched curve and swept protrusion are separate features, with the sweep being a child of the curve. If you aren't going to use the curve for anything else, you might want to ***Hide*** it in the model tree so that it doesn't appear on the screen. This would also prevent you from accidentally picking it as a reference for another feature. There is a better way of dealing with this that we will see later.

In preparation for the next version of a sweep, ***Suppress*** the sweep we just made, keeping the curve in the tree.

Alternate Method for Creating Sweep

In the previous sweep, we launched the sweep tool with the curve pre-selected (highlighted in green). This is not strictly necessary. This time, make sure the curve is not pre-selected (it should just be showing in blue), then select the ***Sweep*** tool. The sweep dashboard opens. Follow the prompts in the message window. Since it was not pre-selected, the first trajectory we pick will form the origin trajectory. Pick anywhere on the blue sketched curve. Make sure the start point is at the left end (click on it if necessary) and that the ***Solid*** button is selected. This gets us to the same place we were before, with the origin trajectory highlighted in bold green, with the start point at the correct end. We won't proceed any farther than this, since the procedure is the same as before. Don't bother creating the section; instead, ***Cancel*** the current feature with the X in the dashboard. ***Resume*** the previously created sweep.

Extending the Trajectory

You will recall that at each end of the origin trajectory there was a "0" parameter. What are those for?

To see a potential problem with the current sweep definition, edit the dimensions of the rounded corners to ***55***. ***Regenerate*** the part and zoom in on the junction of the sweep to the block. You will see small cracks on either side (Figure 10). The sweep is obviously not meeting up with the block on the rounded surfaces. There are three ways to fix this problem.

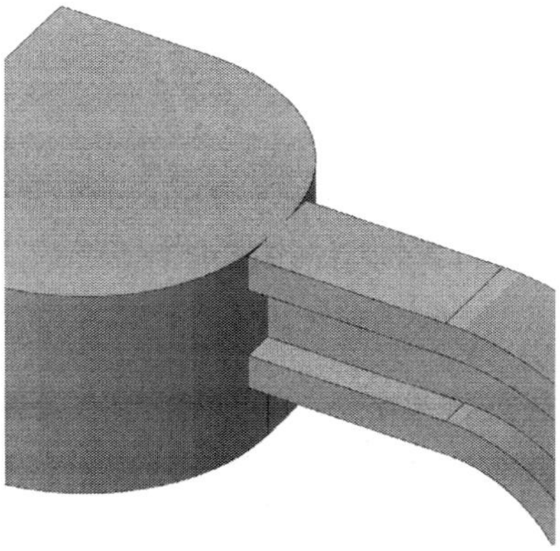

Figure 10 "Cracks" formed at junction of sweep and block

In the model tree, select the sweep and then (in the pop-up) ***Edit Definition***. This opens the sweep dashboard and shows the preview of the sweep, including the trajectory and the parameters (currently ***0***). Double click the one at the block end of the sweep and change it to ***10***. The sweep extends tangent to the existing trajectory by the new value. See Figure 11.

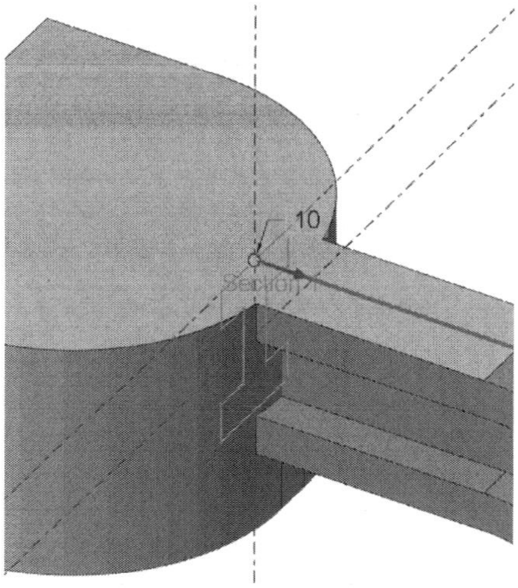

Figure 11 Extending the trajectory

The change in the trajectory "buries" the end of the sweep in the block. If you accept the feature, the cracks will be eliminated (Figure 12). This obviously wouldn't work if the block wall had a thickness less than 10 since the sweep would stick out the other side.

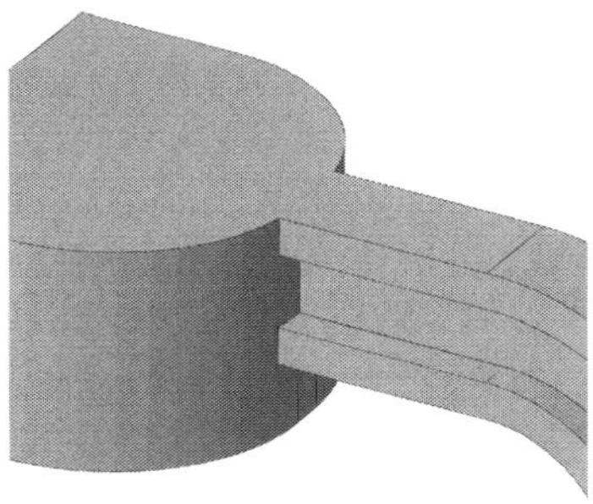

Figure 12 Cracks eliminated!

As an alternative to extending the trajectory by a specified (arbitrary?) amount, try this. Select the sweep and ***Edit Definition***. Pick the round marker at the start point and select ***Extend To...*** in the RMB menu. Pick on the vertical face on the opposite (left) side of the block. This will extend the trajectory all the way across the block (and removes the parameter value). Accept the sweep, then double click on the sweep to see where the section sketch has been moved. This definition will automatically adapt to changes in the block geometry. What do you think would happen if the left face of the block was not normal to the origin trajectory? Come back later and try it!

To return the origin trajectory to its original shape, use ***Edit Definition*** again. Use the RMB pop-up when selecting the square marker at the start point and select ***Trim At...***. Carefully hover the mouse at the original sweep start location and use pre-selection to pick the short straight section of the edge between the two rounds. The sweep origin trajectory is now defined the same as it was before, with the two cracks.

A third method to deal with the cracks is as follows. Open the ***Options*** panel and check ***Merge Ends*** (notice the contents of the pop-up tool tip - exactly what we are looking for). Accept the feature[2]. This is perhaps the cleanest way to deal with the junction of sweeps and solids[3]. Find out where the section sketch is.

Alternate Method #2 for Creating Sweep with Curve "On-the-fly"

Previously, we had a sketched curve that was selected for the origin trajectory either just before or after launching the sweep tool. This leaves the sketched curve in the model tree as a separate feature. You may have wondered if the curve could be created within the sweep itself. This would be "on-the-fly" as we did for make datums. Indeed it can.

Suppress the current sweep and sketch. Now launch the ***Sweep*** tool again; the first thing required is a curve for the origin trajectory. At the right end of the dashboard, launch the ***Sketch*** tool in the **Datum** group. Create the sketch of Figure 5 again. When the sketch is accepted it will automatically become the origin trajectory. Complete the definition of the sweep as before. When you are finished, check out the configuration of the sweep in the model tree. Where has the trajectory sketch shown up?

Another way to get this layout in the model tree is as follows. **Delete** the sweep we just made (with the curve "on-the-fly"). ***Resume*** the previous sketched curve and sweep. In the model tree, left click on the sketch and drag it onto the sweep. It will appear in the

[2] If this doesn't work, you may have to go to ***File ➤ Prepare ➤ Model Properties*** and change the accuracy setting to use relative accuracy.

[3] Beware that in a multibody part, you cannot merge with a surface on a different body.

sweep (along with the section sketch) and like a make datum will be automatically hidden.

Save the part for future experimentation. For example, find out how large a crack can be filled in with the ***Merge Ends*** option. Come back later and try out the other option to change the sketch ***Placement Point***. You will need to insert a datum point somewhere on the trajectory before selecting this option. Can that be done on-the-fly?

Before we leave this sweep, you should note the following:

- It is not strictly necessary for the cross section to lie exactly on or touch the trajectory. If the section is offset from the trajectory at the start point, then the sweep will be offset.
- You have to be careful that during the sweep, the cross section doesn't pass through itself - this can occur when the radius of a trajectory corner is very small (relative to the section size), and the section is on the inside of the curve. Remember in a default sweep, the section remains normal to the trajectory.
- You can sweep a closed section around either an open or closed trajectory.
- The trajectory can also be formed as a three-dimensional spline.
- The trajectory does not have to be a sketched curve - any chain of edges will do (depending on the type of sweep). Try selecting the edge chain around the top surface of the block. This includes both tangent and square corners.
- The trajectory need not be formed of tangent edges. If there are corners in the trajectory, for some sweeps Creo will produce mitered corners in the solid, as shown at the right.

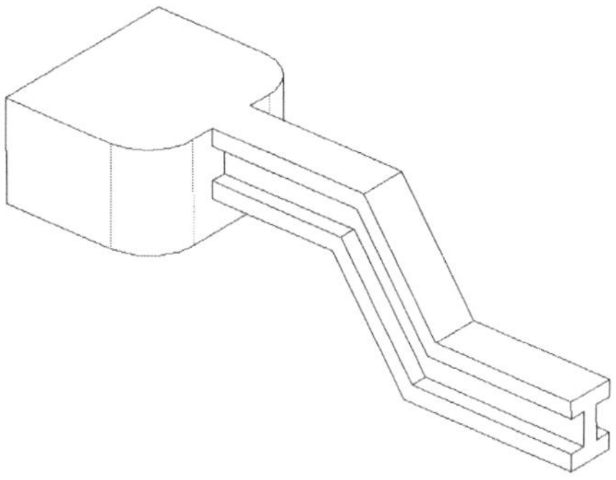

Figure 13 Mitered corners

Let's move on to the next type of sweep. Close the **s_brack** window and erase it from session.

Sweep #2: The Lawn Sprinkler

This version of the sweep command will be used to create the part shown in Figure 14. This part has four features: the hub (revolved protrusion), an arm (extruded) that is patterned, the sweep used to contour the top edge around the part, and a sketched hole down the central axis of the hub. A detailed view of the arm cross section is shown in Figure 15 showing the stepped contour created with the sweep.

Start a new part called **sprinkler** using the mmNs template. Create the hub using a revolved protrusion. The sketch plane is FRONT, and the sketch is shown in Figure 16.

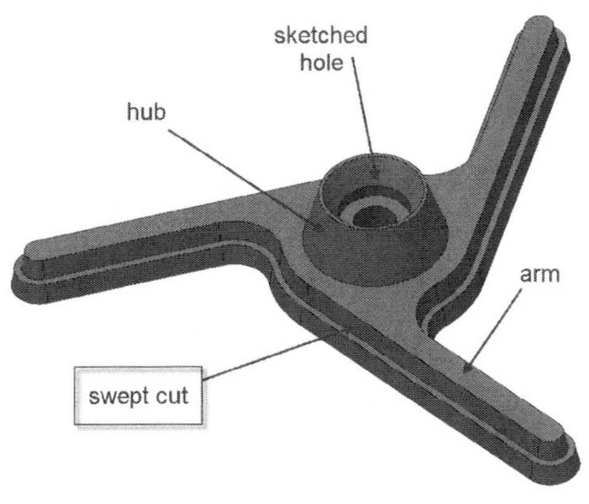

Figure 14 The Lawn Sprinkler

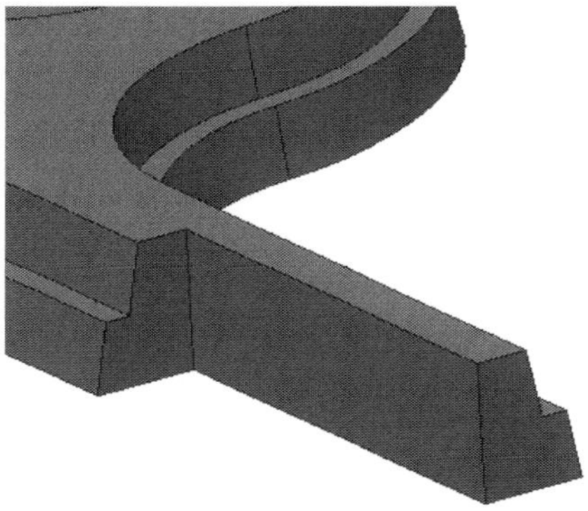

Figure 15 Close-up of lawn sprinkler cross section

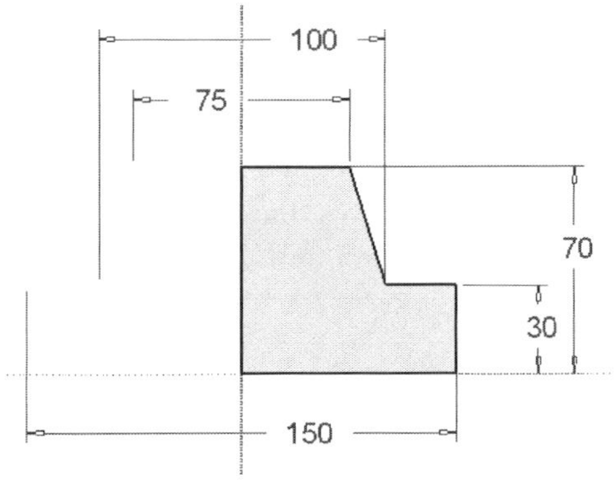

Figure 16 Sketch for the sprinkler hub (constraints off)

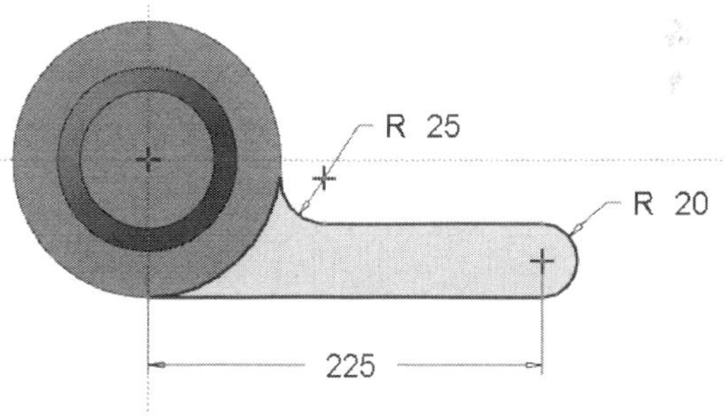

Figure 17 Sketch for the first sprinkler arm (pattern leader) (constraints off)

Now create the first arm using ***Extrude***. The sketch (on **TOP** datum plane) is shown in Figure 17. Use the ***Project*** command in sketcher to select the outer circular edge of the hub. Make sure you have ***Tangent*** constraints where the arm meets the hub - these will be important for the sweep that is coming later. For the depth of the extrusion, pick ***To Reference*** from the RMB pop-up menu and pick on the middle horizontal surface of the hub.

Creating an *Axis* Pattern

We will now create a pattern of arms using a different variation of the ***Pattern*** command. This is a radial pattern and in a previous lesson we created this by using a datum plane through the pattern axis at an angle to a fixed reference. We created a *Dimension* pattern by incrementing this dimension. An ***Axis*** pattern is a bit quicker to create, it does not require a specific dimension to increment to make the pattern, and will work fine for simple radial patterns, although it does not have the extreme flexibility offered by a dimension pattern. You can think of an **Axis** pattern as a multiple-instance rotated copy.

Highlight the existing arm and in the pop-up menu select ***Pattern***. In the dashboard pull-down list, change the pattern type from **Dimension** to **Axis**. The dashboard changes. The first thing to do is to specify the pattern axis. This can be either a datum axis, feature axis (like a hole), or any straight edge in the model. Pick the axis of the hub.

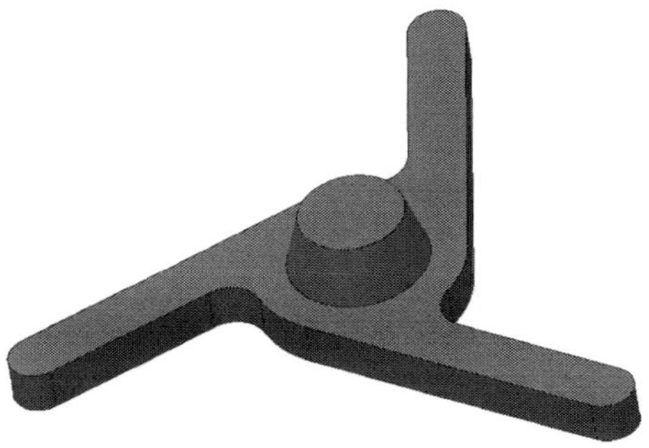

Figure 18 The pattern of arms

In the **1st Direction** we have two ways to specify the pattern: the number of members and the increment between members (default 4 and 90°, respectively), or the number of members and the total included angle. We will do the latter here - set the number of members to 3 and select ***Angular Extent*** that activates the total angle setting (default 360°). The members will automatically be evenly spaced. Axis patterns can create arrays of members by defining a **2nd Direction** radially outward. We will not need that here.

Our pattern definition is completed, but before leaving the dashboard, check out what is available in the **Options** pull-down panel. The usual options for creating patterns (*Variable*, *General*) are there, referring to restrictions on the shapes and references of the members. The other option controls whether the members rotate as they are created. The default, as we want here, is that the orientation will change, following the rotation around the axis to create cyclic symmetry. If you want the members to maintain the same orientation relative to the part, uncheck this box. Come back later to try this out to see the effect.

Accept the pattern. It should look like Figure 18.

Now we will create the swept cut along the top edge of the arms. Launch the ***Sweep*** command from the ribbon. Make sure that ***Solid*** is selected, and also set the button to ***Remove Material***. The default will be a constant cross section. In the **References** panel select the ***Details*** button. This opens the **Chain** window, Figure 19. If you leave the default ***Standard*** button checked, you can go around the top edge of the sprinkler and (with CTRL) add all the edges. This is laborious! Instead, click the ***Rule-based*** button, make sure that ***Tangent*** is selected, and pick on one of the top edges on an arm (all tangent edges will show in orange), then ***OK***. The tangent setting is why we required the tangencies in the sketch of the arm back in Figure 17. The top edge all the way around the model will be highlighted in green, and the trajectory start point will be indicated. See Figure 20. The two arrows at the start point are to indicate the direction of the sweep along the origin trajectory, and the orientation of the section sketch. Leave the latter in the default (up) position.

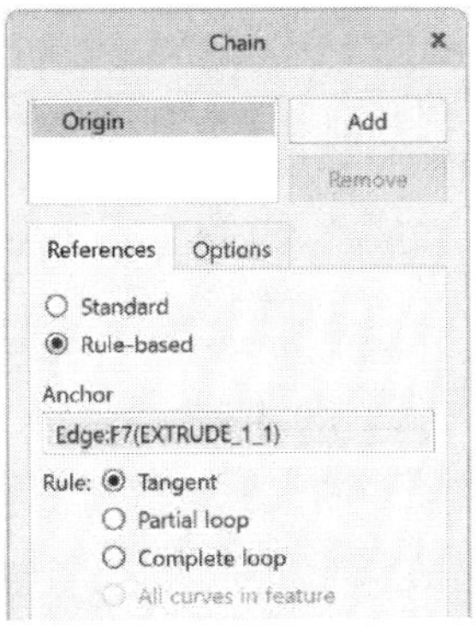

Figure 19 Creating the Origin trajectory as a tangent chain

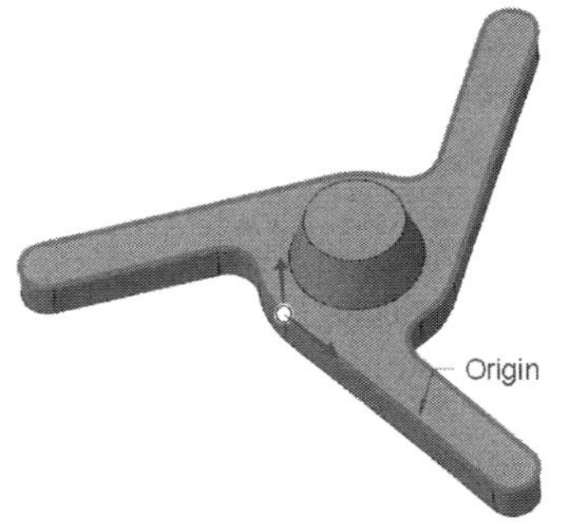

Figure 20 The Origin trajectory defined

Notice the defaults in the **References** panel (*Normal to Trajectory*). We can now define the sweep section by using the ***Sketch*** button in the dashboard, or in the pop-up. The window will re-orient so that you are looking normal to the start point sketching plane, with a cross-hair showing the origin. You might like to give the model a little spin to make sure you understand the orientation. Using ***References*** in the RMB pop-up, select the bottom surface of the part as an additional sketching reference, then create the sketch shown in Figure 21. Accept the sketch, and the cut will preview. Notice there is an arrow on the section sketch showing whether material will be removed from inside or outside the sketch. Obviously, it should be inside. The feature preview is shown in Figure 22. Accept the feature.

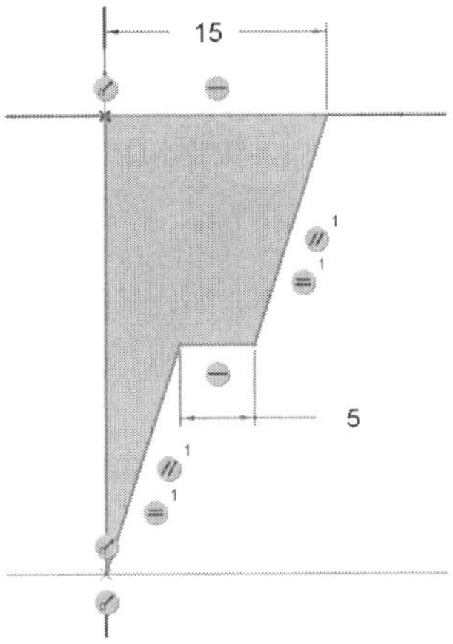

Figure 21 Section sketch for the swept cut

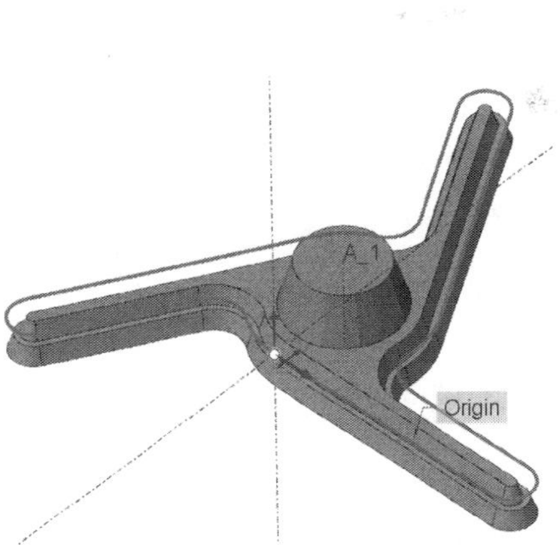

Figure 22 Swept cut previewed

Creating a Sketched Hole

Finally, create a hole on the axis of the hub. This time, instead of a straight or standard hole, we will specify a special cross sectional shape for the hole. This is called a *sketched hole.* This type of hole is essentially a revolved cut that is automatically revolved through 360 degrees. We provide the cross sectional shape or profile of the hole using Sketcher. The placement references are the same as a straight hole. Normally, users will use either a *Coaxial* placement or define a placement point. Sketched holes are handy (especially if the sketch is stored in a library) if you frequently make hole shapes with non-standard cutting tools, like combinations of tapered, stepped, counterbored, and/or countersunk geometries. If you are only going to do this once, you might be tempted to just make a revolved cut, although the placement references for a sketched hole are somewhat clearer and easier to set up.

Select the ***Hole*** tool in the ribbon. The **Hole** dashboard opens. Select the ***Sketched*** button. The dashboard changes to offer two additional options - reading an existing sketch profile (say for a specially shaped cutting tool that you use frequently), or creating a new one. Select the latter with the ***Sketch*** button. In the Sketcher window that opens up, create the sketch shown in Figure 23. You must create the axis of revolution (use the RMB pop-up), and you must also close the sketch down the centerline. When you accept the sketch, you're back in the hole dashboard. We now specify the placement references for the hole. First, select the axis of the hub for the primary reference. The **Coaxial** placement type is then automatic. Open the **Placement** panel to see this. (If we pick the placement surface first, the placement would by default become **Linear** and we could use the datum planes but this means that if the hub moved, the hole would not go with it.) Observe the **Placement** panel - it shows an additional reference is needed (the placement plane). Hold down the CTRL key while you select the top surface of the hub. Since the placement plane is the top surface of the hub, Creo will take the top edge of the sketched hole and align it automatically with the placement plane, with the axis of the hole coinciding with the axis of the hub. The hole is now previewed. Open the **Shape** panel - it shows the shape of the sketched hole.

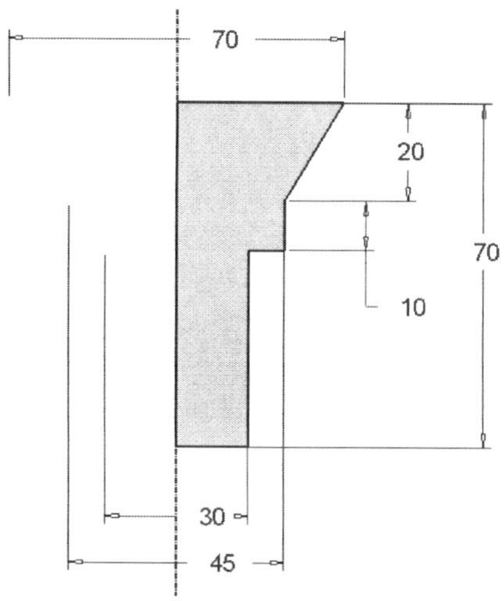

Figure 23 The ***Sketched Hole*** profile

Notice that depth options such as ***Through All*** are not available. That is why the dimension scheme for the hub was chosen (it would be easy to set up a relation between the depth of the hole and the height of the hub above the lower surface of the sprinkler).

Accept the hole feature. That completes our lawn sprinkler part. Save it now.

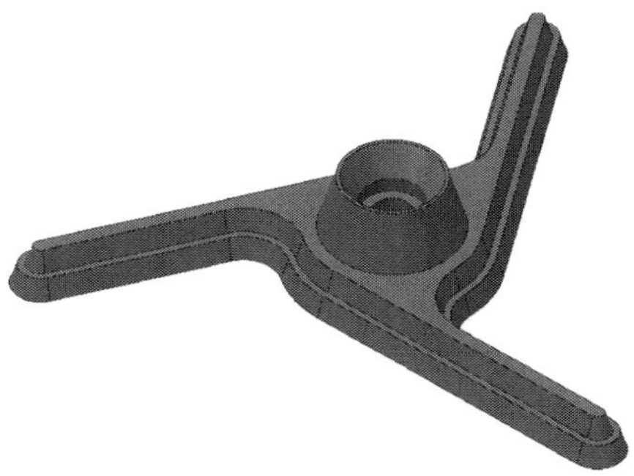

Figure 24 Completed sprinkler

So, that's the end of sweeps! As you can see, these are quite complicated, packing a lot of geometric information into a single feature. For further information on sweeps, there is extensive discussion in the on-line help. You might like to go back and edit any of the dimensions of the sweeps to see what happens. Be aware that arbitrary modifications might make the sweep illegal, so save your part before you try anything drastic. If you are using the dashboard interface, it is very easy and convenient to experiment with different options since you always have a preview of the feature on the screen - if the sweep is illegal, it will not preview.

Close the sprinkler part and remove it from the session.

Blends

A blend is like an extrusion with a changing cross section. Some systems call these "lofts." The different cross sections are specified using a number of sketches or edge chains. These can be created either inside or outside the blend feature. The distance between cross sections is then specified either with references or explicit dimensions. The dimensions can be either a linear distance (forming a *parallel blend*) or an angular distance (forming a *rotational blend*), or a combination of these (a *general blend*). You can maybe imagine what a *swept blend* would involve. In the following, we will look at the first two of these. A blend can be used to create a protrusion or a cut. Some restrictions apply:

- At least two sections are required.
- Sections can be either selected from existing sketched curves, or created from chains of edges.
- Each section must have the same number of vertices; normally this means the same number of line (or arc) segments. This rule can be overridden (for example a blend from a square section to a triangular section) using a *blend vertex* (see the on-line help for information on this).
- Each section has a starting point (one vertex on the sketch) - these must be defined to align all the sections properly or else the resulting geometry will be twisted.
- Parallel blends can use projected sections (obtained by projecting existing sketches or curves onto different planes).
- For a rotational blend, the axis of rotation is common to all sections, and all section planes must intersect the axis.

The sections of the blend can be connected either with *straight* (i.e. ruled) surfaces (see Figure 25)[4], or with *smooth* surfaces (Figure 26).

---

[4] Note that this does not mean *flat* surfaces. A ruled surface is one where from every point along the curve defining one section there exists a straight line or edge to a matching point on the adjacent section.

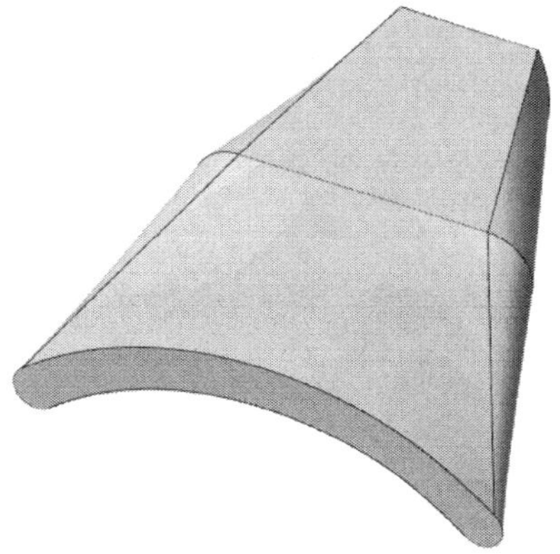

Figure 25 A *Straight* parallel blend

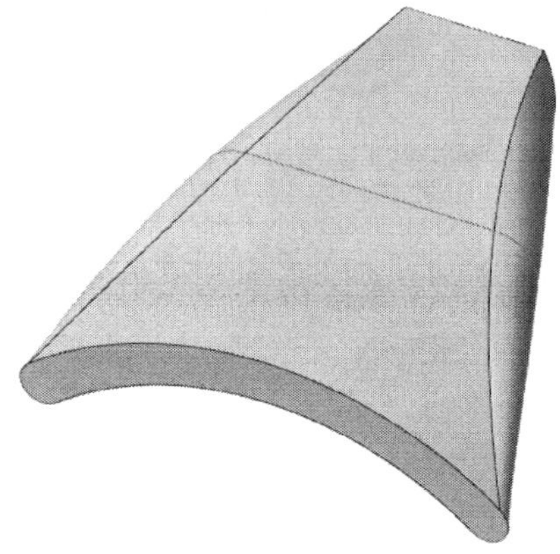

Figure 26 A *Smooth* parallel blend

Parallel Blend

This is the simplest form of a blend. We will create the part shown in Figures 25 and 26. This blend has three sections as seen in Figure 31 created using separate sketched curves on three permanent datum planes, but make datums would do just as well. This construction method has the advantage of making it easier to visualize the final blend shape.

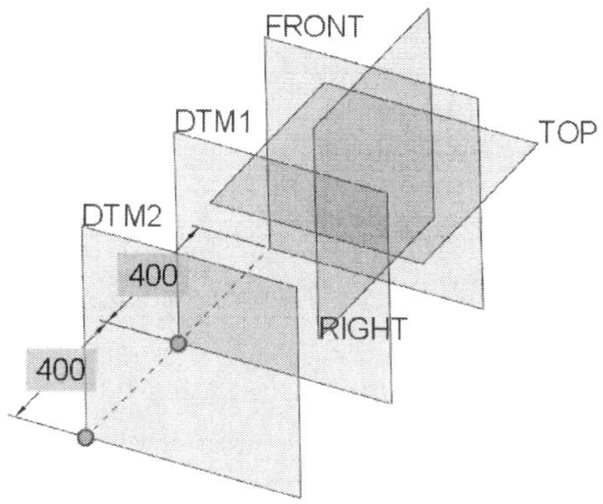

Figure 27 Datum planes to define section sketches

Start up a new part called **blend1** using the mmNs template.

Create a couple of datum planes using the ***Plane*** tool. For the first, click on the FRONT datum. The default is an **Offset** datum, so just enter a value of **400**. Select ***OK*** or middle click.

With DTM1 the active feature (green highlight), you can create DTM2 with two mouse clicks:

Plane

then middle click (since ***Offset*** and **400** are defaults).

Now create our three sketched curves that will define the blend sections. Select ***Sketch*** and pick FRONT as the sketching plane. The default RIGHT reference is fine. The first sketch is shown in Figure 28. The sketch is symmetrical about RIGHT and the lower edge is along TOP. Notice that the sketch has four vertices. This is important, since each section in the blend must have the same number of vertices.

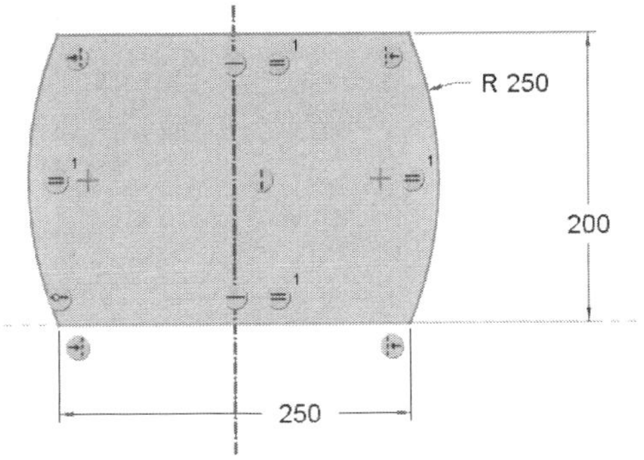

Figure 28 First sketch for blend (on FRONT)

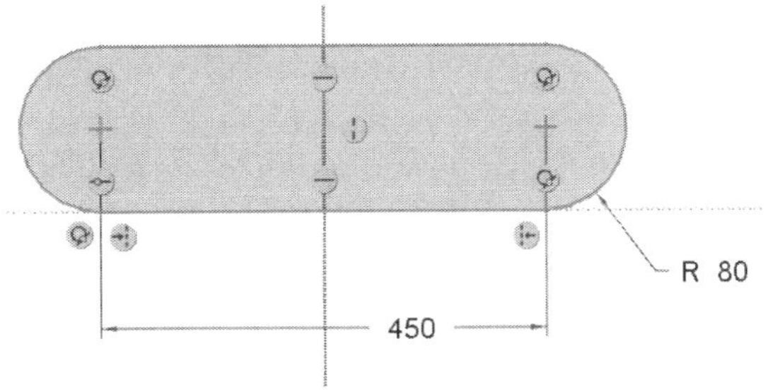

Figure 29 Second sketch for blend (on DTM1)

The second sketch (Figure 29) is created on the second datum plane DTM1. It is also symmetric about RIGHT, and contains four vertices. Note the lower edge is on TOP.

Finally, create the third section sketch on DTM2. This sketch is shown in Figure 30. Again, it has only four vertices. When complete, your model should look like Figure 31.

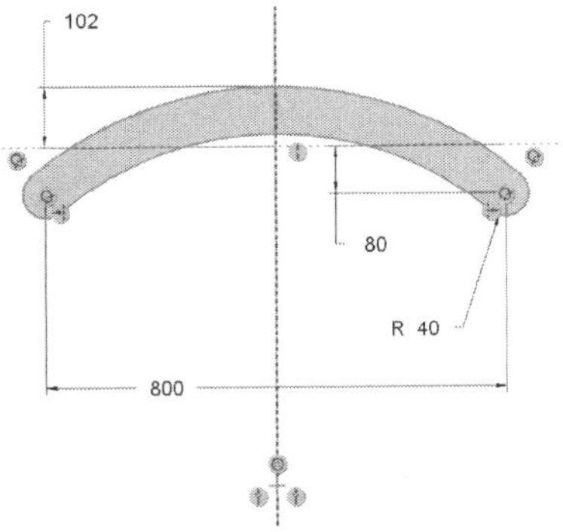

Figure 30 Third sketch for blend (on DTM2)

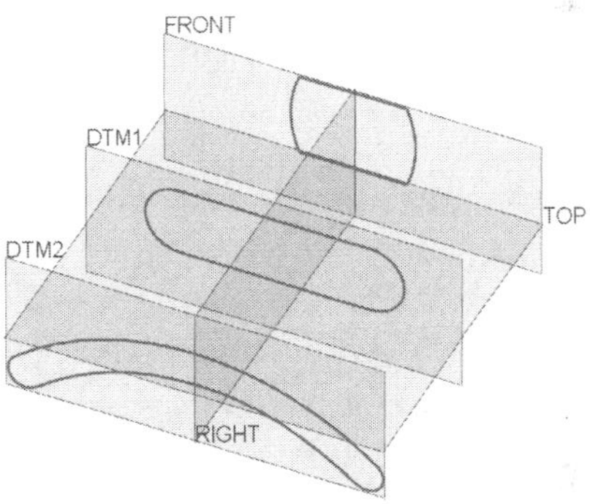

Figure 31 Three curves to define blend sections

Now we can launch the Blend command. In the **Model** ribbon, **Shapes** group, select

Blend

This opens the **Blend** dashboard. The four buttons in two columns on the left in the dashboard are:

- ***Solid*** (default)
- ***Surface***
- ***Sketched Sections*** - create the sections using sketches inside the blend feature
- ***Selected Sections*** - use previously created sketches to define the sections

Leave ***Solid*** selected and pick the ***Selected Sections*** button now. Pick on the sketched curve on the FRONT datum. It will highlight in bold green and a white dot with a magenta arrow will appear on one of the vertices. This is the *Start Point* of the section. It is important that the start points of each section are chosen correctly, as we will see in a moment. Click on the start point and drag it to the top left vertex, as shown in Figure 32. We want the direction arrows on each section to go around the sketch in the same direction - in this case clockwise. If it is pointing the other way, just click on it (or use the RMB menu and select ***Flip Chain Direction***).

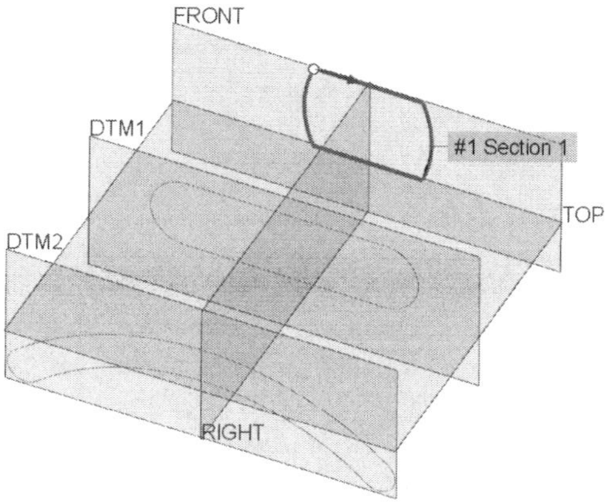

Figure 32 Blend section #1 selected. Note the position of the Start Point.

Open the **Sections** panel in the dashboard and select ***Add*** (or use ***Add Section*** in the dashboard) Now click on the next sketch (on DTM1). It will highlight in bold green, and if it is legal the blend between the two sections will preview. Once again, a white dot will appear on one of the vertices of the sketch. We need the start point to again be on the top left vertex in the section, as shown in Figure 33, to align the sequence of vertices in the second section with those in the first. If the vertex is not at the proper vertex, the blend will be warped - just drag the white dot to the desired vertex, and click on the magenta arrow to make sure the sequence direction is correct.

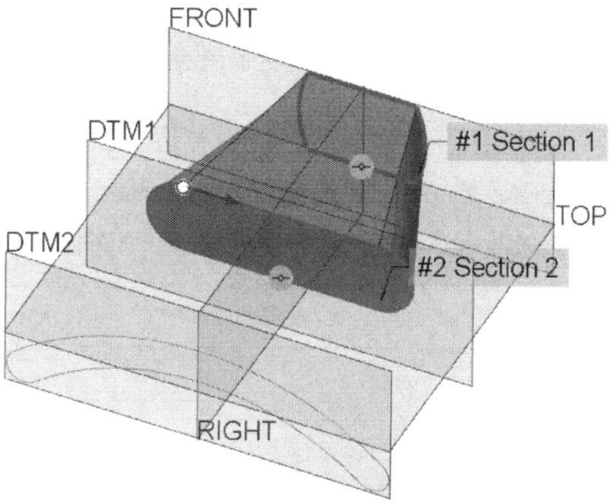

Figure 33 Blend section #2 selected

Once again select ***Add Section***. Click on the third sketch (on DTM2). The sketch highlights in green. As before, if the start point is not on the right vertex we will get a twisted shape (Figure 34). If so, drag the start point to the correct vertex, as in Figure 35.

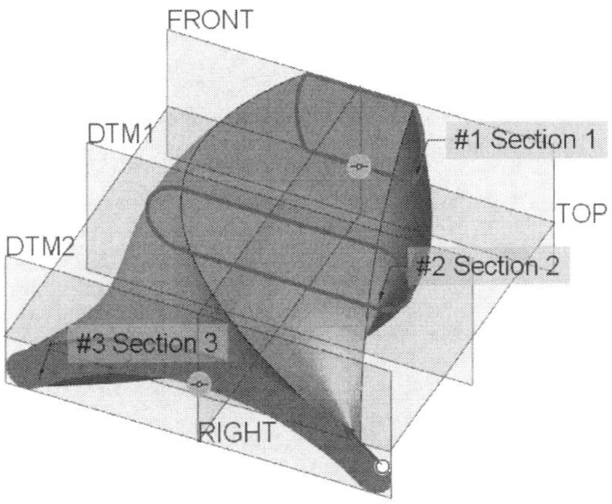

Figure 34 A warped section caused by misplaced start point

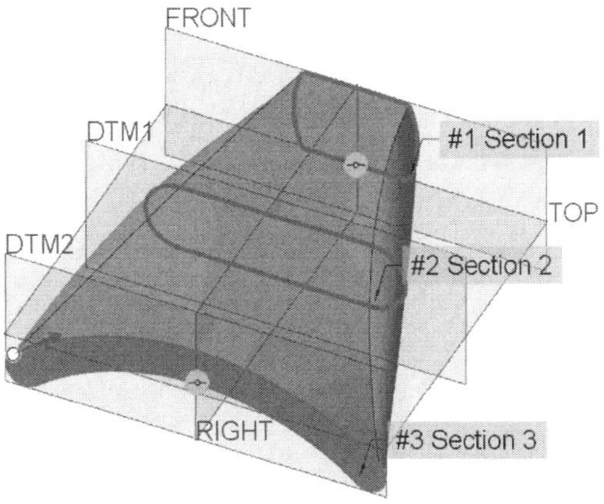

Figure 35 Blend section #3 with correct start point selected

Accept the feature with a middle click; the blend will appear as in Figure 26 above. In the model tree, note the icon used for a blend feature. Select the feature and in the pop-up select ***Edit Definition***. As usual, this brings us back to the feature dashboard.

Open the **Sections** panel in the dashboard. This lists the three sections (which highlight on the model if you select one in the list) and the number of vertices in each section. In the **Options** panel, you can select between a **Smooth** blend, or a **Straight** blend (Figure 25). The **Tangency** panel gives access to controls for how the blend will merge with adjacent geometry at each end of the blend[5].

You might like to try to edit the dimensions of the blend. These are driven by the three sketched curves and the datum planes. Selecting these and using the ***Edit*** command in the pop-up will let you change blend dimensions.

The *Shell* Command

Just for fun, here is a feature creation command we haven't mentioned before. Pre-select the blend feature (edges highlighted in green) and in the **Engineering** group, pick the ***Shell*** tool Shell. This defaults to hollowing out the highlighted object (shelling it) using a default thickness for the remaining shell wall (see the dashboard). A flip button allows you to create the shell on the inside or outside of the existing model surfaces.

In addition to shelling the model, we want to remove the two end surfaces. The **References** panel contains two panes. The **Removed Surfaces** pane is selected so pick on the front and back surfaces (using CTRL) as shown in Figure 36. These will highlight

[5] Move the insert point before the blend, and create two extrudes using the first and third sketches (extrude from sketch #1 going back, sketch #3 coming forward). Resume the blend and experiment with the tangency option to the extrudes coming off each end of the blend.

in green as they are picked. The interior faces are shown in preview orange. Change the shell thickness to ***10***. The toggle beside the thickness value lets you create the shell on the outside of the current solid. Try it! The other pane in the **References** panel allows us to set different thicknesses for different regions of the shell. Accept the shell feature. The part looks like Figure 37. Open the model tree and observe the icon used for the shell.

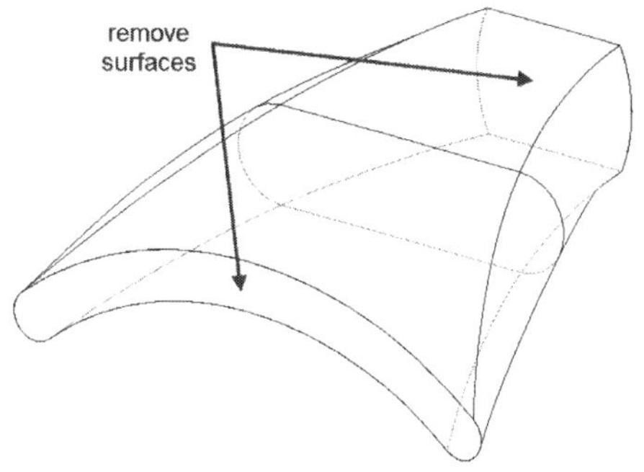

Figure 36 Surfaces to be removed during Shell creation

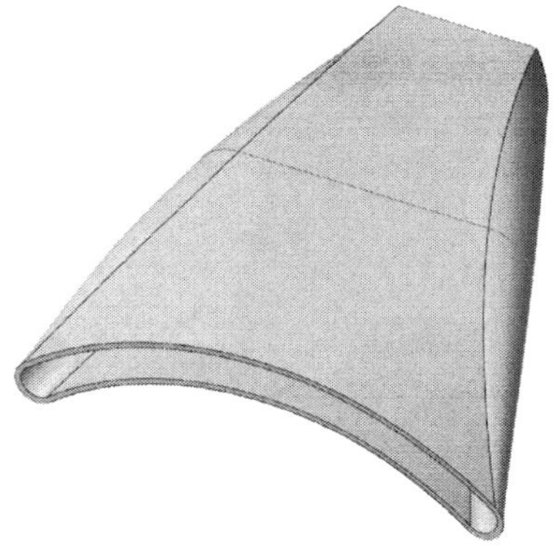

Figure 37 The *Shell*ed part

The ***Shell*** command can handle quite complicated geometry but is not foolproof. You may find that the command will fail sometimes if the geometry is too complicated or physically impossible (typically if a shelled surface has to pass through itself). Normally, the shell command is used fairly early in the model before this situation can develop. Settings in the **Options** panel can often solve these problems.

Let's move on to our last new feature in this lesson. Save the current part (for later experimentation) and remove it from session.

Rotational Blend

A rotational blend is set up by specifying the cross sections on a number of sketching planes that have been rotated around a common axis - thus the sections are not parallel. We are going to make the part shown in Figure 38. Rotational blends can create either protrusions or cuts, and can be either solid or surface features. The usual restrictions apply as to the number of vertices and the start point defined on each section. For some rotational blends, consecutive sections can be no more than 120 degrees apart. The spacing between consecutive sections does not have to be the same. Sections can either be selected from existing sketches (as we will do here), or sketched internally in the feature.

Start a new part called **blend2** using the mmNs template. You can delete the default coordinate system. It will be handy to turn on the display of the datum plane and axis tags.

We are going to create the four rectangular sketches shown in Figure 39. The first sketch will be on FRONT. The next three will be on datum planes created using angle offsets to the previous sketch plane. These datum planes share a common axis.

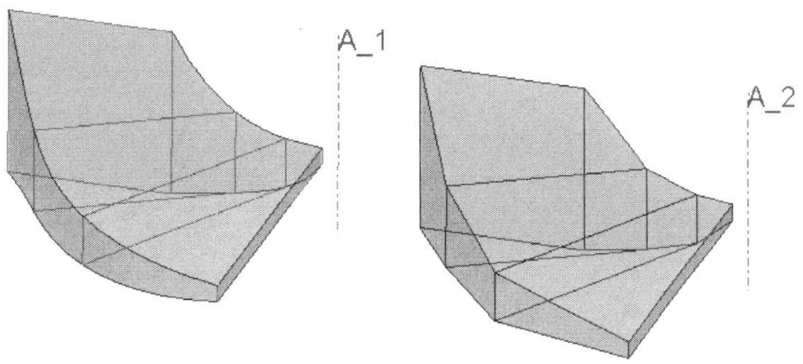

Figure 38 Two rotational blends: smooth (left), and straight (right)

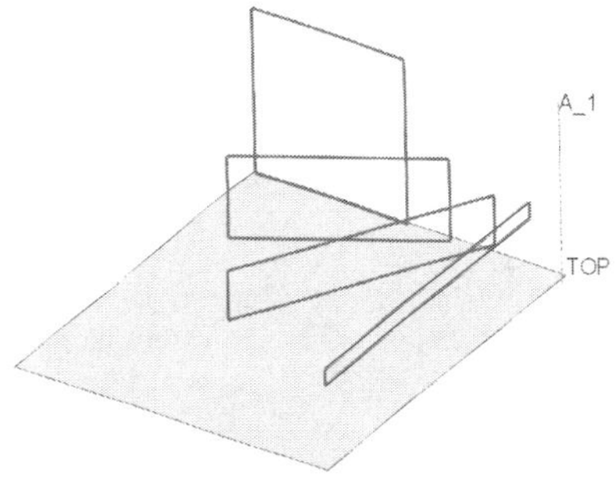

Figure 39 Completed section sketches for rotational blend

Create the axis A_1 by selecting (with CTRL) the RIGHT and FRONT datum planes, then the ***Axis*** tool in the **Datum** group (or pop-up menu).

Create DTM1 by selecting the axis A_1 and FRONT, then ***Plane***. Set an offset of **30**. This will be counterclockwise looking down from the top - you may have to enter –30 to get the angle you want, with the positive side of the new datum facing more-or-less to the front.

Create DTM2 by selecting the axis A_1 and DTM1, again with an offset of **30**. Finally, create DTM3 by selecting the axis A_1 and DTM2, with an offset of **30**. This will line up with the RIGHT datum, which you might as well ***Hide*** now to avoid confusion.

We have set up the section planes this way so that we can easily adjust the angle between successive sections. If you use a parameter and some relations, you can set these all equal, or make them some function of other geometry in the part. Or, of course, you could use a dimension pattern of datum planes.

Create the four sketched curves shown in Figures 40 through 43 on the datum planes indicated. In each sketch, the horizontal reference is TOP, and the second sketching reference is the axis A_1. If the screen gets a little cluttered after the first couple of sketches, go ahead and hide them. You can imagine that using references to existing geometry in the new section sketches could complicate matters here if we want to make a very robust feature.

Launch the ***Rotational Blend*** command in the **Shapes** group. The dashboard opens, looking very similar to the dashboard for the parallel blend we did earlier. There is one new button for specifying the rotation axis. Pick the button for ***Selected Sections***.

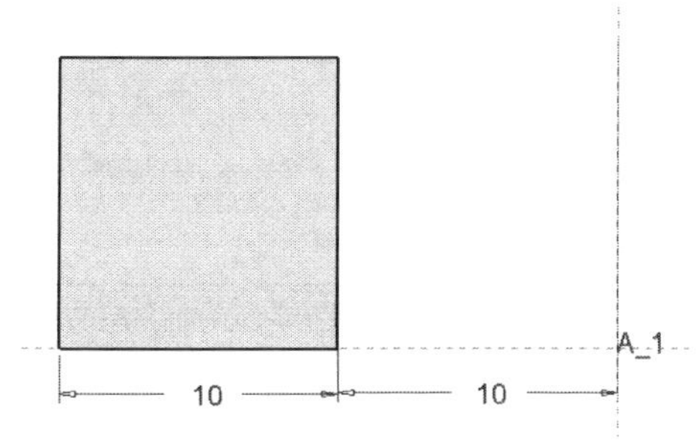

Figure 40 Section 1 on FRONT (constraints off)

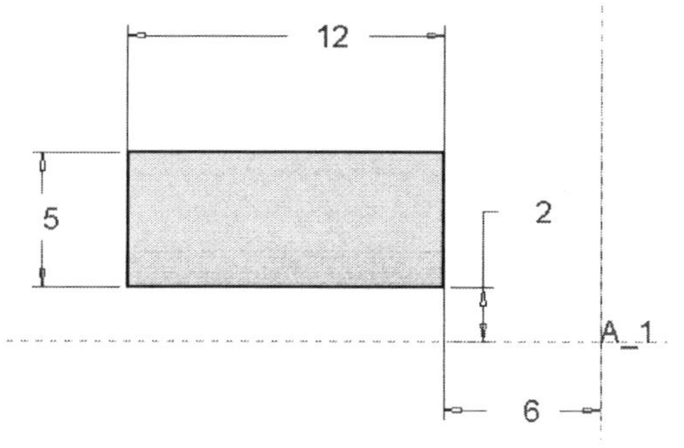

Figure 41 Section 2 on DTM1 (constraints off)

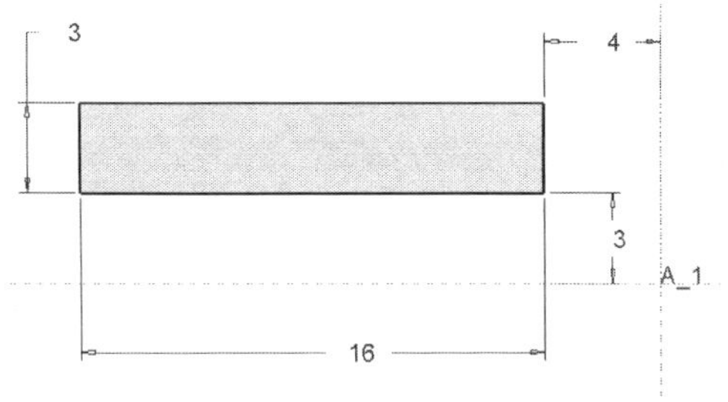

Figure 42 Section 3 on DTM2 (constraints off)

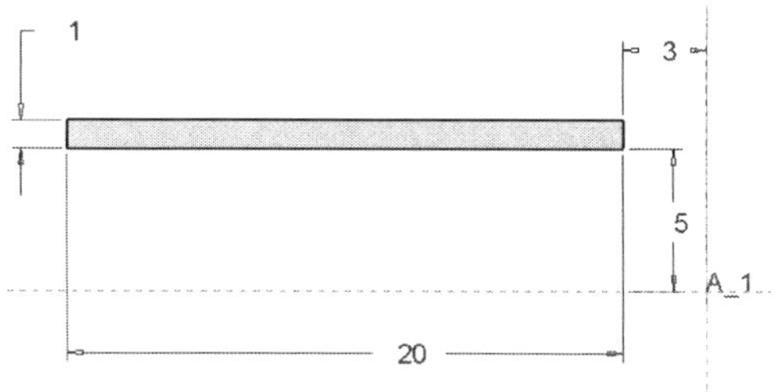

Figure 43 Section 4 on DTM3 (constraints off)

Select the first sketch (on FRONT). Observe the start point, and drag it to the top left corner if necessary, as shown in Figure 44.

Now select ***Add Section***. Pick on the second sketch. The sketch will highlight and the blend will preview as in Figure 45. If the start point is on the wrong vertex, the blend will be twisted.

Continue using ***Add Section*** to select the third and fourth blend sections, making sure the start points all line up. The final previewed feature is shown in Figure 46.

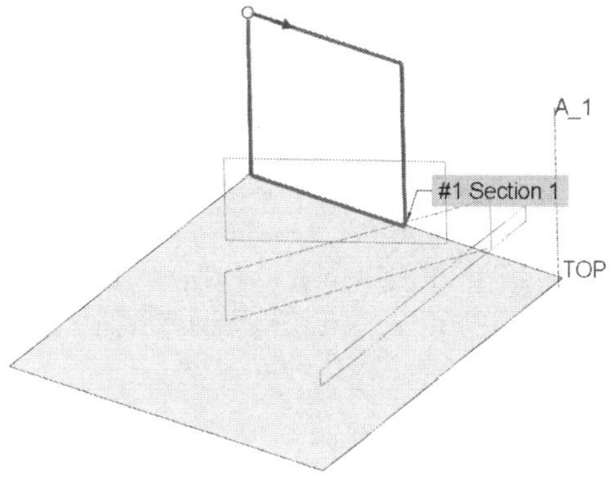

Figure 44 First section selected

Notice that the rotation blend axis was chosen automatically in this case. For some options of the command you must select that yourself, in which case you can use any axis or straight edge. Of course, you only do this once, since the same axis is used for all sections. If you open the **Sections** tab, you will see the same options we saw before (radio buttons for **Sketched** or **Selected** sections, listed sections, number of vertices, etc.). In the **Options** tab, you can select either ***Smooth*** or ***Straight*** blends. There is an additional option here for connecting the end and start sections of the rotational blend[6]. This creates a closed feature. Try it! The **Tangency** options operate the same as the parallel blend, and the **Properties** tab allows you to specify a name for the feature.

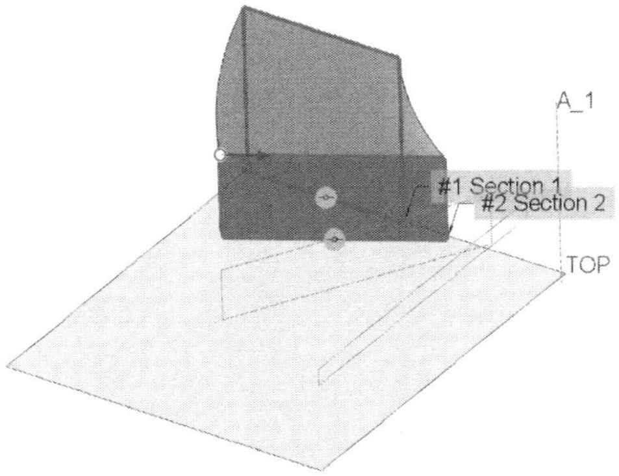

Figure 45 Second section selected

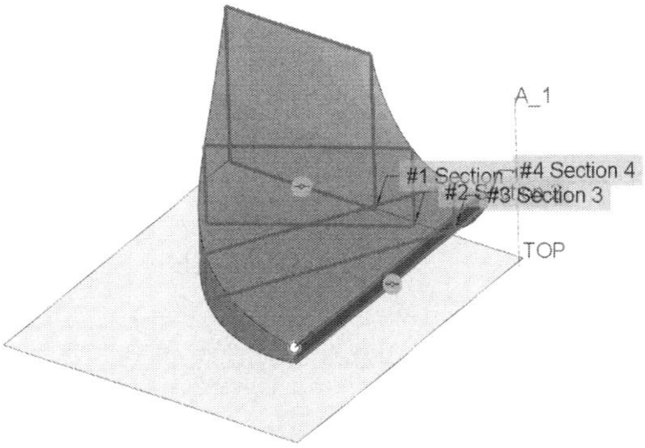

Figure 46 Final section selected

Accept the feature and close the dashboard. Experiment with the dimensions of the feature (the datum plane offsets, sketch dimensions, etc.). The variety of geometry you can create is quite remarkable (Figure 47), including features that can pass through themselves[7].

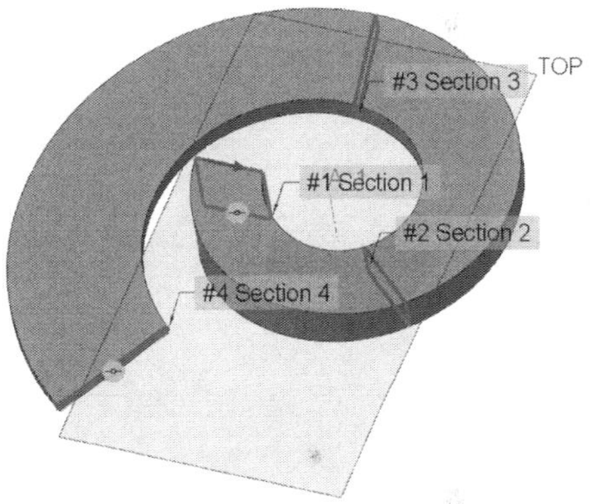

Figure 47 Sections at 140° offset with some section dimension changes to prevent overlap

That completes our limited presentation of blends. As you can see, blends contain a lot of geometric information and are therefore a bit more difficult to set up. However, they offer considerable flexibility and can create very complex shapes not attainable with the simpler features. There are advanced features (swept blends and helical blends, for example) that offer even more complexity/flexibility. Consult the on-line help for information about these.

---

[6] This may not work for all possible blends. For example, in this case the ends cannot be connected if you select a **Straight** blend.

[7] If you would like a challenge some time, try to create a Möbius strip.

Conclusion

Well, we have reached the end of this series of Creo Parametric lessons. We have gone over the fundamentals of creating basic parts, assemblies, and drawings. Much of the material has been presented only once. It is likely that you will have to repeat some of these lessons to get a better grasp on Creo, and it is almost certain that you will need much more practice to be proficient. In some instances, we have only scratched the surface of Creo functionality and it is up to you to explore deeper into the commands and options. The more you know and are comfortable with, the easier it will be to perform modeling tasks. You may find that you will also begin to develop a different way of thinking about part design. As your modeling tasks get more complex, the need to plan ahead will become more important.

You should also remember that what we have covered is only the first step (remember the words of Lao Tzu from Chapter 1!) in the integrated task of design and manufacturing. From here, you can use Creo to head off in a number of directions: engineering analysis using Finite Element Modeling, mechanism kinematics and dynamics, manufacturing analysis, mold design, sheet metal operations, piping layout, and much more. Good luck on your journey of a thousand miles and have fun!

Questions for Review

1. What happens in a *Normal to Trajectory* sketch if the trajectory contains a mix of tangential and non-tangential connections?
2. How many of the vertices of the swept section have to align with or be on the sweep origin trajectory?
3. What problem may arise if the swept section is "large" and the sweep trajectory has a "small" radius arc in it?
4. What happens if the sweep trajectory has discontinuities (kinks) in it rather than being composed of smooth "tangential" transitions?
5. What is the difference in the model between creating the trajectory curve sketch first and then launching the sweep tool, or doing these in the reverse order?
6. When first entering the Sketcher window to define the section for a sweep, it is often difficult to understand the orientation of the view. How can you determine the location and orientation of the section with respect to the trajectory?
7. Find out what happens if you put the section sketch point in the middle of an open trajectory.
8. In the exercises above, the swept section was normal to the trajectory. Is it possible to create a sweep where the section is oriented at an angle to the trajectory?
9. What does the parameter at each end of the origin trajectory do? What is an alternative, and what does it do?
10. Can you change the geometry of the trajectory independently of the geometry of the section?
11. What happens if you try to sweep an open section?
12. What is the difference in the appearance (both wire frame and shaded) in the graphics window between a sweep made using the *Solid* and *Surface* options?
13. Can you have multiple non-overlapping closed sections in a single sketch for a sweep?
14. Can a sweep trajectory intersect itself (like a figure eight)?
15. What are the essential common characteristics of all sections in a parallel blend?
16. When creating a blend, what is meant by the "start point" of the sketch?
17. What is meant by a ruled surface?
18. In a straight, parallel blend must the sections all be centered on a common axis or point?
19. Are all parallel blends symmetrical?
20. In the model tree, can you drag the section sketches into the blend feature, as was done for the sweep?
21. What are the essential common characteristics of all sections in a rotational blend?
22. How can you change the dimensions of a section in a sweep or blend?
23. What happens if you try to delete one of the sections in a blend?
24. How can you easily edit the section dimensions in a parallel or rotational blend?
25. Can a rotational blend loop back on itself? What is a ***Closed*** rotational blend?
26. How do you change a straight blend into a smooth blend?
27. What does ***Flip*** do in the **Shell** command?

Exercises

Here are some parts to try out using the commands in this lesson.

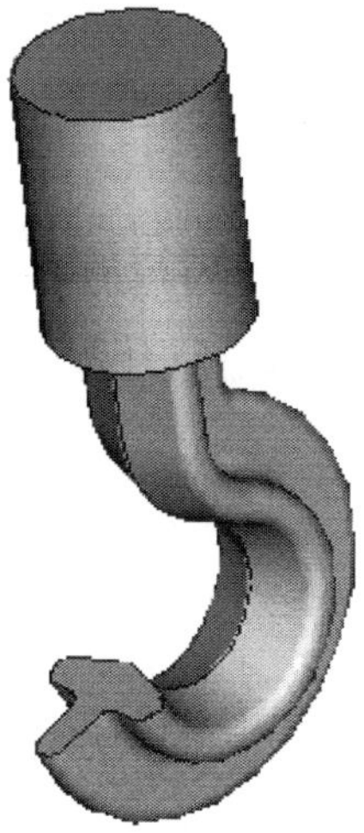

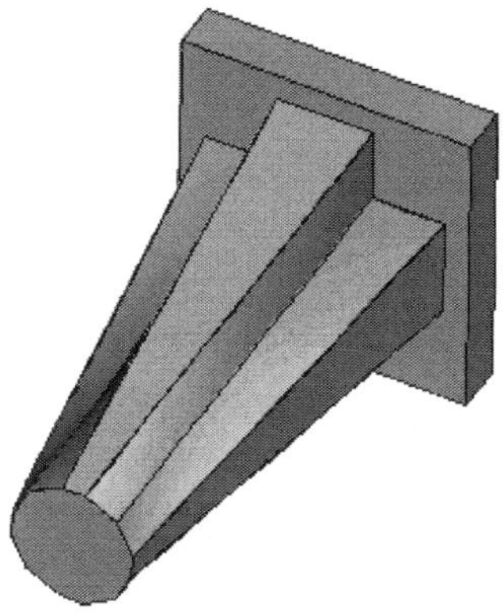

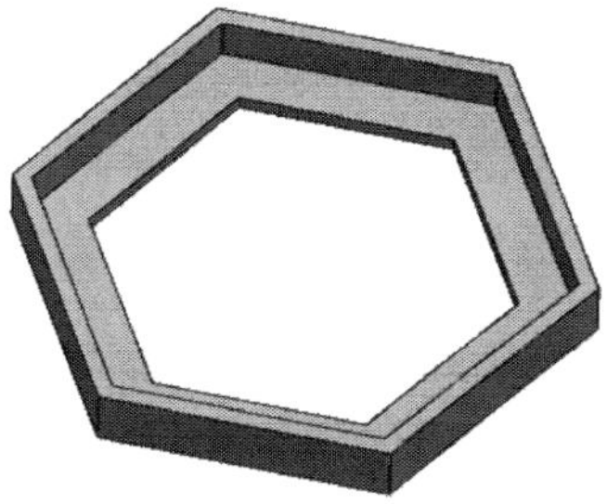

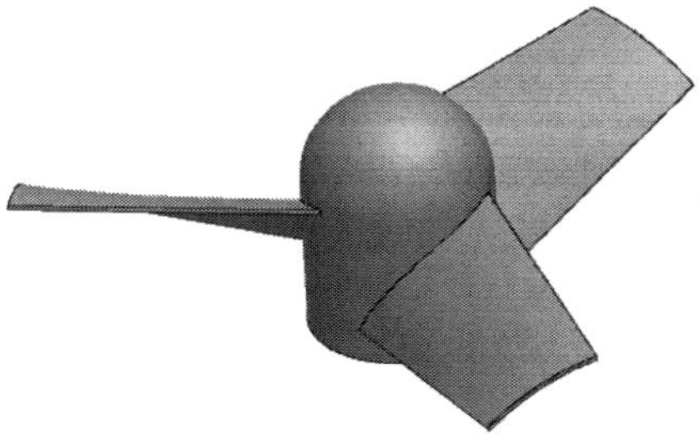

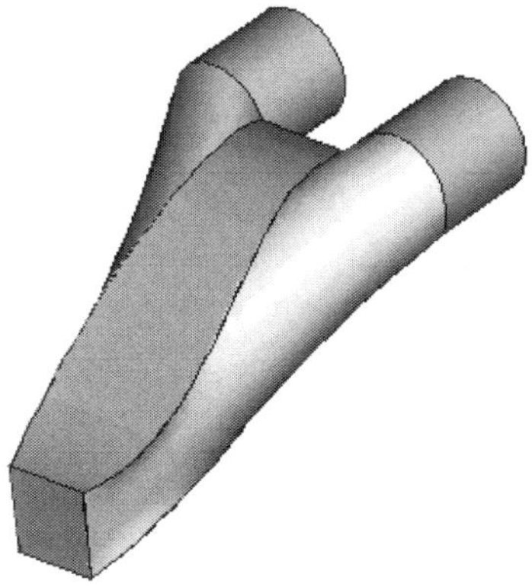

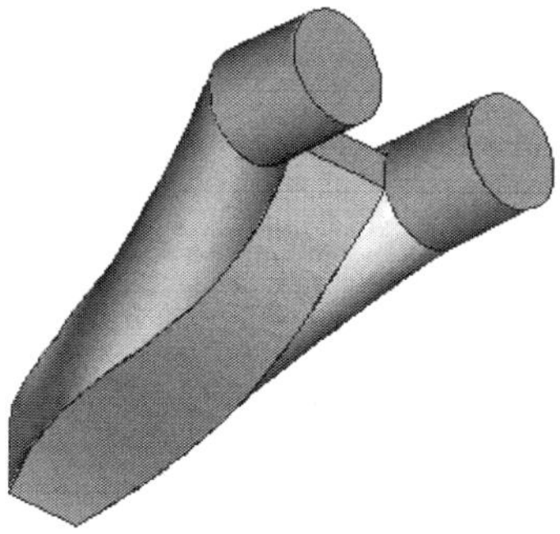

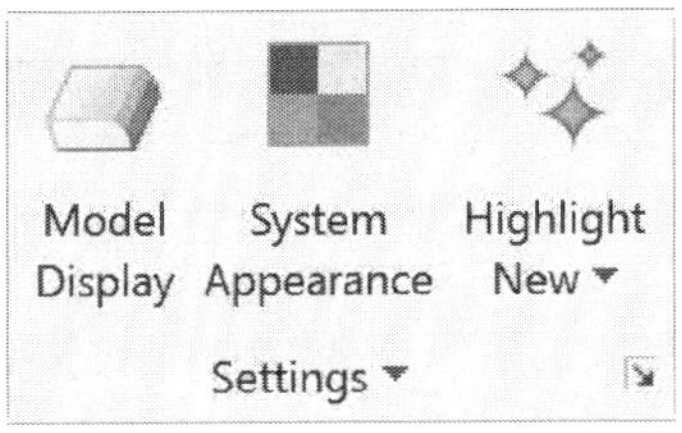

Appendix

Creo Parametric Customization

Synopsis

Configuration settings and *config.pro*; customizing the screen toolbars and ribbons.

Overview

The Creo Parametric interface is infinitely customizable - colors, appearance, ribbon contents, display style, toolbars, and more. By now, you should be familiar with the commands and environment settings that are available in the graphics and Quick Access toolbars and the ribbons. These aspects of the Creo working environment (and much more!) can be controlled using settings stored in configuration files (*config* files for short). Creo has several hundred configuration settings. All settings have default values that will be used if not specifically set in a *config* file. You may also find that you use some command buttons much more frequently than others, and you would like to have them closer to hand. In that case, they can be placed in either of the toolbars. Or, you can create your own ribbon tabs and groups, and populate them with whatever commands you like.

This appendix introduces some of these tools for customizing your Creo working environment. The most important of these is the use of a configuration file, called *config.pro*, that contains settings that control the operation and appearance of the program. This file is automatically read at start-up. In addition, many settings can be changed during a session, with the option to save these in the configuration file for use in a later session. We will only scratch the surface here. Additional information is available in the on-line help.

Configuration Settings

Launch Creo, or if it is already up erase everything currently in session and set your working directory to your normal start-up directory. You should now be looking at the **Home** ribbon, with nothing loaded.

The **Settings** group in the **Home** ribbon is shown in Figure 1. This group is available only in the **Home** ribbon. Select one of these commands ***Model Display***, ***System***

Appearance, or Highlight New Functionality), or the arrow in the lower right corner, and you will see that each is going directly to one of the option areas in the same **Creo Parametric Options** window. At all other times (i.e. when you have a part or assembly loaded), this window is available by selecting

File ➤ Options

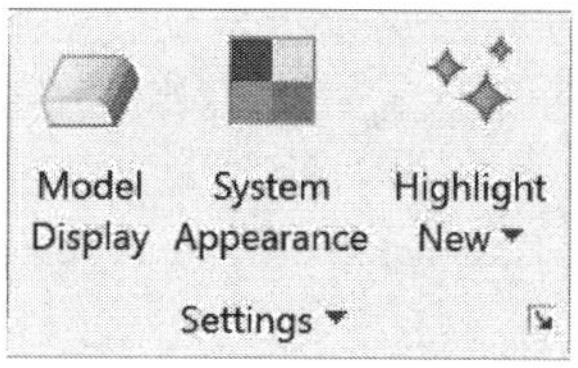

Figure 1 The **Settings** group in the **Home** ribbon

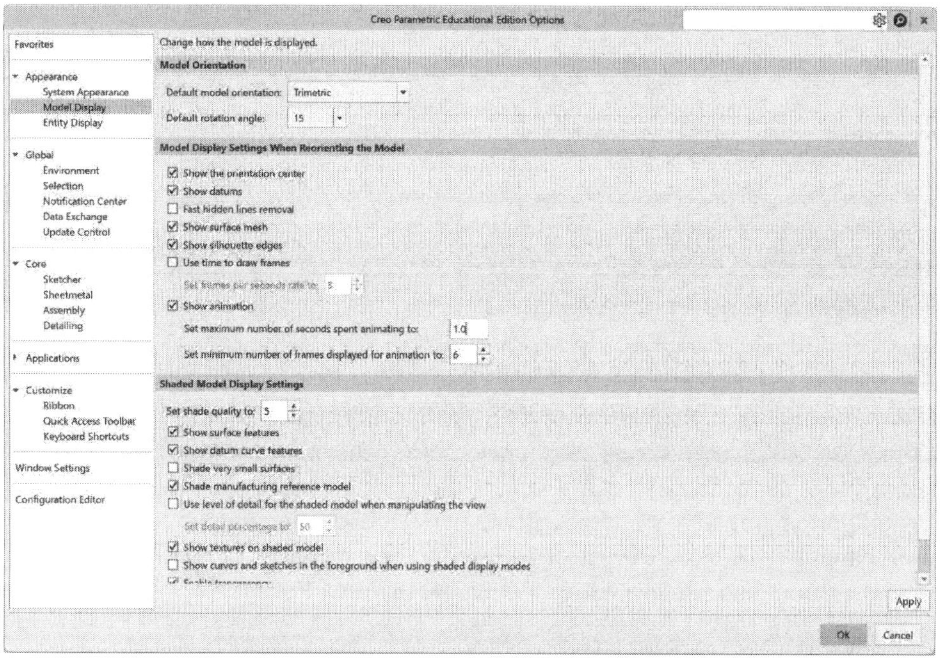

Figure 2 The **Model Display** panel in the **Creo Parametric Options** window

The **Options** window is the main location for most of the customization tools. Select the **Model Display** category as in Figure 2. The panel on the right shows all the settings related to the chosen category. Have a look at some of the other categories and options available.

Any setting changes made will be valid for the current session unless you explicitly save them. This is done by looking at the bottom of some of the panels for the ***Import/Export*** button—this allows you to read in a file (***Import*** a file) of previously created settings or write a file (***Export*** a file) for use in future sessions.

For example, go to the top menu ***File*** pulldown menu and enter ***File ➤ Options ➤ Configuration Editor***. A window will open like the one shown ahead in Figure 4. On the bottom right of the Configuration Editor, you are presented with the Option to ***Import/Export*** the configuration file called *config.pro*. This will open the dialog window shown in Figure 3. If you don't want the settings read in at start-up, but do want them available later for a specific project, you can save the configuration file anywhere, or else don't use the name ***config.pro***. It can then be loaded at any time using the ***Import*** function.

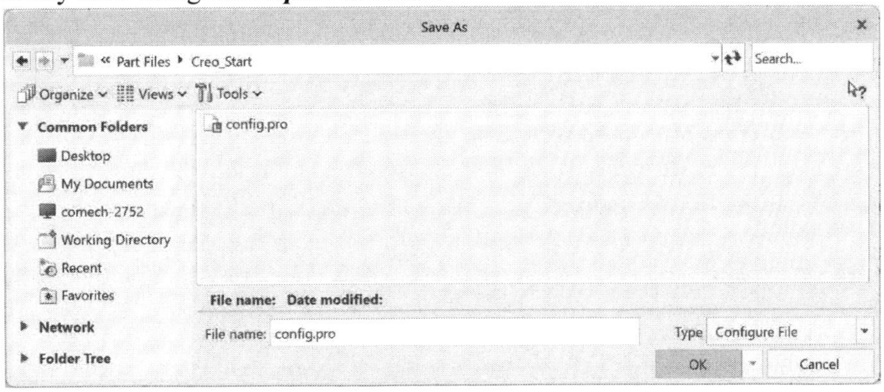

Figure 3 Saving the configuration settings

Now let's have a closer look at the contents of the *config.pro* file itself.

Configuration Files (*config.pro*)

As mentioned above, the most important configuration setting file is a special file called *config.pro* that is automatically read from the startup directory when you launch a new session. You can also read in (and/or change) additional configuration settings at any time during a session. For example, you may want to have one group of settings for one project you are working on, and another group for a different project that you switch to during a single session. Since that is rarely done, in this tutorial we will deal only with the use of the single configuration file, *config.pro*, loaded at start-up.

Several copies of *config.pro* might exist on your system, and they are read in the following order when Creo is launched:

- *config.sup* - this is a protected system file which is read by all users but is not usually available for modification by users. Settings made here cannot be overridden by the user. Your system administrator has control of this file.
- Creo loadpoint\...\text directory - the *config.pro* file stored here is read by all users and would usually contain common settings determined by the system administrator such as search paths, formats, libraries, and so on. This file cannot normally be altered by individual users.
- user home directory - unique for each user
- startup directory - the working directory when Creo starts up[1]. This is highlighted in the Navigator on start-up.

---

[1] In Windows, right click on the Creo icon on the desktop (if it exists), select **Properties ➤ ShortCut** and examine the **Start In** text entry field.

Settings made in the first copy (*config.sup*) cannot be overridden by users. This is handy for making configuration settings to be applied universally across all users at a Creo installation site (search paths for part libraries, for instance).

During the installation procedure for the software, a copy of a stripped-down *config.pro* file was created in the **../text** directory of the software loadpoint. The contents of this file (for example, setting default templates) depend on which system of units was chosen (English or metric) during software installation. If metric was chosen, a number of options are set to reflect ISO standards. If English was chosen, these options are set to ASME and ANSI standards.

An individual user can modify entries in the last two copies of *config.pro* to suit their own requirements. If the same entry appears more than once, the last entry encountered in the start-up sequence is the one the system will use. After start-up, additional configuration settings can be read in at any time. These might be used to create a configuration unique to a special project, or perhaps a special type of modeling (sheetmetal, for example). Be aware that some options may not take effect until Creo is restarted. So it doesn't make sense to try to change these options in the middle of a session by loading a new config file. This is discussed more a bit later.

The Configuration File Editor

As mentioned earlier, you can access your current configuration file using

File ➤ Options ➤ Configuration Editor

This brings up the **Options** window with the existing *config.pro* contents shown in the right panel. See Figure 4.

The first column shows its name, and the second column shows its current value. A value with an asterisk indicates a default value. The third column indicates its status (a solid green circle means the option has been read from the *config* file, rather than the system default), followed by a one-line description in the fourth column. Note that you can resize the column widths by dragging on the vertical column separator bars at the top of the display area.

In the ***Show*** pull-down at the right, select ***All Options***. A complete list of all the Creo configuration options will appear. Browse down through the list. There are a lot of options here (about 1000!). Note that the options are arranged alphabetically. This is because of the setting in the **Sort** pull-down menu in the top-left corner. Change this to ***By Category***. This rearranges the list of options to group them by function. For example, check out the settings available in the **Environment** and **Sketcher** groups. The list of options is a bit overwhelming. Fortunately, there are a couple of tools to help you find the setting you're looking for. Let's see how they work.

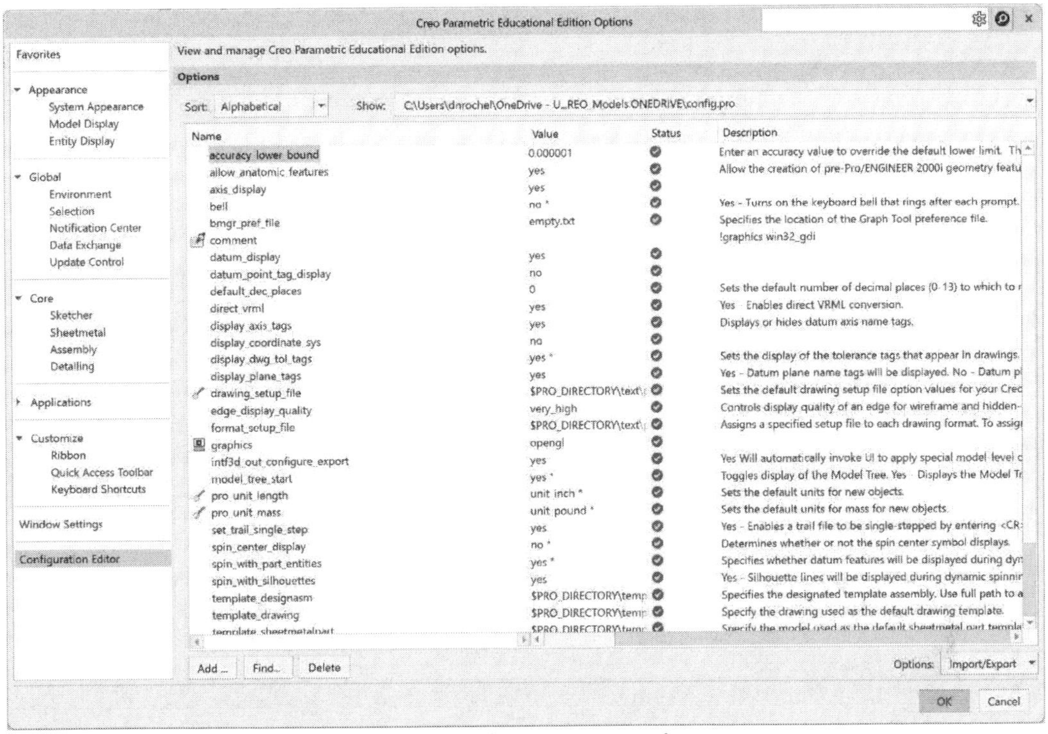

Figure 4 The *config.pro* file layout showing current settings

Open the **Show** pull-down again and select ***Current Session***, then ***Sort(Alphabetical)***.

Adding Settings to *config.pro*

Let's create a couple of useful settings. At the bottom of the **Options** window are two buttons to ***Add*** and ***Find*** options. If you know the name of the option, you can just select ***Add*** and type it in the top box shown in Figure 5.

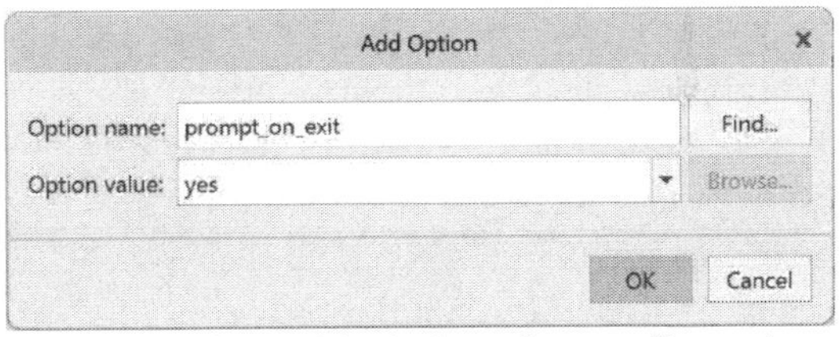

Figure 5 The ***Add*** window for config options

For new users, a useful setting is the following. In the name box enter the option name ***prompt_on_exit***. As you type this in, notice that Creo anticipates the rest of the text based on the letters you have typed in. After typing enough characters (up to the "x" in "exit" - it takes Creo that long to find out you don't want ***prompt_on_erase***), the rest of the desired option will appear. In the pull-down list beside **Value**, select ***Yes*** (or just type "y"). Note that the option name is not case sensitive and the default value is indicated by an asterisk in the pull-down list. Now select the ***OK*** button. The entry now appears in the data area. A bright green star in the **Status** column indicates that the option has been defined but has not yet taken effect.

Now enter a display option. The default part display mode in the graphics window is **Shaded**. Some people prefer to work in shaded mode with part edges shown - let's make it the default on start-up. Once again, we will enter the configuration option name using ***Add*** and pick the value from a drop-down list. The option name and value we want are

display **shadewithedges**

Select ***OK*** as before. Add the following option to control how datum planes are displayed during dynamic spinning with the mouse

spin_with_part_entities **yes**

Another common setting is the location of the Creo trail file. As you recall, the trail file contains a record of every command and mouse click during a Creo session. The default location for this is the start-up directory. Theoretically, trail files can be used to recover from disastrous crashes, but this is a tricky operation. Most people just delete them. It is handy, therefore, to collect trail files in a single directory, where they can be easily removed later. There is an option for setting the location of this directory. Suppose we don't know the configuration option's specific name. Here is where a search function will come in handy.

At the bottom of the **Options** window, click the ***Find*** button (you may have noted that this is also available in the **Add** window, Figure 5). This brings up the **Find Option** window (Figure 6).

Type in the keyword ***trail*** and leave the default ***Look in(ALL_CATEGORY)*** (note the option to search the descriptions for the keyword; we don't need that in this case) then select ***Find Now***.

Several possibilities come up. The option we want is listed as **trail_dir** - scroll the description to the right to confirm this. Select this option and then pick the ***Browse*** button at the bottom to identify a suitable location on your system for the value. Perhaps something like ***c:\temp***. Then select ***Add/Change***. The new entry appears in the **Options** window. In the **Find Option** window, select ***Close***.

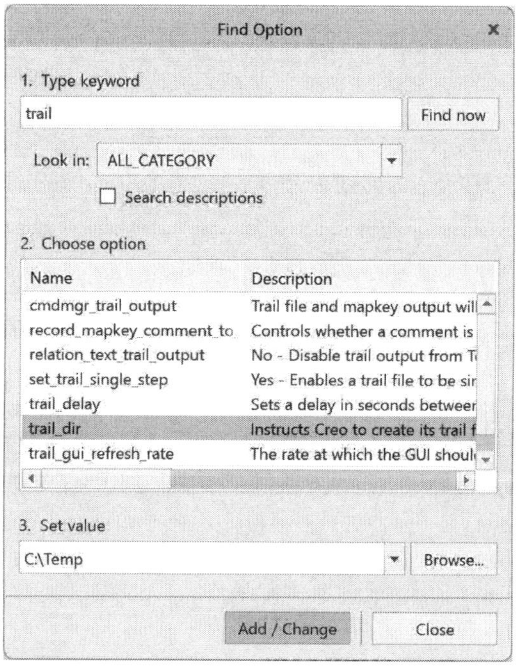

Figure 6 Using ***Find*** to locate an option

For some options, the value is numeric (e.g. setting a default tolerance, number of digits, or the color of entities on the screen). In these cases, you can enter the relevant number (or numbers separated by either spaces or commas).

We have now specified four options. The green stars in the **Status** column indicate they have not taken effect yet.

For practice, enter the following options, either directly or using ***Find***. (HINT: use the

RMB pop-up menu anywhere in the option table and select ***Add.***) The order that the configuration options are declared does not matter. Feel free to add new settings to your file (for search paths, libraries, default editors, default decimal places, import/export settings, and so on).

allow_anatomic features	**yes**
default_dec_places	**1**
display_full_object_path	**yes**
sketcher_starts_in_2D	**yes**
sketcher_lock_modified_dims	**yes**

Notice the icons in the first column beside the option names. These mean the following:

(lightning) - option takes effect immediately

(wand) - option will take effect for the next object created

(screen) - option will take effect the next time Creo is started

(illegal) - option is illegal or not recognized

If you are using a *config* file from a previous version of Creo you may see the red illegal symbol, which means that the option is no longer used.

Try to add an illegal option name. For example, in a previous release there was an option **system_hidden_color**. Try to add that and set it to **Yes**. When you try to set a value for this, it will not be accepted (the ***OK*** button stays gray). Creo only recognizes valid option names! Thus, if you mistype or enter an invalid name, this is indicated by not being able to enter a value for it.

Saving Your *config.pro* Settings

To store the settings we have just created, select the ***Export Configurations*** button in the **Options** window. In the Save As window that opens, the default name for the file ***config.pro*** is already entered. Check the line at the top to make sure this is going into the startup directory, then select ***OK***.

Loading a Configuration File

You may have noticed (in the Import/Export menu) the Import Configuration command. Select that now and pick the desired file and then ***Open***. Note that these settings will be read in but not activated immediately (note the green star).

Deleting Configuration Options

Highlight one of the options (maybe an illegal one) and use the RMB pop-up to select

Delete. Click ***OK*** to close the window.

Creating Configuration Favorites

If you find that you are frequently modifying some of the configuration settings, they can be added to a Favorites list. Select one of the options and in the RMB menu pick ***Add to Favorites***. The list can be viewed and options changed by clicking **Favorites** in the top of the category pane on the left. Note that it doesn't make much sense to add options here that require a restart of the system.

Checking Your Configuration Options

Because some settings will not activate until Creo is restarted (mainly dealing with window properties like fonts and colors), many users will exit after making changes to their *config.pro* file and then restart, just to make sure the settings are doing what they are supposed to. Do that now. This is not quite so critical since the **Options** window shows you with the lightning/wand/screen icons whether an option is active. However, be aware of where Creo will look for the *config.pro* file on start-up, as discussed above. If you have saved *config.pro* in another working directory than the one you normally start in, then move it before startup. On the other hand, if you have settings that you only want active when you are in a certain directory, keep a copy of *config.pro* there and load it once Creo has started and you have changed to the desired directory. To keep things simple, and until you have plenty of experience with changing the configuration settings, it is usually better to have only one copy of *config.pro* in your startup directory.

Note that it is probably easier to make some changes to the environment for a single session using ***File ➤ Options***. Also, as is often the case when learning to use new computer tools, don't try anything too adventurous with *config.pro* in the middle of a critical part or assembly creation session - you never know when an unanticipated effect might clobber your work!

Customizing the Toolbars

For the following, you will have to load one of the parts you have previously made. Do that now.

Click the right mouse button on the Graphics toolbar. This brings up a menu like the one shown in Figure 7. This allows you to select which commands appear on the toolbar, where the toolbar is located, the size of the toolbar icons, or reset to the default toolbar.

Now use the RMB pop-up on the Quick Access toolbar at the top. A pop-up will allow you to specify the location (relative to the ribbon) or remove it entirely, or modify the commands. If you select

Customize Quick Access Toolbar

you will be taken to the regular Creo Parametric Options window. Here you may select from any command in the left pane and ***Add*** it to the toolbar list on the right. You can change its position relative to the other commands using the up/down arrows at the right side of the window, by dragging, or using the RMB menu. As usual, for this to "stick" you must save the settings using ***Export***. Put these in the default file in your startup directory (*creo_parametric_customization.ui*). Notice that there is a button that will reset to all default values. Also, be aware that some commands will not be available unless you are in relevant modes (for example, the Model Colors command is not available unless a model is loaded).

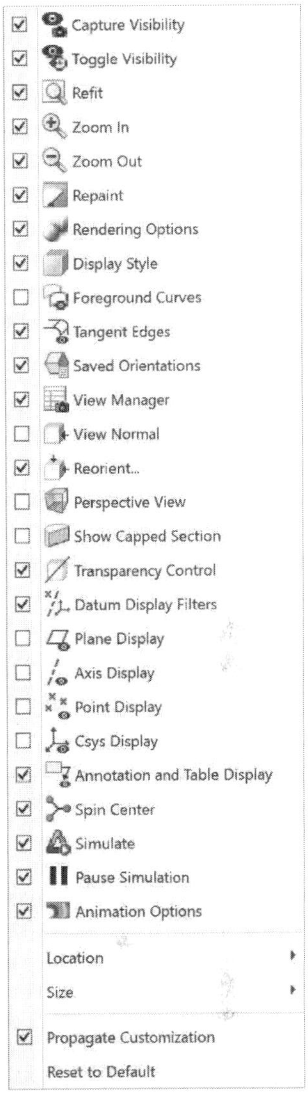

Figure 7 Customizing the **Graphics** Toolbar

Customizing Ribbon Tabs and Groups

With the mouse on a ribbon, use the RMB and select ***Customize the Ribbon***. This opens the usual **Options** window arranged as shown in Figure 8. The Creo commands are listed in the left pane, and the ribbon structure is shown in the right pane. Each of the tabs and groups can be expanded to show the commands it contains.

We cannot modify the command structure of the existing ribbons, tabs, and groups. However, we can do a lot to modify the appearance. First, each tab and group has a checkbox to specify whether it should be displayed. Within each group, we can modify the appearance of the command icons. For example, select the ***Mirror*** command in the ***Editing*** group of the ***Model*** tab. Now you can select the ***Modify*** command at the bottom to change the size and appearance of the command. It is also possible to do this a bit more easily by just selecting the button in the ribbon and using the RMB pop-up menu. Depending on how much space there is on your screen, you may find that the appearance of other commands in other groups may change (mainly the command label). If you right click on a button in the ribbon you can turn on/off the command label.

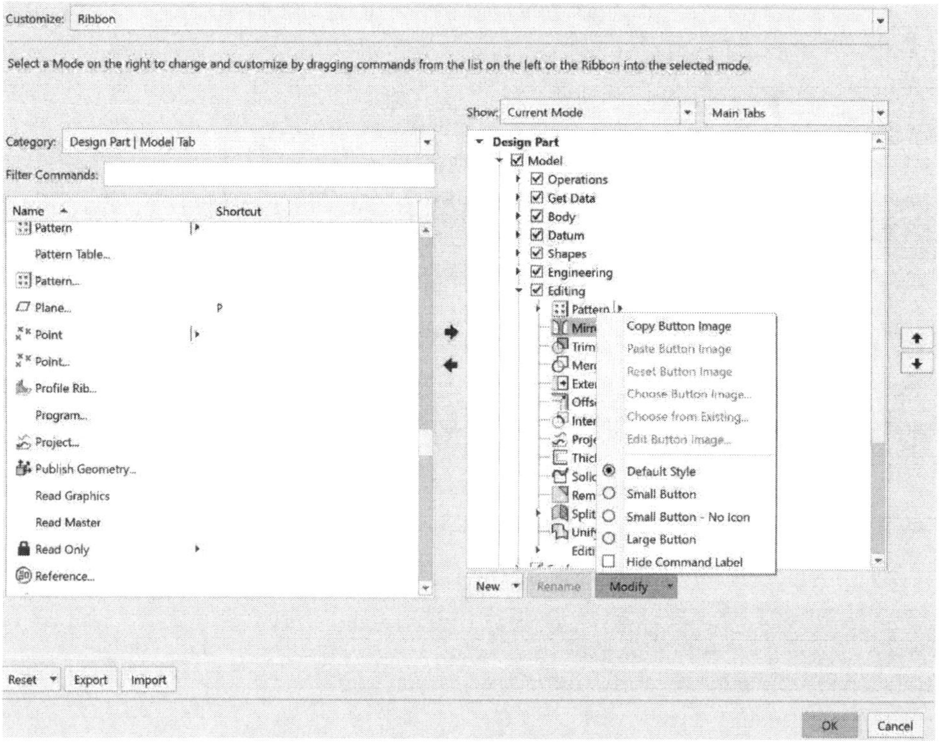

Figure 8 Customizing the Creo Parametric ribbons

Select the **Model** tab, then the ***New Group*** button (either in the RMB pop-up or in the ***New*** pulldown list below the pane). We can specify a new name "My Group" with ***Rename***, also on the RMB menu. Now with the new group selected, we can move the commands from the left pane into the new group. This allows you to collect commands that you use often that are located on different ribbons. Create the group with ***Model Colors***, ***Relations*** and ***Switch Dimensions*** shown in Figure 9. You can remove commands, groups, and tabs using the RMB pop-up menu and selecting ***Remove***.

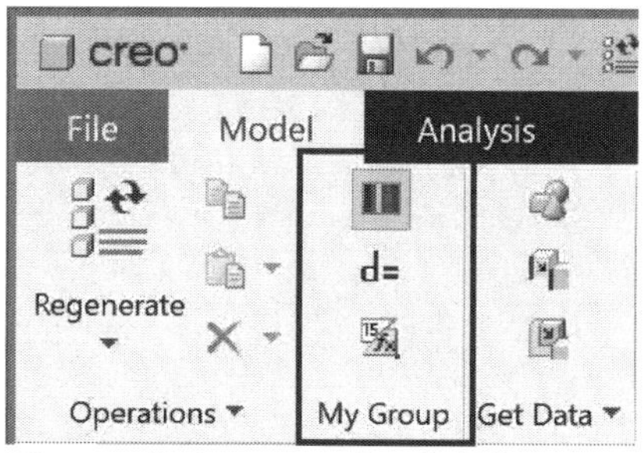

Figure 9 A customized ribbon group

Helpful Hint

It is tempting, if you are blessed with a lot of screen space, to overpopulate the toolbars and ribbons by trying to arrange every commonly used command on the screen at once. Before you do that, you should work with Creo for a while. You will find that Creo will generally bring up the appropriate toolbars for your current program status automatically. For example, if you are in Sketcher, the Sketcher short-cut buttons will appear. Thus, adding these buttons permanently to any toolbar is unnecessary and the buttons will be grayed out when you are not in Sketcher anyway - you are introducing screen clutter with no benefit. Furthermore, most commands are readily available in the pop-up and RMB menus. Since these are context sensitive, you will only have to choose from commands that are useful at that moment.

There is lots more information in the on-line help. A good place to start is to open the following path in the Help Center:

Fundamentals
  Fundamentals
    Configuring Fundamentals
      About Configuring Creo Parametric

You can also do a search in the Help pages for *config.pro*.

This concludes our short introduction to configuration files and customized toolbars. Hopefully, this has given you enough information to explore customization of the Creo interface on your own.

This page left blank.

Index

Commands are shown in ***Bold Italic***.

D

E

F

G

H

I

L

M

T

U

V

W

This page left blank.

PGIL2026USA